Biology Now

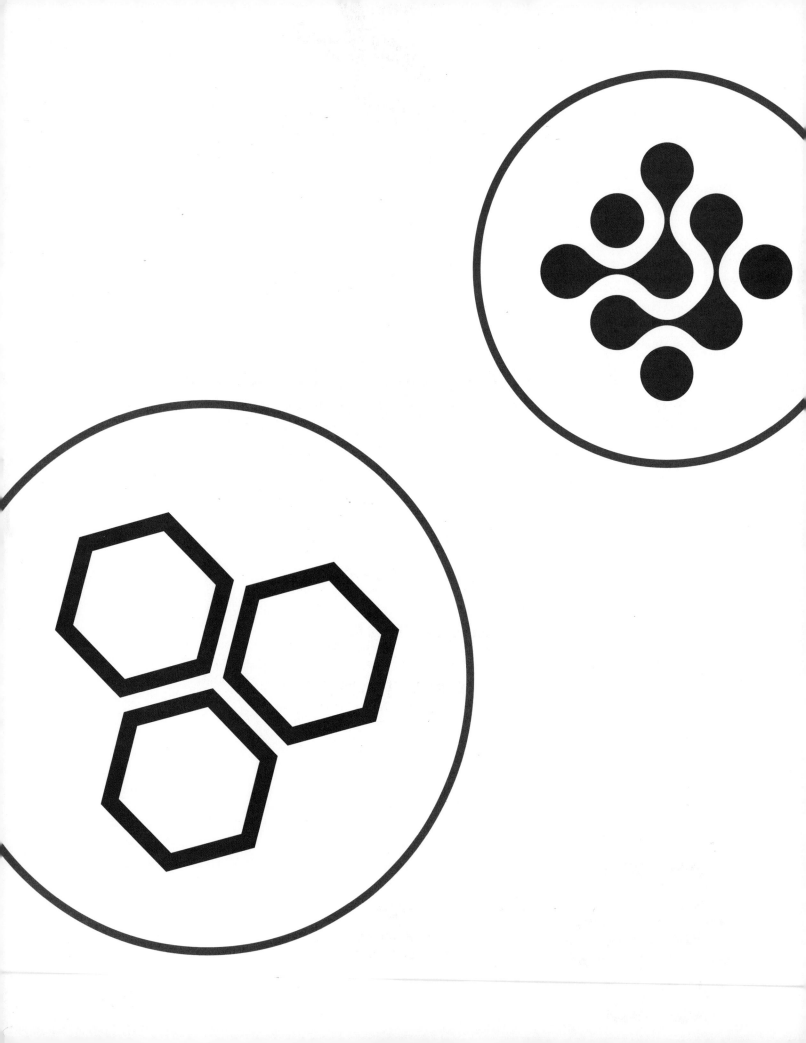

Biology Now

ANNE HOUTMAN
ROSE-HULMAN INSTITUTE OF TECHNOLOGY

MEGAN SCUDELLARI
SCIENCE JOURNALIST

CINDY MALONE
CALIFORNIA STATE UNIVERSITY, NORTHRIDGE

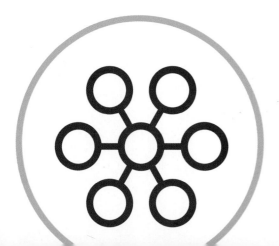

W. W. NORTON
NEW YORK · LONDON

W. W. Norton & Company has been independent since its founding in 1923, when William Warder Norton and Mary D. Herter Norton first published lectures delivered at the People's Institute, the adult education division of New York City's Cooper Union. The firm soon expanded its program beyond the Institute, publishing books by celebrated academics from America and abroad. By midcentury, the two major pillars of Norton's publishing program—trade books and college texts—were firmly established. In the 1950s, the Norton family transferred control of the company to its employees, and today—with a staff of four hundred and a comparable number of trade, college, and professional titles published each year—W. W. Norton & Company stands as the largest and oldest publishing house owned wholly by its employees.

Editor: Betsy Twitchell
Assistant Editor: Taylere Peterson
Developmental Editor: Andrew Sobel
Project Editor: Christine D'Antonio
Manuscript Editor: Stephanie Hiebert
Managing Editor, College: Marian Johnson
Managing Editor, College Digital Media: Kim Yi
Production Manager: Ashley Horna
Media Editor: Kate Brayton
Associate Media Editor: Cailin Barrett-Bressack
Media Project Editor: Jesse Newkirk
Media Editorial Assistant: Gina Forsythe
Marketing Manager, Biology: Todd Pearson
Design Director: Hope Miller Goodell
Photo Editor: Ted Szczepanski
Photo Researcher: Fay Torresyap
Permissions Manager: Megan Schindel
Permissions Clearer: Elizabeth Trammell
Composition by MPS North America LLC
Illustrations by Dragonfly Media Group
Manufacturing: Quad Graphics—Versailles, Kentucky

Permission to use copyrighted material is included in the backmatter of this book.

ISBN 978-0-393-63178-4

W. W. Norton & Company, Inc., 500 Fifth Avenue, New York, NY 10110-0017
wwnorton.com

W. W. Norton & Company Ltd., 15 Carlisle Street, London W1D 3BS

1 2 3 4 5 6 7 8 9 0

Brief Contents

Contents

 INTRODUCTION

UNIT 1: CELLS

CHAPTER 3: Chemistry of Life 39

Ingredients for Life

CHAPTER 4: Life Is Cellular 59

Engineering Life

CHAPTER 5: HOW CELLS WORK 79

Rock Eaters

CHAPTER 6: Cell Division 97

Toxic Plastic

UNIT 3: EVOLUTION

UNIT 4: BIODIVERSITY

UNIT 5: ECOLOGY

About the Authors

ANNE HOUTMAN is Provost and Vice President of Academic Affairs at Rose-Hulman Institute of Technology, where she is also a full professor of biology. Anne has over 20 years of experience teaching nonmajors biology at a variety of private and public institutions, which gives her a broad perspective of the education landscape. She is strongly committed to evidence-based, experiential education and has been an active participant in the national dialogue on STEM (science, technology, engineering, and math) education for over 20 years. Anne's research interests are in the ecology and evolution of hummingbirds. She grew up in Hawaii, received her doctorate in zoology from the University of Oxford, and conducted postdoctoral research at the University of Toronto.

MEGAN SCUDELLARI is an award-winning freelance science writer and journalist based in Boston, Massachusetts, specializing in the life sciences. She has contributed to *Newsweek*, *Scientific American*, *Discover*, *Nature*, and *Technology Review*, among others, and she was a health columnist for the *Boston Globe*. For five years she worked as a correspondent and later as a contributing editor for *The Scientist* magazine. In 2013, she was awarded the prestigious Evert Clark/Seth Payne Award in recognition of outstanding reporting and writing in science. She has also received accolades for investigative reporting on traumatic brain injury and a feature story on prosthetics bestowed with a sense of touch. Megan received an MS from the Graduate Program in Science Writing at the Massachusetts Institute of Technology and worked as an educator at the Museum of Science, Boston.

CINDY MALONE began her scientific career wearing hip waders in a swamp behind her home in Illinois. She earned her BS in biology at Illinois State University and her PhD in microbiology and immunology at UCLA. She continued her postdoctoral work at UCLA in molecular genetics. She is currently a distinguished educator and a professor at California State University, Northridge, where she is the director of the CSUN-UCLA Stem Cell Scientist Training Program funded by the California Institute for Regenerative Medicine. Her research is aimed at training undergraduates and master's degree candidates to understand how genes are regulated through genetic and epigenetic mechanisms that alter gene expression. She has been teaching nonmajors biology for almost 20 years and has won teaching, mentorship, and curriculum enhancement awards at CSUN.

Preface

A good biology class can improve the quality of students' lives. Biology is a part of so many decisions that students will need to make as individuals and as members of society. It helps parents to see the value of vaccinating a child, because they will understand what viruses are and how the immune system works. It helps homeowners in Texas, Florida, and Puerto Rico as they decide how to respond to the ongoing cleanup from 2017's Hurricanes Harvey, Irma, and Maria, because they understand how an ecosystem functions. It helps students make more informed decisions about their own nutrition because they understand the effects of fat, cholesterol, and vitamins, and minerals on our health. The examples are endless. Making informed decisions on these real-world issues requires students to be comfortable with scientific concepts and the process of scientific discovery.

How do we instill that capability in students? The last decade has seen an explosion of research on how students learn best. In a nutshell, they learn best when they see the relevance of a subject to their lives, when they are actively engaged in their learning, and when they are given opportunities to practice critical thinking.

In addition, most faculty who teach nonmajors biology would agree that our goal is to introduce students to both the key concepts of biology (for example, cells, DNA, evolution) and the tools to think critically about biological issues. Many would add that they want their students to leave the class with an appreciation for the value of science to society, and with an ability to distinguish between science and the nonscience or pseudoscience that bombards them on a daily basis.

How can a textbook help combine the ways students learn best with the goals of a nonmajors biology class? At the most basic level, if students don't read the textbook, they can't learn from it. When students read them, traditional textbooks are adept at teaching key concepts, and they have recently begun to emphasize the relevance of biology to students' lives. But students may be intimidated by the length of chapters and the amount of difficult text, and they often cannot see the connections between the story and the science. More important, textbooks have not been successful at helping students become active learners and critical thinkers, and none emphasize the process of science or how to assess scientific claims. It was our goal to make *Biology Now* relevant and interactive, and to be sure that it emphasized the process of science in short chapters that students *want* to read, while still covering the essential content found in other nonmajors biology textbooks.

Following the model of the first edition, each chapter in our book covers a current news story about people *doing* science, reported firsthand by Megan, an experienced journalist who specializes in reporting scientific findings in a compelling and accurate way, and fleshed out with a concise introduction to the science by Anne and Cindy. For this second edition we decided to direct our energies toward writing five current stories that will help instructors keep their courses grounded in real world events, and toward adding content requested by our first-edition adopters. Specifically, we've added a full unit—comprising two new chapters and two revised chapters—on the amazing diversity of life on planet Earth. Not only was more substantial coverage of this topic a common request in feedback about the first edition; it is also essential material for non-biology-major students, for it is partly through an appreciation of the diversity of life that students develop a personal relationship with the natural world.

Finally, we are thrilled for our book to be part of the online-assessment revolution! The second edition is accompanied by two excellent online homework platforms: a formative system called InQuizitive, and a summative system called Smartwork5. We no longer worry that our students aren't seeing the forest for the trees when they read the textbook. These systems are a rich learning environments for students and automatically graded assignment platforms for instructors.

We sincerely hope you enjoy the fruits of our long labors.

Anne Houtman
Megan Scudellari
Cindy Malone

What's New in the Second Edition?

- New chapter stories on current, fun, and unexpected topics like the Zika virus outbreak, the human microbiome, and the discovery of a CRISPR gene editing technology. New stories include:

Chapter 5: How Cells Work—Rock Eaters

Unusual electricity-"eating" microbes could someday provide a new way to store and produce energy as "bacterial batteries."

Chapter 9: What Genes Are—Pigs to the Rescue

CRISPR is perhaps one of the most exciting discoveries of the last century. Chapter 9 describes one application of the CRISPR genome editing technology: creating organs for transplant . . . in pigs.

Chapter 15: Bacteria and Archaea—Navel Gazing

A team at North Carolina State University leads a citizen science project to sequence the human belly button microbiome and gets some surprising results.

Chapter 16: Plants, Fungi, and Protists—The Dirt on Black-Market Plants

Poaching is illegal, and trafficking of tropical plants such as orchids threatens their survival. A group of scientists is tracking illegal plants from the United States to their source.

Chapter 19: Growth of Populations—Zika-Busting Mosquitoes

The spread of Zika throughout the Americas quickly became a health crisis. Genetically modifying mosquitoes is one of the ways that scientists are using to try to control Zika's spread.

- A new unit on biodiversity, which significantly expands coverage of the vast diversity of life on Earth, with two completely new chapters and two significantly revised chapters. Instructors who wish to continue teaching a brief introduction to biodiversity can do so with the "overview" chapter (Chapter 14). But for those wishing to spend time exploring life on Earth, Chapters 15, 16, and 17 provide thorough science coverage and lively stories.
- New, earlier placement of the chapter on applying science to making critical choices. The "capstone" final chapter in the second edition is now Chapter 2: Evaluating Scientific Claims. Introducing the concept of scientifically literate evaluation of scientific claims early in the book gives students the maximum amount of time to benefit from that skill.
- A new end-of-chapter question type—Challenge Yourself—which encourages students to think critically about the chapter's important biological concepts.
- New animation, interactive, and visually based questions in Smartwork5 and InQuizitive that promote critical thinking, interaction with data, and engagement with biology in the real world.
- New resources in the *Ultimate Guide to Teaching with Biology Now*, which will be accessible through the online Interactive Instructor's Guide platform, providing instructors with the ability to easily search and sort for active learning resources by topic, objective, and type of resource.

The perfect balance of science and story

Every chapter is structured around a story about people doing science that motivates students to read and stimulates their curiosity about biological concepts.

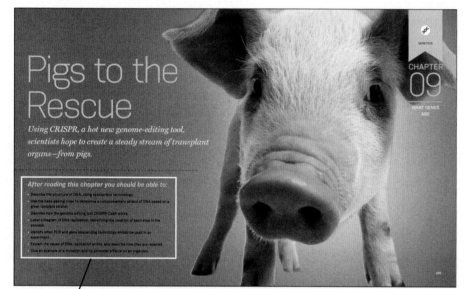

Dynamic chapter-opening spreads inspired by each chapter's story draw students in to the material.

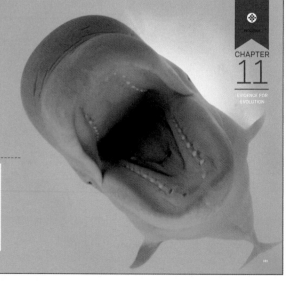

"After reading this chapter you should be able to" introduces learning outcomes that preview the concepts presented in each chapter.

Cast-of-character bios
highlight the scientists, researchers, and professors at the center of each story.

GORDON LARK

A geneticist at the University of Utah in Salt Lake City, Gordon Lark initiated the Georgie Project in 1996 to study the genetics of Portuguese water dogs. The national research project has led to valuable knowledge about the genetic basis of health and disease in humans and dogs.

J. G. M. "HANS" THEWISSEN

Paleontologist and embryologist J. G. M. "Hans" Thewissen is a professor and whale expert at Northeast Ohio Medical University in the Department of Anatomy and Neurobiology. He and his lab study ancestral whale fossils and modern whale species.

LISA COOPER

Lisa Cooper is an assistant professor at Northeast Ohio Medical University in the Department of Anatomy and Neurobiology. She earned her PhD in Thewissen's lab.

MICHAEL HELLBERG AND CARLOS PRADA

Michael Hellberg (right) is an evolutionary biologist at Louisiana State University who studies how species evolve in marine environments. Carlos Prada (left) was a graduate student in Hellberg's lab, and is now a postdoctoral researcher at Penn State studying how organisms cope with changes in the environment.

XU XING

Xu Xing is a paleontologist at the Chinese Academy of Sciences in Beijing. He has discovered more than 60 species of dinosaurs and specializes in feathered dinosaurs and the origins of flight.

An inquiry-based approach that builds science skills—asking questions, thinking visually, and interpreting data.

Most **figures** in the book are accompanied by three questions that promote understanding and encourage engagement with the visual content. Answers are provided at the back of the book, making the questions a useful self-study tool.

Q1: Is a zygote haploid or diploid?

Q2: Which cellular process creates a baby from a zygote?

Q3: If a mother or father was exposed to BPA prior to conceiving a child, how might that explain potential birth defects in the fetus?

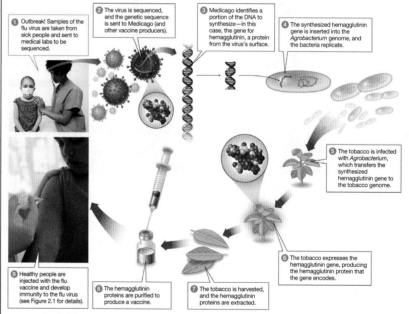

Q1: In which of the step(s) illustrated here does DNA replication occur? In which step(s) does gene expression occur?

Q2: Why do vaccine producers not simply replicate the entire viral genome, instead isolating the gene for one protein and replicating only that gene?

Q3: What role do the bacteria play in this process? Why are they needed?

Engaging, data-driven **infographics** appear in every chapter. Topics range from global renewable energy consumption (Chapter 5) to genetic diseases affecting Americans (Chapter 8) and many more. The infographics expose students to scientific data in an engaging way.

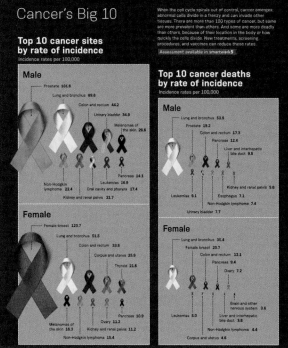

Extensive end-of-chapter review ensures that students see the forest for the trees.

Reviewing the Science
identifies each chapter's key science concepts, providing students with a guide for studying.

End-of-chapter questions follow
Bloom's taxonomy, moving from review (The Basics), to synthesis (Try Something New), to critical thinking (Challenge Yourself), to application (Leveling Up).

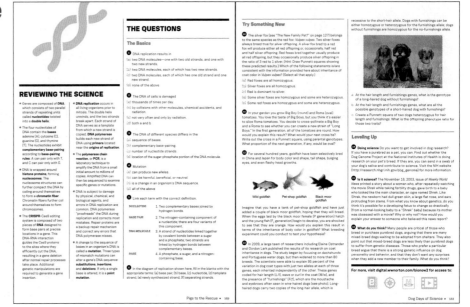

Leveling Up questions, based on
questions the authors use in their classrooms, prompt students to relate biology concepts to their own lives. The questions focus on one of the following themes: "Doing science," "Is it science?," "Life choices," "Looking at data," "What do *you* think?," and "*Write Now* biology."

Powerful resources for teaching and assessment

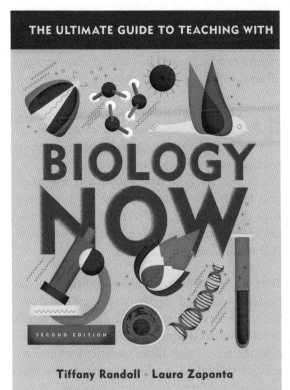

Laura Zapanta, University of Pittsburgh
Tiffany Randall, John Tyler Community College

The **Ultimate Guide** helps instructors bring *Biology Now*'s inquiry-based approach into the classroom through a wealth of resources, including activities useful in a variety of classroom sizes and setups, suggested online videos with discussion questions, clicker questions, sample syllabi, and suggested lecture outlines. The second-edition *Ultimate Guide* has been thoroughly reviewed and updated with new activities, Leveling Up rubrics, and descriptions of animations with discussion questions.

The **Interactive Instructor's Guide** is a searchable database of all the valuable teaching and active learning resources available in the *Ultimate Guide*. Instructors can easily filter by chapter, phrase, topic, or learning objective to find activities with downloadable handouts, streaming video with discussion questions, animations with discussion questions, lecture PowerPoints, and more.

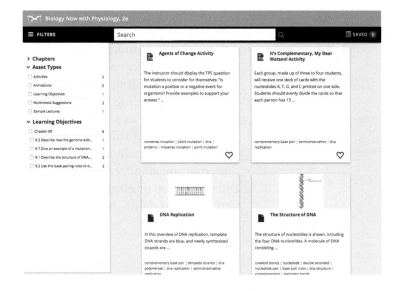

Other presentation tools for instructors

 InQuizitive InQuizitive is Norton's easy-to-use adaptive-learning and quizzing tool that improves student understanding of important learning objectives. Students receive personalized quiz questions on the topics they need the most help with. When instructors assign InQuizitive, students come better prepared to lectures and exams. The second-edition course includes new animation questions, story-based questions, and critical-thinking questions.

Smartwork5 Smartwork5 delivers engaging, interactive online homework to students, helping instructors and students reach their teaching and learning goals. The second edition features:

- New infographic questions, which promote interaction with data and engagement with biology in the real world, while making this popular visual feature of the text an assignable activity.
- New story-based questions, which help students to learn and understand the science behind the stories in the text.
- New critical-thinking questions, which prompt students to think critically about important concepts in biology.
- New animation questions, which engage students with the book-specific animations covering biology concepts.

Coursepacks Norton's free coursepacks offer a variety of concept-based opportunities for assessment and review. The Leveling Up questions from the text are available as writing activities, accompanied by grading rubrics, making them easy to assign. Also included are reading quizzes that contain modified images from the text and animation questions, infographic quizzes that help students build skills in reading charts and graphs, and flashcards for student self-study of key terms.

 Ebook Norton ebooks give students and instructors an enhanced reading experience at a fraction of the cost of a print textbook. Students are able to have an active reading experience and can take notes, bookmark, search, highlight, and even read offline. Instructors can even add their own notes for students to see as they read the text. Norton ebooks can be viewed on—and synced among—all computers and mobile devices.

Animations Key concepts and processes are explained clearly through high-quality, ADA-compliant animations developed from the meticulously designed art in the book. These animations are available for lecture presentation in the Interactive Instructor's Guide, PowerPoint outlines, and the coursepacks, as well as within our ebook, InQuizitive, and Smartwork5.

Test Bank The test bank is based on an evidence-centered design that was collaboratively developed by some of the brightest minds in educational testing. Each chapter's test bank now includes 75 or more questions structured around the learning objectives from the textbook and conforms to Bloom's taxonomy. Questions are further classified by text section and difficulty, and are provided in multiple-choice, fill-in-the-blank, and short-answer form. New infographic questions in every chapter help test student interpretation of charts and graphs.

Art Files All art and photos from the book are available, in presentation-ready resolution, as both JPEGs and PowerPoints for instructor use.

Lecture Slides Comprehensively revised by book author Cindy Malone, complete lecture PowerPoints thoroughly cover chapter concepts and include images and clicker questions to encourage student engagement.

Acknowledgments

We could not have created this textbook without the enthusiasm and hard work of many people. First and foremost, we'd like to thank our indefatigable editor, Betsy Twitchell, for her keen eye to the market, terrific visual sense, and endless author-wrangling skills. Andrew Sobel has done far more than ought to be required of a developmental editor to ensure that our book is both accurate and readable (not to mention his tireless work on the eye-catching infographics you'll see in these pages), and for that he has our eternal gratitude.

Thank you to our supremely focused and talented project editor, Christine D'Antonio, for creating such a superior layout and for keeping our chapters moving. Thank you to our talented copy editor, Stephanie Hiebert, for being so meticulous with our manuscript, and so pleasant to work with.

We are grateful to photo researcher Fay Torresyap for her reliable and creative work, and to Ted Szczepanski for managing the photo process. Production manager Ashley Horna skillfully oversaw the translation of our raw material into the beautiful book you hold in your hands; she, too, has our thanks. Special thanks to book designer Hope Miller Goodell and cover designer Jennifer Heuer for creating such an extraordinary and truly gorgeous book.

Media editor Kate Brayton, associate editor Cailin Barrett-Bressack, and media assistant Gina Forsythe worked tirelessly to create the instructor and student resources accompanying our book. Their determination, creativity, and positive attitude resulted in supplements of the highest quality that will truly make an impact on student learning. Jesse Newkirk's commitment to quality as media project editor ensured that every element of the resource package meets Norton's high standards. Likewise, assistant editor Taylere Peterson contributed in myriad ways, large and small, and for that she has our thanks.

We appreciate the tireless enthusiasm of marketing manager Todd Pearson and his colleagues, director of marketing Steve Dunn and marketing director Stacy Loyal. We thank director of sales Michael Wright and every single one of Norton's extraordinary salespeople for spreading the word about our book. Finally, we thank Marian Johnson, Julia Reidhead, Roby Harrington, Drake McFeely, and everyone at Norton for believing in our book.

Thank you to our accuracy reviewers Erin Baumgartner and Mark Manteuffel. We would be remiss not to thank also all of our colleagues in the field who gave their time and expertise in reviewing, class testing, and contributing to *Biology Now* and its many supplements and resources. Thank you all.

Reviewers

Second Edition

Anne Artz, Preuss School, UC San Diego
Allan Ayella, McPherson College
Erin Baumgartner, Western Oregon University
Joydeep Bhattacharjee, University of Louisiana, Monroe
Rebecca Brewer, Troy High School
Victoria Can, Columbia College Chicago
Lisa Carloye, Washington State University, Pullman
Michelle Cawthorn, Georgia Southern University
Craig Clifford, Northeastern State University

Beth Collins, Iowa Central Community College

Julie Constable, California State University, Fresno

Gregory A. Dahlem, Northern Kentucky University

Danielle M. DuCharme, Waubonsee Community College

Robert Ewy, SUNY Potsdam

Clayton Faivor, Ellsworth Community School

Michael Fleming, California State University, Stanislaus

Kathy Gallucci, Elon University

Kris Gates, Pikes Peak Community College

Heather Giebink, Pennsylvania State University

Candace Glendening, University of Redlands

Sherri D. Graves, Sacramento City College

Cathy Gunther, University of Missouri

Meshagae Hunte-Brown, Drexel University

Douglas P. Jensen, Converse College

Ragupathy Kannan, University of Arkansas–Fort Smith

Julia Khodor, Bridgewater State University

Jennifer Kloock, Garces Memorial High School

Karen L. Koster, University of South Dakota

Dana Robert Kurpius, Elgin Community College

Joanne Manaster, University of Illinois

Mark Manteuffel, St. Louis Community College

Jill Maroo, University of Northern Iowa

Tsitsi McPherson, SUNY Oneonta

Kiran Misra, Edinboro University of Pennsylvania

Jeanelle Morgan, University of North Georgia

Lori Nicholas, New York University

Fran Norflus, Clayton State University

Christopher J. Osovitz, University of South Florida

Christopher Parker, Texas Wesleyan University

Brian K. Paulson, California University of Pennsylvania

Carolina Perez-Heydrich, Meredith College

Thomas J. Peri, Notre Dame Preparatory School

Kelly Norton Pipes, Wilkes Early College High School

Gordon Plague, SUNY Potsdam

Benjamin Predmore, University of South Florida

Jodie Ramsey, Highland High School

Logan Randolph, Polk State College

Debra A. Rinne, Seminole State College of Florida

Michael L. Rutledge, Middle Tennessee State University

Celine Santiago Bass, Kaplan University

Steve Schwendemann, Iowa Central Community College

Sonja Stampfler, Kellogg Community College

Jennifer Sunderman Broo, Saint Ursula Academy

J. D. Swanson, Salve Regina University

Heidi Tarus, Colby Community College

Larchinee Turner, Central Carolina Technical College

Ron Vanderveer, Eastern Florida State College

Calli A. Versagli, Saint Mary's College

Mark E. Walvoord, University of Oklahoma

Lisa Weasel, Portland State University

Derek Weber, Raritan Valley Community College

Danielle Werts, Golden Valley High School

Elizabeth Wright, Athenian School

Steve Yuza, Neosho County Community College

First Edition

Joseph Ahlander, Northeastern State University

Stephen F. Baron, Bridgewater College

David Bass, University of Central Oklahoma

Erin Baumgartner, Western Oregon University

Cindy Bida, Henry Ford Community College

Charlotte Borgeson, University of Nevada, Reno

Bruno Borsari, Winona State University

Ben Brammell, Eastern Kentucky University

Christopher Butler, University of Central Oklahoma

Stella Capoccia, Montana Tech

Kelly Cartwright, College of Lake County

Emma Castro, Victor Valley College

Michelle Cawthorn, Georgia Southern University

Jeannie Chari, College of the Canyons

Jianguo Chen, Claflin University

Beth Collins, Iowa Central Community College

Angela Costanzo, Hawai'i Pacific University

James B. Courtright, Marquette University

Danielle DuCharme, Waubonsee Community College

Julie Ehresmann, Iowa Central Community College

Laurie L. Foster, Grand Rapids Community College

Teresa Golden, Southeastern Oklahoma State University

Sue Habeck, Tacoma Community College

Janet Harouse, New York University

Olivia Harriott, Fairfield University

Tonia Hermon, Norfolk State University

Glenda Hill, El Paso Community College

Vicki J. Huffman, Potomac State College, West Virginia University

Carl Johansson, Fresno City College

Victoria Johnson, San Jose State University

Anthony Jones, Tallahassee Community College

Hinrich Kaiser, Victor Valley College

Vedham Karpakakunjaram, Montgomery College

Dauna Koval, Bellevue College

Maria Kretzmann, Glendale Community College

MaryLynne LaMantia, Golden West College

Brenda Leady, University of Toledo

Lisa Maranto, Prince George's Community College

Roy B. Mason, Mt. San Jacinto College

Gabrielle L. McLemore, Morgan State University

Paige Mettler-Cherry, Lindenwood University

Rachel Mintell, Manchester Community College

Kiran Misra, Edinboro University of Pennsylvania

Lori Nicholas, New York University

Louise Mary Nolan, Middlesex Community College

Fran Norflus, Clayton State University

Brian Paulson, California University of Pennsylvania

Carolina Perez-Heydrich, Meredith College

Ashley Ramer, University of Akron

Nick Reeves, Mt. San Jacinto College

Tim Revell, Mt. San Antonio College

Eric Ribbens, Western Illinois University

Kathreen Ruckstuhl, University of Calgary

Michael L. Rutledge, Middle Tennessee State University

Brian Sato, UC Irvine

Malcolm D. Schug, University of North Carolina at Greensboro

Craig M. Scott, Clarion University of Pennsylvania

J. Michael Sellers, University of Southern Mississippi

Marieken Shaner, University of New Mexico

David Sheldon, St. Clair County Community College

Jack Shurley, Idaho State University

Daniel Sigmon, Alamance Community College

Molly E. Smith, South Georgia State College, Waycross

Lisa Spring, Central Piedmont Community College

Steven R. Strain, Slippery Rock University of Pennsylvania

Jeffrey L. Travis, SUNY Albany

Suzanne Wakim, Butte College

Mark E. Walvoord, University of Oklahoma

Sherman Ward, Virginia State University

Lisa Weasel, Portland State University

Jennifer Wiatrowski, Pasco-Hernando State College

Rachel Wiechman, West Liberty University

Bethany Williams, California State University, Fullerton

Satya M. Witt, University of New Mexico

Donald A. Yee, University of Southern Mississippi

Focus Group Participants

Michelle Cawthorn, Georgia Southern University
Marc Dal Ponte, Lake Land College
Kathy Gallucci, Elon University
Tamar Goulet, University of Mississippi
Sharon Gusky, Northwestern Connecticut Community College
Krista Henderson, California State University, Fullerton
Tara Jo Holmberg, Northwestern Connecticut Community College
Brenda Hunzinger, Lake Land College
Jennifer Katcher, Pima Community College
Cynthia Kay-Nishiyama, California State University, Northridge
Kathleen Kresge, Northampton Community College
Sharon Lee-Bond, Northampton Community College
Suzanne Long, Monroe Community College
Boriana Marintcheva, Bridgewater State University
Roy B. Mason, Mt. San Jacinto College
Gwen Miller, Collin College
Kimo Morris, Santa Ana College
Fran Norflus, Clayton State University
Tiffany Randall, John Tyler Community College
Gail Rowe, La Roche College
J. Michael Sellers, University of Southern Mississippi
Uma Singh, Valencia College
Patti Smith, Valencia College
Bethany Stone, University of Missouri
Willetta Toole-Simms, Azusa Pacific University
Bethany Williams, California State University, Fullerton

Class Test Participants

Bruno Borsari, Winona State University
Jessica Brzyski, Georgia Southern University
Beth Collins, Iowa Central Community College

Christopher Collumb, College of Southern Nevada
Jennifer Cooper, University of Akron
Julie Ehresmann, Iowa Central Community College
Michael Fleming, California State University, Stanislaus
Susan Holecheck, Arizona State University
Dauna Koval, Bellevue College
Kiran Misra, Edinboro University of Pennsylvania
Marcelo Pires, Saddleback College
Michael L. Rutledge, Middle Tennessee State University
Jack Shurley, Idaho State University
Uma Singh, Valencia College
Paul Verrell, Washington State University
Daniel Wetzel, Georgia Southern University
Rachel Wiechman, West Liberty University

Instructor and Student Resource Contributors

Holly Ahern, SUNY Adirondack
Steven Christenson, Brigham Young University–Idaho
Beth Collins, Iowa Central Community College
Julie Ehresmann, Iowa Central Community College
Jenny Gernhart, Iowa Central Community College
Julie Harless, Lone Star College
Janet Harouse, New York University
Vedham Karpakakunjaram, Montgomery College
Dauna Koval, Bellevue College
Brenda Leady, University of Toledo
Boriana Marintcheva, Bridgewater State University
Paige Mettler-Cherry, Lindenwood University
Lori Nicholas, New York University
Christopher Osovitz, University of South Florida
Tiffany Randall, John Tyler Community College
Lori Rose, Sam Houston State University
Suzanne Wakim, Butte College
Bethany Williams, California State University, Fullerton

This book wouldn't have happened without Anne's husband, Will, who took care of every single other thing in her life so that she could write. His support, and that of her children, Abi and Ben, are what keep her going every day. With great patience, Megan's husband, Ryan, bore many dinner conversations about bats, algae, wolves, and more, and for that he has her thanks. To Megan's children: May you read this book and share your mother's joy about all things biology. Cindy thanks her husband, Mike; children, Ben and Lily; and their numerous pets for the chaotic lifestyle that inspired her to step up her game. Also, Cindy thanks her friends and students who laugh at her jokes and keep her grounded in reality.

Perhaps most of all, we are indebted to the many scientists and individuals who shared their time and stories for these chapters. To the men and women we interviewed for this book, we cannot thank you enough. Your stories will inspire the next generation of biologists.

Biology Now

Caves of Death

Scientists scramble to identify a mysterious scourge decimating bat populations.

After reading this chapter you should be able to:

- Caption a diagram of the scientific method, identifying each step in the process.
- Develop a hypothesis from a given observation and suggest one or more predictions based on that hypothesis.
- Design an experiment using appropriate variables, treatments, and controls.
- Give specific examples of a scientific fact and a scientific theory.
- Create a graphic showing the levels of biological organization.
- Determine whether something is living or nonliving based on the characteristics of living things.

CHAPTER
01

THE NATURE OF SCIENCE

Every spring for 30 years, Alan Hicks laced up his hiking boots, packed his camera, and set out to count bats in caves in upstate New York. A biologist with the New York State Department of Environmental Conservation, Hicks leads one of the few efforts in the country to collect annual data on bat populations. Since 1980, he had never missed the annual cave trip—until March 17, 2007.

"That day, of all days in my entire career, I stayed at my desk," recalls Hicks, who had remained behind to write a report for his supervisor. A couple of hours after his crew left to inspect some local caves, 15 miles from the Albany office, Hicks's cell phone rang.

"Hey, Al. Something weird is going on here," said a nervous voice. "We've got dead bats. Everywhere."

The line went quiet. "What are we talking here?" asked Hicks. "Hundreds of dead bats?"

"No," said the voice. "Thousands."

At first, Hicks conjectured that the bats had died in a flood, which had happened in that particular cave before. But the next day, a young volunteer who had been out with the team told Hicks to check his e-mail. The volunteer had sent him a picture taken the day before of eight little brown bats (*Myotis lucifugus*) hanging from a cave outcropping. Each one had a fuzzy white nose. This was a surprise because little brown bats do not have white noses.

Hicks e-mailed the picture to every bat researcher he knew. The fuzzy white material looked like a fungus, but there was no previous record of a fungus killing bats. As scientist after scientist looked at the picture, they all replied the same way: "What is that?" Hicks resolved to find out what was killing the bats and whether the white fuzz was involved.

Why was Hicks so interested in saving the bats? And why should any of us care, apart from valuing the preservation of all of Earth's creatures? For one thing, bats help us by devouring insects that would otherwise destroy agricultural crops and forests (see "Bug Zappers" on page 15). And mosquitoes, which bats eat, are the world's most deadly animal to humans: through malaria transmissions, mosquitoes kill hundreds of thousands of people each year.

As a biologist, Hicks took a scientific view of the world—logical, striving for objectivity, and valuing evidence over other ways of discovering the truth. **Science** is a body of knowledge about the natural world, but it is much more than just a mountain of data. Science is an evidence-based process for acquiring that knowledge.

- Science deals with the natural world, which can be detected, observed, and measured.
- Science is based on evidence that can be demonstrated through observations and/or experiments.
- Science is subject to independent validation and peer review.
- Science is open to challenge by anyone at any time on the basis of evidence.
- Science is a self-correcting enterprise.

To gather knowledge, Hicks would apply the **scientific method** (**Figure 1.1**). The scientific method is not a set recipe that scientists follow in a rigid manner. Instead, the term is meant to capture the core logic of how science works. Some people prefer to speak of the **process of science** rather than the scientific method. Whatever we call it, the practices that produce scientific knowledge can be applied across a broad range of disciplines—including bat biology.

Keep in mind that, as powerful as the scientific method is, it is restricted to seeking natural causes to explain the workings of our world. There are other areas of inquiry that science cannot address. The scientific method cannot tell us what is morally right or wrong. For example, science can inform us about the differences between humans and other animals, but it cannot identify the morally correct way to act on that information. Science also cannot speak to the existence of God or any other supernatural being. Nor can it tell us what is beautiful or ugly, which poems are most lyrical, or which paintings are most inspiring. So, although science exists comfortably alongside different belief systems—religious, political, and personal—it cannot answer all questions.

ALAN HICKS

Alan Hicks is a retired bat specialist who began the investigation of a mysterious bat illness while working for the New York Department of Environmental Conservation.

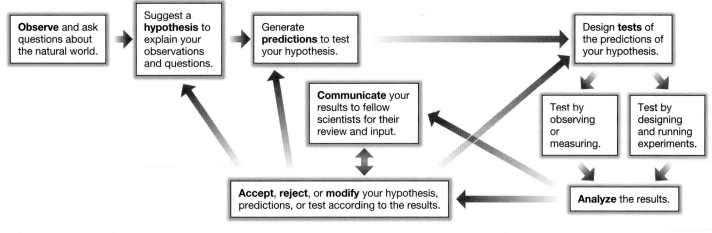

Figure 1.1

The scientific method

The scientific method is a logical process that helps us learn more about the natural world. ▶

Q1: What were the original observation and question of the scientists studying the sick bats?

Q2: At what point in the scientific method would a scientist decide on the methods she should use to test her hypothesis?

Q3: How might you explain the scientific method to someone who complains that "scientists are always changing their minds; how can we trust what they say?"

But science is the best way to answer questions about the natural world. The first two steps of the scientific method are to *gather observations* and *form a hypothesis*. Hicks didn't waste a moment of time before applying the scientific method to the question of the white fuzz. Bats were dying. "Bats are part of the planet and vital members of the ecosystem," says Hicks. "They play an important role in the environment in which we live."

Bat Crazy

On March 18, the day after the first dead bats were discovered, Hicks entered the cave to make observations—a key part of the scientific process. An **observation** is a description, measurement, or record of any object or phenomenon. Hicks's team observed that the sick bats had not only white noses, but also depleted fat reserves, meaning that the bats did not have enough stored energy to get through the winter. The bats also had white fuzz on their wings with scarred and dying wing tissue, and they were behaving abnormally, waking up early from hibernation and leaving the cave when it was still too cold outside to hunt.

Hicks's team also observed that the illness cut across species—many different types of bats were getting sick—and the bats exhibited a high rate of death: in some cases, up to 97 percent of infected bats died. Hicks and others began to call the illness white-nose syndrome (WNS). They still didn't know what caused the syndrome, but its characteristics led them to the assumption that the cause was a living organism (see "The Characteristics of Living Organisms" on page 6).

"For the first few years, we were just sleuthing," says Paul Cryan, a research biologist with the U.S. Geological Survey (USGS), and one of the scientists who received the original e-mailed picture from Hicks. From that first picture, Cryan was involved in trying to pinpoint the cause. "We were trying to understand something that had never happened before in a group of animals that was poorly understood."

In the caves, Hicks began collecting dead bats and sending them to laboratories around the nation. In those labs, technicians scraped samples from the bats' noses and wings, rubbed the samples into petri dishes (shallow glass or plastic plates containing a nutrient solution used to grow microorganisms), and watched to see whether the white fuzz would grow. Time after time, many different types of bacteria and fungi grew on the dishes, speckling them with dots of different-colored colonies, but none of the samples were unusual. Nothing special or dangerous appeared to be present on the bats.

One researcher, a young microbiologist named David Blehert, decided to try something different. Blehert worked at the USGS National Wildlife Health Center in Madison, Wisconsin. In December 2007, Hicks called Blehert. Blehert listened carefully as Hicks described how WNS was spreading. "He said, 'We have a major problem on our hands,'" recalls Blehert. "It turns out he was 100 percent right."

Hicks described to Blehert the conditions under which the bats lived during hibernation—caves in upstate New York, where the temperature was often between 30°F and 50°F. Blehert realized that most of the laboratories, including his, were trying to grow the samples

The Characteristics of Living Organisms

All living things share certain features that characterize life.

1. *They are composed of one or more cells.* The **cell** is the smallest and most basic unit of life; all organisms are made of one or more cells. Larger organisms are made up of many different kinds of specialized cells and are known as *multicellular organisms*.

2. *They reproduce autonomously using DNA.* All living organisms are able to **reproduce**, to make new individuals like themselves. **DNA** is the genetic material that transfers information from parents to offspring. A segment of DNA that codes for a distinct genetic characteristic is called a *gene*. Life, no matter how simple or how complex, uses this inherited genetic code to direct the structure, function, and behavior of every cell.

3. *They obtain energy from the environment to support metabolism.* All organisms need **energy** to survive. Organisms use a wide variety of methods to capture this energy from their environment. The capture, storage, and use of energy by living organisms is known as *metabolism*.

4. *They sense the environment and respond to it.* Living organisms **sense** many aspects of their external environment, from the direction of sunlight to the presence of food and mates. All organisms gather information about the environment by sensing it, and then respond appropriately.

5. *They maintain a constant internal environment.* Living organisms sense and respond to not only the external environment, but also their internal conditions. All organisms maintain constant internal conditions—a process known as **homeostasis**.

6. *They can evolve as groups.* **Evolution** is a change in the genetic characteristics of a group of organisms over generations. When a characteristic becomes more or less common across generations, evolution has occurred within the group.

	Rock	Virus	Fungus	Plant	Animal
Composed of one or more cells	✗	✗	✓	✓	✓
Autonomously reproduce themselves	✗	✗	✓	✓	✓
Obtain energy from their environment	✗	✗	✓	✓	✓
Sense their environment and respond to it	✗	✗	✓	✓	✓
Maintain a constant internal environment (homeostasis)	✗	✗	✓	✓	✓
Can evolve as groups	✗	✓	✓	✓	✓
Living	✗	?	✓	✓	✓

from the bats at room temperature—a method conducive to the growth of many fungi. But in the caves, any living thing would have to grow at cold temperatures, so Blehert and his technicians took samples from dead bats, put them on petri dishes, and placed the dishes in the fridge.

At the same time, Melissa Behr, an animal disease specialist at the New York State Health Department, accompanied Hicks on a trip to a local cave (**Figure 1.2**). Behr swabbed a sample of the white fuzz directly from a bat in the cave, immediately spread it onto a glass slide, and looked at it under a microscope. A unique fungus was on the plate. The fungus was visible in little white fuzzy patches of cells, and up close, the individual spores of the fungus appeared crescent-shaped—different from all the other "normal" microbes growing on the bats' skin, and different from any fungus known to the researchers.

But Behr's single observation wasn't enough evidence to convince anyone that the strange-looking fungus was the cause of WNS. To be of use in science, an observation must be repeatable, preferably by multiple techniques. Independent observers should be able to see or detect the same object or phenomenon, at least some of the time.

In this case, Blehert was able to reproduce Behr's results by an independent technique. After letting his plates sit in the fridge for a few weeks, Blehert removed them and observed white patches of the same strange, crescent-shaped fungal spores. "OK, we now have in laboratory culture what Melissa captured when she collected white material in the caves," thought Blehert. "We've got it."

Prove Me Wrong

In science, just as in everyday life, observations lead to questions, and questions lead to potential explanations. For example, if you flip on a light switch but the light does not turn on, you wonder why, and then you look for an explanation: Is the lamp plugged in? Has the lightbulb burned out? You then identify one of these explanations as the most likely hypothesis for why the light did not turn on.

Figure 1.2

Preparing to enter the bat cave
Scientists suit up to collect more observations on the infected bats and the environmental conditions in the bats' roosting cave.

Q1: Which step(s) in the scientific method does this photograph illustrate?

Q2: What types of environmental data might the researchers have collected?

Q3: Why do you think the researchers are wearing protective gear?

A scientific **hypothesis** (plural "hypotheses") is an informed, logical, and plausible explanation for observations of the natural world. From the start, Hicks hypothesized that a new, cold-loving fungus was the primary cause of death in the bats. After observing the unique crescent-shaped fungal spores, Behr and Blehert agreed with this hypothesis. "It was the simplest

DAVID BLEHERT

A microbiologist and branch chief of the Wildlife Disease Diagnostic Laboratories at the National Wildlife Health Center, David Blehert studies a variety of fungal and bacterial pathogens that are harmful to bats, humans, and other species.

solution," says Blehert. "We had bats with a white fungus that nobody had ever seen before growing on them, so that was the most likely thing that was doing it."

But other scientists disagreed. A fungus itself is rarely deadly to a mammal; more often, a fungus causes an annoying, but not lethal, skin infection or is a secondary response after an animal gets sick from a viral or bacterial infection. So scientists proposed other hypotheses for the cause of WNS. Some suggested the fungus was a secondary effect of an underlying condition, such as a viral infection. Others hypothesized that an environmental contaminant, such as a pesticide, was the cause of death. "There were so many different hypotheses," says Cryan. "But that's what is beautiful about the scientific process. You observe as much as you can, and from those observations you can form multiple hypotheses. Science doesn't proceed by just landing on the right hypothesis the first time."

One of the joys, and challenges, of the scientific method is that after scientists suggest competing hypotheses, they then test their own hypotheses against those of others. A scientific hypothesis must be constructed in such a way that it is potentially **falsifiable**, or refutable. In other words, it must make predictions that can be clearly determined to be true or false, right or wrong (**Figure 1.3**). A well-constructed hypothesis is precise enough to make predictions that can be expressed as "if . . . then" statements.

For example, *if* WNS is caused by a transmissible fungus, *then* healthy bats that hibernate in contact with affected bats should develop the condition. *If* the fungus is secondary to an underlying condition, *then* the infection will occur in bats only after the primary underlying condition is present. *If* an environmental contaminant is the cause, *then* bats with WNS symptoms will have elevated levels of that contaminant in their blood or on their skin.

In each "if . . . then" case, it is possible to design tests able to demonstrate that a prediction is right or wrong. Although predictions can be shown to be true or false, the same is not true of hypotheses. Hypotheses can be *supported*, but no amount of testing can *prove* a hypothesis is correct with complete certainty (**Figure 1.4**).

The reason a hypothesis cannot be proved is that there might be another factor, unmeasured or unobserved, that explains why the prediction is true. For example, consider the first prediction stated in the previous paragraph—that healthy bats hibernating in contact with affected bats will develop WNS. If this is true, the reason might be that the healthy bats were infected by a fungus from their neighbor, supporting the hypothesis that the disease is caused by a transmissible fungus. Alternatively, related bats may tend to hibernate together in the same cave, and the disease, or at least vulnerability to the disease, might be genetically based. The hypothesis that the disease is fungal is *supported* but not *proved* by the correctness of this prediction.

Blehert set out to test the hypothesis that he, Behr, and Hicks had put forward—that a unique, cold-loving fungus was the primary cause of death in the bats. One can test a hypothesis through observational studies or experimental studies. Blehert's first study was observational. Observational studies can be purely **descriptive**— reporting information (**data**) about what is found in nature. Observational studies can also be **analytical**—looking for (analyzing)

1 **Observations and questions:** Bats are observed with white noses. What is causing the white fuzz? These bats are dying at higher rates than bats without white noses. Why?

3 **Predictions:** *If* the white noses are caused by a transmissible fungus, *then* healthy bats that hibernate in contact with affected bats should develop the condition. *If* the white noses are caused by a deadly fungus, *then* healthy bats inoculated with the fungus should develop white noses and die at higher rates.

2 **Hypothesis:** Bats with white noses are infected with a fungus, and this fungus is causing death.

Figure 1.3

From observation to hypothesis to testable prediction

Figure 1.4

Hypotheses are supported or not supported, but never proved
Although the claim of scientifically confirmed mildness in this vintage advertisement for cigarettes seems ridiculous, "science" is still used to sell products today. Most Americans see thousands of advertisements every day, and many of these make "scientific" claims that are exaggerated or inaccurate.

Q1: State the hypothesis that this advertisement is claiming was scientifically tested.

Q2: State a prediction that comes from this hypothesis. Is it testable? Why or why not?

Q3: Explain in your own words why the hypothesis cannot be "proved."

patterns in the data and addressing how or why those patterns came to exist. The tools of **statistics**—a branch of mathematics that can quantify the reliability of data—help scientists determine how well those patterns support a hypothesis. Observational studies usually rely on both descriptive and analytical methods to test predictions made by a hypothesis.

In 2009, Blehert, Behr, and Hicks published a scientific paper in which they described the results from inspecting 117 dead bats. They identified microscopic damage caused by a specific kind of fungus in 105 of the bats, and isolated and identified the fungus from a subset of 10 of them. It was a type of cold-loving fungus belonging to a group of fungi called *Geomyces*. They named this new species *Geomyces destructans*.

Their observational study revealed a correlation between white fungus on the noses of bats and bat illness and death. Observational studies suggest possible causes for a phenomenon, but they do not establish a cause-effect relationship. To demonstrate that the fungus was actually causing the illness—and not just correlated with it—Blehert designed and conducted an experiment. Testing scientific hypotheses often involves both observational and experimental approaches (**Figure 1.5**).

Catching the Culprit

An **experiment** is a repeatable manipulation of one or more aspects of the natural world. Blehert's experiment was to take healthy bats into his laboratory and expose them to the fungus. Like analytical observational studies, experimental studies use statistics to determine whether the experimental results support or refute the hypothesis being tested.

In studying nature, whether through observations, experiments, or both, scientists focus on **variables**, characteristics of any object or individual organism that can change. In a scientific experiment, a researcher typically manipulates

MELISSA BEHR

Melissa Behr, formerly with the New York Department of Health, is now a doctor of veterinary medicine at South Dakota State University. She conducts research on the pathology and biology of bats and teaches at the UW veterinary school.

Descriptive:
Looking for physical evidence of WNS

Analytical:
Measuring the weight of each bat

Experimental:
Injecting a fungicide to determine whether it will protect a bat from WNS

One research team; three approaches

Figure 1.5

Testing hypotheses using multiple approaches

Scientists set up an underground laboratory in Tennessee's New Mammoth Cave to test hypotheses about white-nose syndrome (WNS) using descriptive, analytical, and experimental approaches.

Q1: Give a possible hypothesis that could be tested by weighing the bats.

Q2: State the hypothesis being tested in the photo on the bottom right.

Q3: Explain in your own words why an experimental study is the only way to show a cause-effect relationship.

a single variable, known as the **independent variable**. In this case, Blehert's independent variable was fungal exposure. Some bats were exposed; others were not. A **dependent variable** is any variable that responds, or could potentially respond, to changes in the independent variable. Blehert's dependent variable was any sign of WNS on the healthy bats.

If we think of the independent variable as the cause, then the dependent variable is the effect. In the most basic experimental design, a researcher manipulates a single independent variable and tracks how that manipulation changes the value

of a dependent variable. Blehert manipulated his independent variable—exposing some bats to the fungus but not others—and then tracked his dependent variable, whether the bats showed symptoms of WNS.

Blehert made sure his experiment was a controlled experiment. A **controlled experiment** measures the value of the dependent variable for two groups of subjects that are comparable in all respects, except that one group is exposed to a change in the independent variable and the other group is not. In this case, healthy bats were either exposed to the fungus

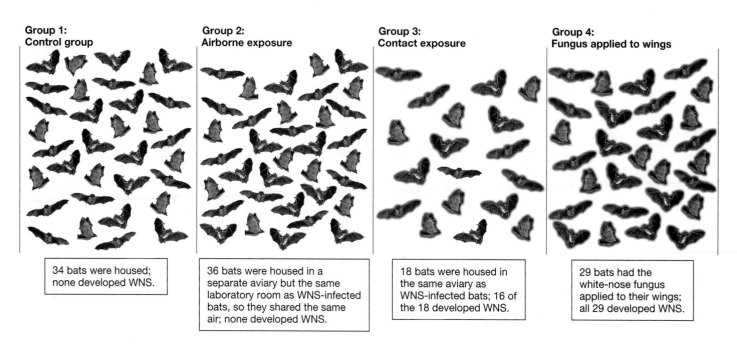

Figure 1.6

Blehert's experimental design

Blehert and his colleagues captured 117 healthy bats and brought them into the laboratory. They divided the bats into control and treatment groups and observed them for 102 days.

Q1: Which is the control group in this experiment, and what are the three treatment groups?

Q2: What is the hypothesis being tested in this experiment?

Q3: In one or two sentences, state the conclusions you can draw from the experiment. Was the hypothesis supported? Why or why not?

or not exposed. Typically, a researcher obtains a sufficiently large sample of study subjects and assigns them randomly to two groups. Randomization helps ensure that the two groups are comparable to start with.

One group, the **control group**, is maintained under a standard set of conditions with no change in the independent variable. Blehert had 34 healthy bats in his control group; he kept these bats in the laboratory, under the same conditions as the other bats, except for one difference: he did not expose them to *Geomyces destructans*.

The other group, known as the experimental or **treatment group**, is maintained under the same standard set of conditions as the control group, but the independent variable is manipulated. Blehert exposed 83 healthy bats to the fungus. Of these, 36 were exposed to the

fungus through the air, 18 were put into close contact with naturally infected bats, and 29 had the fungus applied directly to their wings (**Figure 1.6**).

As noted earlier, when scientists test a prediction of a hypothesis and find it upheld, the hypothesis is said to be supported. Scientists can be relatively confident in a supported hypothesis, but they cannot say that the hypothesis has been proved true. Even well-established scientific ideas can be overturned if new evidence against the prevailing view comes to light. Albert Einstein is famously reported to have said, "No amount of experimentation can ever prove me right; a single experiment can prove me wrong."

After watching both the control group and the treatment group, Blehert saw what he had hypothesized: physical exposure to the fungus caused white-nose syndrome, but exposure through the air did not. Healthy bats that had fungus applied directly to their wings or were caged with naturally infected bats had high rates of WNS by the experiment's end. It was the first direct evidence that the fungus was the primary cause of white-nose syndrome.

When a prediction is not upheld, the hypothesis is reexamined and changed, or it is discarded. Over the years, other hypotheses about the cause of white-nose syndrome have not been upheld. For example, scientists were not able to identify a single environmental contaminant at elevated levels in infected bats. A follow-up study by Blehert and others showed that the fungus not only leads to symptoms of WNS in bats, but is sufficient to cause death (**Figure 1.7**).

The story of the bats with white noses is just a single example of the scientific process. One of the greatest strengths of science is that scientific knowledge is tentative and therefore open to challenge at any time by anyone. An absolute requirement of the scientific method is that evidence be based on observations or experiments, or both. Furthermore, the observations and experiments that furnish the evidence must be subject to testing by others; independent researchers should be able to make the same observations, or obtain the same results, if they use the same conditions. In addition, the evidence must be collected in as objective a fashion as possible—that is, as free of personal or group bias as possible. Blehert's experiment fit all these conditions.

The main mechanism for policing personal or group bias and even outright fraud in science is **peer-reviewed publication**. Peer-reviewed publications are found in scientific journals that publish original research after it has passed the scrutiny of experts who have no direct involvement in the research under review. Before Blehert's research was published, it was reviewed by numerous scientists who had not participated in the experiment. If reviewers have concerns during the peer-review process, such as whether the evidence is strong enough to support a hypothesis, they can ask the paper's authors to address those concerns (for example, by gathering additional evidence) and to resubmit the paper. Blehert's paper passed the peer-review process and was published in the scientific journal *Nature* in 2011. At that point, says Blehert, the evidence that *G. destructans* causes WNS was strong enough that "I think we'd convinced most people." (In 2013, scientists renamed *G. destructans* as *Pseudogymnoascus destructans* when the fungus's genus was reclassified.)

But identifying the cause of WNS did not stop the disease from spreading. By March 2008, just a year after Hicks's team found the thousands of dead bats near Albany, more bats were found dead and dying in caves across Vermont, Massachusetts, and Connecticut. Within a year, the disease had spread as far as Tennessee and Missouri. The spread of WNS is a fact: bats around the United States are dying. In casual conversation, we typically use the term "fact" to mean a thing that is known to be true. A scientific **fact** is a direct and repeatable observation of any aspect of the natural world.

Figure 1.7

Experiments support the hypothesis that the fungus causes WNS in bats

The experiment whose results are plotted here was conducted by some of the same researchers interviewed for this story.

Q1: What is the control group in this experiment, and what are the two treatment groups?

Q2: At day 40, approximately how many individuals were alive in each treatment group? At day 80? At day 100?

Q3: In one or two sentences, state the conclusions you can draw from the experiment. Was the hypothesis supported? Why or why not?

Scientific fact should not be confused with scientific **theory**. Outside of science, people often use the word "theory" to mean an unproven explanation. In science, a theory is a hypothesis, or a group of related hypotheses, that has received substantial confirmation through diverse lines of investigation by independent researchers. Scientific theories have such a high level of certainty that we base our everyday actions on them. For example, the *germ theory of disease*, formally verified by Robert Koch in 1890, is the basis for treating infections and maintaining hygiene in the modern world (**Figure 1.8**).

No End in Sight

According to the U.S. Fish and Wildlife Service, white-nose syndrome has killed more than 6 million bats across the United States since 2007 and shows no signs of slowing. As of this writing, the disease has spread to 29 U.S. states and 5 Canadian provinces. In March 2016, hikers found a bat in Washington State with WNS, and Rhode Island's first case was confirmed two months later. Almost all species of bats that hibernate in these regions have been affected, including little brown bats and

1 **Fact:** Bats with white noses have been observed in the wild.

3 **Theory:** Specific pathogens ("germs") are directly responsible for specific diseases and conditions (germ theory of disease).

2 **Hypothesis:** A fungal infection is responsible for the spread of white-nose syndrome and is associated with higher mortality across bat populations and species.

Pseudogymnoascus destructans is the fungus responsible for white-nose syndrome in bats.

Frog populations worldwide have been decimated by the fungus *Batrachochytrium dendrobatidis*, or Bd.

Tomato blight is caused by the fungus *Phytophthora infestans*.

Figure 1.8

Facts, hypotheses, and theories

It is important to distinguish among facts, hypotheses, and theories when thinking and talking about science.

Q1: Give one *fact* about bats that you learned from this chapter.

Q2: What is another example of evidence for the *germ theory of disease*? (*Hint:* Think about human diseases.)

Q3: Explain in your own words the difference between a fact and a hypothesis, and between a hypothesis and a theory.

endangered Indiana bats, both of which have been particularly hard-hit.

The fungus appears to be related to a type of fungus common in caves in Europe, so a human traveler from Europe most likely carried it across the Atlantic and into the Albany cave, where it infected its first bat in the United States. Researchers continue to explore exactly how the fungus kills the bats. It appears to wake them from hibernation too many times during the winter, so the bats use up their fat reserves too soon and do not survive the months of cold weather. The fungus also eats through bats' delicate wings, which are important not only for flight, but also for maintaining healthy levels of water, oxygen, and carbon dioxide in the bats' bodies. The impact of this fungus on bats is a powerful example of how a microorganism can affect many levels of life, from individual tissues and organs up to whole populations, communities, and even the ecosystem itself (**Figure 1.9**).

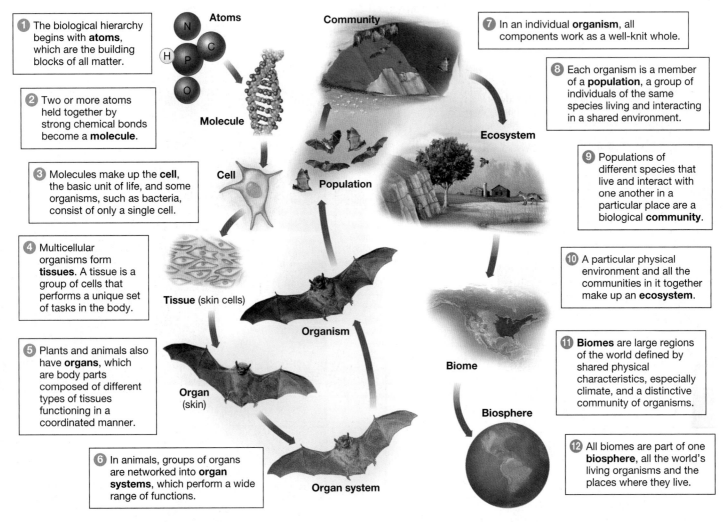

1. The biological hierarchy begins with **atoms**, which are the building blocks of all matter.

2. Two or more atoms held together by strong chemical bonds become a **molecule**.

3. Molecules make up the **cell**, the basic unit of life, and some organisms, such as bacteria, consist of only a single cell.

4. Multicellular organisms form **tissues**. A tissue is a group of cells that performs a unique set of tasks in the body.

5. Plants and animals also have **organs**, which are body parts composed of different types of tissues functioning in a coordinated manner.

6. In animals, groups of organs are networked into **organ systems**, which perform a wide range of functions.

7. In an individual **organism**, all components work as a well-knit whole.

8. Each organism is a member of a **population**, a group of individuals of the same species living and interacting in a shared environment.

9. Populations of different species that live and interact with one another in a particular place are a biological **community**.

10. A particular physical environment and all the communities in it together make up an **ecosystem**.

11. **Biomes** are large regions of the world defined by shared physical characteristics, especially climate, and a distinctive community of organisms.

12. All biomes are part of one **biosphere**, all the world's living organisms and the places where they live.

Atoms — Molecule — Cell — Tissue (skin cells) — Organ (skin) — Organ system — Organism — Population — Community — Ecosystem — Biome — Biosphere

Figure 1.9

The biological hierarchy

The **biological hierarchy** is a way to visualize the breadth and scope of life, from the smallest structures to the broadest interactions between living and nonliving systems.

Q1: Give examples of other kinds of organs that mammals such as bats have. (*Hint:* Think of the organs in your own body.)

Q2: Are bats in California part of the community of bats in upstate New York if they are of the same species? Why or why not?

Q3: Is the soil in a cave where bats live a part of the bats' population, community, or ecosystem? Explain your reasoning.

Bug Zappers

Bats are skilled natural exterminators, consuming billions upon billions of insects each year, including crop pests and mosquitoes. For this reason, bats play a critical role in agriculture and potentially in human health with the onset of increased mosquito-borne viruses such as West Nile and Zika. The loss of bat populations to white-nose syndrome would have a significant impact on farms, forests, and people around the country.

Assessment available in smartwork5

1.3 MILLION

A single colony of **150 big brown bats** (*Eptesicus fuscus*) in Indiana is estimated to eat nearly **1.3 million pest insects each year.**

4–8 GRAMS

A single little brown bat (*Myotis lucifugus*) can consume **4–8 grams of insects each night** during the active season.

660–1,320 METRIC TONS

Extrapolating the diet of a single bat to 1 million bats estimated to have died from WNS, between **660 and 1,320 metric tons of insects** are no longer being consumed each year in WNS-affected areas.

$22.9 BILLION

If bats disappeared entirely from the United States, it would cost the agricultural industry roughly **$22,900,000,000 per year** to save crops by dealing with the insects no longer being eaten by bats.

Today, WNS continues to spread, and scientists have not yet found a way to stop it, though many are trying. Recently, an international team of scientists identified a few populations of bats in Asia and the United States that survived the fungal infection. Evidence from those bats suggests they evolved resistance to the deadly disease. Researchers have also had some luck treating infected bats with a type of bacteria that kills the fungus, but they are still working on ways to deliver the treatment to large groups of bats.

For now, bats continue to die across America every winter. Hicks worries that students, hikers, and tourists will visit caves in the United States, not see any bats, and think that's normal. "In 2006, in one big cave in the Adirondacks, we counted over 185,000 bats. Anywhere you shined a light, there was a bat," says Hicks. "You go in now, and there's not a bat in sight."

REVIEWING THE SCIENCE

- **Science** is both a body of knowledge about the natural world and an evidence-based process for generating that knowledge.

- The **scientific method** represents the core logic of the process by which scientific knowledge is generated. The scientific method requires that we (1) make **observations**, (2) devise a **hypothesis** to explain those observations, (3) generate predictions from that hypothesis, (4) test those predictions, and (5) share the results of the tests for **peer review** by fellow scientists.

- A hypothesis cannot be proved true; it can only be supported or not supported. If the predictions of a hypothesis are not supported, the hypothesis is rejected or modified. If the predictions are upheld, the hypothesis is supported.

- We can test hypotheses by making further observations or by performing **experiments** (controlled, repeated manipulations of nature) that will either uphold the predictions or show them to be incorrect.

- In a scientific experiment, the **independent variable** is manipulated by the investigator. Any variable that can potentially respond to the changes in the independent variable is called a **dependent variable**.

- A scientific **fact** is a direct and repeatable observation of any aspect of the natural world. A scientific **theory** is a major idea that has been supported by many observations and experiments.

- The term **biological hierarchy** refers to the many levels at which life can be studied: atom, molecule, cell, tissue, organ, organ system, organism, population, community, ecosystem, biome, biosphere.

- All living organisms have the following characteristics in common: (1) They are built of **cells**; some are single-celled and some are multicellular. (2) They **reproduce**, using **DNA** to pass genetic information from parent to offspring. (3) They take in **energy** from their environment. (4) They **sense** and respond to their environment. (5) They exhibit **homeostasis**, maintaining constant internal conditions. (6) They can **evolve** as groups.

THE QUESTIONS

The Basics

1 Which of the following is a living organism? (Select all that apply.) If it is not living, which criterion, or criteria, does it not meet?

(a) an oak tree

(b) an influenza virus

(c) the fungus that causes white-nose syndrome in bats

(d) a diamond

(e) your teacher

2 When scientists use the word "theory," they mean

(a) an educated guess.

(b) an overarching explanation of an interrelated set of observations.

(c) wild speculation.

(d) an experimental prediction.

(e) a fact proved by many experiments.

3 Select the correct terms:
The process of science begins with a(n)

(**prediction / observation**) about the natural world. A scientist then proposes a (**hypothesis / prediction**), which is the basis of one or more testable (**observations / predictions**).

4 Place the following steps of the scientific method in the correct order by numbering them from 1 to 7.

_____a. Make observations about the natural world.

_____b. Test the predictions by designing an experiment or collecting observational data.

_____c. Run the experiment and analyze the results.

_____d. Generate predictions to test the hypothesis.

_____e. Share the results with fellow scientists so that they can review and evaluate them.

_____f. Develop a hypothesis to explain the observations.

_____g. Accept, reject, or modify your hypothesis depending on the results.

5 Identify the level of biological organization for each of the following.

_____a. the kidney of a bat

_____b. an oak tree outside a cave in upstate New York

_____c. bats in a cave in upstate New York

_____d. the physical and biological components of a cave in upstate New York

_____e. the respiratory system of a bat

_____f. all the species living and interacting within a cave in upstate New York

Challenge Yourself

6 Describe one observation, one hypothesis, and one experiment from the white-nose syndrome research discussed in this chapter.

7 Which of the following statements is a scientific hypothesis (that is, it makes testable predictions)? (Select only one.)

(a) Even though no one else can see him, the ghost of my dog lives in my backyard.

(b) The Atkins diet helps people lose more weight and keep it off than Weight Watchers does.

(c) People born under the sun sign Aquarius are kinder and cuter than those born under Scorpio.

(d) It is unethical to text while driving.

(e) none of the above

8 Consider an experiment in which subjects are given a pill to test its effectiveness in reducing the duration of a cold. Which of the following is the best way to treat the control group?

(a) Give the control group two pills instead of one.

(b) Do nothing with the control group.

(c) Give the control group a pill that looks like the test pill but does nothing.

(d) Let the control group choose whether or not to take any pills.

(e) Expose the control group to the cold virus.

Try Something New

9 If a virus was discovered that had the ability to reproduce itself without the help of a cell and its machinery, several experiments could be performed in order to show that it should be categorized as "alive." For each experiment, specify which one of the six criteria for being classified as "alive" would be tested.

_____a. Place a set number of viral particles in a nutrient broth and check for an increase in viral particles after several days.

_____b. Place a set number of viral particles in a broth where all the nutrients are corralled to one side of the container and check for viral particle accumulation in the nutrient-rich area of the container.

_____c. Place a set number of viral particles in a nutrient broth and check for waste products (broken-down nutrients) in the broth after several days.

_____d. Place a set number of viral particles in a nutrient broth for several months, adding different nutrients daily. Randomly isolate viral particles and sequence their DNA to determine whether there have been changes in genetic characteristics of this population of viral particles compared to the DNA of the original population before culturing.

_____e. Perform biochemical and microscopic analyses on a number of viral particles to determine whether they have the main components of a living cell, including a plasma membrane, ribosomes, and DNA.

10 Mad cow disease, or bovine spongiform encephalopathy, appears to be caused by a novel infectious agent: a protein that replicates by causing related proteins to modify their structure from a harmless shape to a dangerous one. These prions (short for "proteinaceous infectious particles") also appear to be the cause of several other spongiform encephalopathy diseases, such as scrapie in sheep, and kuru and Creutzfeldt-Jakob disease in humans. Which of the following observations or experiments would *not* support the hypothesis that a prion causes spongiform encephalopathy?

(a) The brains of many sheep with scrapie contain prion proteins, but the brains of most sheep without scrapie do not.

(b) There is a high incidence of kuru in populations of people who consume brain tissue from prion-infected animals.

(c) Coyotes that feed on cows with mad cow disease do not subsequently develop spongiform encephalopathy.

(d) When introduced into sheep brain cells in culture, prions cause the normal proteins to change shape into dangerous prion proteins.

(e) When prions are fed to sheep, most of them subsequently develop scrapie, whereas sheep not fed prions do not develop scrapie.

11 Label each of the following statements as an observation, a hypothesis, an experiment, or a result.

_____ a. People who ingest a liquid contaminated with a rotavirus will subsequently experience acute diarrhea.

_____ b. Seventeen out of twenty students given rotavirus-contaminated chocolate bars subsequently experienced acute diarrhea.

_____ c. Ten students were given a liquid containing a rotavirus, and 10 other students were given uncontaminated liquid.

_____ d. People exposed to rotaviruses often experience acute diarrhea.

_____ e. The latest outbreak of acute diarrhea on campus was due to a rotavirus-contaminated elevator button in the dorms.

12 Review the map showing the confirmed and suspected outbreaks of WNS. Is each statement that follows a testable, properly formulated scientific hypothesis for the spread of WNS to bats in Washington State? Explain why or why not. (*Hint:* Think about the criteria for a scientific hypothesis.)

a. Did bats infected with WNS arrive in Washington State via human smuggling for research purposes?

b. The WNS fungus was transported from the Midwest to Washington State via a bat found in Flight 1701's cargo hold.

c. Bats in Washington State were infected by contaminated clothing and equipment from visiting researchers checking for WNS in these colonies.

d. How did bats end up with WNS in Washington State, so far away from all other sites of contamination?

e. A mysterious cloud of fungal particles must have floated over Washington State in early 2016.

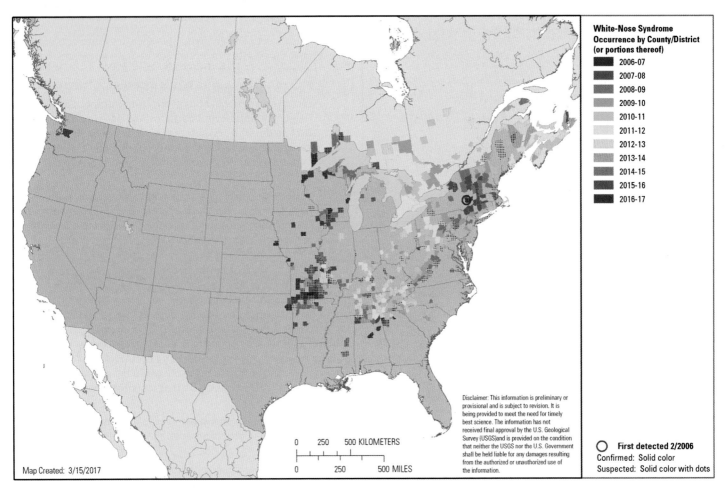

White-Nose Syndrome Occurrence by County/District (or portions thereof)
- 2006-07
- 2007-08
- 2008-09
- 2009-10
- 2010-11
- 2011-12
- 2012-13
- 2013-14
- 2014-15
- 2015-16
- 2016-17

Disclaimer: This information is preliminary or provisional and is subject to revision. It is being provided to meet the need for timely best science. The information has not received final approval by the U.S. Geological Survey (USGS) and is provided on the condition that neither the USGS nor the U.S. Government shall be held liable for any damages resulting from the authorized or unauthorized use of the information.

○ First detected 2/2006
Confirmed: Solid color
Suspected: Solid color with dots

Map Created: 3/15/2017

0   250   500 KILOMETERS
0   250   500 MILES

Map depicts the first time WNS is reported suspect or confirmed in a county or district (or portions thereof); each time period in the legend spans a winter bat hibernation period.

Citation: White-nose syndrome occurrence map - by year (2017). Data Last Updated: 3/15/2017. Available at: https://www.whitenosesyndrome.org/resources/map.

Leveling Up

13 **What do _you_ think?** Insectivorous bats save the agriculture industry billions of dollars each year because they eat insects that damage crops. Other species of bats pollinate some crops. What would you predict about crop yields in areas that are experiencing heavy bat mortality from WNS? Does your prediction affect your level of concern about WNS? Why or why not?

14 **Doing science** Although you are an expert on white-nose syndrome in bats, you have been asked to contribute your scientific expertise to understanding the fungal infection that is decimating frog populations worldwide: _Batrachochytrium dendrobatidis_, a.k.a. Bd.

a. Read the May 2013 _National Geographic_ article on how Bd has been spread around the world: http://news.nationalgeographic .com/news/2013/13/130515-chytrid-fungus-origin-african -clawed-frog-science.

b. The graph provided here, based on data from a scientific paper published in 2013, shows that frog species with higher average body temperatures tend to have lower Bd infection rates. It also shows that, compared to males of the same species, females tend to have higher average body temperatures and lower Bd infection rates. Is this observational study descriptive or analytical? Explain your answer.

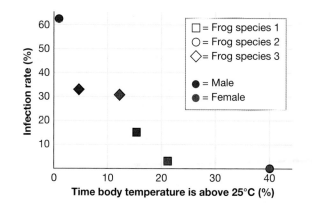

c. Propose a hypothesis to explain the results depicted in the graph, and then identify a testable prediction from that hypothesis. Design an experiment to test the prediction, identifying your independent and dependent variables, the control group, and the treatment conditions. Create a graph to show the results you expect to find (1) if your hypothesis is supported and (2) if your hypothesis is not supported.

For more, visit digital.wwnorton.com/bionow2 for access to:

A Critical Choice

A mother peels back the controversy to decide whether to vaccinate her children.

After reading this chapter you should be able to:

- Evaluate a scientific claim, using the process described in the chapter.

- Explain the importance of scientific literacy for making informed decisions.

- Distinguish between secondary and primary literature, and explain the role of peer review in the latter.

- Compare and contrast basic and applied research, and give an example of each.

- Distinguish between correlation and causation, and give a clear example of each.

- Determine whether a scientific claim is based on real science or pseudoscience.

CHAPTER
02

EVALUATING
SCIENTIFIC
CLAIMS

In 2009, Anna Eaton discovered she was pregnant. Several of her friends were also expecting or had recently had children, so the friends' afternoon chats were soon about all things baby: nursery colors, pediatricians, car seats. Every once in a while, however, one of the mothers broached a subject far more contentious than strollers and baby carriers. One of their hot topics of discussion became vaccination.

Vaccination is the injection of material—typically an inactivated and harmless infectious organism or parts of such an organism (e.g., a single protein)—that stimulates the immune system to protect against future exposure to that pathogen (**Figure 2.1**). When the body's immune system is exposed to a vaccine, it recognizes the inactivated organism (or its parts) as an invader and mounts an attack against it. Upon later contact with the infectious organism, a vaccinated individual's immune system remembers the inactivated organism from the vaccine and is already armed and bristling. The inactivated organism from the vaccine cannot cause disease because it was made harmless by the process of creating the vaccine. The vaccine itself cannot cause disease because it is not a functional infectious organism and therefore cannot replicate.

Over the past 200 years, scientists have developed vaccines to protect against dozens of pathogens, starting with the smallpox virus. In England in 1796, Dr. Edward Jenner first vaccinated his gardener's 8-year-old son against smallpox using the cowpox virus, because it was known that individuals who had previously suffered cowpox were immune to smallpox. Jenner's breakthrough ultimately led to a worldwide smallpox vaccination campaign, which was so successful that in 1980 the disease was officially declared eradicated by the World Health Organization. Vaccines have also been developed for the Tic Tac–shaped bacteria that cause the upper respiratory infection diphtheria, the round and highly contagious rotavirus that causes vomiting and diarrhea, the tiny airborne bacteria that cause pertussis (also known as whooping cough), and many more pathogens.

Before vaccines, children died in large numbers from smallpox, diphtheria, whooping cough, polio, and many other diseases (**Figure 2.2**). The infectious pathogens that cause those diseases still exist today in our environment, but because vaccines protect people, we almost never see those infections. Today, the Centers for Disease Control and Prevention (CDC), the public health branch of the U.S. government, recommends that children receive 10 vaccines, given over a total of 24 doses, between birth and 15 months of age. Additional vaccinations are recommended between the ages of 18 months and 18 years, including the annual influenza ("flu") vaccine, also recommended for all adults each year (see "Flu Shot" on page 24).

Eaton, pregnant with her first baby, was a microbiologist by training. She worked at the Cleveland Clinic in Ohio, an academic medical center, before moving on to teach high school biology and chemistry, as well as bioscience manufacturing at a local community college. Some of her friends were also scientists, and one pregnant friend adamantly commented, "Of course we're going to vaccinate. We're scientists."

Vaccine   Vaccine   Antibodies   Disease organisms   Antibodies

A vaccine with a harmless form of a virus (or other organism) is injected under the skin.

The vaccine stimulates the immune system to produce antibodies (in green) that recognize the virus.

When the individual is exposed to the virus after vaccination, the new antibodies are primed to attack the invader.

Figure 2.1

How vaccines work

A vaccine trains the body's immune system to fight infection.

Q1: Describe in one sentence how a vaccine creates immunity to a virus.

Q2: Why is it impossible to become infected with a virus from a vaccine composed of viral proteins?

Q3: Natural immunity occurs without a vaccine, just by exposure to a particular stimulus, like the chicken pox virus. Explain why people don't get chicken pox twice.

ANNA EATON

Anna Eaton was, at the time of her research into vaccines, a high school and community college science teacher in Cleveland Heights, Ohio.

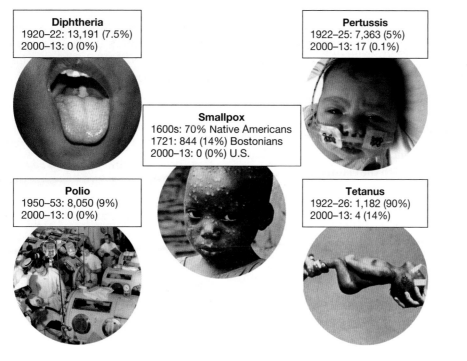

Diphtheria
1920–22: 13,191 (7.5%)
2000–13: 0 (0%)

Pertussis
1922–25: 7,363 (5%)
2000–13: 17 (0.1%)

Smallpox
1600s: 70% Native Americans
1721: 844 (14%) Bostonians
2000–13: 0 (0%) U.S.

Polio
1950–53: 8,050 (9%)
2000–13: 0 (0%)

Tetanus
1922–26: 1,182 (90%)
2000–13: 4 (14%)

Figure 2.2

Vaccines save lives

The numbers in each box represent each disease's average annual number of deaths and percent mortality rate in the United States during prevaccination periods (for example, 1950–53 for polio) and the postvaccination twenty-first century.

Q1: Before vaccinations, which diseases had the highest and lowest mortality rates? What are these mortality rates?

Q2: After vaccinations, which diseases have the highest and lowest mortality rates? What are these mortality rates?

Q3: If there were a sudden outbreak of pertussis at a university where pertussis vaccinations were not required and no one was protected, how many students would die? What is the probability or chance that you would die if infected?

But Eaton was not so confident. She had seen media stories suggesting that vaccines might cause autism spectrum disorders, and when she searched online, she read scary accounts from parents who had had a child vaccinated and then watched as their child developed autism. "I'm on message forums, encountering a hundred different moms with stories," recalls Eaton. "It was frightening."

Stories and anecdotes are powerful. They can change how we feel or think about a subject. But anecdotal evidence is not scientific evidence. Anecdotes are not representative of collected data or collected scientific observations, and therefore they cannot reliably give us the big picture of a subject or phenomenon. Scientific evidence can.

Eaton's scientist friend said there were data backing the safety of vaccines, but Eaton had not seen those data and wasn't about to inject her new baby girl with something she knew nothing about. "Feelings of distrust are in all of us, and possibly the strongest in a new mother," says Eaton. "Even though she was a friend of mine, it was still not enough for me to just go in the office and have my daughter vaccinated."

True or False?

The statement that vaccines cause autism, or the opposite statement that vaccines are safe, is a **scientific claim**, a statement about how the world works that can be tested using the scientific method. We are exposed to scientific claims every day—dozens of them, in fact. They are not all true. In 2004, editors at the magazine *Popular Science* asked one of their journalists to write down every scientific claim he heard in a day and evaluate each one. He recorded a whopping 106 claims and dug into each of them, from Cheerios' claim that it reduces cholesterol (supported by scientific evidence) to the claim that a face cream infused with vitamin A would revitalize skin (not supported). Of the 106 claims, most were bogus.

The majority of scientific claims come from advertisers. Though companies are legally bound to tell the truth, not all of them do. For example, in 2005 an Italian footwear company named Vibram introduced the $100 FiveFingers running shoes, glove-like shoes that mimic the feel of barefoot running (**Figure 2.3**). In a marketing campaign, the company touted the health benefits of these unique shoes, advertising that they could reduce foot injuries and strengthen foot muscles. In 2013, however, two peer-reviewed studies of over a hundred joggers found that the shoes actually increased the likelihood of injury. Soon, Vibram was sued for false advertising,

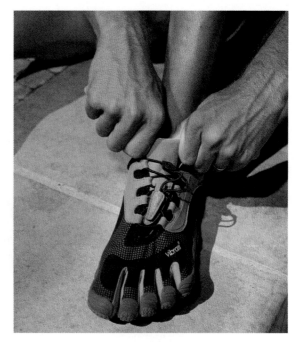

Figure 2.3

Shoes designed to mimic bare feet
FiveFingers running shoes did not live up to the manufacturer's scientific claims.

and in 2014 the company settled that lawsuit for $3.75 million, agreeing to refund FiveFingers shoe purchases and remove all health claims about the shoes from its advertisements.

Scientific claims also come from special-interest groups and organizations that exist to advance certain causes, often for political or religious reasons. These claims include statements about global warming, evolution, and medical care. Again, such claims are often untrue. Therefore, it is important to question the "truth" of a claim when you hear it.

We, the public, are not simply consumers of science and technology. We are participants. By voting on issues that have a scientific underpinning, we shape the course of science and influence which technologies are used, as well as where and how they're used. Although our personal values and political leanings are likely to influence how we vote, the underlying science should also be taken into consideration. **Scientific literacy**, an understanding of the basics of science and the scientific process, enables us to make informed decisions about the world around us and to communicate our knowledge to others. Our hope is that this book will help you become scientifically literate.

Flu Shot

Why do we need a flu shot every year? Why doesn't it protect us for life, like other vaccines do? In fact, each year's flu vaccine *does* protect you for life from a particular strain of the flu virus, but the virus mutates rapidly, and every year's flu strain is likely to be different from the previous year. To the immune system, the new strain looks like a completely new virus. But that's not all! That pesky flu virus can even combine itself with other flu viruses, from animals like birds and pigs, to create deadly flu pandemics like the H1N1 strain, or "swine flu," in 2009.

Each flu season, the CDC uses current flu cases to try to determine what the next season's flu viruses will be. It then creates a vaccine to protect us from the anticipated viruses. Sometimes the experts are extremely accurate and the flu vaccines are highly effective against that season's virus. Sometimes, however, they are not so accurate, and the vaccine is not as protective. It's a viral genetics puzzle and a guessing game, with a little attempt at clairvoyance thrown in. Still, even in years when the vaccine does not provide full coverage for that season, it is likely to protect you from future manifestations of the flu.

Here's the bottom line: The more yearly flu vaccines you receive, the more flu viruses your immune system will be ready to fight, and the less likely you will be to catch the seasonal flu. More important, you will be less likely to succumb to a deadly flu pandemic.

Eaton didn't know the source of the original claim that vaccines cause autism, but she knew that in order to be scientifically literate and make an informed decision for her daughter, she needed to find it. Scientific claims directly affect our lives because we make decisions based on them. Some of these are small decisions: Should I take a multivitamin every morning? How often should I exercise?

Others are larger decisions: Should I vote to support carbon taxes? Does a cell phone cause tumors from radiation? Should I vaccinate my children? The good news is that you can learn to evaluate scientific claims. You can be skeptical about claims and, using critical thinking, make scientifically literate decisions for yourself.

Credentials, Please

Caroline was born in January 2010 (**Figure 2.4**). Eaton had Caroline vaccinated when she was two months old, but then Eaton found herself gripped with fear. "I worried I was doing more harm than good," she says. Because of her rising fears, she did not get Caroline's four-month vaccinations.

Eaton sought out a pediatrician who would discuss her fears about vaccination. Several pediatricians refused to talk with her about a delayed vaccination schedule; some even refused to treat Caroline if she was not vaccinated. Finally, Eaton found a pediatrician who was willing to listen and discuss. After Eaton shared her concerns about vaccine safety, the pediatrician handed Eaton a book, *Vaccines and Your Child*, by a pediatrician named Paul Offit. In the book, Offit laid out how vaccines work, how they are made, and which risks are real and which are false. He also provided documentation, with detailed references to scientific papers, showing that there is no scientific evidence that vaccines cause autism.

Eaton found the book to be well written and informative, but she wondered whether she could trust Offit. She checked his **credentials**—a first step toward assessing the strength of a person's scientific claim. Does the person making a scientific claim have a PhD or MD? Is the degree in the field in which they are making the claim? PhDs in physics do not have training in germ theory, for example, and medical doctors do not have training in atmospheric science.

Figure 2.4

Anna and Caroline Eaton
Anna Eaton had concerns about vaccinations for her first child, Caroline.

Offit had an MD from the University of Maryland and was a professor of vaccinology and pediatrics at the University of Pennsylvania. He was also chief of the Division of Infectious Diseases and the director of the Vaccine Education Center at the Children's Hospital of Philadelphia. In other words, Offit had an advanced degree in an appropriate field and held a job at a reputable university. While good credentials alone do not guarantee that a source is trustworthy, scientists practice for many years in their area of expertise, and their scientific claims tend to be based on that expertise and carefully stated.

In addition to evaluating someone's credentials, it is important to assess whether the person making a scientific claim has an agenda or **bias** (a prejudice or opinion for or against something). Does the person have an ideological, political, or religious belief that will be supported by the scientific claims being made? Does the person have a

PAUL OFFIT

Paul Offit is a pediatrician and chief of the Division of Infectious Diseases at the Children's Hospital of Philadelphia.

conflict of interest? Does he or she stand to make money in any way if others accept those claims?

Doing scientific research almost always requires money, so it is important to take into account where the money comes from. In North America, the vast majority of **basic research** in science is funded by the federal government—that is, by taxpayers. Basic research is intended to expand the fundamental knowledge base of science. In the United States, Capitol Hill appropriates more than $30 billion each year for basic and medical research in the life sciences, including biomedicine and agriculture. Researchers must compete vigorously for the limited funds, and this competition helps ensure that public money goes toward supporting high-quality science. Research funded by the government is normally not considered biased, since the funding comes from taxpayers.

But industries and businesses spend a great deal of money funding science as well, often in areas of **applied research**, in which scientific knowledge is applied to human issues and often commercial applications. In some cases, researchers funded by industry may have a bias in favor of whatever that industry is selling. Funding from industry does not necessarily mean that a scientific claim is incorrect, but the claim should be looked at closely to rule out possible bias.

Offit, Eaton read, had worked with other researchers to develop the RotaTeq vaccine against rotavirus, a virus that can cause death from severe diarrhea. The pharmaceutical company Merck had purchased the vaccine, and Offit had received an unspecified amount of money from that transaction. This financial compensation raised the possibility that Offit might have some bias toward the use of the rotavirus vaccine. Though she had enjoyed Offit's book, Eaton decided she did not want to take his word alone as an answer to her questions about vaccines. "In the end, I decided to just read about it, do a lot of research, and talk to a lot of people," she says.

To the Books

Eaton dove into the secondary literature on vaccines. When investigating a scientific claim, your first stop should be the Internet or the library to get a basic overview of the topic from the **secondary literature**, which summarizes and synthesizes an area of research. Textbooks, review

articles, and popular science magazines such as *National Geographic*, *Popular Science*, and *Scientific American* are good secondary sources.

Eaton went to her local library and came home with a stack of books, including a thick vaccine textbook so that she could learn the underlying science of how a vaccine works inside the body. She read about how vaccines stimulate cells of the immune system to protect a person from a virus or bacteria. Once, while she was reading, Eaton's husband walked into the room and looked at her with surprise. "What are you doing?" he asked. "I'm looking for answers!" she replied.

For secondary literature on the Internet, try to visit sites that are affiliated with the government, a university, or a respected institution like a major hospital or museum. Wikipedia often has overview articles that link to science blogs and review articles in science journals. Like Eaton did, it's important to check the credentials of the person or people behind a resource, especially on the Internet. Anonymous sources are not to be trusted.

When evaluating a scientific claim, you may need more detailed information than is available in the secondary literature, especially if you're dealing with a particularly important life decision or if the area of science involved is changing rapidly. In that case, you should next review the **primary literature**, where scientific research is first published (**Figure 2.5**). Primary sources include technical reports, conference proceedings, and dissertations, but the most important primary sources are peer-reviewed scientific journals such as *Science*, *Ecology*, and the *Journal of the American Medical Association* (*JAMA*).

Pulling from references she found in the secondary literature, Eaton compiled a stack of primary literature on vaccines. If her library didn't carry a particular journal, she acquired a copy through interlibrary loan or, in some cases, even e-mailed the journal directly and got articles for free. She found dozens and dozens of papers about vaccines.

Correlation or Causation?

One of the first papers that Eaton came across in her investigation—one that had made huge waves in the media—was published in the *Lancet* in

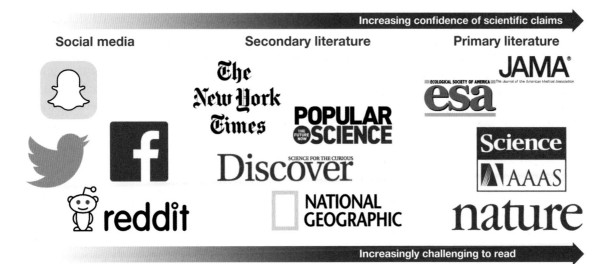

Figure 2.5

Scientific claims in the media and literature
It is easy to find and read scientific claims in social media. However, this is not a good source of scientific information. For help in making important life choices, it is important to go to the secondary literature or even the primary scientific literature for accurate and reliable information.

Q1: Why are we less confident of scientific claims made over social media?

Q2: Where would you place a blog in this figure? Would it matter whether or not it was written by a practicing scientist? Explain your reasoning.

Q3: Give an example of when you would rely on secondary literature to evaluate a scientific claim and an example of when you would go to the primary literature. What is the basis of that decision?

1998. The *Lancet* is a well-known peer-reviewed medical journal and therefore a reputable source. The paper was unremarkably titled "Ileal-Lymphoid-Nodular Hyperplasia, Non-specific Colitis, and Pervasive Developmental Disorder in Children." It was a study of 12 children ranging in age from 3 to 10 years old who had experienced a loss of language skills—a symptom of autism spectrum disorders—as well as diarrhea and abdominal pain. Parents of 8 of the 12 children said the onset of symptoms occurred shortly after the child's immunization with the measles, mumps, and rubella (MMR) vaccine.

The authors of the paper, a team of 12 researchers, concluded that more research was needed to study a possible relation between the observed brain dysfunction, bowel problems, and the MMR vaccine. In a press conference when the paper was published, one of the authors, a British doctor named Andrew Wakefield, stated that he believed single vaccines, rather than the MMR triple vaccine, were likely to be "safer" for children. The study and press conference sparked widespread fear among parents that the MMR vaccine could cause autism.

The study was published around the same time that officials began documenting an increase in rates of autism. Since the early 1970s, when organizations started counting the number of people with autism for the first time, the incidence of autism has increased 20- to 30-fold in the United States and other countries. In 2002, the estimate was that one in 150 children aged eight years (the age of peak prevalence) had the disorder. In 2004, that number had risen to one in 125. In 2006, one in 110. By the year 2012, the most recent year with available data, an estimated one in every 68 eight-year-olds had an autism spectrum disorder. The CDC has said that the rise in rates is likely due to heightened disease awareness, more screening

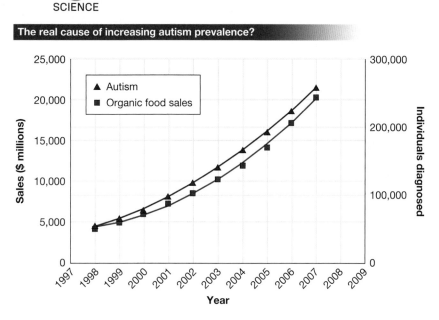

The real cause of increasing autism prevalence?

Figure 2.6

Correlation is not causation: organic food and autism

Reddit user Jasonp55 created a tongue-in-cheek demonstration of why it is important not to assume causation from correlation. Jason used real data on organic food sales and the prevalence of autism from 1998 to 2007. The two are highly correlated, but one does not cause the other.

> **Q1:** How much did organic food sales grow during the period covered in the graph? How much did the incidence of autism grow?
>
> **Q2:** Why might both organic food sales and autism prevalence have increased during this time period? A Reddit user in the original discussion thread suggested that both might be affected by increasing wealth in the United States. How might increased wealth affect these variables?
>
> **Q3:** In what way has the vaccine-autism debate been confused by people misinterpreting correlation as causation?

within schools, and a willingness to label the condition. But after Wakefield suggested that the MMR vaccine might be causing autism, the press and other organizations began to report that the rise in autism rates was caused by the increased use of vaccines.

Linking rising rates of autism with increased use of vaccines is a correlation. **Correlation** means that two or more aspects of the natural world behave in an interrelated manner: if one shows a particular value for one aspect, we can predict a value for the other aspect. But correlation does not establish **causation**, in which a change in one aspect *causes* a change in another. Correlations may suggest possible causes for a

phenomenon, but they do not establish a cause-and-effect relationship. For example, there is also a correlation between organic food sales and the increase in autism. From 1998 to 2007, rates of autism increased hand in hand with organic food sales. They are correlated, but there is no scientific evidence that eating organic food causes autism (**Figure 2.6**).

Another correlation that spurred fears about vaccines is that the onset of autism symptoms occurs at about the same age that children receive vaccinations. Most children receive the MMR vaccine at about 15 months old, which is shortly before the first symptoms of autism are often noticed. Parents of children with autism saw their children begin to exhibit symptoms of the illness after their vaccination and therefore directly observed a correlation between the administration of the vaccine and the onset of an autism spectrum disorder.

Yet, again, correlation does not prove causation. Only scientific experiments can demonstrate causation. So, was Wakefield's claim based on good science? Did the MMR vaccine cause autism?

Real or Pseudo?

Unfortunately, sometimes a claim that superficially looks like science is actually **pseudoscience**. Pseudoscience is characterized by scientific-sounding statements, beliefs, or practices that are not actually based on the scientific method. Asking a few simple questions at each step of the progression toward a claim purported to be "scientific" can help you distinguish science from pseudoscience (**Figure 2.7**).

Using these criteria, Eaton analyzed Wakefield's claim. It was quickly obvious that his study did not live up to the standards of good scientific research (**Figure 2.8**). First, the study was small—only 12 children participated—yet the conclusions were grand: Wakefield suggested that all children should stop receiving the MMR vaccine. Sample size is extremely important in observational studies, where small samples may skew data to one extreme. Large sample sizes are more representative of the population, are less likely to be affected by outliers, and provide the power to draw more accurate conclusions.

A second problem with the study was that it was not made up of a random sample of

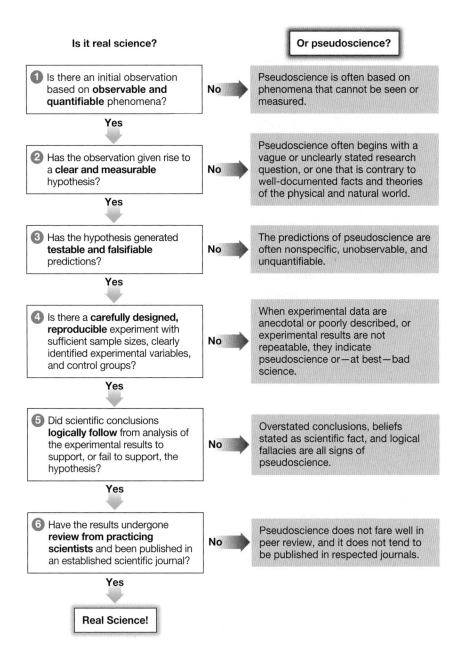

Is it real science?

Or pseudoscience?

1 Is there an initial observation based on **observable and quantifiable** phenomena?

No → Pseudoscience is often based on phenomena that cannot be seen or measured.

Yes ↓

2 Has the observation given rise to a **clear and measurable** hypothesis?

No → Pseudoscience often begins with a vague or unclearly stated research question, or one that is contrary to well-documented facts and theories of the physical and natural world.

Yes ↓

3 Has the hypothesis generated **testable and falsifiable** predictions?

No → The predictions of pseudoscience are often nonspecific, unobservable, and unquantifiable.

Yes ↓

4 Is there a **carefully designed, reproducible** experiment with sufficient sample sizes, clearly identified experimental variables, and control groups?

No → When experimental data are anecdotal or poorly described, or experimental results are not repeatable, they indicate pseudoscience or—at best—bad science.

Yes ↓

5 Did scientific conclusions **logically follow** from analysis of the experimental results to support, or fail to support, the hypothesis?

No → Overstated conclusions, beliefs stated as scientific fact, and logical fallacies are all signs of pseudoscience.

Yes ↓

6 Have the results undergone **review from practicing scientists** and been published in an established scientific journal?

No → Pseudoscience does not fare well in peer review, and it does not tend to be published in respected journals.

Yes ↓

Real Science!

Figure 2.7

Science or pseudoscience?

A series of simple questions based on the scientific method can help you determine whether a "scientific" study is real science or pseudoscience.

Q1: State the hypothesis of the people who believe that vaccines cause autism. What is an alternative hypothesis?

Q2: What part(s) of the figure show(s) where Wakefield's study failed to meet the standards of the scientific method?

Q3: Why does only one arrow point to "real science," whereas multiple arrows point to pseudoscience?

CREDENTIALS: Does the person making a scientific claim have a PhD or MD? Is the degree in the field in which the claim is being made?

BIAS or AGENDA: Does the person making the scientific claim hold a belief that will be supported by the claim? Does he/she stand to make money if others accept the claim?

Evaluating Scientific Claims

PEER REVIEW & PUBLICATION: Has the claim been reviewed by practicing scientists and published in a reputable scientific journal?

REAL SCIENCE or PSEUDOSCIENCE? Does the scientific study purporting to support the claim meet the standards of the scientific method?

Figure 2.8

Evaluating scientific claims

We are constantly bombarded with scientific claims. A few simple questions will help you evaluate which of these claims are valid and which are not.

Q1: Why is it important to know the education and expertise of a person making a scientific claim?

Q2: List at least five possible biases that people making scientific claims might have.

Q3: Describe a situation in which you might not dismiss the scientific claim of a person who did not have appropriate credentials, or who had a bias toward the claim.

children. Instead, the subjects were picked specifically for their symptoms. Third, the study did not have a control group, such as children who had been vaccinated but did not show signs of autism, or children who had autism but had not been vaccinated. Finally, and most important, the finding could not be repeated. In fact, as of October 2016, at least 110 published papers, studying millions of children, have found no evidence of a link between vaccines and autism. Vaccination does not cause autism, and the two are not even directly correlated. In fact, studies first published in March 2014 provide evidence that autism begins in the womb, before birth.

In one of the most recent papers, researchers at the University of Sydney in Australia, who stated no conflicts of interest, performed a **meta-analysis** (work combining results from different studies) of five *cohort studies*—observational studies of a group of people over a certain period of time—involving 1,256,407 children. The researchers also looked at five *case-control studies*—studies comparing patients with a disease to those without—involving 9,920 children. The data showed no relationship between vaccination and autism, autism and MMR, or autism and thimerosal (a mercury-containing preservative in some vaccines). The work, published in the journal *Vaccine*, included data from the United States, United Kingdom, Japan, and Denmark on vaccines for measles, mumps, diphtheria, tetanus, and pertussis.

Many more peer-reviewed studies with good control groups and sufficient sample sizes, conducted by researchers without a conflict of interest, have shown that the number of vaccines given from birth to 15 months of age under the CDC's recommended vaccination schedule is safe. Researchers have also found that delaying the vaccination schedule increases a child's risk of contracting a disease.

Today it is clear that the paper that started it all, Wakefield's 1998 *Lancet* paper, proposed a claim that is incorrect. All told, it's OK for science to be wrong. There is no expectation that every single one of the thousands of studies published each year will be correct. Yet in the case of the *Lancet* paper, there was bias and wrongdoing. Years after publication of the paper, it came to light that Andrew Wakefield had received large amounts of money as a paid expert for lawyers who were suing vaccine manufacturers. Wakefield had also applied for a patent on a vaccine that would rival the most commonly used MMR vaccine.

Because of these conflicts of interest, in 2010 the General Medical Council, an organization that licenses all medical doctors in the United Kingdom, concluded that Wakefield's conduct had been "irresponsible and dishonest" and revoked his medical license. Eventually, 10 of the paper's 11 other authors retracted their support of the study and its conclusions. Ultimately, in February of 2010, the *Lancet* officially retracted the paper—a very rare action for publishers (**Figure 2.9**). A peer-reviewed paper is retracted (withdrawn as untrue or inaccurate) by a publisher or author when its findings are no longer considered trustworthy because of error, plagiarism, a violation of ethical guidelines, or other scientific misconduct.

Fears versus Facts

Wakefield's paper had not yet been retracted when Eaton was making her decision about whether to vaccinate Caroline. But Eaton followed the guidelines to distinguish science from pseudoscience, and she concluded that Wakefield and his colleagues had looked at too small a sample of children. Therefore, the paper "did not make a lasting impression," recalls Eaton. So she looked beyond that one paper to others. The multitude of papers she found, all

The Lancet, Volume 351, Issue 9103, Pages 637 - 641, 28 February 1998
doi:10.1016/S0140-6736(97)11096-0 (?) Cite or Link Using DOI

< Previous Article | Next Article >

This article was retracted

RETRACTED: Ileal-lymphoid-nodular hyperplasia, non-specific colitis, and pervasive developmental disorder in children

Dr AJ Wakefield FRCS [a][✉], SH Murch MB [b], A Anthony MB [a], J Linnell PhD [a], DM Casson MRCP [b], M Malik MRCP [b], M Berelowitz FRCPsych [c], AP Dhillon MRCPath [a], MA Thomson FRCP [b], P Harvey FRCP [d], A Valentine FRCR [e], SE Davies MRCPath [a], JA Walker-Smith FRCP [a]

Summary

Background
We investigated a consecutive series of children with chronic enterocolitis and regressive developmental disorder.

Methods
12 children (mean age 6 years [range 3–10], 11 boys) were referred to a paediatric gastroenterology unit with a history of normal development followed by loss of acquired skills, including language, together with diarrhoea and

Figure 2.9

The paper that precipitated the vaccine-autism scare is debunked and retracted

with controls and larger sample sizes, assured her that vaccination does not cause autism. In the end, her research and scientific literacy led Eaton to fully vaccinate all of her children on the CDC's recommended schedule.

In making her decision, Eaton was aware that vaccines can have side effects. Most are minor, like a sore arm or a low-grade fever. In some very rare cases, however, vaccination can result in a serious reaction. An allergic reaction to the hepatitis B vaccine, for example, is estimated to occur once in 1.1 million doses. A serious allergic reaction to the MMR vaccine is seen less than once in a million doses. The CDC maintains a list of side effects and constantly monitors the safety of vaccines (**Figure 2.10**).

Weighing the benefits of avoiding disease against the small risk of side effects made the choice to vaccinate her children clear, says Eaton. "We [parents] are often so focused on possible effects of the vaccines that we don't think about our kids getting these diseases" that vaccines prevent, she says. The same was not true for your great-grandparents; in the early 1900s, one in five children worldwide died of a vaccine-preventable disease.

When making life choices based on scientific claims, you sometimes need to consider nonscientific aspects of your decision. Your

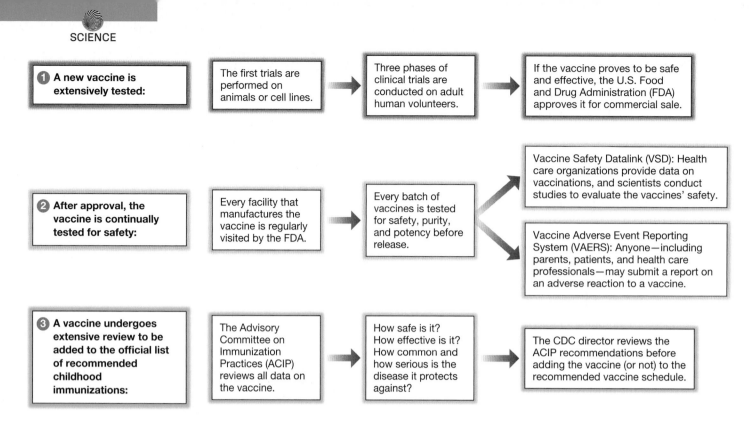

Figure 2.10

Evaluating vaccines: an ongoing process

Vaccines are continually tested and evaluated for effectiveness, safety, and side effects. The severity of the disease a vaccine works against, as well as how likely a child is to be infected with the disease, is also taken into consideration in determining whether to recommend the vaccine.

Q1: Why do vaccine manufacturers begin with tests on animals or cell lines before moving on to adult human subjects?

Q2: What ongoing testing and reporting are vaccines subjected to?

Q3: What do ACIP, FDA, and CDC stand for, and what is the role of each in evaluating vaccines?

values, ethical stances, and religious beliefs will make some choices more acceptable than others. For example, if you believe it is not ethical to eat meat, the scientific finding that leaner meats are healthier than fattier meats won't matter to you.

It is also important to consider how your choices will affect other people. Choosing not to vaccinate a child, for instance, affects the entire community. Rates of vaccination in the United States have been hovering around 75 percent because of misconceptions about the safety of vaccines, and this low level of immunization puts at risk not only the children who are not vaccinated, but also those who *cannot* be vaccinated, such as infants too young for a vaccine

or people who are genetically unable to respond to a vaccine. When a critical portion of a population is vaccinated, then the spread of disease is contained—a concept known as **herd immunity** (**Figure 2.11**). In other words, vaccinating a large number of people keeps germs out of circulation and protects the vulnerable members of the community.

When parents opt not to vaccinate their children, herd immunity disappears. And because of that loss of herd immunity, infectious diseases of the past, some thought to be eliminated, are roaring back. In August 2013, a megachurch in Texas whose founder had spoken out against vaccines made headlines after 21 members of its congregation contracted

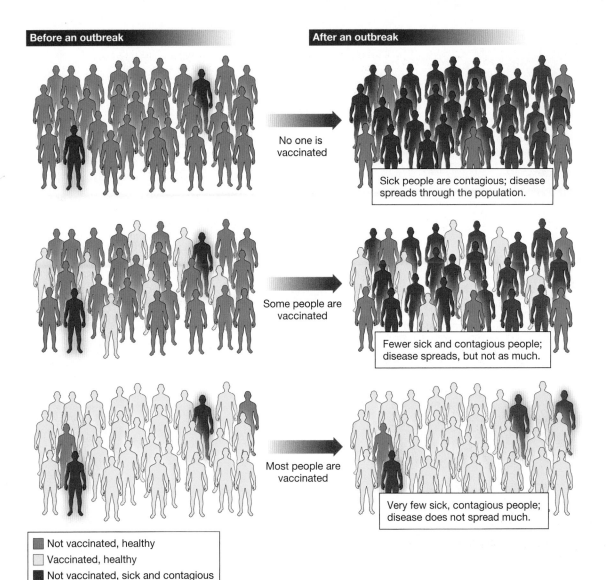

| | Before an outbreak | After an outbreak |

No one is vaccinated

Sick people are contagious; disease spreads through the population.

Some people are vaccinated

Fewer sick and contagious people; disease spreads, but not as much.

Most people are vaccinated

Very few sick, contagious people; disease does not spread much.

Not vaccinated, healthy
Vaccinated, healthy
Not vaccinated, sick and contagious

Figure 2.11

Vaccine prevalence and herd immunity

When the majority of the population is not vaccinated, an outbreak of disease will spread further and may cause disease and death in members of the population who are too young to be vaccinated or have a compromised immune system. However, when most of the population is vaccinated against a contagious disease, the disease will spread little during an outbreak; this tighter containment of a disease is the result of what is known as herd immunity. Image modified from the National Institute of Allergy and Infectious Diseases (NIAID).

Q1: What happens to an immunized person when a disease spreads through a population? (*Hint:* In the graphic, follow an immunized individual before and after a disease spreads.)

Q2: Explain why a disease is less likely to spread to vulnerable members of a population if most people are immunized.

Q3: How does vaccination help an individual person? How does it help that person's community?

Safety in Numbers

In 1999, the United Kingdom began a meningitis vaccination program for children up to the age of 18. Meningitis rates rapidly fell off throughout the population—children and adults alike—thanks to both the effectiveness of the vaccine and the protective effects of herd immunity. Models show that without herd immunity, the number of meningitis cases would have rebounded. The U.S. CDC recommends meningitis vaccination up to the age of 23, as outbreaks have been reported on college campuses in recent years.

Assessment available in smartw●rk5

Meningitis cases in England and Wales

measles. Sixteen of them were unvaccinated. In 2014, measles cases in the United States hit a 20-year high, with 667 reported cases. And in the 2013–14 flu season, 90 percent of the 100 American children who died of complications from the flu had not received the flu vaccine. Decisions based on scientific claims have serious consequences.

Today, Eaton is thankful that all of her children are fully vaccinated (**Figure 2.12**). She encourages her friends, relatives, and colleagues to foster scientific literacy and use their critical-thinking skills to distinguish science from pseudoscience. Not everyone will want to read a vaccine textbook, Eaton admits with a laugh, but anyone can think critically and follow the process of evaluating scientific claims.

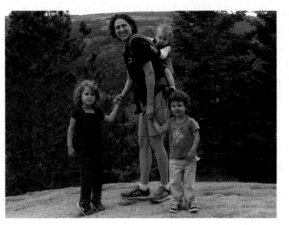

"Look to see what has credible backing and what doesn't. This issue transcends vaccines," says Eaton. "We should all have the tools to become scientifically literate citizens."

Figure 2.12

Anna Eaton and her children
Although Eaton initially had concerns about vaccinations for her first child, Caroline, her research gave her confidence that vaccines are a safe and responsible choice. Caroline and her younger brothers are now all up-to-date on their vaccinations.

REVIEWING THE SCIENCE

- A **scientific claim** is a statement about how the world works that can be tested using the scientific method. To evaluate scientific claims, it is important to look at the **credentials** and **bias** of those making the claim.

- **Scientific literacy** requires a basic understanding of scientific facts and theories, and of the process of science. It is important to be scientifically literate in order to make well-informed life decisions.

- Scientific claims can be found in advertising, in social media, in the popular press, and in scientific publications. The best source for a review or overview of a scientific topic is the **secondary literature**, including the popular press, reputable websites, or review articles in scientific journals.

- The actual experimental or observational results related to a scientific claim are found in the **primary literature**, articles published in scientific journals that have undergone peer review.

- With **correlation**, two or more aspects of the natural world behave in an interrelated manner. With **causation**, a change in one aspect *causes* a change in another.

- **Basic research** explores questions about the natural world and expands the fundamental knowledge base of science. **Applied research** seeks to use the knowledge gained from basic research to address human issues and concerns. It may involve developing commercial applications.

- **Pseudoscience** is characterized by scientific-sounding statements, beliefs, or practices that do not meet the standards of the scientific method.

THE QUESTIONS

The Basics

1 Which of the following should receive the *least* consideration when evaluating a scientific claim?

(a) the scientific credentials of the person making the claim

(b) your personal beliefs and values

(c) whether the study supporting the claim has been published in a peer-reviewed scientific journal

(d) whether the study supporting the claim meets the standards of the scientific method

(e) any possible biases of the person making the claim

2 Scientific literacy means that you

(a) are able to easily read and understand a scientific journal article.

(b) have taken a university-level science course.

(c) understand the process of science and basic scientific facts and theories.

(d) enjoy reading current science news in newspapers and blogs.

(e) are a good critical thinker.

3 "Correlation does not prove causation" means that

(a) if two variables are correlated, one is likely to have caused the other.

(b) if changes in one variable cause changes in another, the variables are not correlated.

(c) only experimental research can answer questions about the natural world.

(d) although two variables are interrelated, changes in one do not necessarily cause changes in the other.

(e) none of the above

4 Link each term with the correct definition.

SCIENTIFIC LITERACY	1. Uses scientific knowledge to address human issues.
BASIC RESEARCH	2. Helps in making informed life choices.
APPLIED RESEARCH	3. Consists of peer-reviewed scientific journal articles.
SECONDARY LITERATURE	4. Gives an overview of scientific findings on a subject.
PRIMARY LITERATURE	5. Contributes to fundamental science knowledge.

5 Select the correct terms:

Evaluating a scientific claim begins with reviewing the (**credentials / fame**) of the person making the claim. It is also important to know whether those making a particular claim have a (**detachment / bias**), a vested interest in whether or not the claim is true. To gain an overview of scientific studies related to the claim, it is helpful to read the (**primary literature / secondary literature**).

6 You are trying to determine whether a scientific claim is based on real science or pseudoscience. Place the following questions you will address in the correct order by numbering them from 1 to 6.

_____ a. Are the study's claims observable and quantifiable?

_____ b. Has the study been reviewed by practicing scientists and published in an established scientific journal?

_____ c. Are the predictions specific, testable, and falsifiable?

_____ d. Is the hypothesis clearly stated, measurable, and aligned with current scientific facts and theories?

_____ e. Are the experimental design and analysis well described, well designed, reproducible, and conducted with a large sample size?

_____ f. Are the study conclusions logical, based on evidence, and justified, given the study results?

Challenge Yourself

7 An example of basic research is

(a) a study of how hummingbirds learn song.

(b) an investigation of how the melting of polar ice caps affects agriculture.

(c) looking at possible genetic contributions to autism spectrum disorder.

(d) designing more effective vaccines for dangerous infectious diseases.

(e) exploring how agricultural waste can be turned into fuel.

8 For each of the following, select the term that best describes the type of literature it represents: primary, secondary, or neither.

_____ a. A research study from Dr. Drake and colleagues on the blood sugar levels of diabetic rats that eat only kale, compared to control diets, is published in a peer-reviewed scientific journal.

_____ b. In the infomercial selling his own health care products, Dr. Horton states, "I believe my personal cure of daily meditation and yoga for diabetes is more effective than any drug I've ever prescribed in 20 years of practicing medicine."

_____ c. Dr. DeBellard insists in her blog that people will lose weight in a healthy manner if they acquire some of her personal parasitic tapeworm.

_____ d. In an article published in the *Annual Review of Nutrition*, a peer-reviewed journal, Dr. Pepper summarizes the last 10 years of basic research on diet and diabetes.

9 Which of the following situations has the greatest potential for biased or inaccurate results in an experimental process?

(a) Ms. Ochoa-Bolton is an outside consultant who is conducting a health and wellness survey for a pharmaceutical company. She does not know the name of the company, nor does she know the name of the drug being tested during the survey.

(b) Ms. Adamian is a research technician surveying a study group on response to a new cold remedy. She knows only the e-mail address of each subject and asks them identical questions by computer.

(c) Dr. Wisidagama is evaluating cancer patients for their responses to a new therapeutic drug. She knows which patients are receiving the placebo and which are receiving the drug.

(d) Dr. Waters is analyzing biopsy samples from rats that have been given either a placebo or an experimental drug believed to reduce inflammation. Each sample is identified by a code number such that she cannot tell which treatment each rat received.

(e) Ms. Nuno is conducting a survey of weight loss regimens as part of her master's degree project. Her online survey is anonymous and asks each participant the same questions.

Try Something New

10 Determine whether each of the following statements is likely to represent real science or pseudoscience. In each case of likely pseudoscience, identify which scientific standard is not met.

_____ a. Dr. Oz says that green coffee beans will burn fat, so you can lose weight without dieting.

_____ b. The *Wall Street Journal* reports that climate scientists have conspired to exaggerate the effects of global climate change.

_____ c. A group of researchers reports at a national scientific conference that they have found a genetic link to autism.

_____ d. Many astrologers agree that people born under the sun sign Aquarius are more intelligent than those born under Scorpio.

_____ e. A study published in the scientific journal *Diabetes* finds that sleeping in a cooler room may increase metabolic rate and insulin sensitivity.

11 The following graphs illustrate the incidence of pertussis (whooping cough) cases in the United States. The first graph organizes the data by year from 1922 to 2012, with the inset showing a zoomed-in view of 1990–2012. The second graph organizes the data by age group from 1990 to 2012. (The 2013 data are not complete in either graph.) DTP, Tdap, and DTaP are different formulations of the vaccine that covers tetanus, diphtheria, and pertussis.

a. Describe what the first graph shows. What do each of the axes represent? What does any point on the line show? What general trend is seen, if all the data shown in the graph are considered?

b. Why do you think there was an increase in cases of pertussis in the first decade of the twenty-first century?

c. Describe what the second graph shows. What do each of the axes represent? What does any point on each of the different lines show? What general trend is seen, if all the lines are considered? What differs among the lines?

d. Compare the incidence of pertussis cases in children under 1 year old and people over the age of 20.

e. Summarize your reflections on reviewing these graphs and the relative risks of pertussis and the pertussis vaccine. What would you recommend to someone trying to decide whether to vaccinate a child?

12 Human papillomavirus (HPV) is the most common sexually transmitted disease in the United States; almost all sexually active people are infected with HPV at some point. HPV can be contagious even when an infected person shows no symptoms, and symptoms may not appear until years after infection. Some but not all strains of HPV can cause genital warts and cancer. The HPV vaccine is very effective and is recommended for all young people from the ages of 11 to 26 years. What steps would you go through to decide whether to vaccinate yourself or your child with the HPV vaccine? (*Hint:* Use the method for evaluating scientific claims, including assessing real science versus pseudoscience.)

Leveling Up

13 **Life choices** Most American universities require students to have the following immunizations: MMR (measles, mumps, and rubella), varicella (chicken pox), and Tdap (tetanus, diphtheria, and pertussis/whooping cough). Most also recommend or require hepatitis A and B, meningococcal conjugate (meningitis), HPV (human papillomavirus), poliovirus, and the annual flu vaccine.

a. Which of these vaccines have you received?

b. If you have not received one or more of them, why not?

c. Some school districts allow parents to "opt out" of vaccines because of their belief system. Do you think this option should be allowed for university students? Why or why not? What are the possible consequences of allowing people to opt out of vaccinations?

14 *Write Now* **biology: evaluating scientific claims** John Oliver, in a 2014 episode of *Last Week Tonight*, discussed a poll finding that one in four Americans is skeptical of global climate change. He dismissed the poll results and compared it to a poll asking, "Which number is bigger, 15 or 5?" or "Do owls exist?" Why does Oliver feel that what the American public believes about climate change is not relevant? (*Hint:* It could be argued that each of the people polled was making a scientific claim, either in support of or against the scientific consensus on climate change.)

15 **Is it science?** Watch the 2011 movie *Contagion*, which depicts the spread of a fictional virus. While watching the movie, make a list of the scientific concepts that are presented, and note whether you think they represent real science or pseudoscience. Then use your textbook and Internet research to determine whether you were correct or incorrect in each case. If you evaluated any of the concepts incorrectly, reflect on how you came to your initial conclusions.

16 **What do *you* think?** Read Peg Kehret's novel *Small Steps: The Year I Got Polio*, the author's true account of contracting polio as a seventh-grader. Keep a reflective journal while reading the book, documenting your emotional reactions to each chapter in this story. While reading, also note where you learned something new that you did not know previously about the poliovirus, the symptoms of polio, and the treatment options (or lack thereof) in the late 1940s. In your final entry, please reflect on how this novel did or did not change your feelings toward vaccination.

For more, visit digital.wwnorton.com/bionow2 for access to:

Ingredients for Life

A dusty box, rediscovered after 50 years, sparks a scientific treasure hunt for the origins of life on Earth.

After reading this chapter you should be able to:

Compare and contrast the types of bonds used to form molecules.

Diagram the hydrogen bonds that form between water molecules and explain how those bonds produce the unique properties of water.

Explain the difference between hydrophilic and hydrophobic molecules, and between acidic and basic molecules.

Predict whether a solution has a high or low concentration of free hydrogen ions by using its pH number.

Describe the chemical qualities of carbon that make it the basis of life on Earth.

Create a graphic depicting how each of the four main classes of biomolecules and their subunits are related.

CHAPTER
03

CHEMISTRY OF
LIFE

Jim Cleaves hauled yet another box to the dumpster. The young, dark-haired researcher at UC San Diego had the unenviable task of cleaning out the laboratory of his PhD adviser, the recently retired chemist Stanley Miller. It was 2003. Cleaves tossed box after box—20 years' worth of chemical samples from experiments done by Miller and his former students. As the lab emptied, a small cardboard box high on a shelf caught Cleaves's eye. It was labeled "Electric Discharge Samples."

"That box looked like something I'd really regret throwing out," says Cleaves. So instead of making a trip to the trash, Cleaves gave the box to another former student and friend of Miller's, marine chemist Jeffrey Bada at the Scripps Institution of Oceanography. Once again, the box was relegated to a shelf, where it sat, unopened, for years. Neither Bada nor Cleaves had any idea there was a scientific treasure trove inside, preserved for them by Miller.

In 1952, Miller was a thin, bespectacled graduate student at the University of Chicago, looking for a thesis idea. He approached Harold Urey, a Nobel Prize–winning chemist who, a year and a half earlier, had proposed a radical hypothesis about Earth's early atmosphere and the origins of life. Scientists knew that several key types of **matter** (anything that has mass and occupies a volume of space) existed on the early Earth, but they debated which types were necessary for the emergence of life 3–4 billion years ago. One type of matter is an **element**, a pure substance that has distinct physical and chemical properties, and that cannot be broken down into other substances by ordinary chemical methods. There are 98 natural elements known to us, and another 20 have been created in laboratories.

An **atom** is the smallest unit of an element that retains the element's distinctive properties. Atoms make up all common materials, including this book, the air, and you. Every atom has a dense core called a **nucleus** (plural "nuclei") made up of positively charged **protons** and electrically neutral **neutrons**. A cloud of negatively charged **electrons** surrounds the nucleus (**Figure 3.1**). Electrons have significantly less mass than protons and neutrons have: if an electron weighed as much as a 1-liter bottle of water, a proton or neutron would be as heavy as a car.

The number of protons in an atom's nucleus is called its **atomic number** and is unique to that element. **Isotopes** of an element have the same number of protons but different numbers of neutrons. The sum of the number of protons and the number of neutrons is the **atomic mass number** of an isotope. The atomic mass number is how much mass is in an element or, in other words, how much it weighs. For example, the most common isotope of carbon has 6 protons and 6 neutrons, giving it an atomic mass number of 12, and we call it carbon-12 (^{12}C). The isotope of carbon with 6 protons and 8 neutrons is carbon-14 (^{14}C).

Atoms interact with other atoms via electrons; they can donate electrons, accept electrons, and even share electrons. When two atoms share electrons, they form a **covalent bond**. Atoms linked by covalent bonds form **molecules**. Molecules that include at least one carbon atom are referred to as **organic molecules**. Urey and others suspected that gas molecules in early Earth's atmosphere combined to form the earliest organic molecules, which later assembled into the first living organism.

Urey proposed that Earth's early atmosphere resembled that of other planets in our solar system, including Jupiter, Saturn, and Uranus. He suggested that, like the atmospheres of those planets, our atmosphere was once rich with important **chemical compounds**, molecules that contain atoms from two or more elements. Earth's early atmosphere, he hypothesized, was made up of mainly four compounds: the gases methane (CH_4), ammonia (NH_3), hydrogen (H_2), and water vapor (H_2O).

The young Miller was intrigued by Urey's hypothesis, and he asked Urey if he could perform an experiment to test it, to see whether such a combination of gases could be used to create other simple compounds. Urey dismissed the idea, saying that it was too difficult for a student's thesis and that, if it didn't work, Miller would have nothing to show in order to graduate. But Miller pestered him until at last Urey

STANLEY MILLER

Stanley Miller was an American chemist who designed the first experiment to mimic Earth's early atmosphere and pioneered the study of the origins of life. He died in 2007 at the age of 77.

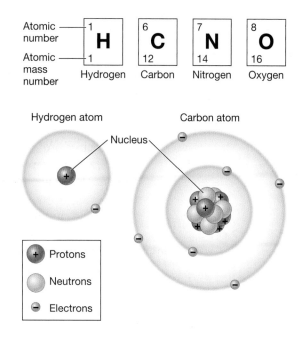

Atomic number — [1 **H** 1] [6 **C** 12] [7 **N** 14] [8 **O** 16]
Atomic mass number —

Hydrogen   Carbon   Nitrogen   Oxygen

Hydrogen atom   Carbon atom

Nucleus

+ Protons

◯ Neutrons

⊖ Electrons

Figure 3.1

Atomic structure

The electrons, protons, and neutrons of these hydrogen and carbon atoms are shown greatly enlarged in relation to the size of the whole atom.

Q1: How many protons, electrons, and neutrons does the hydrogen atom shown here have? What are the atomic number and the atomic mass number of the hydrogen atom?

Q2: What are the atomic number and the atomic mass number of the carbon isotope shown?

Q3: Nitrogen-11 is an isotope of nitrogen that has 7 protons and 4 neutrons. What are the atomic number and atomic mass number of nitrogen-11?

agreed to let Miller work on the project for six months, a year at the most. As it turned out, all Miller needed was a few weeks.

Miller mixed the four gases together in a large glass apparatus. He then zapped the swirling cloud of compounds with a continuous electrical spark to mimic lightning strikes on early Earth, which he surmised would break apart the compounds (**Figure 3.2**). The process of breaking existing chemical bonds and creating new ones is known as a **chemical reaction**. The **reactants** (the gases, in this case) undergo a chemical change and form new molecules, called **products**. Some chemical reactions, like the one Miller devised, require energy in order to proceed. Others release energy.

After the first day of the experiment, a pink liquid pooled in the bottom of the apparatus. By the end of a week, it had turned "deep red and turbid," Miller reported. The color changes, he later discovered, reflected the formation of new products through chemical reactions.

To Miller and Urey's pleased surprise, the resulting red broth contained several biologically significant compounds, including five **amino acids**, small molecules important to life.

HAROLD UREY

Harold Urey was an American chemist who won the Nobel Prize in Chemistry in 1934 for the discovery of an isotope of hydrogen. He was the first to speculate that Earth's early atmosphere was composed of ammonia, methane, hydrogen, and water vapor.

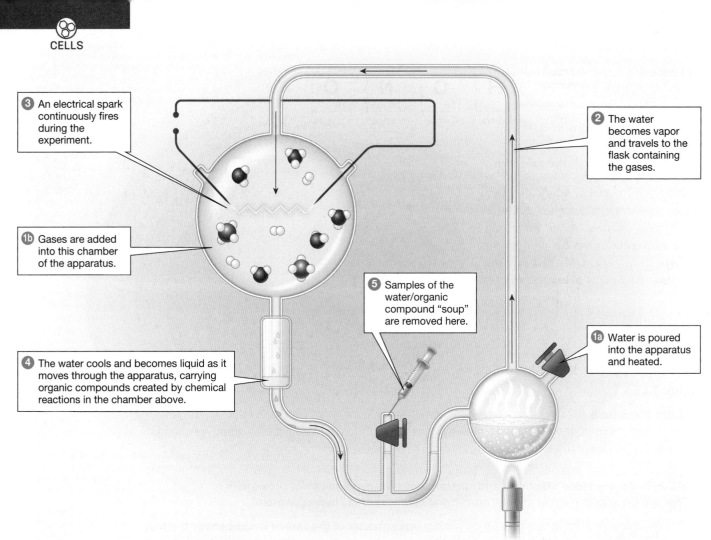

3 An electrical spark continuously fires during the experiment.

2 The water becomes vapor and travels to the flask containing the gases.

1b Gases are added into this chamber of the apparatus.

5 Samples of the water/organic compound "soup" are removed here.

4 The water cools and becomes liquid as it moves through the apparatus, carrying organic compounds created by chemical reactions in the chamber above.

1a Water is poured into the apparatus and heated.

Figure 3.2

Miller's original spark discharge experiment

When Stanley Miller zapped a mixture of methane (CH_4), ammonia (NH_3), hydrogen (H_2), and water vapor (H_2O) with electrical sparks, he produced compounds called amino acids.

Q1: Before the experiment was run, the apparatus was sterilized and then carefully sealed. Why was this an important thing to do?

Q2: Why is inclusion of methane in the gas flasks an essential part of the hypothesis that complex organic molecules were formed in the early atmosphere of Earth? (*Hint:* What makes a molecule organic?)

Q3: Answer this question after reading about Miller's "steam injection" experiments: Where was the steam injected in the experimental apparatus?

There are hundreds of known amino acids, but only 20 are the building blocks for proteins, a major class of molecules found in the cells of living things. Of the five amino acids that Miller isolated, three were part of that group of 20. Thus, from a simple mixture of gases Miller had indeed created some, but not all, of the amino acids used in life on Earth.

At a crowded seminar, Miller presented his results. Urey sat in the front row. Colleagues at the meeting recalled that after the presentation, the famed physicist Enrico Fermi turned to Urey and said, "I understand that you and Miller have demonstrated that this is one path by which life might have originated. Harold, do you think it was the way?" Urey looked at Fermi and replied, "Let me put it this way, Enrico. If God didn't do it this way, he overlooked a good bet!"

Miller and Urey's experiment, which became an instant classic in the scientific community, demonstrated that the basic chemicals for life could arise under natural conditions. Miller continued to perform spark discharge experiments, tweaking the experimental conditions and types of gases in the hope of producing more amino acids. But for an unknown reason, perhaps lack of time, Miller never published or followed up on many of his results—that is, until the mysterious box sitting on Jeffrey Bada's shelf was finally opened.

One Picture, a Thousand Experiments

In 2007, Bada was visiting the University of Texas to give a lecture. His talk was scheduled immediately after a talk by another close friend of Miller's, Antonio Lazcano, a biologist at the School of Sciences at the National Autonomous University of Mexico in Mexico City. The two men agreed to review each other's lecture slides to make sure their talks didn't overlap.

One of Lazcano's slides caught Bada's eye. It was a picture of a small glass vial, labeled as containing a residue from Miller's early experiments. Bada asked Lazcano about it. Lazcano explained that during a visit to his friend, Miller had pulled a cardboard box off a shelf, lifted out the vial, and let Lazcano take a picture. At the time, Miller told Lazcano that it was one of the leftover samples of his spark discharge experiments. Miller had saved them all.

"I was flabbergasted," says Bada. "I'd known Stanley since 1965, and he never once mentioned it." It dawned on Bada that the box might still exist. He called his lab and asked if anyone had seen the box that Cleaves had given him four years earlier. As soon as Bada returned to the lab, he found and opened the box. It was like a scientist's Christmas; the box was full of carefully labeled plastic boxes containing thin glass

Figure 3.3

Electric discharge samples from Miller's experiments

vials, many with films of dried brown, tar-like gunk in the bottom. "It was on the order of 200–300 vials," says Bada (**Figure 3.3**). "It was extracts from experiments throughout the course of his life."

Luckily, Miller kept notebooks detailing the specific contents of each vial. He had performed two other experiments shortly after his original spark discharge work, using variations on the original apparatus. In one, a different method generated the spark. In another, which caught Bada's attention, hot steam was injected directly into the spark chamber.

HENDERSON (JIM) CLEAVES

Jim Cleaves is an organic geochemist at the Carnegie Institution for Science in Washington, DC. As a graduate student of Stanley Miller in 2003, he discovered old vials from Miller's 1950s experiments. Today, Cleaves continues to study how life arose on Earth.

Bada decided to analyze the contents of the vials. He sent the samples to Jason Dworkin, also one of Miller's former students, at NASA's Goddard Space Flight Center in Maryland. The Goddard Center is home to an advanced mass spectrometer, an instrument that measures the weights of tiny amounts of matter as a way to identify them.

The results were stunning. In 1953, Miller had identified just five amino acids, three of which were common constituents of proteins. In 2008, analyzing the same samples with more sophisticated techniques, Bada's team identified 14 amino acids in the original experiment and a whopping 22 in the unpublished steam experiment. Bada was intrigued. Why did the second experimental apparatus result in more kinds of amino acids?

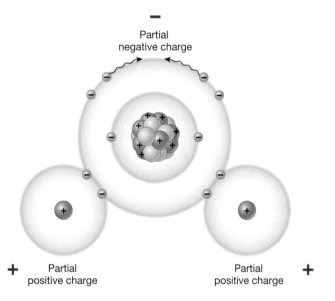

Partial negative charge

Partial positive charge

Partial positive charge

Figure 3.4

Water molecules are polar

The polarity of water molecules is the cause of water's unique properties.

Q1: Where are the covalent bonds in this figure?

Q2: This figure shows a water molecule (H_2O). A hydrogen molecule (H_2) consists of two hydrogen nuclei that share two electrons. Draw a simple diagram of a hydrogen molecule indicating the positions of the two electrons.

Q3: When table salt (sodium chloride, NaCl) dissolves in water, it separates into a sodium ion (Na^+) and a chloride ion (Cl^-). Which portion of a water molecule would attract the sodium ion, and which portion would attract the chloride ion?

The World of Water

The major difference between the two experiments was steam—hot water vapor shooting through the gas chamber. Water is essential for life because of its unique chemical properties, which enable molecules to dissolve and interact in special ways through chemical reactions.

A water molecule is made up of two hydrogen atoms and one oxygen atom held together by shared electrons—that is, by covalent bonds. Electrons moving around an atom's nucleus have different energy levels, so we can think of the electrons as segregated into rings or shells, each of which can contain up to a fixed number of electrons. The innermost shell, closest to the nucleus, can hold up to two electrons. Moving outward, the next two shells can hold up to eight electrons each.

A shell needs to be full to be stable, so in an effort to fill their shells, atoms may form covalent bonds. In a covalent bond, atoms share electrons in their outermost shell of electrons, also called the valence shell. In a water molecule, the oxygen atom uses an electron from each of two hydrogen atoms to fill its outer shell, increasing its count of electrons from six to eight. The hydrogen atoms also benefit from the bond: they fill their outer (and only) shell with the two electrons needed to be stable.

However, the electrons are not shared equally; they spend more time near the oxygen atom than near the hydrogen atoms. Because electrons are negatively charged particles, the oxygen end of a water molecule therefore takes on a slightly negative charge, and the hydrogen ends become slightly positively charged. This lopsided electron sharing means that water is a **polar molecule** (**Figure 3.4**).

In addition to covalent bonds, two other types of **chemical bonds** attach atoms to one another: ionic bonds and hydrogen bonds. Atoms that have lost or gained electrons are called **ions**. Since electrons are negatively charged, an atom that has gained an electron is a negative ion, and an atom that has lost an electron is a positive ion. When a negatively charged ion and a positively charged ion are in the same vicinity, they will chemically attract each other and form an **ionic bond**. Common table salt, sodium chloride (NaCl), is composed of sodium and chlorine held together by ionic bonds. Unlike a covalent bond, no electrons are shared in an ionic bond.

The third means of attaching atoms to one another is called a **hydrogen bond**. Hydrogen bonds are weak electrical attractions between a hydrogen atom with a partial positive charge and a neighboring atom with a partial negative charge. Molecules of water bind to each other through hydrogen bonds because the negatively charged oxygen end of one water molecule weakly attracts one of the positively charged hydrogen ends of another water molecule. A single hydrogen bond is about 20 times weaker than a covalent bond, but water makes up for that lack of strength with sheer quantity. The collective cross-linking of many, many water molecules through hydrogen bonds amounts to a potent force.

The polarity of water molecules and hydrogen bonding explain nearly all of the special properties of water, which were critical in Miller's experiments. The foremost of these was that water was able to break apart the compounds in the flask. As you may have noticed the last time you soaked a dirty dish, water has an incredible ability to dissolve other materials. This is because water molecules form hydrogen bonds with other polar molecules, like sugars or, in Miller's experiment, ammonia. The formation of hydrogen bonds with polar molecules causes those compounds to dissolve in water. Such compounds are said to be **soluble**; that is, they mix completely with the water.

A **solution** is any combination of a **solute** (a dissolved substance, such as sugar) and a **solvent** (the fluid, such as water, into which the solute has dissolved). Water is called the "universal solvent" because it successfully dissolves so many substances. The polar nature of water molecules, however, means that they will *not* interact with uncharged or nonpolar substances, such as fat or oil. Molecules that are soluble in water (such as salt) are called **hydrophilic** ("water-loving"); molecules that don't dissolve well in water (such as oil) are called **hydrophobic** ("water-fearing"). **Figure 3.5** shows these processes in action.

When Bada and his colleagues published their results from reanalyzing Miller's vials, some scientists proposed that Miller's second experiment, in which he shot a jet of steam

Oil molecules are hydrophobic. They are excluded from water and tend to clump together.

Olive oil

Vinegar molecules are hydrophilic. They are held in solution by water molecules.

Vinegar

Figure 3.5

Hydrophilic substances dissolve in water, but hydrophobic substances do not

Q1: Describe what will happen to the molecules of olive oil if you shake the bottle and then leave it alone for an hour. What about the molecules of vinegar?

Q2: What would happen if you added another fat to the bottle, such as bacon grease, and shook it?

Q3: Given how sugar behaves when it is mixed into coffee or tea, would you predict that it is hydrophobic or hydrophilic?

into the spark, had resulted in more amino acids because the hot water enabled a wider variety of chemical reactions. Whether or not that was the case, water was central to Miller's success.

JEFFREY BADA

Jeffrey Bada is a chemist at the Scripps Institution of Oceanography at UC San Diego. With Jim Cleaves, he closely analyzed and then duplicated the Miller-Urey experiments. He is also a leading scientist studying organic compounds beyond Earth, including in meteorites and on Mars.

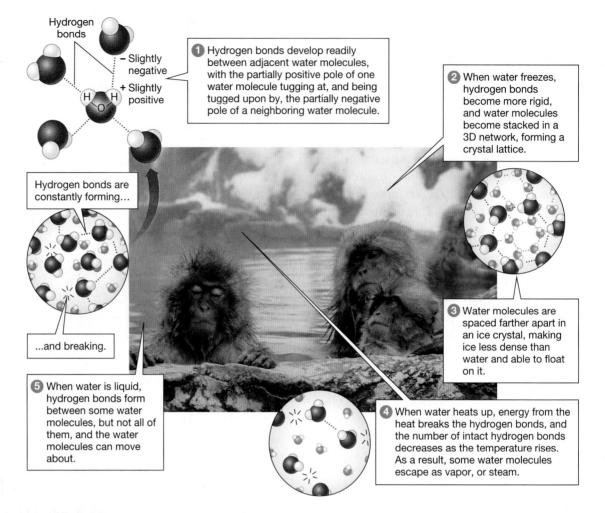

Hydrogen bonds

– Slightly negative

+ Slightly positive

1 Hydrogen bonds develop readily between adjacent water molecules, with the partially positive pole of one water molecule tugging at, and being tugged upon by, the partially negative pole of a neighboring water molecule.

2 When water freezes, hydrogen bonds become more rigid, and water molecules become stacked in a 3D network, forming a crystal lattice.

Hydrogen bonds are constantly forming...

...and breaking.

3 Water molecules are spaced farther apart in an ice crystal, making ice less dense than water and able to float on it.

5 When water is liquid, hydrogen bonds form between some water molecules, but not all of them, and the water molecules can move about.

4 When water heats up, energy from the heat breaks the hydrogen bonds, and the number of intact hydrogen bonds decreases as the temperature rises. As a result, some water molecules escape as vapor, or steam.

Figure 3.6

Water molecules change state as hydrogen bonds increase or decrease
Japanese snow macaques escape the cold with a daily dip in natural hot springs. Water can be seen here in its liquid, solid, and gas states.

Q1: Identify where in the picture water can be seen in its liquid, solid, and gas states.

Q2: In the gas state, water molecules move too rapidly and are too far apart to form hydrogen bonds. Compare the volumes occupied by an equal number of water molecules in the liquid, solid, and gas states.

Q3: Explain in your own words how ice floats on water.

Water can exist in all three states of matter: liquid, solid, and gas. Hydrogen bonds explain the physical properties of water in these three states. Though water is composed of two elements that are gases at room temperature (hydrogen and oxygen), it forms a liquid at room temperature because hydrogen bonds stick water molecules together, keeping them close. Those hydrogen bonds are constantly forming and breaking in water, creating a nonstop jostling that gives water its liquid form (**Figure 3.6**, left).

When water chills, its molecules cannot move about as vigorously (**Figure 3.6**, right). As water turns into ice at 0°C (32°F), a stable network of hydrogen bonds emerges. The molecules become spaced farther apart, locked into an orderly pattern known as a crystal lattice. That spacing is the reason ice occupies more space than liquid water. Normal ice is 9 percent less dense than liquid water (density is mass divided by volume), which explains why ice floats on water. This property of water is quite unusual—most substances are *more* dense in the solid state than in the liquid state—but it has helped shape life as we know it. If ice did not float on water, then each time a lake or river froze in the winter, the frozen top layer would sink to the bottom and the new, liquid top layer would freeze. This process would repeat until the entire lake or river was frozen solid, no matter the depth. All the aquatic creatures and plants would be killed in the process, and terrestrial plants and animals would no longer have access to the freshwater.

To boil water, as Miller did to inject steam into his experiments, requires the addition of a significant amount of energy to snap the water molecules' network of hydrogen bonds before the molecules can move fast enough to escape as steam (**Figure 3.6**, bottom). That transition from the liquid to the gas state is a phenomenon known as **evaporation**. By the opposite reaction, as water vapor cools, molecules slowly re-form hydrogen bonds and return to the liquid state—a process known as **condensation**. Without evaporation and condensation, both of which are essential parts of the water cycle, life on Earth could not exist (see Figure 18.12 for more details).

Figure 3.7

Miller's experiments were like volcanic eruptions
An erupting volcano releases all of the gases included in Miller's mixtures, and the volcanic ash contains iron and other metals. Steam is produced when the magma (superheated rock) comes in contact with groundwater, and lightning bolts lance through the cloud of gas and ash.

> **Q1:** Suppose you were going to repeat Miller's experiments. How would you decide how much of each gas to include in the chamber?
>
> **Q2:** Why did the addition of steam to the gases in Miller's second set of experiments increase the yield of amino acids?
>
> **Q3:** Miller used electrical energy in his experiment. What other forms of energy were present in the early atmosphere of Earth that could have led to the formation of complex molecules?

The Smell of Success

Water was key to the success of Miller's spark discharge experiments, but another compound would turn out to be almost as important. In 2011, three years after reanalyzing the first of Miller's samples, Bada, Cleaves, and a group of scientists teamed up again to analyze another set of vials in the cardboard box.

In 1958, Miller had performed spark discharge experiments that included a new gas: hydrogen sulfide (H_2S). You may know hydrogen sulfide

gas as the cause of the awful smell released by rotten eggs, but it is also released by volcanoes and so may have been present in Earth's early atmosphere (**Figure 3.7**). Miller's lab notes show that he performed the experiment and isolated the amino acids, but then never analyzed them. "Why he never analyzed it, I don't know," says Bada. Bada's team finished the job, identifying 22 amino acids in Miller's test tube, including six that contained sulfur. One of those sulfur-containing amino acids, methionine, initiates the construction of all proteins in cells.

Miller's experiments support the hypothesis that volcanoes—a major source of hydrogen sulfide today—coinciding with lightning, which is often focused around volcanoes, may have played a role in making large and varied quantities of biologically crucial molecules, setting the stage for the evolution of life on Earth. But this is not the only hypothesis about the origins of life. There are other ideas, including one that has gained traction in recent years: that the building blocks for life arrived from space.

Although proteins use only 20 amino acids, scientists have identified 90 amino acids in meteorites, suggesting that the first organic materials could have come to Earth from outer space. At NASA's Ames Research Center in California, astrophysicist Scott Sandford and colleagues have been able to make amino acids from gases using conditions like those that exist in interstellar space.

In several experiments, Sandford's team froze different combinations of gases to simulate ices found on comets and in interstellar clouds (where new stars and planets form), and then bombarded those ices with radiation. And what do their icy-hot recipes create? "We make amino acids all the time," says Sandford. In many cases, they identified amino acids similar to those that Miller's experiment created. "The universe seems to be an organic chemist," says Sandford. "It is hardwired to turn simple molecules into more complicated ones. The consequence of that is that it makes a whole host of products, some of which are biologically interesting."

On October 7, 2008, an asteroid entered Earth's atmosphere over Africa, and a NASA satellite photographed its impact (**Figure 3.8**). Initially 6–15 feet in diameter, the asteroid exploded over eastern Africa, and its fragments landed in Sudan. NASA scientists were able to find pieces of the asteroid on the ground because the dark rocks stood out against the light-colored desert sand. These fragments are called the "Almahata Sitta" or "Station Six" meteorites, named after a train station near the location where pieces were recovered.

When the fragments were analyzed, scientists found 19 different amino acids. The extraterrestrial origin of these amino acids is confirmed by a particular characteristic of amino acid molecules: They have two forms, left-handed and right-handed, that are mirror images of each other. All of the amino acids produced by living organisms on Earth are of the left-handed form; right-handed amino acids are produced only in laboratories. The amino acids in the Almahata Sitta meteorites are a mixture of left-handed and right-handed forms. The presence of right-handed amino acids shows that the amino acids in the fragments really came from space, not from terrestrial organic contamination after the fragments landed.

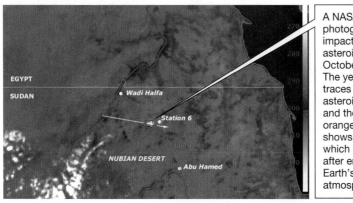

A NASA satellite photographed the impact of the asteroid on October 7, 2008. The yellow arrow traces the path the asteroid followed, and the reddish-orange blob shows the point at which it exploded after entering Earth's atmosphere.

The asteroid broke into fist-sized pieces that fell into the Nubian Desert in Sudan. The dark color of the fragments made them conspicuous on the desert sand, and many of the fragments were recovered by NASA scientists in February 2009.

Figure 3.8

Asteroid 2008 TC3 contains amino acids that formed in outer space

Q1: How did the NASA scientists find the fragments of the meteorite that exploded over eastern Africa?

Q2: What piece of evidence suggests that amino acids found in the meteorite fragments originated in outer space?

Q3: Speculate on the significance of finding extraterrestrial amino acids.

Getting the Right Mix

Miller never stopped trying to make complex molecules from simple ones. Though his 1953 experiment was originally met with fanfare, scientists later began to dispute the usefulness of his experiments. Methane and ammonia, they argued, didn't exist in large amounts on the early Earth. Instead, new evidence suggested that the atmosphere contained nitrogen gas (N_2) and carbon dioxide (CO_2). "Most people agree now [that Miller and Urey] didn't have the right composition," says Sandford. So, in 1983, 30 years after his original experiment, Miller repeated it using nitrogen and carbon dioxide. But instead of a deep red/brown broth, the liquid produced was clear and seemingly barren. The experiment looked like a failure.

In 2007, Bada and Cleaves decided to revisit that experiment to see what had gone wrong. Instead of just reanalyzing samples, they redid the experiment and discovered that the reactions between the new gases were producing chemicals called nitrites, which destroy amino acids. The solution also became acidic because of the presence of nitrous acid (HNO_2). An **acid** is a hydrophilic compound that dissolves in water and loses one or more hydrogen ions (H^+). By donating H^+ ions to water, acids increase the concentration of free H^+ ions in an aqueous solution. H^+ ions are extremely reactive and can disrupt or alter other chemical reactions. The acidic solution, Bada and Cleaves realized, was preventing amino acids from forming.

To counteract that acidity, Bada added a **base**. Acids and bases are chemical opposites. Unlike acids, bases *accept* hydrogen ions from aqueous surroundings. Because a base removes H^+ ions from solution, it has the overall effect of *reducing* the concentration of free H^+ ions in an aqueous solution (strong bases, like strong acids, can be dangerous because they disrupt chemical reactions important to life). Acids react with bases to have an overall neutralizing effect, reducing the concentration of reactive H^+ ions.

Hydrogen ion concentration is commonly expressed on a scale from 0 to 14, where 0 represents an extremely high concentration of free H^+ ions and 14 represents the lowest concentration. This scale, called the **pH scale**, is logarithmic: each pH unit represents a 10-fold increase or decrease in the concentration of hydrogen ions (**Figure 3.9**). Pure water is said to be neutral at pH 7, in the middle of the pH scale. The addition of acids to pure water raises the concentration of free hydrogen ions, making the solution more acidic and pushing the pH below

Figure 3.9

The pH scale indicates hydrogen ion concentration

Q1: Which has a higher concentration of free hydrogen ions: vinegar, pH 2.8; or milk, pH 6.5?

Q2: What happens to the concentration of free hydrogen ions in your stomach when you drink a glass of milk?

Q3: Black coffee has a pH of 5. Does adding coffee to water (pH 7) increase or decrease the concentration of free hydrogen ions in the liquid?

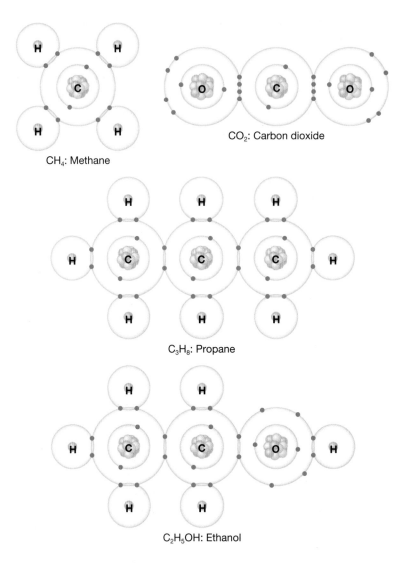

CH₄: Methane

CO₂: Carbon dioxide

C₃H₈: Propane

C₂H₅OH: Ethanol

Figure 3.10

Versatile carbon

All organic compounds are built from carbon atoms interacting with atoms of other elements through strong covalent bonds. A carbon atom can bond with up to four other atoms because it has four valence electrons in its outer shell. Recall that atoms must fill their outer shells to be stable and that covalent bonds require the sharing of valence electrons. Count the electrons circling each of the atoms in the molecules shown; all the outer shells are full.

Q1: In methane gas, how many electrons is each hydrogen atom sharing? How many is the carbon atom sharing?

Q2: In carbon dioxide, how many electrons is each oxygen atom sharing? How many is the carbon atom sharing?

Q3: Draw a molecule of formaldehyde (CH_2O). How many electrons is the oxygen atom sharing with the carbon atom? How many is the carbon atom sharing with the oxygen atom and with each hydrogen atom?

the neutral value of 7. Adding a base lowers the concentration of free hydrogen ions in the solution, making the resulting solution more basic and raising the pH above 7.

Bada added a simple base, calcium carbonate, to the experiment in order to raise the pH of his solution. This time, the resulting brew was bursting with amino acids, suggesting that the building blocks of life could indeed have originated on Earth and not been solely delivered to Earth by meteorites or comets. Today, many scientists agree that it is likely that both processes—amino acids arriving from space and originating on Earth—contributed to life as we know it. But no matter where amino acids came from, the logical follow-up question is one that continues to stump scientists to this day: What happened next?

Life's First Steps

"No one really knows how life got started," says Sandford. But researchers do agree on what life requires. If all the water in any living organism were removed, four major classes of large organic molecules, or **biomolecules** (sometimes called "macromolecules"), would remain, all of them critical for living cells: proteins, carbohydrates, nucleic acids, and lipids.

Each of these biologically important molecules is built on a framework of covalently bonded carbon atoms. Carbon is the predominant element in living systems, partly because it can form large molecules that contain thousands of atoms. A single carbon atom can form strong covalent bonds with up to four other atoms (**Figure 3.10**). Carbon atoms can also bond to other carbon atoms, forming long chains, branched molecules, and even rings. No other element is as versatile as carbon in the sheer diversity of complex molecules that can be assembled from it. In fact, while there are only about 4,500 known naturally occurring inorganic molecules (molecules that do not contain a carbon atom) on Earth, the number of known organic molecules is in the range of millions, and the number of those that are not yet uncharacterized is likely several orders of magnitude larger.

Proteins, carbohydrates, and nucleic acids are **polymers**, long strands of repeating units of small molecules called **monomers**. Amino acids are the monomers making up proteins, simple sugars are the monomers in carbohydrates, and nucleotides are the basis of nucleic acid polymers.

All three of these polymers have essential functions for every life-form. **Proteins** are known to be the most numerous and versatile of the biomolecules. Different combinations of the twenty amino acid monomers allow for countless proteins that vary in size and shape, and therefore function (**Figure 3.11**, top left). For example, enzymatic proteins, like polymerases, enable us to copy our DNA (see Chapter 9 for more on DNA replication). Structural proteins give our cells shape. Hormone and receptor proteins, like insulin and its receptor, allow our cells to take up sugars for use as energy. Other equally important protein categories include membrane transport proteins that help move substances into and out of cells, antibody molecules that protect us from disease, storage proteins like LDL and HDL (low-density and high-density lipoproteins, respectively) that carry cholesterol, and venoms and toxins such as the tetanus toxoid.

Carbohydrates are the next-most-versatile biomolecules. They range in size from simple sugar monomers (monosaccharides) and two-monomer sugars (disaccharides) to complex carbohydrates that may contain thousands of monomers (**Figure 3.11**, top right). Simple sugars are the cell's direct fuel to make ATP (adenosine triphosphate), the molecular energy source essential for all cellular work (see Chapter 4 for more on ATP). Other carbohydrates, such as glycogen in animals and starch in plants, are used for energy storage. Three additional complex carbohydrates provide structural support to cells: cellulose, also known as fiber, helps plants to grow tall; chitin forms a hard outer covering to protect organisms without an internal skeleton, such as insects, spiders, and crustaceans; and peptidoglycan is a major component of bacterial cell walls.

The third and most crucial category of polymer, the **nucleic acids**—DNA (deoxyribonucleic acid) and RNA (ribonucleic acid)—form the basis of life itself. Nucleic acids are polymers of nucleotide monomers: DNA is composed of deoxyribonucleotides, and RNA is composed of ribonucleotides (**Figure 3.11**, bottom left). DNA provides living organisms with long-term, stable genetic information storage in a form that is easily copied and passed on to future generations (see Chapter 9). Our genes are DNA. That sounds important, but what about RNA? Without RNA,

Proteins

Proteins are polymers of amino acids and perform many functions in the cell.

Some proteins join other proteins and form a complex.

Tetanus toxoid | Antibodies

Human antibodies and the bacterial tetanus toxoid are both proteins.

Foods rich in protein include meats, fish, dairy, and beans (legumes).

The chain folds further into a 3D shape that supports the protein's function.

This chain then forms coils or sheets.

There are four levels of protein structure. The simplest level is a chain of amino acids.

Supplements advertising amino acids contain these building blocks of proteins.

The 20 amino acids differ only at the "R" side group.

Amino acid

H_3N^+—C—C—OH
with H above and R below, =O

Hydrogen

Phosphorus

Carbon

Nucleic acids

Nucleic acids are polymers of nucleotides and function to store and transmit genetic information.

Nucleotide

The four different ribonucleotides make up RNA.

RNA is made up of one strand of nucleotides.

Nitrogen

The four different deoxyribonucleotides make up DNA.

DNA is built of two strands of joined nucleotides.

A eukaryotic plant cell.

Figure 3.11

Biomolecules are critical for life

A handful of monomers (*mono*, "one") can be assembled into a great variety of polymers (*poly*, "many"). Each of these biologically important molecules is built on a framework of repeating monomers made up of some, and in several cases all, of the atoms shown in the center circle—with the exception of the lipids, which are instead characterized by repeating atoms of carbon and hydrogen. ▶

Protein
mRNA Ribosome
Cytoplasm
Cell

RNA is read by ribosomes to create proteins.

DNA
RNA
mRNA
Nucleus
Cytoplasm
Cell

The genetic information in DNA is transcribed into RNA.

A human has 46 chromosomes.

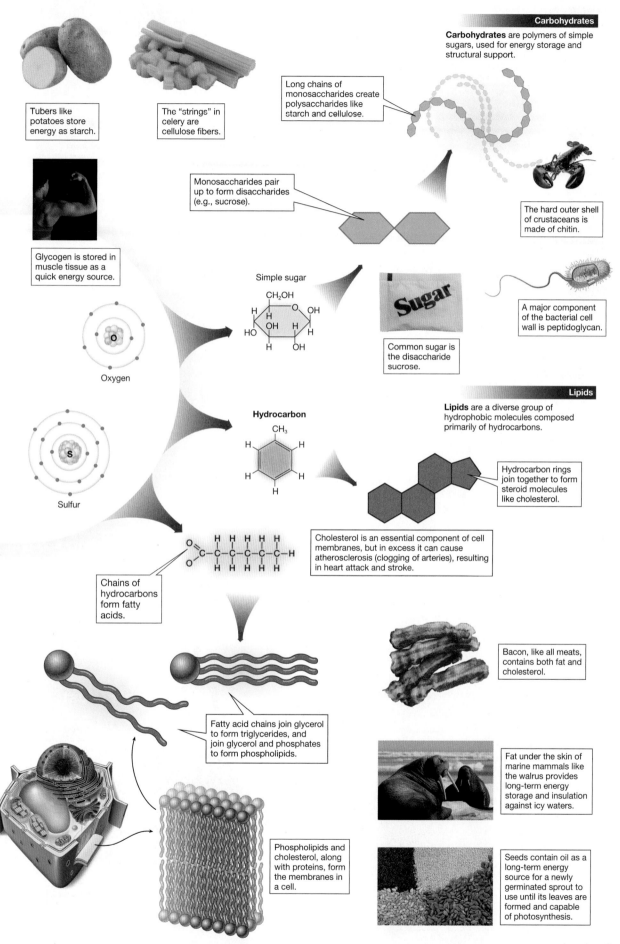

Tubers like potatoes store energy as starch.

The "strings" in celery are cellulose fibers.

Glycogen is stored in muscle tissue as a quick energy source.

Oxygen

Sulfur

Carbohydrates

Carbohydrates are polymers of simple sugars, used for energy storage and structural support.

Long chains of monosaccharides create polysaccharides like starch and cellulose.

The hard outer shell of crustaceans is made of chitin.

Monosaccharides pair up to form disaccharides (e.g., sucrose).

Simple sugar

CH₂OH

A major component of the bacterial cell wall is peptidoglycan.

Common sugar is the disaccharide sucrose.

Lipids

Lipids are a diverse group of hydrophobic molecules composed primarily of hydrocarbons.

Hydrocarbon

CH₃

Hydrocarbon rings join together to form steroid molecules like cholesterol.

Cholesterol is an essential component of cell membranes, but in excess it can cause atherosclerosis (clogging of arteries), resulting in heart attack and stroke.

Chains of hydrocarbons form fatty acids.

Bacon, like all meats, contains both fat and cholesterol.

Fatty acid chains join glycerol to form triglycerides, and join glycerol and phosphates to form phospholipids.

Fat under the skin of marine mammals like the walrus provides long-term energy storage and insulation against icy waters.

Phospholipids and cholesterol, along with proteins, form the membranes in a cell.

Seeds contain oil as a long-term energy source for a newly germinated sprout to use until its leaves are formed and capable of photosynthesis.

What's It All Made Of?

Everything in the universe is composed of matter—from ordinary matter, made of atoms, to dark matter, which may consist of unknown types of particles. Here, we stick with what we know and describe the common elements that compose the world around us.

Assessment available in smartwork5

Earth's Atmosphere

- **78%** Nitrogen
- **21%** Oxygen
- **<1%** Argon
- **<<1%** Other Elements

The Universe

- **75%** Hydrogen
- **23%** Helium
- **2%** Other Elements

The Human Body

- **65%** Oxygen
- **18%** Carbon
- **10%** Hydrogen
- **3%** Nitrogen
- **2%** Calcium
- **1%** Phosphorus
- **1%** Other Elements

Earth's Crust

- **46%** Oxygen
- **28%** Silicon
- **8%** Aluminum
- **5%** Iron
- **4%** Calcium
- **3%** Sodium
- **2%** Potassium
- **2%** Magnesium
- **2%** Other Elements

All proportions are by mass except Earth's atmosphere, which is by volume the information stored in our genes would be stuck there, like a blueprint in a foreign language that no one can decipher. RNA comes in many forms and plays many roles, but its most important job is providing a readable genetic language that enables genes to be expressed as proteins.

A final valuable type of biomolecule is the **lipids**, which are better known as fats, oils, and steroids. Lipids are not polymers, because their structure is not composed of a chain of monomers (**Figure 3.11**, bottom right). Nevertheless, the lipids form a diverse group of biomolecules made up of combinations of hydrocarbons (carbon- and hydrogen-only molecules), fatty acids, or glycerol molecules. For example, triglycerides—composed of three fatty acid molecules linked to a glycerol molecule—provide long-term energy storage in both plants (as oils) and animals (as fats), and serve as insulation against the cold in animals. And phospholipids, which have a hydrophilic half and a hydrophobic half, are the main component of all cellular membranes (see Chapter 4 for more on phospholipids).

Miller, Bada, and other scientists have not yet been able to identify any biomolecules in their prebiotic soups. "It's still very much of an unknown how we go from simple to complex molecules," says Bada. Lipids and carbohydrate-like compounds, however, have been found in meteorites, and Sandford's team at NASA is now attempting to create carbohydrates using its space ice simulations. Other scientists are trying to re-create lipids and nucleic acids in the lab (see Chapter 4).

Fifty More Years

Sadly, Miller died in 2007, never having seen the results of Bada's reanalysis. Today, researchers continue to perform spark discharge experiments with equipment that is essentially the same as in Miller's original design. "We're not done with it," says Jim Cleaves. "There are just so many variations to look at—different combinations of gases, changing the pH, adding metals, et cetera."

"This experiment lives on," agrees Bada. "Not many people can say that an experiment done over 50 years ago is still being investigated today, and finding a wealth of new information." And with better and better instruments to detect the results, who knows what scientists might discover over the next 50 years?

REVIEWING THE SCIENCE

- The physical world is composed of **matter**, which is anything that has mass and occupies space. There are 98 distinct **elements** found naturally on Earth. An **atom** is the smallest unit of an element that maintains its unique properties; it contains positively charged **protons**, uncharged **neutrons**, and negatively charged **electrons**.

- The chemical interactions that cause atoms to associate with each other are known as **chemical bonds**. When an atom loses or gains electrons, it becomes, respectively, a positively or negatively charged **ion**. Ions of opposite charge are held together by **ionic bonds**.

- **Covalent bonds** are formed by the sharing of electrons between atoms. A **molecule** contains at least two atoms that are held together by covalent bonds.

- **Hydrogen bonds** are weak associations between two molecules such that a partially positive hydrogen atom within one molecule is attracted to a partially negative region of the other molecule.

- Partial electrical charges result from the unequal sharing of electrons between atoms, giving rise to **polar molecules**. The polarity of individual water molecules and the hydrogen bonding across water molecules explain nearly all of the special properties of water.

- A **solution** is any combination of a dissolved substance, known as the **solute**, and a fluid into which the solute has dissolved, known as the **solvent**.

- Ions and polar molecules are **hydrophilic**; they readily dissolve in water. Nonpolar molecules cannot associate with water and are therefore **hydrophobic**.

- In **chemical reactions**, bonds between atoms are formed or broken. The participants in a chemical reaction (**reactants**) are modified to give rise to new ions or molecules (**products**).

- The concentration of free hydrogen ions in water is expressed by the **pH scale** and reflects whether a solution is **acidic**, **basic**, or neutral.

- **Chemical compounds** are molecules that contain atoms from at least two different elements. Carbon atoms can link with each other and with other atoms to generate a great diversity of chemical compounds called **organic molecules**. The four main types of large organic molecules, or **biomolecules**, are **proteins**, **carbohydrates**, **nucleic acids**, and **lipids**.

THE QUESTIONS

The Basics

1 The atomic number of an atom is determined by the number of _____ in the atom.

(a) protons

(b) neutrons

(c) electrons

(d) electrons plus neutrons plus protons

2 The atomic mass number of an atom is determined by the sum of the number of _____ in the atom.

(a) protons plus electrons plus neutrons

(b) protons plus electrons

(c) neutrons plus electrons

(d) neutrons plus protons

3 Link each term with the correct definition.

ION	1. The smallest subunit of an element.
MATTER	2. A molecule made up of repeating monomers.
SOLUTION	3. An atom that has gained or lost an electron.
ELEMENT	4. Consists of two or more atoms chemically bonded together.
CHEMICAL COMPOUND	5. Consists entirely of atoms with the same atomic number.
MOLECULE	6. Anything that has mass and occupies space.
ISOTOPE	7. Has the same atomic number but a different atomic mass than its original element.
POLYMER	8. A molecule that contains atoms from two or more different atoms.
ATOM	9. A combination of a solvent and a solute.

4 The partial negative charge at one end of a water molecule is attracted to the partial positive charge of another water molecule. What is this attraction called?

(a) a hydrogen bond

(b) a van der Waals interaction

(c) an ionic bond

(d) a covalent bond

(e) a hydrophilic bond

5 Select the correct terms:

Proteins are (**polymers / monomers**) of amino acids. A common carbohydrate is (**sugar / fat**). Nucleic acids are composed of (**nucleotides / DNA**). Lipids (**are / are not**) polymers. All of these organic molecules contain (**carbon / nitrogen**).

Challenge Yourself

6 You are asked to determine the classification of an unknown biomolecule. You are told that it can be broken down into only carbon, oxygen, and hydrogen atoms, and nothing else. Using your knowledge of biomolecules and their components, identify which of the following statements about your unknown sample are true. (Select all that apply.)

(a) It is an organic compound.

(b) It contains amino acids.

(c) It contains sugar.

(d) It is a nucleic acid.

(e) It is a carbohydrate.

7 Explain why life on Earth is carbon-based rather than, for example, hydrogen- or oxygen-based.

8 Why did Miller's addition of hydrogen sulfide (H_2S) to the mixture of molecules in the spark discharge experiments increase the number of amino acids he found?

Try Something New

9 In the accompanying figure, a carbon atom resides at each unlabeled corner of the hexagons and wherever a "C" is shown. Each "O" is an oxygen atom, and each "H" is a hydrogen atom. What is this structure?

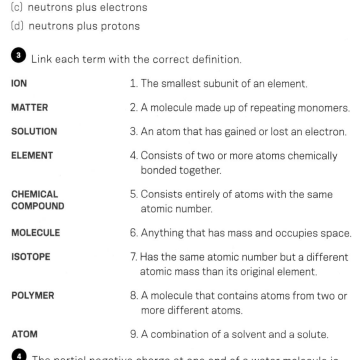

(a) a fatty acid monomer

(b) two simple sugar molecules linked together

(c) a nucleotide polymer

(d) a hydrocarbon ring polymer

(e) an amino acid

10 Lipoproteins are relatively large, combined clumps of both protein and lipid molecules that circulate in the blood of mammals. They come in two forms, called HDL and LDL, and they act like suitcases to move cholesterol, fatty acid remnants, triglycerides, and phospholipids from one place to another through the bloodstream. (LDL recirculates lipids throughout the body, while HDL takes lipids to the liver to excrete them in feces.) Given that lipids are hydrophobic and proteins can be hydrophilic, which of the following statements is correct?

(a) The lipid portion of LDL does not dissolve in the bloodstream, while the lipid portion of HDL does.

(b) The protein portions of both LDL and HDL can dissolve or interact with the water molecules in the bloodstream.

(c) Neither the protein nor the lipid portions of LDL molecules can interact with water molecules in the bloodstream.

(d) Both the protein and the lipid portions of HDL molecules can interact with water molecules in the bloodstream.

(e) none of the above

11 You have found an unknown cleaning solution in your roommate's under-sink cabinet. You have several spring cleaning projects to take care of and would like to use this product if it has the correct pH for the job. You know you need an extremely basic solution to clean your oven, but an acidic solution like vinegar to clean your coffee maker. You borrow a few pH indicator strips from your biology lab TA and test the solution. For each of the following scenarios, select the best answer.

a. The solution tested at a pH of 11; you will be able to clean your (**oven / coffee maker / both / neither**).

b. The solution tested at a pH of 2; you will be able to clean your (**oven / coffee maker / both / neither**).

c. The solution tested at pH 7; you will be able to clean your (**oven / coffee maker / both / neither**).

d. The solution tested at pH 9; you will be able to clean your (**oven / coffee maker / both / neither**).

e. The solution tested at pH 4; you will be able to clean your (**oven / coffee maker / both / neither**).

12 Laundry detergent molecules have a short polar end and a long nonpolar end, enabling them to bind to water (on the polar end) and oils (on the nonpolar end). Suppose you spill some salad dressing on your shirt. Explain how washing your shirt with detergent will help remove the dressing.

Leveling Up

13 **What do *you* think?** Read the articles "Two Miles Underground, Strange Bacteria Are Found Thriving" in *News at Princeton* (http://www.princeton.edu/main/news/archive/S16/13/72E53/index.xml?section=newsreleases) and "Martian Underground Could Hold Clues to Life's Origins" (http://www.studentnews.eu/s/3229/68257-Articles/4014087-Martian-underground-could-hold-clues-to-lifes-origins.htm). Then answer the following questions.

a. How are bacteria able to stay alive below Earth's surface or (potentially) below the surface of Mars?

b. How does the discovery of bacteria living 2 miles beneath Earth's surface relate to the importance of the McLaughlin Crater on Mars?

c. What research do you think should next occur to answer the questions brought up by these studies?

d. From what you've read, what do you think is the probability of life on other planets? What would you expect that life to look like?

14 **Is it science?** Life on Earth evolved as carbon-based. Silicon shares many of the chemical properties of carbon, yet it is not a building block of life on Earth. Watch the episode of the TV series *The X-Files* titled "Firewalker" (http://www.hulu.com/watch/158588), or read through its story line (https://en.wikipedia.org/wiki/Firewalker_%28The_X-Files%29). While doing so, make a list of the scientific concepts presented, and note whether you think they represent real science or pseudoscience. Then use your textbook and Internet research to determine whether you were correct or incorrect in each case. If you evaluated any of the concepts incorrectly, reflect on how you came to your initial conclusions.

For more, visit digital.wwnorton.com/bionow2 for access to:

Engineering Life

In 2003, scientists began trying to build an artificial cell from scratch. Today, they're closer than you might think.

- Explain cell theory and why it is central to the study of life.
- Describe the differences in the structures of viruses, prokaryotes, and eukaryotes.
- Diagram a plasma membrane, showing how the structure allows some substances in and keeps others out.
- Compare and contrast passive and active transport of materials into and out of cells.
- Differentiate between exocytosis and the three types of endocytosis.
- Describe the role of any given organelle in a eukaryotic cell.
- Identify the main differences between a plant cell and an animal cell.

CHAPTER
04

LIFE IS
CELLULAR

The sky was still dark when Daniel Gibson hurried into the J. Craig Venter Institute (JCVI) in La Jolla, California. At 5:00 a.m., his footsteps echoed through the empty halls of the building. He reached a laboratory door and slipped inside. There, Gibson peered into a warm incubator, his eyes scanning rows of palm-sized petri dishes. His stomach was in knots. For 3 months the experiment had failed. Would this day—Monday, March 29, 2010—be any different?

Gibson is part of a team at the JCVI with a single, audacious goal: to create life. For more than a decade, this team of scientists and engineers has attempted to build a synthetic, or human-made, **cell**. Cells are the smallest and most basic unit of life—microscopic, self-contained units enclosed by a protective membrane (**Figure 4.1**). The human body is composed of approximately 100 trillion (10^{14}) cells. On that day in 2010, however, the JCVI was trying to synthesize just one cell—a single-celled bacterium.

Gibson's team had sequenced a bacterium's complete genetic information, its **genome**; built a synthetic version of that genome using basic laboratory chemicals; and, finally, replaced the natural DNA of another species of bacterium with the synthetic DNA. **DNA** (deoxyribonucleic acid) is a large and complex molecule that acts as a set of instructions for building an organism, like a blueprint. Almost every cell of every living organism contains DNA. DNA transfers information from parents to offspring, which is why it is essential for reproduction. Life, no matter how simple or how complex, uses this inherited genetic code to direct the structure, function, and behavior of every cell. DNA is made up of many nucleotides held together in a structure called the double helix, a ladderlike assembly twisted along its length into a spiral (see Figure 3.11).

Gibson's boss, the famous geneticist J. Craig Venter, worked for more than 15 years and spent millions of dollars to construct a synthetic DNA helix from chemicals in the laboratory, but Gibson and the team had been unable to get that synthetic DNA to work inside a cell. Every Friday for 3 months, they transplanted the synthetic DNA into a bacterial cell whose own DNA had been removed. The synthetic DNA included a gene to make the cells turn bright blue, so every Monday, Gibson hurried to the incubator and checked the petri dishes for a colony of blue cells. But Monday after Monday, the dishes were barren. "We did the genome transplantation again and again," he recalls, "but nothing was working."

Then, in mid-March, Gibson identified an error in a single gene in the synthetic DNA. A **gene** is a segment of DNA that codes for a distinct genetic characteristic, such as having O-type blood or a dimpled chin. In Gibson's bacterium, the gene with the error was responsible for DNA replication. When it wasn't working, the bacterium couldn't replicate its DNA, and it died. So in late March, Gibson fixed the DNA error, transplanted the genome yet again, and waited.

Salmonella is a single-celled bacterium that is a common cause of food poisoning.

This is one cell of the multicellular plant *Arabidopsis*, which is used extensively in genetic studies.

Yeasts are single-celled, but more complex than bacteria. Some species are critical for making bread and beer, while others are pathogens.

Humans are multicellular animals with many specialized cells, such as these neurons within the central nervous system.

Figure 4.1

An individual organism may consist of a single cell or many cells

All of these photos of cells were taken using electron microscope technology. Color has been added to the images to differentiate structures within the cells.

Life, Rewritten

The JCVI is just one institution among a large group of universities and companies pursuing synthetic biology, a field that aims to design and construct new biological entities with novel and useful functions. Scientists are working to create algae that digest trash and produce energy, microbes that use light and water to create hydrogen gas, and bacteria that produce new kinds of antibiotics to treat infections. With synthetic biology "we can harness what nature has made, but repurpose it," says James Collins, a synthetic biologist at Boston University. "We can reprogram organisms and endow them with novel functions." In addition to useful tools, the pursuit of artificial life sheds light on the very origins of life. "It's going to be a big challenge to create a totally synthetic cell," says Collins, "but it's fundamentally intriguing to explore how life may have arisen on the planet."

To make their own cell, scientists are pushing the boundaries of **cell theory**, one of the unifying principles of biology. Cell theory has two main parts: every living organism is composed of one or more cells, and all cells living today came from a preexisting cell. By trying to engineer a cell in the laboratory, Venter, Gibson, and others are challenging the second part of the definition.

Starting Small

The JCVI's first step toward a synthetic cell was a small one. In 2003, Venter's team flexed its scientific muscles by synthesizing the 11-gene, 5,386-base-pair genome of phiX174, a virus that infects bacteria. A **virus** is a small, infectious agent that can replicate only inside a living cell. Most viruses are little more than stripped-down genetic material wrapped in proteins, yet these pathogens attack and devastate organisms in every kingdom of life, from bacteria to plants and animals. Though the JCVI team successfully created a virus with a synthetic genome, it was not considered the first synthetic life, because scientists debate whether viruses are alive. (For more on this debate, see "Viruses—Living or Not?," page 66.)

Next, Venter and his colleagues moved up to a bacterium, a living organism that consists of a single cell. In 2010, they sequenced and built the genome of a bacterium called *Mycoplasma mycoides* (*M. mycoides*), an organism that can cause the mammary glands of goats to swell. They constructed the genome—a 1.1-million-base-pair DNA sequence—using four necessary ingredients: the nucleotides **adenine** (**A**), **thymine** (**T**), **guanine** (**G**), and **cytosine** (**C**), the building blocks of DNA. A, T, G, and C, organized in different combinations, carry all the instructions for everything a cell does.

The team used a machine to read the nucleotide sequence of the *M. mycoides* genome and then "print out" little bits of that code, creating strands of DNA about 50–80 bases long. They then strung these pieces together using living cells as factories, inserting the short segments into yeasts and *Escherichia coli*—small, single-celled organisms. These organisms interpreted the strands as broken pieces of DNA and stitched them together, creating longer and longer sequences. It was like building the Eiffel Tower from a massive box of Legos, constructing a single support beam at a time. The effort—with many mistakes along the way—took years. "It was very complex," said Venter. "It was a long, involved process."

Congratulations, It's a . . . Cell

Once the DNA sequence of *Mycoplasma mycoides* was complete and intact, it was up to the JCVI team to transfer it into another species and make it work. This was the experiment that almost drove Gibson crazy. The researchers removed all the DNA from a cell of a closely related bacterium, *Mycoplasma capricolum*, and replaced it with the *M. mycoides* synthetic DNA.

After months of trying, on that Monday morning at 5:00 a.m., Gibson cautiously scanned the

J. CRAIG VENTER

J. Craig Venter is an American biologist and founder and CEO of the J. Craig Venter Institute. He led a team to fully sequence and publish the human genome in 2001, and initiated the effort to create the first cell constructed with synthetic DNA.

petri dishes. There, on a single dish, was a group of bright-blue cells—proof that the *M. capricolum* cell had "booted up" the *M. mycoides* DNA and transformed itself into an *M. mycoides* cell (**Figure 4.2**). Gibson was ecstatic. Moments later, he sent a text message to Venter, waking him up. Within the hour, Venter was in the lab with a camera, taking pictures of the tiny blue dollop in the dish. "How does it feel to create life?" Venter asked Gibson. They opened champagne and toasted their success.

Over the following weeks, Gibson repeated the experiment hundreds of times to make sure the blue cells were not an accident or a fluke, confirming that they contained only

Two colonies of *M. mycoides*, transformed from *M. capricolum*.

100 µm

Individual *M. mycoides* cells.

500 nm

Figure 4.2

The first synthetic organism

A colony of bacteria was transformed from *M. capricolum* by the insertion of synthetic DNA of a closely related species of bacterium, *M. mycoides*. The investigators inserted a gene into the synthetic DNA that codes for blue pigment.

Q1: What was the purpose of inserting the gene that codes for blue pigment into the synthetic DNA?

Q2: What part of the transformed bacterium is synthetic?

Q3: Did this experiment create life?

M. mycoides DNA. Every time, the cells with a synthetic genome survived. The team had done it—created the first synthetic cell. "They are living cells," Venter told *The Scientist* magazine when the research was published 2 months later. "The only difference is that they have no natural history. Their parents were the computer."

Once the research was published, the reaction from the academic community, captured in *Nature* magazine, was swift and divided. Some called it a significant advance: "We now have an unprecedented opportunity to learn about life," said Mark Bedau, a professor of philosophy at Reed College in Oregon. Arthur Caplan, a bioethicist at the University of Pennsylvania, said, "Venter's achievement would seem to extinguish the argument that life requires a special force or power to exist. In my view, this makes it one of the most important scientific achievements in the history of mankind."

Others were more hesitant. "Has [Venter] created 'new life'?" asked George Church, a prominent geneticist at Harvard Medical School. "Not really. . . . Printing out a copy of an ancient text isn't the same as understanding the language." Gibson and Venter agree that they did not create life from scratch, but they argue that they did create new life from existing life.

A Different Approach

While Venter and Gibson were making headlines, other scientists were quietly pursuing a different approach to building a cell, working from the bottom up rather than from the top down. One young scientist in California decided to start by building one of the simplest, yet most vital, components of a cell: the layer of molecules that surrounds it.

In the chemistry department at UC San Diego, assistant professor Neal Devaraj was fascinated by the idea of building life and approached synthetic biology from the perspective of chemistry. "Most molecular biologists study what exists," says Devaraj, "but when you're a chemist and constantly make new compounds, you want to engineer something from scratch." So instead of taking a cell apart and determining how

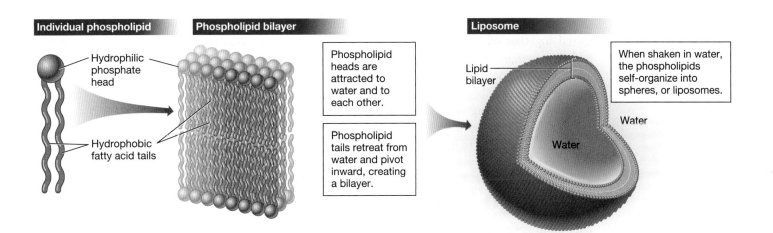

Individual phospholipid

Hydrophilic phosphate head

Hydrophobic fatty acid tails

Phospholipid bilayer

Phospholipid heads are attracted to water and to each other.

Phospholipid tails retreat from water and pivot inward, creating a bilayer.

Liposome

Lipid bilayer

When shaken in water, the phospholipids self-organize into spheres, or liposomes.

Water

Water

Figure 4.3

Liposomes form when phospholipids and water are shaken together

When you shake a mixture of phospholipids and water, the phospholipid bilayers bend and link together to form spheres called liposomes. This simple structure is remarkably similar to the basic structure of a cell.

Q1: Why is it important that the phosphate head of a phospholipid is hydrophilic?

Q2: What essential component of a cell do liposomes lack, and why is that omission important?

Q3: Could the tendency of phospholipid bilayers to spontaneously form spheres have played a role in the origin of life? (*Hint:* Refer to "The Characteristics of Living Organisms" on page 6 of Chapter 1.)

it works, Devaraj decided to try to build a cell artificially, using materials not typically found in nature. "If you want to really understand the principles by which life operates and evolves, the best way to do so is to build a cell from the ground up," he says.

Scientists suspect that one of the first events at the beginning of life on Earth was the formation of a **plasma membrane**, a barrier separating a cell from its external environment. A plasma membrane is made of two layers of **phospholipids**, organic molecules with a water-loving, or *hydrophilic*, head, and a water-fearing, or *hydrophobic*, tail. In water, these molecules form a double layer with heads out and tails in, a barrier that separates the contents of the cell from what lies outside the cell. Thus, the membrane is a **phospholipid bilayer**, a mostly impermeable barrier. When a phospholipid bilayer forms a sphere, or **liposome**, the fluid inside the liposome can have a different composition from the fluid outside (**Figure 4.3**). The ability to maintain an internal environment separate from the external environment is one

of the most critical functions of the plasma membrane of a cell.

Given a container of phospholipids, anyone can make a membrane, says Devaraj. "Making membranes is almost a trivial thing," he says. "You take natural or synthetic phospholipids, add water, and they form membranes." Yet researchers had been unable to form a membrane from scratch—without using preexisting phospholipids. In nature, new phospholipids are created by enzymes embedded in the cell.

Instead of trying to engineer new phospholipids, Devaraj wanted to start with something simpler. He worked with graduate student Itay Budin, who was then at Harvard University and is now a postdoctoral researcher at UC Berkeley,

NE AL DEVARAJ

JNeal Devaraj is a biochemist at UC San Diego who is working to make an artificial cell from the bottom up, starting with the membrane and then building other organelles.

to create a self-assembling membrane. They first mixed together oil and a detergent. Then they added copper, a metal ion, as a catalyst to spark a chemical reaction. With the addition of copper, sturdy membranes begin to bud off the oil; these were self-assembling structures. "There's no equivalent whatsoever in nature," says Devaraj. "Our goal was simply to mimic biology."

Through the Barrier

Devaraj admits that his artificial membrane is far simpler than a real cell's plasma membrane, which is dotted with numerous proteins, including transport proteins. **Transport proteins** are gates, channels, and pumps that allow molecules to move into and out of the cell, making it selectively permeable (**Figure 4.4**). **Selective permeability** means that some substances can cross the membrane, others are excluded, and still others can pass through the membrane when aided by transport proteins.

All movement of substances through the plasma membrane occurs by either active or passive transport. Some transport proteins facilitate **active transport**, the movement of a substance that requires an input of energy (**Figure 4.4**, left). Molecules move across the plasma membrane by active transport when they need to move from a region of *lower* concentration to a region of *higher* concentration. In contrast, **passive transport** is the movement of a substance without the addition of energy (**Figure 4.4**, middle and right). Movement via passive transport is spontaneous.

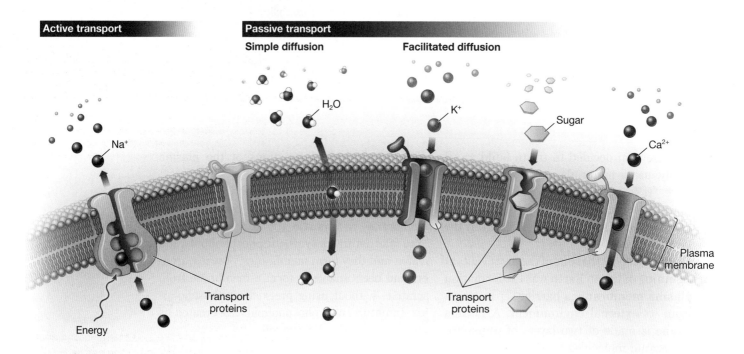

Figure 4.4

The plasma membrane is a barrier and a gatekeeper

The plasma membrane moves substances in a highly selective fashion, determined in large part by the types of membrane proteins embedded in the phospholipid bilayer.

> **Q1:** In what ways is the plasma membrane a barrier, and in what ways is it a gatekeeper?
>
> **Q2:** Why can't ions (such as Na⁺) cross the plasma membrane without the help of a transport protein?
>
> **Q3:** If no energy were available to the cell, what forms of transport would not be able to occur? What forms of transport could occur? (*Hint:* Look ahead at Figures 4.5 and 4.6.)

A primary type of passive transport is **diffusion**, the movement of a substance from a region of *higher* concentration to a region of *lower* concentration (**Figure 4.5**). Water, oxygen, and carbon dioxide usually enter and leave cells by **simple diffusion**: These small, uncharged molecules slip between the large molecules in the phospholipid bilayer without much hindrance.

Water moves in and out of cells (and compartments inside cells) by **osmosis**. Osmosis is a form of simple diffusion because the water molecules are moving from areas of higher concentration to areas of lower concentration (**Figure 4.6**). Osmosis is critical for cellular processes because most cells are at least 70 percent water, and nearly all cellular processes take place in a water-rich environment.

Cells must maintain a stable internal water concentration to function properly, but the concentration of water in most cells changes from moment to moment. For example, when salt molecules move into a cell, the concentration of water in the cell decreases because additional molecules have been added. In response, water

At first, the molecules of food coloring are concentrated in one region.

Net movement of the food coloring is from regions of high concentration to regions of low concentration.

Diffusion ceases when food-coloring molecules are evenly distributed. At equilibrium, just as many molecules move into any given region as leave that region, so there is no net change in concentration.

Sugar has just been added to a beaker of water. This beaker is divided by a semipermeable membrane—that is, a membrane with pores large enough to allow water molecules to pass through, but too small for sugar molecules to pass.

After a period of time, water molecules have moved by osmosis from the right side of the membrane (which had a higher concentration of water) to the left side of the membrane. The concentration of water on the two sides of the membrane is now the same, and the movement of water molecules is now the same in both directions.

Figure 4.5

Food coloring in water illustrates diffusion

> **Q1:** Is the dye at equilibrium in any of these glasses? Describe how the first glass will look when the dye is at equilibrium with the water.
>
> **Q2:** Will diffusion mix the molecules of dye evenly through the water, or is it necessary to shake the container to get a uniform mixture?
>
> **Q3:** Will diffusion mix the dye faster in hot water than in cold water? Why or why not? (*Hint:* Review the discussion of the behavior of water molecules at different temperatures in Chapter 3.)

Figure 4.6

In osmosis, water diffuses across a semipermeable membrane

Osmotic movement of water between a cell and the external environment is critical for maintaining a constant water concentration in the cell, which it needs to function properly.

> **Q1:** What would the second diagram look like if the pores in the semipermeable membrane were too small to allow water molecules to pass through?
>
> **Q2:** What would the second diagram look like if the pores were large enough to let both water molecules and sugar molecules through?
>
> **Q3:** The fluid in an IV bag is isotonic to blood. What change would you see in the red blood cells of a patient if a bag of a hypertonic solution was used in error?

molecules immediately move by osmosis—that is, they diffuse—across the plasma membrane into the cell until the concentration of water inside the cell is the same as the concentration on the outside. On the other hand, when salt molecules move out of a cell, the water molecules become more concentrated, and osmotic movement of water out of the cell then restores the concentration of water in the cell to its correct level.

You can imagine, then, that the concentration of solutes within a cell in relation to the concentration *outside* the cell is of critical importance. When cells are surrounded by fluid with the same solute concentration as the cell interior, the extracellular and intracellular environments are said to be **isotonic** to each other (*iso*, "equal"). If the extracellular environment has a higher solute concentration, it is **hypertonic** to the cell interior (*hyper*, "more"). If it has a lower solute concentration than the cell's interior, it is said to be **hypotonic** to it (*hypo*, "less").

The careful balance of concentrations is particularly vital in human blood. Red blood cells are typically in an isotonic solution in our bloodstream, where the solute concentrations inside and outside the cells are the same. Doctors and nurses are taught to administer an IV drip of saline, a solution of salt in water, to dehydrated patients. If a dehydrated patient was administered water instead of a saline solution, the water would dilute the patient's blood, making it hypotonic. Osmosis would then occur, causing water to rapidly diffuse into red blood cells to the point where the cells could burst and die.

Most hydrophobic molecules, even fairly large ones, can pass through the plasma membrane via simple diffusion because they mix readily with the hydrophobic tails that form the core of the phospholipid bilayer. But hydrophilic substances such as sodium ions (Na^+), hydrogen ions (H^+), and larger molecules, including sugars and amino acids, cannot cross the plasma membrane without assistance. These substances move across the plasma membrane by **facilitated diffusion**, a type of passive transport that requires transport proteins (**Figure 4.4**, right).

Devaraj's artificial membrane does not currently contain any transport proteins, he says, so it is impermeable to large hydrophilic molecules. But small molecules such as water can pass through his membrane via simple diffusion.

The plasma membrane also contains **receptor proteins**, which are sites where molecules released by other cells can bind. The binding of a molecule to a receptor protein starts a chain of

Viruses—Living or Not?

You've heard their names: Ebola virus, Zika virus, H5N1, dengue virus. Viruses—microscopic, noncellular infectious particles—are perhaps the smallest biological agents with the greatest impact on human health. Like living organisms, viruses reproduce and evolve, yet they lack some of the key characteristics of life—which is why most scientists today regard viruses as nonliving. For one thing, viruses are not made up of cells. A virus is much simpler than a cell, usually consisting of a small piece of genetic material (for example, DNA) that is wrapped in a protein coat. Some viruses also have an envelope, a lipid layer usually stolen from a cell's plasma membrane, enclosing the central core of genetic material and protein.

Protein coat
RNA
Envelope

Another difference, compared to living organisms, is that viruses lack the many structures within cells that are necessary for critical cellular functions such as homeostasis, autonomous reproduction, and metabolism. To gain these functions, they become "body snatchers": Viruses use their genetic material to make the cells of the organisms they infect do their work for them. They accomplish this feat by invading cells, releasing their genetic material into the cell interior, and "hijacking" the host cell's machinery. Viruses multiply to huge numbers, and viral offspring escape from a host cell either by causing it to burst open or by budding off from the cell, wrapped in a layer of the host cell's plasma membrane.

Uniquely, unlike the case with living organisms, the genetic material that viruses pass from one generation to the next is not always DNA; sometimes it is RNA. Viruses are generally classified by the type of genetic material they possess (type of DNA or RNA molecule), their shape and structure, the type of organism (host) they infect, and the disease they produce. The variant forms of a particular type of virus are called **viral strains** or serotypes. Viruses evolve new strains within a host so quickly that sometimes an antiviral drug or vaccine developed to fight an older strain becomes useless against a new strain.

events inside the cell that causes the cell to do something. For example, the receptors in nerve cells receive molecular signals from other nerve cells that cause the cells to fire. Receptor proteins are key components of a cell's communication system, enabling it to respond appropriately to changes in the organism.

Another Way Through

In addition to transport proteins, there is another way that molecules move into and out of a cell. Sections of the plasma membrane can bulge inward or outward to form packages called **vesicles**. Vesicles move molecules from place to place inside a cell but also transport substances into and out of the cell (**Figure 4.7**).

Cells expel materials in vesicles via **exocytosis** (**Figure 4.7**, top left). The substance to be exported from the cell is packaged into a vesicle, and as the vesicle approaches the plasma membrane, a portion of the vesicle's membrane fuses with the plasma membrane. The inside of the vesicle then opens to the exterior of the cell, discharging its contents.

Endocytosis (**Figure 4.7**, right and lower left) is the opposite of exocytosis. In this process, a section of plasma membrane bulges inward to form a pocket around extracellular fluid,

Exocytosis

Exocytosis is used to eject substances from the cell. Here, a cell ejects waste material into the outside environment.

Receptor-mediated endocytosis

Receptor-mediated endocytosis is a selective process in which only certain molecules bind to receptor proteins. Here, low-density lipoprotein (LDL) particles bind to LDL receptors and are transported to the cell interior.

Endocytosis

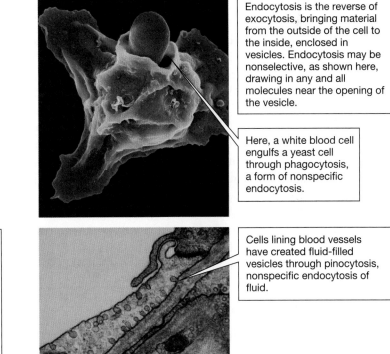

Endocytosis is the reverse of exocytosis, bringing material from the outside of the cell to the inside, enclosed in vesicles. Endocytosis may be nonselective, as shown here, drawing in any and all molecules near the opening of the vesicle.

Here, a white blood cell engulfs a yeast cell through phagocytosis, a form of nonspecific endocytosis.

Cells lining blood vessels have created fluid-filled vesicles through pinocytosis, nonspecific endocytosis of fluid.

Figure 4.7

Substances move into cells by endocytosis and out of cells by exocytosis

Q1: If endocytosis itself is nonspecific, how does receptor-mediated endocytosis bring only certain molecules into a cell?

Q2: What sorts of molecules could be moved by endocytosis or exocytosis, but not by diffusion?

Q3: How does the fluid that enters a cell via pinocytosis differ from the fluid that enters by osmosis?

molecules, or particles. The pocket deepens until the opening in the membrane pinches off and the membrane breaks free as a closed vesicle, now wholly contained within the cell. Endocytosis can be nonspecific or specific. In nonspecific endocytosis, all of the material in the immediate area is surrounded and enclosed; in specific endocytosis, one particular type of molecule is enveloped and imported.

There are three types of endocytosis. **Receptor-mediated endocytosis** (**Figure 4.7,** bottom left) is a form of endocytosis in which receptor proteins embedded in the membrane recognize specific surface characteristics of substances to be incorporated into the cell. For example, our cells use receptor-mediated endocytosis to take up cholesterol-containing packages called low-density lipoprotein (LDL) particles. **Phagocytosis** (**Figure 4.7,** top right), or "cellular eating," is a large-scale version of endocytosis in which particles considerably larger than biomolecules are ingested. Specific cells in the immune system use phagocytosis to ingest an entire bacterium or virus. **Pinocytosis** (**Figure 4.7,** bottom right) is a form of endocytosis that is often described as "cellular drinking" because it involves the capture of fluids. However, the cell does not attempt to collect particular solutions. Pinocytosis is nonspecific: the vesicle budding into the cell contains whatever happened to be dissolved in the fluid when the cell "drank."

There is a long way to go before an artificial plasma membrane will perform processes like endocytosis and exocytosis, says Neal Devaraj. "But just because a research problem is difficult doesn't mean we shouldn't tackle it," he adds.

Prokaryotes versus Eukaryotes

Devaraj's team continues to pursue that ideal. Prior synthetic membranes, their own included, did not have the ability to grow by adding new phospholipids, but in 2015 the team succeeded in designing and synthesizing an artificial membrane that sustained continual growth, just like a living cell. Membranes are important not only because they form the structure of a cell, but because they compartmentalize processes within eukaryotic cells. Depending on the fundamental structure of their cells, all living organisms can be sorted into one of two groups: **prokaryotes** or **eukaryotes** (**Figure 4.8**). *Mycoplasma mycoides* and all other bacteria are prokaryotes, but virtually all the organisms you see every day, including all plants and animals, are eukaryotes.

Eukaryotic cells are larger and more complex than prokaryotic cells: they are roughly ten times wider, with a cell volume about a thousand times greater. Unlike prokaryotic cells, eukaryotic cells have a membrane-enclosed **nucleus** (plural "nuclei") that contains the organism's DNA, and they have a variety of membrane-enclosed subcellular compartments called **organelles**. Through specialization and division of labor, these organelles act like cubicles in a large office, allowing the cell to localize different processes in different places. In contrast, prokaryotic cells are like an open floor plan: they lack a cell nucleus or any membrane-encased organelles.

Though the first fully artificial cell will most likely be a simple prokaryotic cell, synthetic biologists aspire to build a eukaryotic cell. "We're not quite there yet, but it's interesting to think about making complex structures like organelles," says Devaraj. Having achieved a self-assembling and growing artificial membrane, his team is working to create membranes inside preexisting vesicles, mimicking organelles like mitochondria that are made up of membranes within membranes. It is the first step toward creating the variety of organelles inside eukaryotic cells.

What's in a Cell?

The nucleus is the control center of the cell. It contains most of the cell's DNA and may occupy up to 10 percent of the space inside the cell (**Figure 4.9**, top left). Inside the nucleus, long strands of DNA are packaged with proteins into a remarkably small space. The boundary of the nucleus, called the **nuclear envelope**, is made up of two concentric phospholipid bilayers. The nuclear envelope is speckled with thousands of small openings called **nuclear pores**. These pores allow chemical messages to enter and exit the nucleus.

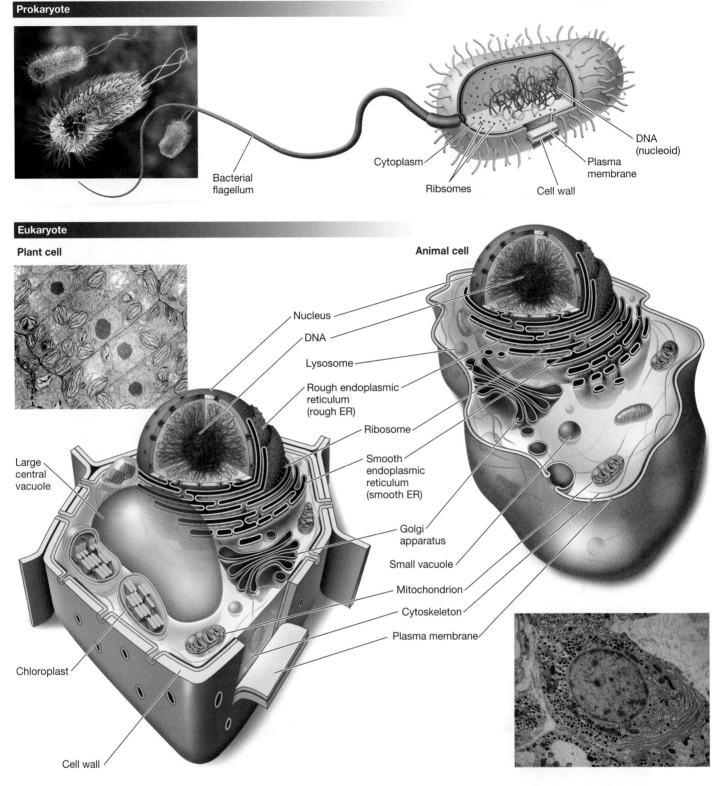

Prokaryote

Bacterial flagellum

Cytoplasm

Ribosomes

DNA (nucleoid)

Plasma membrane

Cell wall

Eukaryote

Plant cell

Animal cell

Nucleus

DNA

Lysosome

Rough endoplasmic reticulum (rough ER)

Ribosome

Smooth endoplasmic reticulum (smooth ER)

Golgi apparatus

Small vacuole

Mitochondrion

Cytoskeleton

Plasma membrane

Large central vacuole

Chloroplast

Cell wall

Figure 4.8

Prokaryotic and eukaryotic cells

A prokaryotic cell, like all cells, contains DNA, cytoplasm, ribosomes, and a plasma membrane. Many prokaryotes also have a cell wall that serves as a kind of exoskeleton. To enable movement, some bacteria possess a flagellum, or several flagella. The components of eukaryotic cells are described in the text and in Figure 4.9. ▶️

Q1: What structures do prokaryotic and eukaryotic cells have in common?

Q2: What cellular processes occur in both prokaryotic and eukaryotic cells?

Q3: Both plants and animals are eukaryotes, but there are differences in their cellular structure. What are those differences?

Nucleus

All eukaryotes contain their DNA within a nucleus, the control center of the cell.

Nuclear envelope

Nuclear pores

DNA

Smooth ER is a site for lipid and hormone manufacture.

Endoplasmic reticulum (ER)

Smooth ER

Rough ER

Ribosomes

With its embedded ribosomes, rough ER is a site for protein production.

Lysosomes

Lysosomes function as cellular "recycling centers," breaking down molecules no longer needed by the cell.

Proteins and lipids are chemically modified as they move from one part of the Golgi apparatus to the next.

Golgi apparatus

Transport vesicle being formed

Free transport vesicle

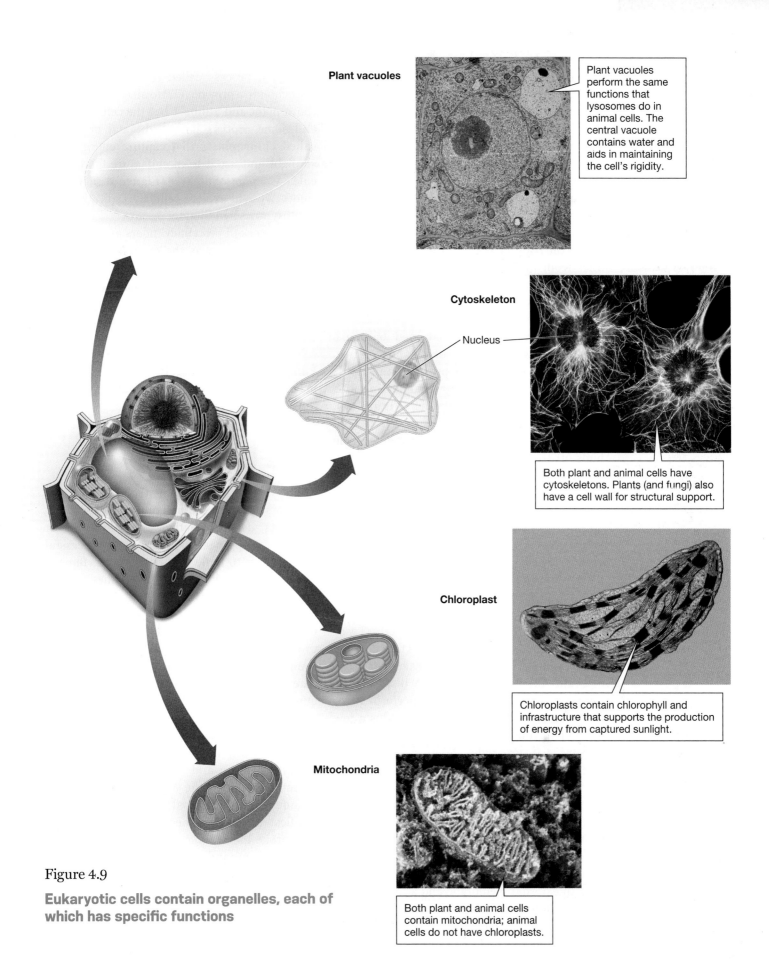

Plant vacuoles

Plant vacuoles perform the same functions that lysosomes do in animal cells. The central vacuole contains water and aids in maintaining the cell's rigidity.

Cytoskeleton

Nucleus

Both plant and animal cells have cytoskeletons. Plants (and fungi) also have a cell wall for structural support.

Chloroplast

Chloroplasts contain chlorophyll and infrastructure that supports the production of energy from captured sunlight.

Mitochondria

Both plant and animal cells contain mitochondria; animal cells do not have chloroplasts.

Figure 4.9

Eukaryotic cells contain organelles, each of which has specific functions

The **endoplasmic reticulum** (**ER**) is an extensive and interconnected network of sacs made of a single membrane that is continuous with the outer membrane of the nuclear envelope (**Figure 4.9**, middle left). The membranes of the ER are classified into two types based on their appearance: smooth and rough. Enzymes associated with the surface of the **smooth ER** manufacture lipids and hormones. In some cell types, smooth-ER membranes also break down toxic compounds. **Ribosomes** embedded in the **rough ER** give it the knobby appearance from which it gets its name. Ribosomes on the rough ER assemble proteins that will be inserted into the cell's plasma membrane or its organelles.

Resembling a pile of flattened balloons, the **Golgi apparatus** is like a post office, packaging and directing proteins and lipids produced by the ER to their final destinations either inside or outside of the cell (**Figure 4.9**, bottom left). Each molecule is first packaged into a transport vesicle. The transport vesicle buds off from the ER membrane and delivers the cargo to its destination by fusing with the membrane of the target compartment. The Golgi apparatus targets these destinations by adding a specific chemical tag to each molecule it receives, like attaching a shipping label to a package.

In animal cells, transport vesicles bring large molecules that will be discarded to **lysosomes**, organelles that act as garbage and recycling centers (**Figure 4.9**, bottom far left). Lysosomes contain a variety of enzymes that degrade biomolecules and release the breakdown products into the cell interior to be discarded or reused. In plant cells, **vacuoles** perform functions similar to those of lysosomes, plus a few additional functions, such as water storage (**Figure 4.9**, top right). Some plant vacuoles stockpile noxious compounds that can deter herbivores from eating plants.

In most eukaryotic cells, the main source of energy is the **mitochondrion** (plural "mitochondria"), a tiny power plant that fuels cellular activities (**Figure 4.9**, bottom right). Mitochondria are made up of double membranes—a smooth external membrane and a folded internal membrane—that form a mazelike interior. Mitochondria use chemical reactions to transform the energy of food molecules into ATP (adenosine triphosphate), the universal cellular fuel, in a process called *cellular respiration*. Mitochondria provide ATP to all eukaryotic cells (both plant and animal), but the cells of plants and some protists have additional organelles called **chloroplasts** that capture energy from sunlight and use it to manufacture food molecules via *photosynthesis* (**Figure 4.9**, middle right). We will explore cellular respiration and photosynthesis in Chapter 5.

A network of protein cylinders and filaments collectively known as the **cytoskeleton** forms the framework of a cell (**Figure 4.9**, middle right). The cytoskeleton organizes the interior of a eukaryotic cell, supports the intracellular movement of organelles such as transport vesicles, and enables whole-cell movement in some cell types. It also gives shape to wall-less cells. Fungi and plants separately evolved cell walls to maintain cell structure, but animals rely only on the cell's cytoskeleton.

Life Goes On

From Venter's complex synthetic genome to Devaraj's simple self-assembling membrane, scientists debate whether the top-down or bottom-up approach will be more successful in the effort to build an artificial cell. But the two can be complementary, says James Collins. "I'm not sure one will win out. I think they bring different things to the table," he notes. Gibson agrees: "I would like to one day be able to combine all of a cell's parts from nonliving components, including the genome, and incubate them, and see if we can get life out of those nonliving components," he says. "It would help us better understand how cells work."

Today, the work on both ends continues. While Devaraj makes his membranes more complex, Gibson and his team (**Figure 4.10**) recently simplified the *Mycoplasma mycoides* genome. With the goal of determining the smallest set of genes needed to maintain life, they broke the *M. mycoides* genome into eight DNA segments and mixed and matched them to see which combinations would produce viable cells. Eventually, they narrowed the genome down to just 473 genes capable of sustaining life. Amazingly, the team could not identify the function of 149 of the 473 genes. "We don't

Sizing Up Life

Cells dramatically range in size, and viruses are even smaller. The amoeba, a single-celled eukaryote that eats smaller unicellular organisms, is visible under a light microscope. But to observe a virus, one would have to use an electron microscope, which relies on a high-voltage beam of electrons to magnify an image.

Assessment available in smartwork**5**

|←———— 500 μm ————→|

Amoeba proteus

|←———— 130 μm ————→|

Human egg cell
(shown next to an Amoeba proteus)

|←———— 30 μm ————→|

Skin cell
(shown next to a human egg cell)

|←———— 8 μm ————→|

Red blood cell
(shown next to a skin cell)

|←———— 3 μm ————→|

***E. coli* bacterium**
(shown next to a red blood cell)

|←———— .22 μm ————→|

Measle virus
(shown next to an E. coli bacterium)

|←———— .13 μm ————→|

Influenza virus
(shown next to a measle virus)

|←———— .03 μm ————→|

Rhinovirus
(shown next to an influenza virus)

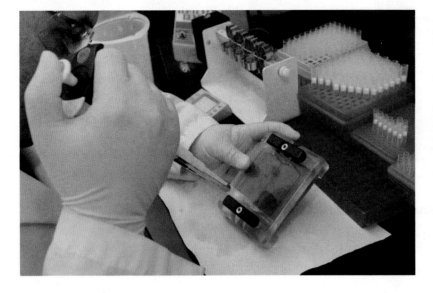

Figure 4.10

Scientist in Action
Laboratory technician Javier Quinones is setting up the beginning of the sequencing procedure used to sequence the *M. mycoides* genome in the sequencing laboratory at the J. Craig Venter Institute in Rockville, Maryland.

know about a third of essential life, and we're trying to sort that out now," Venter told *Nature* magazine in 2016.

Knowing both the identities and function of a core set of genes necessary for life will make it easier to build other synthetic cells with specific functions, says Gibson. "There's not a single organism in the world where we understand what every gene does," he notes. If they can reach such understanding with those 473 genes in the *M. mycoides* genome, "this would be the first example of that."

REVIEWING THE SCIENCE

- **Cells** are the basic units of all living organisms. **Cell theory** states that all living things are composed of one or more cells, and that all cells living today came from a preexisting cell.

- Every cell is surrounded by a **plasma membrane** that separates the chemical reactions inside the cell from the surrounding environment. The plasma membrane is **selectively permeable** and formed by a **phospholipid bilayer** embedded with proteins that perform a variety of functions.

- In **passive transport**, substances move across the plasma membrane without the direct expenditure of energy. **Active transport** by cells requires energy. **Diffusion** is the passive transport of a substance from a region where it is at a higher concentration to a region where it is at a lower concentration.

- **Osmosis** is the diffusion of water across a selectively permeable membrane. In a **hypotonic** solution, a cell gains water. In a **hypertonic** solution, a cell loses water. In an **isotonic** solution, there is no net movement of water. Cells can actively balance their water content through osmosis.

- Cells export materials by **exocytosis** and import materials by **endocytosis**.

- In **receptor-mediated endocytosis**, **receptor proteins** in the plasma membrane recognize and bind the substance to be brought into the cell.

- **Prokaryotes** are single-celled organisms lacking a nucleus and complex internal compartments. **Eukaryotes** may be single-celled or multicellular, and their cells typically possess many membrane-enclosed compartments.

- By volume, eukaryotic cells can be a thousand times larger than prokaryotic cells. They require internal compartments, or **organelles**, that concentrate and organize cellular chemical reactions.

- The **nucleus** of a eukaryotic cell contains DNA. It is bounded by the **nuclear envelope**, which has **nuclear pores** that allow communication between the nucleus and the cell interior.

- Lipids are made in the **smooth endoplasmic reticulum**. Some proteins are manufactured in the **rough endoplasmic reticulum**.

- The **Golgi apparatus** receives proteins and lipids, sorts them, and directs them to their final destinations.

- In animals, **lysosomes** break down biomolecules such as proteins into simpler compounds that can be used by the cell. Plant **vacuoles** are similar to lysosomes but also store ions and molecules and lend physical support to plant cells.

- **Mitochondria** produce chemical energy for eukaryotic cells in the form of ATP.

- **Chloroplasts** harness the energy of sunlight to make sugars through photosynthesis.

- Eukaryotic cells depend on the **cytoskeleton** for structural support and for the ability to move and change shape. Plants and fungi also have a cell wall that provides structural support.

THE QUESTIONS

The Basics

1 To be able to recognize a colony of bacteria that had grown from cells of *Mycoplasma capricolum* in which the DNA had been replaced by a synthetic DNA of *Mycoplasma mycoides*, Daniel Gibson added a _____ that coded for a blue pigment.

(a) cell

(b) chromosome

(c) gene

(d) bacterium

2 A phospholipid is a biomolecule composed of a phosphate group that is bonded to two lipid chains. Which of the following correctly describes the nature of those two components?

(a) The phosphate group is hydrophobic, and the lipid chains are hydrophilic.

(b) Both the phosphate group and the lipid chains are hydrophilic.

(c) Both the phosphate group and the lipid chains are hydrophobic.

(d) The phosphate group is hydrophilic, and the lipid chains are hydrophobic.

3 Link each process with the correct definition.

RECEPTOR-MEDIATED ENDOCYTOSIS	1. A cell ingests a large particle, such as a bacterial cell.
PHAGOCYTOSIS	2. Receptor proteins embedded in the membrane recognize specific surface characteristics of substances.
PINOCYTOSIS	3. A transport vesicle inside the cell approaches the plasma membrane of the cell, fuses with it, and releases its contents to the outside of the cell.
EXOCYTOSIS	4. A vesicle containing whatever molecules are in solution outside the cell bulges inward, pinches off, and enters the cell.

4 Link each structure with the correct function.

CHLOROPLAST	1. Location of the cell's DNA.
GOLGI APPARATUS	2. Site of protein synthesis.
LYSOSOME	3. Site of lipid synthesis.
MITOCHONDRION	4. Adds chemical tags to newly synthesized proteins to direct them to their correct location.
NUCLEUS	5. Breaks down biomolecules by enzymatic action.
ROUGH ENDOPLASMIC RETICULUM	6. Site of cellular respiration.
SMOOTH ENDOPLASMIC RETICULUM	7. Site of photosynthesis.

5 In the table below, indicate whether the specified cellular component is found in each type of cell by placing an "X" in the relevant columns.

Component	Prokaryotes	Eukaryotes Animals	Eukaryotes Plants
Plasma membrane			
Cellulose cell wall			
Nucleus			
Endoplasmic reticulum			
Golgi apparatus			
Ribosomes			
Cytoskeleton			
Mitochondria			
Chloroplasts			

Challenge Yourself

6 How does the phospholipid bilayer of a liposome differ from the phospholipid bilayer of the plasma membrane of a cell?

(a) The phospholipid bilayer of a liposome contains only phospholipids, without the proteins that are embedded in the plasma membrane of a cell.

(b) The phospholipid bilayer of a liposome contains two bilayers of phospholipid molecules, whereas the plasma membrane of a cell contains only one.

(c) The phospholipid bilayer of a liposome completely envelops the liposome, whereas the plasma membrane of a cell does not completely envelop the cell.

(d) The phospholipid molecules in the phospholipid bilayer of a liposome are oriented with the lipid ends on the outside of the bilayer and the phosphate groups on the inside.

7 Examine the graph presented here. Using your knowledge of how the processes work, label each curve as representative of either simple diffusion or facilitated diffusion. Explain how you determined which curve was which.

8 Which of the following statements about transport through the plasma membrane of a cell is correct?

(a) Both passive and active transport require the input of energy.

(b) Passive transport does not require the input of energy, and active transport does.

(c) Passive transport does require the input of energy, and active transport does not.

(d) Neither active nor passive transport requires the input of energy.

Try Something New

9 Picture a beaker divided by a semipermeable membrane like the one in Figure 4.6. Imagine that you put 5 grams of sugar in the left side of the beaker and 10 grams of sugar in the right side, and then add water to bring both sides to the same depth. Select the correct terms in the following description of what happens next: The water will move by osmosis to the (**left side / right side**) of the beaker because there are (**more / fewer**) molecules of solute and (**more / fewer**) molecules of water on that side.

10 When the chemical oligomycin is added to cells, the production of ATP is blocked. Which of the following processes would be most affected by oligomycin?

(a) simple diffusion

(b) facilitated diffusion

(c) active transport

(d) All of these processes would be equally affected by oligomycin.

(e) None of these processes would be affected by oligomycin.

11 Use the accompanying image to select the correct terms in the statements below each of the cells shown.

(a) This red blood cell is in a(n) (**hypertonic / hypotonic / isotonic**) solution.

(b) In this environment, normal red blood cells (**gain / lose / neither gain nor lose**) water.

(c) The total solute concentration outside the cell is (**lower than / higher than / equal to**) the total solute concentration inside it.

(d) This red blood cell is in a(n) (**hypertonic / hypotonic / isotonic**) solution.

(e) In this environment, normal red blood cells (**gain / lose / neither gain nor lose**) water.

(f) The total solute concentration outside the cell is (**lower than / higher than / equal to**) the total solute concentration inside it.

(g) This red blood cell is in a(n) (**hypertonic / hypotonic / isotonic**) solution.

(h) In this environment, normal red blood cells (**gain / lose / neither gain nor lose**) water.

(i) The total solute concentration outside the cell is (**lower than / higher than / equal to**) the total solute concentration inside it.

Leveling Up

12 **What do *you* think?** Viruses display many of the characteristics of living organisms. In particular, they reproduce, creating new virus particles. During reproduction, viruses make copies of their genetic material, and some of the copies contain mutations that are beneficial to the virus. For example, HIV (human immunodeficiency virus), the virus that causes the disease AIDS (acquired immunodeficiency syndrome) mutates so often that its surface proteins change faster than we can develop antiviral drugs. New drug-resistant strains of HIV are appearing constantly. However, viruses can reproduce only after entering a living cell of an organism, because viruses hijack the cell's machinery and use it to produce new viruses. Where does that combination of characteristics place viruses on the scale of nonlife to life? Are viruses living organisms? Nonliving? If neither of those categories fits the properties of viruses, how should they be classified?

13 **What do *you* think?** Did Venter create "new life" by inserting synthetic DNA copied from the genome of *Mycoplasma mycoides* into a cell of *Mycoplasma capricolum* from which the DNA had been removed? Be prepared to support your opinion in a class discussion.

14 **Doing science** Go to the National Science Foundation website (http://www.nsf.gov), and search for and read "Biologists Replicate Key Evolutionary Step in Life on Earth" (Press Release 12-009, January 16, 2012). State the hypothesis that the biologists were testing, describe their experimental design, and explain their results. Was their hypothesis supported? What experiment do you think they should try next? State a hypothesis that you would test using their multicellular yeast samples.

For more, visit digital.wwnorton.com/bionow2 for access to:

Rock Eaters

Unusual "electric" microbes could provide the energy storage solution we've been looking for.

After reading this chapter you should be able to:

- Explain how photosynthesis and cellular respiration support life on Earth.
- Differentiate between anabolic and catabolic metabolic processes.
- Detail the role of ATP in a cell.
- Describe the importance of enzymes in metabolic pathways.
- Label a diagram of the light reactions and the Calvin cycle of photosynthesis.
- Define the three stages of cellular respiration and explain the function of each.
- Compare and contrast photosynthesis and cellular respiration.

CHAPTER
05

HOW CELLS
WORK

Annie Rowe stands knee-deep in cold, clear salt water. Sweating, she pushes a shovel into the ground and lifts a mound of heavy mud. It's a sunny day on Santa Catalina Island, a rocky retreat off the coast of California, where Rowe has come to collect samples. Once the bucket is full, she sits on a rock and retrieves a mesh filter to sieve out any small invertebrates, such as snails and worms, from the sand. They are not part of the experiment.

A postdoctoral research fellow at the University of Southern California, Rowe pours the filtered sediment into one of her five 10-gallon aquariums. When each aquarium is about two-thirds full, she and a graduate student transport the tanks to a local laboratory on the island. There, they hook each aquarium up to a steady stream of seawater from the harbor.

To the untrained eye, the aquariums appear sterile, empty. In reality, both the water and sand in the tanks are teeming with microscopic organisms, or *microbes*, and Rowe is aiming to find a rare, strange one. Deep in the sand of each aquarium, she buries a metal electrode buzzing with electrical current to attract her quarry: a microbe that eats electricity (**Figure 5.1**).

"I went into microbiology because there's so much diversity in things that microbes can do," says Rowe. As a PhD student, Rowe began studying unusual microbes with the hope of someday putting them to use toward one of the greatest human challenges of our time: developing sources of alternative energy.

Humans rely heavily on fossil fuels such as oil and petroleum to power our cars, trains, planes, and more. Hundreds of millions of years ago, single-celled organisms used carbon dioxide and sunlight to make cell materials, which then became buried and concentrated into oil in Earth's crust. Burning that oil releases that ancient carbon dioxide back into the

Figure 5.1

Annie Rowe burying electrodes in her sediment aquariums

atmosphere, and a consequence of this increase in carbon dioxide is that the surface of our planet has begun to warm. To prevent further climate change and lessen our dependence on fossil fuels—many of which we pay a premium to import from other countries—scientists and engineers have developed clean, renewable ways to capture energy, such as solar, wind, and water power.

Yet to switch from a fossil fuel–based economy to renewable energy technologies, we need efficient and inexpensive ways to store renewable energy. With the right chemical stabilizers, oil can be left in a barrel for decades, but a direct current of electricity from a wind farm or solar panels needs to be used immediately or fed into the electrical grid. Today's batteries aren't a good storage option; they are bulky, expensive, and not terribly efficient.

Now imagine the ability to siphon electricity from a wind farm or solar panel and transform it into a liquid fuel. That's where electricity-eating microbes come in.

Energy for Life

All living cells require energy. Organisms use energy for growth, reproduction, and defense, and to manufacture the many chemical compounds that make up living cells. They must obtain energy from the living or nonliving components of their environment, and at the very core of making and storing energy is the transfer of *electrons*—subatomic particles with a negative charge. Electrons play an essential role in electricity, magnetism, and many other physical phenomena that shape the world we know. The success of computers, solar cells, cell phones, and other devices is due to our ability to shape and control the flow of electrons.

But long before electrons flowed through computers, they moved through cells. The first law of thermodynamics says that energy cannot be created or destroyed, but it can be changed from one form to another. In other words, cells cannot create energy from nothing, so they must utilize one form of energy and change it to another form. However, organisms can't simply perform the biological equivalent of plugging into an electrical socket and sucking up electricity. That's because the cell membrane is an electrically neutral zone that prevents charged particles from sneaking through, including electrons. So, to move electrons into and out of a cell, living organisms attach electrons to molecules. Plants, for example, smuggle electrons into the cell via water molecules. Humans and other organisms obtain electrons from food (sugars, proteins, fats, etc.). As the Nobel Prize–winning physiologist Albert Szent-Györgyi reportedly said, "Life is nothing but an electron looking for a place to rest."

Cells use and store energy by transferring electrons among molecules via chemical reactions. Thousands of different types of chemical reactions are required to sustain life in even the simplest cell. The term **metabolism** describes all the chemical reactions that occur inside living cells, including those that store or release energy. Most chemical reactions in a cell occur in chains of linked events known as **metabolic pathways**. Metabolic pathways produce key biological molecules in a cell, including important chemical building blocks like amino acids and nucleotides.

Two metabolic pathways drive most of the life around us. The sun is the ultimate source of energy for most living organisms, and in the first process, known as **photosynthesis**, organisms capture energy from the sun and use it to create sugars from carbon dioxide and water (**Figure 5.2**). In this way, photosynthetic organisms such as plants transform light energy into chemical energy stored in the covalent bonds of sugar molecules. These sugar molecules, for example, glucose, fuel the cell's activities, and some are converted to fatty acids to help build cell membranes and to store energy for future needs.

The second important process is **cellular respiration**, a process reciprocal to photosynthesis. During cellular respiration, the cell breaks down sugars into usable energy

ANNETTE ROWE

Annette ("Annie") Rowe is a postdoctoral research associate at the University of Southern California, where she studies microbes that take up electrons from inorganic surfaces.

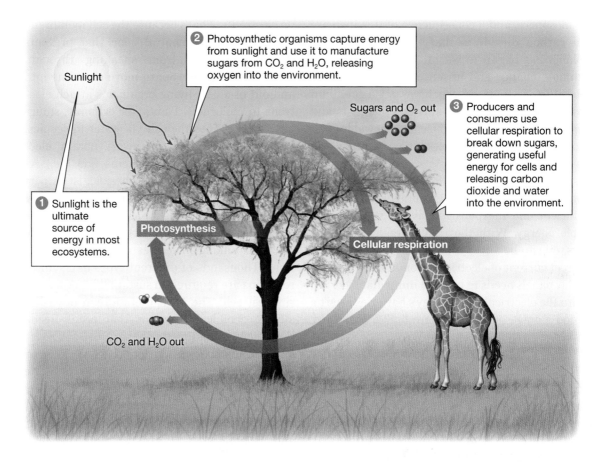

Figure 5.2

From sunlight to usable energy

Photosynthesis transforms sunlight into sugar molecules within the cell, releasing oxygen as a by-product. Without photosynthesis, we would not have access to the sun's energy, and we wouldn't have oxygen in our atmosphere to support life. In a complementary process, cellular respiration breaks down these sugar molecules, allowing organisms to access the energy stored in them.

Q1: Why is photosynthesis called "primary production"?

Q2: How does animal life depend on photosynthesis?

Q3: Explain how photosynthesis and cellular respiration are "complementary" processes.

(**Figure 5.2**). Plants break down the glucose made during photosynthesis, whereas nonphotosynthetic organisms, including animals, eat plants or other animals and break down sugars in that food to release energy. Electrons are transferred back and forth during these two metabolic processes. Metabolism, at its core, is a feat of juggling electrons.

Rowe spent a week on Catalina setting up her tanks and then let them sit for 3 months, untouched. During that time, her instruments began to detect evidence of life: the negative charge flowing through the electrodes in the sand began to steadily increase. Something was stealing the electrons, consuming more and more electricity.

An Unusual Pathway

Scientists in the 1980s first discovered two types of bacteria that expel—breathe, actually—electrons onto metal. Because it was assumed

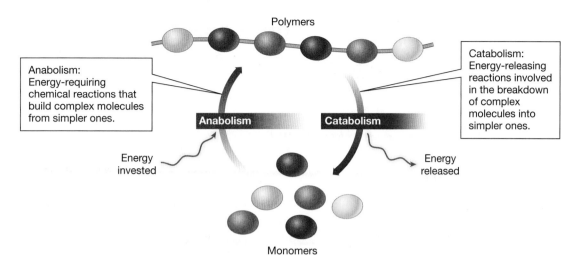

Polymers

Anabolism:
Energy-requiring
chemical reactions that
build complex molecules
from simpler ones.

Catabolism:
Energy-releasing
reactions involved
in the breakdown
of complex
molecules into
simpler ones.

Anabolism

Catabolism

Energy
invested

Energy
released

Monomers

Figure 5.3

Anabolism builds biomolecules; catabolism breaks them down

Metabolic reactions either build or break down molecules. Molecule-building reactions (anabolism) cost energy, and molecule-breakdown reactions (catabolism) release energy.

Q1: What source of energy would plants use for anabolic reactions? Would an animal use the same kind of energy?

Q2: What source of energy would plants release in catabolic reactions? Would an animal release the same kind of energy?

Q3: Create a mnemonic or jingle that helps you remember the difference between anabolism and catabolism.

that electrons couldn't cross the cell membrane, the discovery was a big surprise. For decades, however, the way the electron transfer worked remained a mystery. Then, in 2006, a scientific team identified a trio of proteins that form a bridge in the bacterial cell membrane, allowing the transfer of electrons from the inside to the outside of the cell.

Inspired by these "rock breathers" that expelled electrons onto a mineral surface, scientists began looking for "rock eaters." If nature produced microbes that spit out electrons, why not microbes that directly ingest electrons from minerals?

All living cells have two main types of metabolism: anabolism and catabolism (**Figure 5.3**). **Anabolism** refers to metabolic pathways that create complex molecules from simpler compounds—a process used by all cells to make the basic building blocks of the cell. These processes often require an energy input. Photosynthesis is an example of anabolism where the energy is provided by sunlight: as plants sit in the noonday sun, the cells in their green leaves capture energy to make glucose from carbon dioxide; a complex molecule is assembled from simpler ones. **Catabolism**, unsurprisingly, refers to the opposite: metabolic pathways that release chemical energy in the process of breaking down complex molecules. Plant cells rely on cellular respiration, a catabolic process, to break down the glucose made by photosynthesis.

All of this metabolic activity requires energy, so cells need **energy carriers** to deliver usable "on-demand" energy. Every living cell uses **ATP** (adenosine triphosphate), a small, energy-rich organic molecule, to carry energy from one part of the cell to another. ATP powers almost all activities in the cell, such as moving molecules and ions in and out of the cell, sending nerve impulses, triggering muscle contractions, and moving organelles around inside the cell. In addition to powering a cell's activities, ATP fuels metabolic reactions and most other enzymatic reactions. If a cell exhausts its ATP supply, it will die.

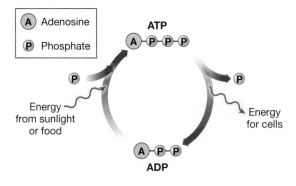

Ⓐ Adenosine
Ⓟ Phosphate

ATP

Energy from sunlight or food

Energy for cells

ADP

Figure 5.4

Cells store and deliver energy using ATP

Loading a high-energy phosphate (P) on ADP transforms the otherwise sedate molecule into the live wire that is ATP. But turning ADP and phosphate groups into ATP—the universal energy currency of cells—takes metabolic energy.

> **Q1:** Define ATP in your own words.
>
> **Q2:** How is ATP involved in anabolism and catabolism? (*Hint:* Review Figure 5.3.)
>
> **Q3:** Arsenic disrupts ATP production. Why would this characteristic cause it to be a potent poison?

Much of the usable energy in ATP is held in its energy-rich phosphate bonds (**Figure 5.4**). Energy is released when a molecule of ATP loses its terminal phosphate group, breaking into a molecule of **ADP** (<u>a</u>denosine <u>di</u>phosphate) and a free phosphate. Converting ADP and phosphate back into ATP takes metabolic energy.

ATP is not the only energy carrier that cells rely on; NADPH, NADH, and FADH$_2$ are also energy carriers. Each one is a specialist in terms of the amount of energy it carries and the types of chemical reactions to which it supplies energy and from which it receives energy. ATP, because of its phosphate bonds, carries the most energy.

ALFRED SPORMANN

Alfred Spormann is a professor of civil and environmental engineering, and of chemical engineering, at Stanford University. He investigates novel microbial metabolism, including how microbes transport electrons between themselves and surfaces.

Into the Light, Part 1

In 2009, environmental engineers at Pennsylvania State University identified the first elusive rock eaters by exposing a mix of microbes to a negatively charged electrode (**Figure 5.5**). One of the species, *Methanobacterium palustre*, survived on the electricity, taking in electrons and using them in a metabolic pathway to convert carbon dioxide to methane. One news article summed up the experiment with the headline BUG EATS ELECTRICITY, FARTS BIOGAS. The Penn State engineers proposed, but were unable to prove, that the bacteria ("bugs") were ingesting naked electrons straight from the surface of the electrode.

A few years later, Alfred Spormann launched his own investigation into bacteria that seemed to be "eating" electrons. Spormann, a microbiologist at Stanford University, has long been interested in outliers, microbes with unusual metabolisms. Tall and thin with a throaty German accent, Spormann has spent more than 25 years studying such microbes, from those in the human gut that contribute to irritable bowel syndrome to bacteria that could be used to help decontaminate groundwater. "I'm really interested in exceptions to the rules, and in understanding the plasticity of microbial metabolism," says Spormann. Typically, the microbes he studies are **anaerobic**, growing in places with little or no oxygen. These cells don't need oxygen

Figure 5.5

A negatively charged electrode used to grow "rock eaters"

for metabolism; in fact, some are even poisoned by it.

Oxygen plays an important role in photosynthesis, considered by many to be the most important life process on Earth; it is the way our planet stores energy from the sun and produces oxygen for animals to breathe. In the cells of algae and all plants, photosynthesis takes place inside chloroplasts, organelles that look like green, oval gumballs when viewed under a light microscope (**Figure 5.6**). Chloroplasts contain an extensive network of structures called thylakoids, piled up like stacks of pancakes, that contain enzymes needed for photosynthesis. Also embedded in those membranes is a green pigment, called **chlorophyll**, that is specialized for absorbing light energy.

Photosynthesis takes place in two principal stages: the *light reactions* and the *light-independent reactions*, or the *Calvin cycle* (**Figure 5.7**). During the **light reactions**, chlorophyll molecules absorb energy from sunlight and use that energy to split water (**Figure 5.7**, left). The splitting of water produces oxygen gas (O_2)—the oxygen that we breathe—as a by-product that is released into the atmosphere. More important for photosynthesis, electrons and protons (H^+) from the light reactions are handed over to other molecules via the **electron transport chain**, an elaborate chain of chemical events that ultimately generates ATP and NADPH. The light reactions depend on protein complexes embedded in the chloroplast membrane, including photosystems I and II, and ATP synthase.

In the next stage, the **light-independent reactions**, or **Calvin cycle**—a series of chemical reactions—convert carbon dioxide (CO_2) into sugar, using energy delivered by ATP, and electrons and hydrogen ions donated by NADPH (**Figure 5.7**, right). Enzymes catalyze these reactions at each step; the enzyme needed in the first step—and the most abundant enzyme on the planet—is **rubisco**. This process is also known as **carbon fixation**. By capturing inorganic carbon atoms from CO_2 gas and converting them into glucose, the Calvin cycle reactions make carbon from the nonliving world available to the photosynthetic organisms and eventually to other living organisms, including us.

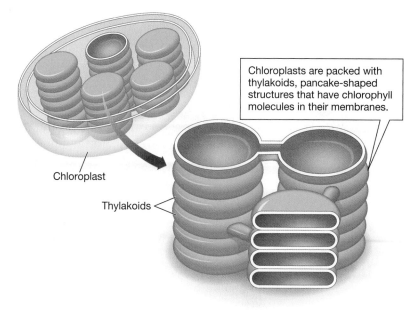

Chloroplasts are packed with thylakoids, pancake-shaped structures that have chlorophyll molecules in their membranes.

Chloroplast

Thylakoids

Figure 5.6

Chloroplasts are the site of photosynthesis in eukaryotes

Most plants and algae have multiple chloroplasts to contain their chlorophyll. Photosynthetic bacteria embed chlorophyll directly into the plasma membrane.

Q1: Is chlorophyll found only within chloroplasts?

Q2: What could be an advantage of concentrating chlorophyll molecules in the membranes of chloroplasts?

Q3: What is the advantage of having multiple chloroplasts per cell?

Catalyzing Reactions

In about 2014, Spormann's team began studying an "electric" microbe that lives in water, *Methanococcus maripaludis*, to see whether it was capable of directly ingesting electrons. It wasn't. The researchers found that the microbe actually excretes an enzyme onto the surface of an electrode to do its dirty work. Nearly all metabolic reactions are facilitated by enzymes. **Enzymes** are biological catalysts—molecules that speed up chemical reactions. Without the action of enzymes—most of which are proteins—metabolism would be extremely slow, and life as we know it could not exist. Enzymes work by positioning **substrates**—molecules that will react to form new products—in

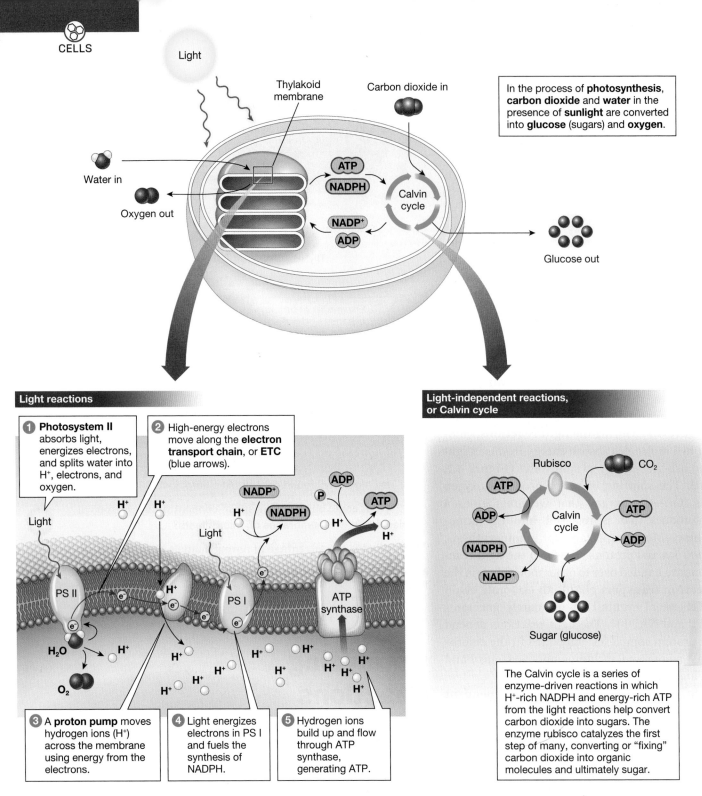

Light

Thylakoid membrane

Carbon dioxide in

Water in

Oxygen out

ATP
NADPH
NADP⁺
ADP

Calvin cycle

In the process of **photosynthesis**, **carbon dioxide** and **water** in the presence of **sunlight** are converted into **glucose** (sugars) and **oxygen**.

Glucose out

Light reactions

1 Photosystem II absorbs light, energizes electrons, and splits water into H⁺, electrons, and oxygen.

2 High-energy electrons move along the **electron transport chain**, or **ETC** (blue arrows).

Light-independent reactions, or Calvin cycle

Rubisco   **CO₂**

Calvin cycle

Sugar (glucose)

The Calvin cycle is a series of enzyme-driven reactions in which H⁺-rich NADPH and energy-rich ATP from the light reactions help convert carbon dioxide into sugars. The enzyme rubisco catalyzes the first step of many, converting or "fixing" carbon dioxide into organic molecules and ultimately sugar.

3 A proton pump moves hydrogen ions (H⁺) across the membrane using energy from the electrons.

4 Light energizes electrons in PS I and fuels the synthesis of NADPH.

5 Hydrogen ions build up and flow through ATP synthase, generating ATP.

Figure 5.7

Photosynthesis occurs in two stages

The light reactions generate energy carriers; the light-independent Calvin cycle reactions create sugars. Rubisco contributes to fixing carbon during the Calvin cycle, but enzymes are critical throughout both stages of photosynthesis. ▶

Q1: What is the source of the carbon dioxide used for photosynthesis?

Q2: Which products of the light reactions of photosynthesis does the Calvin cycle use?

Q3: What are the two major products of photosynthesis?

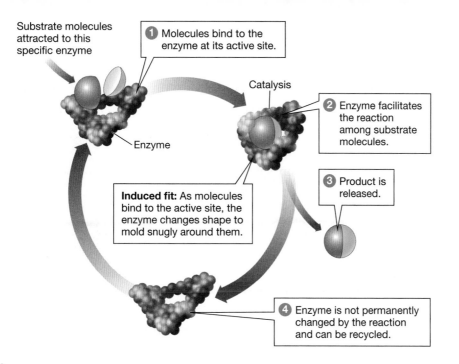

Substrate molecules attracted to this specific enzyme

① Molecules bind to the enzyme at its active site.

Catalysis

② Enzyme facilitates the reaction among substrate molecules.

Enzyme

③ Product is released.

Induced fit: As molecules bind to the active site, the enzyme changes shape to mold snugly around them.

④ Enzyme is not permanently changed by the reaction and can be recycled.

Figure 5.8

Enzymes are molecular matchmakers

Enzymes dramatically increase the rate of chemical reactions by positioning molecules so that they more easily form or break chemical bonds.

Q1: Why is it important that enzymes are not permanently altered when they bind with substrate molecules?

Q2: How would a higher temperature or higher salt concentration make it more difficult for an enzyme to function effectively?

Q3: If a cell was unable to produce a particular enzyme necessary for a metabolic pathway, describe how the absence of that enzyme would affect the cell.

an orientation that favors the making or breaking of chemical bonds (**Figure 5.8**).

Each enzyme binds only to a specific substrate or substrates and catalyzes a specific chemical reaction, such as rubisco catalyzing the first step of the Calvin cycle. An enzyme's function is based on its chemical characteristics and the three-dimensional shape of its **active site**—the location within the enzyme where substrates bind. When molecules bind to the active site, the enzyme changes shape—a process called **induced fit**. The enzyme's shape, and therefore its activity, can be affected by temperature, pH, and salt concentration. Because the enzyme's active site is not permanently changed as reactions are catalyzed, an enzyme, like rubisco, is used over and over.

The enzyme that Spormann identified was unusual because it was excreted to the exterior of the cell. Most enzymes work on the inside of cells. A multistep metabolic pathway can proceed rapidly and efficiently because the required enzymes are physically close together and the products of one enzyme-catalyzed chemical reaction serve as the basis for the next reaction in the series, as is especially true for metabolic pathway reactions.

In the case of *M. maripaludis*, the excreted enzyme's role was to grab an electron from the metal, pair it with a proton from water, and create a hydrogen atom—a familiar food for microbes, and easily passed through the cell membrane. "They [*M. maripaludis* bacteria] found a way to produce a compound that is easily metabolized by the cells," says Spormann. And although this particular microbe wasn't eating naked electrons, Spormann has since isolated a microbe that does directly take up electrons. His team has yet to publish the details.

Into the Light, Part 2

As Spormann was unraveling the mechanisms of how electron-eating life works, Rowe's prospecting trip to Catalina uncovered more species of rock eaters (**Figure 5.9**). From her tanks, Rowe identified 30 new varieties of microbes sucking electrons from the electrodes. She was even able to grow a few of them on plates in the laboratory, using minerals like sulfur and iron as electron sources for the microbes to munch. Not all the species relied exclusively on electrons as a food source; that is, they didn't just live off mineral sediments. Like eukaryotic cells, most microbial cells use the traditional route of cellular respiration for obtaining energy, converting sugars into energy through catabolic reactions.

During cellular respiration, the carbon-carbon bonds in glucose molecules are broken,

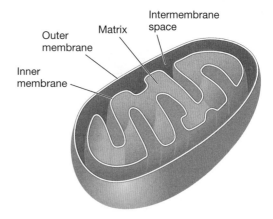

Figure 5.10

Mitochondria are the site of cellular respiration in eukaryotes

and each carbon atom is released into the environment in a molecule of CO_2, with water (H_2O) as a by-product. Thus, cellular respiration is the reciprocal process of photosynthesis, as illustrated earlier, in **Figure 5.2**. In most eukaryotes, cellular respiration is a multistep process that occurs in the cell's mitochondria (**Figure 5.10**). Cellular respiration is the main way that animal cells, which do not perform photosynthesis, obtain energy. We digest glucose from food sources; break it down into carbon dioxide, hydrogen, and electrons; and then process those captured electrons for energy.

Simply put, cellular respiration consists of three stages: glycolysis, the Krebs cycle, and oxidative phosphorylation (**Figure 5.11**). The first stage, **glycolysis**, takes place in the cytoplasm of the cell. During glycolysis, sugars (mainly glucose) are split to make a three-carbon compound called pyruvate. This process results in two useful molecules of ATP and two molecules of NADH for each glucose molecule that is split (**Figure 5.11**, top left). In other words, glycolysis converts some of the chemical energy of glucose into the chemical energy of NADH and ATP.

From an evolutionary standpoint, glycolysis was probably the earliest means of producing ATP from food molecules, and it is still the primary means of energy production in many prokaryotes, including some of Rowe's microbes. However, because glucose is only partially broken down through this process, the

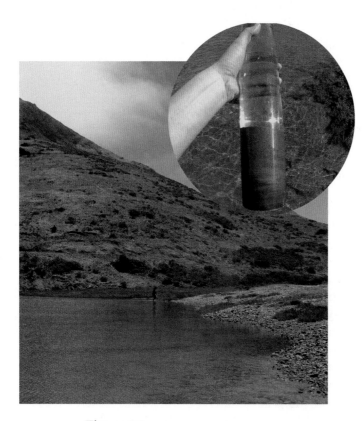

Figure 5.9

Sediment containing rock eaters, collected off the coast of Santa Catalina Island

Figure 5.11

Cellular respiration is highly efficient in eukaryotes

In glycolysis, each six-carbon glucose molecule is converted into two molecules of pyruvate. The next two stages of cellular respiration require oxygen. The Krebs cycle (bottom left) releases carbon dioxide and generates high-energy molecules. Oxidative phosphorylation (bottom right), the last stage in cellular respiration, produces more ATP than any other metabolic pathway.

Q1: What are the products of cellular respiration?

Q2: Considering the inputs and products of each process, why is cellular respiration considered the reciprocal process to photosynthesis?

Q3: Which of the three stages of cellular respiration—glycolysis, the Krebs cycle, or oxidative phosphorylation—could organisms have used 4 billion years ago, before photosynthesis by cyanobacteria released oxygen into the atmosphere?

energy yield from glycolysis is small. For most eukaryotes, glycolysis is just the first step in extracting energy from sugars, and the second and third stages of cellular respiration, which occur in the mitochondria, help to generate much more ATP than is possible through glycolysis alone.

Glycolysis is an anaerobic process; that is, it does not require oxygen. Cells experiencing low oxygen levels—such as our muscle cells during intense exercise or cancer cells in the interior of a tumor mass—use glycolysis alone because there is not enough oxygen available for the later stages of cellular respiration to proceed.

But when oxygen is present, eukaryotic cells rely on all three stages of cellular respiration. After pyruvate is made in the cytoplasm during glycolysis, it enters the mitochondria, where it is broken down during the second stage of cellular respiration: a sequence of enzyme-driven reactions known as the **Krebs cycle** or **citric acid cycle**. The carbon backbone of the pyruvate molecule is taken apart, releasing carbon dioxide (**Figure 5.11**, bottom left). The breakdown of carbon backbones by the Krebs cycle produces a large bounty of energy carriers, including ATP, NADH, and $FADH_2$. Essentially, during the Krebs cycle the remaining chemical energy of glucose is completely converted to the chemical energy of these energy carriers.

The largest output of ATP, however, is generated during the third and last stage of cellular respiration: **oxidative phosphorylation**. The electrons and hydrogen atoms are removed from NADH and $FADH_2$ and handed over to molecular O_2 through an electron transport chain, creating water (H_2O). In the process, a large amount of ATP is generated (**Figure 5.11**, bottom right). During oxidative phosphorylation, the chemical energy of NADH and $FADH_2$ is converted into the chemical energy of ATP. In fact, oxidative phosphorylation can generate 15 times as much ATP as can glycolysis alone.

ATP production in mitochondria is crucially dependent on oxygen; that is, the Krebs cycle and oxidative phosphorylation are strictly **aerobic** processes. Highly aerobic tissues, like your muscles, have high concentrations of mitochondria and a rich blood supply to deliver the large amounts of O_2 needed to support their activity. Even so, intense exercise exhausts this oxygen

supply, at which point your cells switch to glycolysis, as mentioned already. But how do the cells extract sufficient energy via glycolysis alone?

Under anaerobic conditions, these muscle cells use a fermentation pathway. **Fermentation** begins with glycolysis, followed by a special set of reactions whose only role is to help perpetuate glycolysis. This process enables organisms to generate ATP through glycolysis alone, when aerobic ATP production is constrained by low oxygen levels (**Figure 5.12**). If glycolysis cannot be perpetuated, the cells will run out of ATP and die.

Many species of bacteria that live in places like oxygen-deficient swamps, sewage, or deep layers of soil never use oxygen or are actually poisoned by oxygen. Similar to rock-breathing microbes, many microbes can utilize other electron acceptors, such as iron or sulfur compounds. When these run out, anaerobic organisms use the fermentation pathway to regenerate molecules needed to perpetuate glycolysis, and exclusively generate ATP through glycolysis without ever using oxygen.

Most eukaryotic cells can perform both aerobic and anaerobic ATP production, depending on the oxygen levels in their environment. Because the Krebs cycle and oxidative phosphorylation produce magnitudes more ATP than glycolysis and the fermentation pathway yield, these cells always use oxygen and aerobic respiration when levels permit.

Together, photosynthesis and cellular respiration enable cells to store and utilize energy from the sun. They are two sides of the same coin; the products of one reaction are the ingredients for the other. Photosynthesis, an anabolic process, requires energy (sunlight) and CO_2, and releases O_2 and glucose. Cellular respiration, a catabolic process, requires O_2 and glucose, and releases CO_2 and energy.

Unlike the situation with photosynthesis and cellular respiration—metabolic processes that have been well studied for decades—researchers don't yet know the details of how rock eaters store and use energy. "There isn't really any pathway worked out yet," says Rowe. "Truthfully, it's amazing that it works." She's eager to find out how. Of the 30 species Rowe has isolated, she is currently focused on 8, working to develop tools to identify which enzymes are involved in their metabolism.

Figure 5.12

Fermentation produces energy in the absence of oxygen

When the oxygen supply is inadequate to support ATP production through cellular respiration, fermentation enables ATP production through glycolysis alone.

Q1: Which product released by fermentation accounts for the bubbles in beer?

Q2: Bakers of yeast breads also rely on fermentation, allowing bread to "rise" before baking. Describe what is occurring with the yeast as the bread rises.

Q3: Explain in your own words why lactic acid builds up in your muscles during strenuous physical activity.

Making Way
for Renewables

Microbes that "eat" electricity may someday provide an efficient way to store renewable energy. That's good, because we need such storage technologies as soon as possible. Technologies that produce renewable energy are no longer "alternative"—they're mainstream and compete financially with fossil fuel producers. From solar power to biofuels, the fuel economy is rapidly changing toward more sustainable solutions.

Assessment available in smartwork**5**

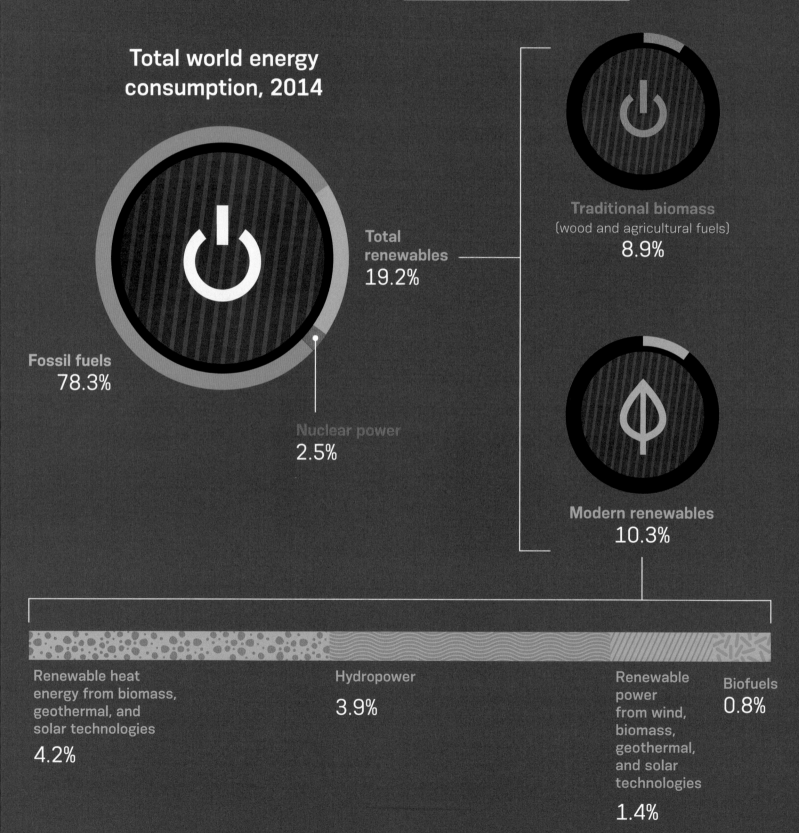

Total world energy consumption, 2014

Total renewables — 19.2%

Fossil fuels 78.3%

Nuclear power 2.5%

Traditional biomass (wood and agricultural fuels) 8.9%

Modern renewables 10.3%

Renewable heat energy from biomass, geothermal, and solar technologies 4.2%

Hydropower 3.9%

Renewable power from wind, biomass, geothermal, and solar technologies 1.4%

Biofuels 0.8%

Bacterial Batteries

Because cell membranes are insulated, researchers long believed that cells could not transport electrons directly across their membranes, but now it appears they can. In 2010, in fact, Rowe's postdoctoral adviser, Mohamed El-Naggar, a pioneer in the rock breather field, found that the tiny hairlike structures called pili (singular "pilus") on some bacteria are conductive (**Figure 5.13**)—just as conductive, in fact, as silicon, the basis of most electronics. One single nano-sized pilus, which researchers refer to as a bacterial "nanowire," can transport about 106 electrons per second—enough to sustain the respiration of a whole cell.

Today, Rowe, Spormann, and others are focusing on how to use these unique bacteria in "bacterial batteries." Energy absorbed during the day from a solar panel currently has to be either instantly used or siphoned off into the electrical grid. But what if it could be passed into a tank of bacteria instead, where the microbes would gobble up the electricity and store it in chemical bonds in the form of methane or another natural gas? In that case the gas could be stored until needed.

Figure 5.13

The rock eater *Methanococcus maripaludis*, with electrically conductive hairlike pili

"If you could convert electrical energy, through microbial processes, to a liquid fuel, that could be a really valuable way of storing energy," says Rowe. This electricity-to-fuel scenario could also be run in reverse, using microbes to digest a fuel source such as raw sewage and convert it into electricity.

There's also the intriguing scenario that someday our electronic devices will be powered by microorganisms that manipulate the flow of electrons. In this rapidly evolving field, the possibilities are endless. And so the hunt for rock eaters continues. "We don't know how many organisms do or don't do it," adds Spormann. "It's entirely possible that rock eating is more widespread than previously thought. We just don't know because it's so new."

REVIEWING THE SCIENCE

- The sun is the source of energy fueling most living organisms. Plants, algae, and some bacteria gain energy from their environment through **photosynthesis**. Most organisms use **cellular respiration** to extract usable energy from sugar molecules. In chemical terms, photosynthesis is the opposite of cellular respiration.

- All the many chemical reactions involved in the capture, storage, and use of energy by living organisms are collectively known as **metabolism**. Energy-releasing breakdown reactions like cellular respiration are **catabolism**; energy-requiring synthesis reactions like photosynthesis are **anabolism**.

- A **metabolic pathway** is a multistep sequence of chemical reactions, with each step catalyzed by a different enzyme.

- **Energy carriers** store energy and deliver it for cellular activities. **ATP** is found in all cells and is the most commonly used energy carrier.

- Photosynthesis takes place in chloroplasts and occurs in two stages. In the **light reactions**, energy is absorbed using pigment molecules that include

chlorophyll as electrons flow along the **electron transport chain**. The light reactions create the energy carriers ATP and NADPH, splitting water molecules and releasing oxygen gas.

- The energy carriers are then used to convert carbon dioxide into sugar molecules during the **light-independent reactions**, or **Calvin cycle**. In the first of the reactions, the enzyme **rubisco** catalyzes the fixation of CO_2.

- **Enzymes** are biological catalysts, usually small proteins, that speed up chemical reactions. An enzyme's function is based on its chemical characteristics and the three-dimensional shape of its **active site**.

- Cellular respiration occurs in three stages: (1) **glycolysis**, which yields small amounts of ATP and NADH; (2) the **Krebs cycle**, which releases carbon dioxide and produces NADH, $FADH_2$, and ATP; and (3) **oxidative phosphorylation**, which generates many molecules of ATP.

- In the absence of oxygen, **fermentation** breaks down the products of glycolysis into alcohol or lactic acid.

THE QUESTIONS

The Basics

1 Metabolic pathways

(a) always break down large molecules into smaller units.

(b) only link smaller molecules together to create polymers.

(c) are often organized as a multistep sequence of reactions.

(d) occur only in mitochondria.

2 Enzymes

(a) speed up reactions that would otherwise occur much more slowly.

(b) spur reactions that would otherwise never occur.

(c) provide energy for anabolic but not catabolic pathways.

(d) are consumed during the reactions they catalyze.

3 The most common energy-carrying molecule in all organisms is

(a) carbon dioxide.

(b) water.

(c) ATP.

(d) rubisco.

4 The major product of photosynthesis is

(a) lipids.

(b) sugar.

(c) amino acids.

(d) nucleotides.

5 Which of the following statements is *not* true?

(a) Glycolysis is the first stage of cellular respiration.

(b) Glycolysis can proceed under low oxygen levels with the assistance of fermentation.

(c) Glycolysis produces less ATP than does either the Krebs cycle or oxidative phosphorylation.

(d) Glycolysis produces most of the ATP required by aerobic organisms like us.

6 Select the correct terms:

Photosynthesis and cellular respiration are chemically (**identical / opposite**) processes. Cellular respiration is an example of (**anabolism / catabolism**), which (**produces / expends**) energy.

7 In the diagram of photosynthesis shown here, fill in each blank with the appropriate term: (a) Sunlight; (b) CO_2; (c) Oxygen; (d) H_2O; (e) ATP and NADPH; (f) ADP and $NADP^+$; (g) Sugar; (h) Light reactions; (i) Calvin cycle.

8 Place the following steps of cellular respiration in the correct order by numbering them from 1 to 4.

_____ a. The Krebs cycle produces the energy carriers NADH, $FADH_2$, and ATP.

_____ b. If oxygen levels are adequate, pyruvate is transported into the mitochondrion. If oxygen levels are very low, fermentation proceeds.

_____ c. Glucose is broken down to produce ATP and NADH.

_____ d. An electron transport chain produces ATP from ADP.

Challenge Yourself

9 The Calvin cycle reactions are sometimes called the "light-independent reactions" or "dark reactions" to contrast them with the light (or light-dependent) reactions. Can the Calvin cycle be sustained in algae that are kept in total darkness for several days? Why or why not?

10 After strenuous exercise, you may notice that your muscles burn and feel sore the next day. Which statement best explains this phenomenon?

(a) Proteins in muscle cells are being digested to provide energy.

(b) Carbon dioxide is building up in muscle cells and changing their pH.

(c) Spontaneous combustion occurs during strenuous exercise, so avoid it at all costs!

(d) ATP is accumulating in muscle cells, causing a burning sensation.

(e) Without adequate oxygen, muscle cells are fermenting pyruvate into lactic acid.

11 The graph presented here depicts the activation energy, or the amount of energy needed for a reaction to proceed, with and without an enzyme.

a. Which reaction requires more energy to proceed—the one with or without an enzyme? How do you know?

b. Is this reaction anabolic or catabolic? How do you know?

Try Something New

12 In 2012, an Illinois man was killed by cyanide poisoning after he won a million dollars in the lottery. Cyanide is a lethal poison because it interferes with the electron transport chain in mitochondria. What effect would cyanide have on cellular respiration?

(a) Glycolysis, the Krebs cycle, and oxidative phosphorylation would all be inhibited.

(b) The Krebs cycle would be inhibited, but oxidative phosphorylation would not.

(c) Oxidative phosphorylation would be inhibited.

(d) Glycolysis, the Krebs cycle, and oxidative phosphorylation would all be stimulated.

13 Plants in the genus *Ephedra* have been harvested for their active substance ephedrine for centuries. Ephedrine is used to reduce the symptoms of bronchitis and asthma, as a stimulant and study aid, and as an appetite suppressant. It is also the main ingredient in the illegal production of methamphetamine. Since 2006, the sale of ephedrine and related substances has been limited and monitored in the United States. One effect of ingesting ephedrine is greatly increased metabolism, which has been known to kill users of ephedrine. How might an increased metabolic rate cause death?

14 You have been transported into the future, where a nanosensor can be inserted into a living cell and subsequently travel into any organelle. The sensor relays information stating that it has lodged itself in a compartment of the mitochondrion, where there is a concentration gradient of hydrogen ions (H^+) and levels of ATP production are high. From what you know about cellular respiration in mitochondria, where is the sensor lodged?

(a) inner mitochondrial membrane

(b) mitochondrial matrix

(c) chloroplast stroma

(d) cell cytoplasm

(e) plasma membrane

Leveling Up

15 **What do *you* think?** What would happen if a virus destroyed all photosynthetic organisms on Earth?

16 *Write Now* biology: calorie-burning fat Your friend has e-mailed you a link (see below) to a *New York Times* article on "brown" fat. He has been trying to lose weight and wants to know whether you think it would be a good idea for him to spend more time in the cold, rather than continuing to exercise regularly. He is also interested that the article mentions ephedrine's ability to stimulate brown fat, and he asks if you think he should begin to take ephedrine supplements. Compose an reply to your friend addressing the following points (using one to two paragraphs for each one). [*Note:* You may need to do further reading to answer (b) and (c).]

a. Explain in detail how brown fat burns calories when someone is chilled.

b. Explain how ephedrine affects metabolism and what its possible side effects are.

c. Contrast (a) and (b) with the effect of exercise on metabolism, both in the short term and in the longer term by increasing muscle mass.

d. In your final paragraph, advise your friend as to whether he should begin spending time in the cold to increase weight loss or take ephedrine supplements, and whether he should continue to exercise regularly. Justify your opinion with data and logic.

To research your answer, begin with the following *New York Times* article, published on April 8, 2009: "Calorie-Burning Fat? Studies Say You Have It" (http://www.nytimes.com/2009/04/09/health/research/09fat.html). Consult and reference at least two additional sources in your answer.

17 **Is it science?** Wouldn't it be amazing if humans could manufacture their own food through photosynthesis like plants do? Use Internet research to determine whether this idea is real science or science fiction. Is there evidence that animals can use the sun's energy directly and/or perform photosynthesis in nature? In the lab? If so, how would that capability affect life as we know it?

For more, visit digital.wwnorton.com/bionow2 for access to:

Toxic Plastic

Two ruined experiments expose the health risks of chemicals in everyday products.

After reading this chapter you should be able to:

- Label a figure of the major stages of the cell cycle, and explain the processes that occur during each of these stages.
- Compare and contrast cell division by binary fission, mitosis, and meiosis.
- Distinguish between sister chromatids and homologous chromosomes.
- Diagram, using the appropriate terms, the steps in mitosis and in meiosis.
- Explain the importance of the checkpoints in the cell cycle and the consequences of bypassing those checkpoints.
- Identify the ways in which meiosis and fertilization together produce genetically diverse offspring.

CHAPTER

06

CELL DIVISION

It began as a run-of-the-mill experiment. In 1989, biologists Ana Soto and Carlos Sonnenschein at Tufts University in Massachusetts were studying how the hormone estrogen regulates the growth of cells in the female reproductive system. For their research, the duo developed an experimental setup consisting of human breast tumor cells growing in plastic bottles called cell culture flasks (**Figure 6.1**, top). The flasks were filled with a liquid containing a factor that prevented the cells from multiplying. But when estrogen was added to the flasks, the cells grew. When estrogen was absent, they didn't.

One day, suddenly and surprisingly, cells in the flasks began growing even when estrogen hadn't been added. "What had worked for years didn't work anymore," says Sonnenschein. The two scientists immediately stopped their experiments and began searching for the cause. "It smacked of contamination," recalls Soto, as if estrogen had somehow gotten into the flasks. But after weeks of searching, Soto and Sonnenschein still couldn't identify a source of contamination. They became so paranoid that they suspected someone was entering the lab at night and secretly dripping estrogen into their flasks.

Almost 10 years later, in August 1998, geneticist Patricia Hunt at Case Western Reserve University in Ohio stared dumbfounded at another experimental anomaly. Hunt was studying why older women are at increased risk of having children with chromosomal abnormalities, like Down syndrome, in which an individual has 47 **chromosomes**—the tiny, stringlike structures in cells that contain genes—instead of the usual 46 (see Chapter 8 for more on chromosomes). She hypothesized that hormone levels have an impact on that increased risk. To test her hypothesis, Hunt

Figure 6.1

Two research models, two unusual results

Breast tumor cells (top) and mouse oocytes (bottom) grew in unexpected ways, leading scientists to study them more deeply.

raised groups of mice with varying levels of hormones and checked their egg cells for abnormal numbers of chromosomes (**Figure 6.1**, bottom).

The experiment was almost complete when Hunt went in to check on the control mice one last time. A control population is a necessary baseline for comparison against an experimental population; in this case, the control was a group of healthy mice whose hormone levels had not been altered. Using a light microscope, Hunt examined mouse oocytes—precursors to egg cells—at the moment just before the cells

ANA SOTO AND CARLOS SONNENSCHEIN

Ana Soto is a biologist at Tufts University who studies how cell division is affected by sex steroids. Carlos Sonnenschein, also a professor at Tufts University, studies chemicals that disrupt hormone systems in mammals.

undergo a specialized type of cell division that produces the eggs. She was shocked. The cells were a mess, the chromosomes scrambled. A whopping 40 percent of the resulting eggs had chromosomal defects. "The controls were completely bonkers," says Hunt. "One week they were fine, the next week they weren't. That's when we knew something was going on."

Like Soto and Sonnenschein, Hunt scrutinized every method and every piece of lab equipment used in the experiment, looking for the culprit. But as weeks passed, she couldn't figure out what had ruined her experiment. Soto, Sonnenschein, and Hunt didn't know it at the time, but their botched experiments would change the course of their scientific careers forever. The three would spend the next decade identifying, tracking, and investigating a toxic chemical that pervades our environment.

Divide and Conquer

In Soto and Sonnenschein's experiment, the breast cells were multiplying under the wrong circumstances. In Hunt's experiment, the mouse oocytes were not producing egg cells correctly. In both cases, something was interfering with the ability of the cells to divide—disrupting the cell cycle. The **cell cycle** is a sequence of events that make up the life of a typical eukaryotic cell, from the moment of its origin to the time it divides to produce two daughter cells. The time it takes to complete a cell cycle depends on the organism, the type of cell, and the life stage of the organism. Human cells, for example, typically have a 24-hour cell cycle, while Hunt's mouse oocytes can take days to complete a cycle. Some fly embryos, on the other hand, have cell cycles that are only 8 minutes long.

There are two main stages in the cell cycle of eukaryotes: *interphase* and *cell division*—each marked by distinctive cell activities (**Figure 6.2**). **Interphase** is the longest stage of the cell cycle; most cells spend 90 percent or more of their life span in interphase. During this phase, the cell takes in nutrients, manufactures proteins and other substances, expands in size, and conducts special functions depending on the cell type. Neurons in the brain, for example,

① The cell has grown and is ready to replicate its DNA.

② DNA replication happens here.

③ Checks are made to ensure that conditions are suitable for cell division.

④ The replicated DNA is separated with precision.

⑤ DNA molecules are divided equally between the two daughter cells.

⑥ Many cells in the human body exit the cell cycle and remain in a resting stage for some time. Some cells never leave this phase.

Interphase

Cell division

G_0 phase

G_1 phase

S phase

G_2 phase

Cytokinesis

Mitosis

Figure 6.2

The cell cycle

The eukaryotic cell cycle consists of two major stages: interphase and cell division.

Q1: When is DNA replicated during the cell cycle?

Q2: When in the cell cycle does DNA separate into the two genetically identical daughter cells?

Q3: If a cell is not destined to separate into daughter cells, what phase does it enter? Is this part of the cell cycle?

transmit electrical impulses, while beta cells in the pancreas release insulin.

Interphase can be divided into three main intervals: G_1, S, and G_2. The **G_1 phase** (for "gap 1") is the first phase in the life of a newborn cell. In cells that are destined to divide, preparations for cell division begin during the **S phase**

("S" stands for "synthesis"). A critical event during the S phase is the copying, or *replication*, of all the cell's DNA molecules, which contain the organism's genetic information. The G_2 **phase** (for "gap 2") occurs after the S phase but before the start of cell division.

Early cell biologists bestowed the term "gap" on the G_1 and G_2 phases because they believed those phases to be less significant periods in the life of a cell than are the S phase and cell division. We now know that the "gap" phases are often periods of growth during which both the size of the cell and its protein content increase. Furthermore, each "gap" phase serves as a checkpoint to prepare the cell for the phase immediately following it, ensuring that the cell cycle does not progress unless all conditions are suitable.

Cell division is the last stage in the cell cycle of an individual cell. As cell division begins, the cell contains twice the usual amount of DNA because of DNA replication during the S phase.

Not all cells complete the cell cycle. Many types of cells—neurons and beta cells, for example—become specialized shortly after entering G_1, and they pull out of the cell cycle to enter a nondividing state called the G_0 phase. The G_0 **phase** can last for a period ranging from a few days to the lifetime of the organism.

Cells begin cell division for two basic reasons: (1) to reproduce and (2) to grow and repair a multicellular organism. Most single-celled organisms use cell division to produce offspring through **asexual reproduction**. Asexual reproduction generates *clones*, offspring that are genetically identical to the parent.

Cell division occurs in all living organisms (eukaryotes and prokaryotes) and involves copying the DNA in the parent cell, then delivering one copy of it to each of two daughter cells. Most prokaryotes carry their genetic material in just one loop of DNA. These cells reproduce through **binary fission**, a type of cell division in which a

Parent cell
Cell wall
Plasma membrane
Circular DNA molecule

DNA replication and segregation
DNA is replicated, giving rise to two circular DNA molecules.

Cell division
The cell expands, and a partition is created that isolates the two DNA molecules into separate compartments.
New cell wall

Cell separation

Two daughter cells

Figure 6.3

Cell division in a prokaryote

Many prokaryotes reproduce asexually in a type of cell division known as binary fission.

Q1: Name one similarity between cell division in prokaryotes and eukaryotes.

Q2: Why is binary fission referred to as "asexual reproduction"?

Q3: Name one difference between cell division in prokaryotes and eukaryotes.

PATRICIA HUNT

Patricia Hunt is a reproductive biologist at Washington State University in Pullman, Washington, who studies meiosis in mammalian cells.

cell simply copies the circular chromosome and each daughter cell receives one copy of the DNA loop—resulting in two cells that are genetically identical (**Figure 6.3**).

Soto, Sonnenschein, and Hunt, however, were studying eukaryotic cells, where cell division is more complicated than binary fission. Eukaryotic DNA forms multiple, distinct linear chromosomes wrapped around proteins and coiled into fibers that have to be unwound, replicated, and equally distributed between the two daughter cells. In addition to the multiple-chromosome problem, eukaryotic DNA lies inside a nucleus, enclosed by a double layer of membranes that make up the nuclear envelope. In most eukaryotes, the nuclear envelope must be disassembled in the dividing cell and then reassembled in each of the daughter cells toward the end of cell division. To make things even more complicated, eukaryotic cells undergo two types of division, depending on cell type: asexual reproduction through *mitosis*, and sexual reproduction—the production of sperm and eggs—through *meiosis*.

Trade Secret

Back in Massachusetts, Soto and Sonnenschein spent 4 months trying to figure out why their experiment had stopped working—how unknown estrogen was getting into their cell culture flasks and causing the cells to divide. By trial and error, they determined that a compound seemed to be shedding from the walls of the plastic tubes in which they stored the liquid being added to the cell culture flasks.

They called the tube manufacturer, who confirmed that an ingredient was added to make the tubes more impact resistant. But the company refused to reveal the identity of the "trade secret" ingredient. Soto and Sonnenschein spent a year purifying the secret ingredient and finally identified a compound called nonylphenol, a chemical used to make detergents and hard plastics. The reason for all their problems became clear: nonylphenol mimics the action of estrogen.

Like estrogen, nonylphenol activates **mitotic division**, a type of cell division that generates two genetically identical daughter cells from a single parent cell in eukaryotes. Mitotic division consists of two steps that occur after interphase: *mitosis* and *cytokinesis*. The first step, **mitosis**, refers to the division of the copied chromosomes in the nucleus. Mitosis is divided into four main

Cancer: Uncontrolled Cell Division

Cancer accounts for almost 600,000 deaths in the United States each year—nearly one in every four deaths. Only heart disease kills more people. Over the course of a lifetime, an American male has a nearly one in two chance of being diagnosed with cancer; American women fare slightly better, with a one in three chance of developing cancer. There are more than 200 different types of cancer, but the big four—lung, prostate, breast, and colon cancers—combine to account for about half of all cancers. More than 15 million Americans alive today have been diagnosed with cancer and are either in remission or undergoing treatment. The National Cancer Institute estimates that the collective price tag for the various forms of cancer is more than $100 billion per year.

Every cancer begins with a single rogue cell that starts dividing without the checkpoints of a normal cell. This runaway cell division rapidly creates a cell mass known as a **tumor**. Tumors that remain confined to one site are **benign**. Because benign tumors can usually be surgically removed, they are generally not a threat to the patient's survival. However, an actively growing benign tumor is like a cancer-in-training. Because these tumor cells are not subject to the monitoring that occurs at checkpoints during the cell cycle of a normal cell, their descendants can become increasingly abnormal—changing shape, increasing in size, and ultimately ceasing normal cell functions. As tumor cells progress toward a cancerous state, they begin secreting substances that cause **angiogenesis**, the formation of new blood vessels. The resulting increase in blood supply to the tumor is important for delivering nutrients to it and whisking waste away from it, allowing the tumor to grow larger.

Most cells in the adult animal body are firmly anchored in one place and will stop dividing if they are detached from their surroundings—a phenomenon known as **anchorage dependence**. But some tumor cells may acquire anchorage *in*dependence, the ability to divide even when released from their attachment sites. When tumor cells gain anchorage independence and start invading other tissues, they are transformed into **cancer cells**, also known as **malignant cells**. Cancer cells may break loose from their attachment sites and enter blood or lymph vessels to emerge in distant locations throughout the body, where they form new tumors. The spread of a disease from one organ to another is known as **metastasis**. Metastasis typically occurs at later stages in cancer development. Once a cancer has metastasized to form tumors in multiple organs, it may be very difficult to fight.

Cancer cells multiply rapidly wherever they establish themselves, overrunning neighboring cells, monopolizing oxygen and nutrients, and starving normal cells in the vicinity. Without restraints on their growth and migration, cancer cells steadily destroy tissues, organs, and organ systems. The normal function of these organs is then seriously impaired, and cancer deaths are ultimately caused by the failure of vital organs.

Interphase

Mitosis

Prophase

Metaphase

Unpacked chromosomes

Mitotic spindle

Nuclear envelope

Plasma membrane

Fragments of nuclear envelope

Replicated, condensed chromosomes

Sister chromatids

| Chromosomes are replicated during the S phase, before mitosis begins. | **1** The replicated chromosomes condense and attach to the **mitotic spindle**, which is made up of cytoskeletal proteins. The nuclear envelope breaks down. | **2** The mitotic spindle aligns the replicated chromosomes at the center of the cell at the metaphase plate. |

Onion root tip

Figure 6.4

Cell division in a eukaryote

Mitotic cell division is composed of two main stages: mitosis (with four substages) and cytokinesis. ▶▸

Q1: Do all cells in an organism enter each stage of mitosis at the same time? (*Hint*: See image of onion root tip at far left in the figure.)

Q2: What happens between the end of interphase and early prophase that changes the appearance of the chromosomes?

Q3: Explain in your own words the role of the mitotic spindle in mitosis.

phases: *prophase, metaphase, anaphase,* and *telophase.* Each phase is defined by easily identifiable events (**Figure 6.4**).

A parent cell sets up for an upcoming mitotic division by replicating its DNA during the S phase of interphase, well before mitosis gets under way. Note that DNA in the nucleus is not tightly packaged during the gap and synthesis phases. This is because the DNA must be accessible for replication and for conducting the business of the cell. Then, as cell division begins, each long, double-stranded DNA molecule is attached to proteins that help pack it for cell division into a chromosome. This packing is

necessary because every DNA molecule is enormously long, even in the simplest eukaryotic cells. When a chromosome is replicated, two identical DNA molecules called **sister chromatids** are produced. These sister chromatids are firmly attached at a central region of the chromosome called the **centromere** and do not separate until metaphase (**Figure 6.5**).

One of the main objectives of mitosis is to separate those sister chromatids, pulling them apart at the centromere and delivering one of each to the opposite ends of the parent cell. Eukaryotic cells have evolved an elaborate choreography to minimize the risk of mistakes during

Anaphase

Telophase

Interphase

Cell division

Nuclear
envelope
forming

Chromosomes
decondensing

New chromosomes

3 The sister chromatids are separated and become individual chromosomes. These new chromosomes are moved toward the poles of the cell by the shortening of the mitotic spindle.

4 The new chromosomes reach the poles. The nuclear envelopes re-form, and the chromatin decondenses.

Overlapping with the end of mitosis, the cell physically divides into two daughter cells through cytokinesis.

---

the equal and symmetrical partitioning of the replicated genetic material. Normally, no daughter cell winds up short a chromosome, nor does it acquire duplicates. Unless an error occurs, each daughter cell inherits the same genetic information that the parent cell possessed in the G_1 phase of its life.

After the replicated DNA has been divided in two, half to each end of the parent cell, the cytoplasm is divided by a process called **cytokinesis** ("cell movement"), like pulling apart a ball of Silly Putty into two halves. Cytokinesis gives rise to two self-contained daughter cells that are clones of each other.

Mitotic division can serve both the eukaryotic organism's need to replace itself (to reproduce) and its need to add new cells to its body. Many multicellular eukaryotes use mitotic division to reproduce asexually, including seaweeds, fungi, and plants, and some animals, such as sponges and flatworms. All multicellular organisms also rely on mitotic division for the growth of tissues and organs and the body as a whole, and for repairing injured tissue and replacing worn-out cells. Mitosis is why children grow taller and why skin closes over a cut.

Good Cells Gone Bad

Cell division is not always a good thing. Runaway cell division can create a tumor (see "Cancer: Uncontrolled Cell Division" on page 101). In a developing organism, it can also cause an organ such as the heart or liver to form incorrectly and not function properly. It is little wonder, then, that the cell cycle is carefully controlled in healthy individuals. The decision to divide a cell is made during the G_1 phase of the cell cycle in response to internal and external signals. In humans, external signals that influence the commitment to divide include hormones and proteins called growth factors. Some hormones and growth factors act like the gas pedal in a car and push a cell toward cell division; others act like a brake and prevent cell division.

After a cell enters the cell cycle, special *cell cycle regulatory proteins* are activated. These proteins "throw the switch" that enables the cell to pass through critical checkpoints and proceed from one phase of the cell cycle to the

During gap phase | Synthesis phase | At the beginning of mitosis

One replicated chromosome

Centromere

Two sister chromatids

Each chromosome is replicated before mitosis begins.

Centromere

One replicated chromosome

Two sister chromatids

As mitosis begins, the chromatin packs even more tightly, and the chromosome becomes easily visible with a light microscope.

Figure 6.5

Chromosomes are copied and condensed in preparation for cell division
Chromosomes spend the majority of the cell cycle unpackaged (left). They are copied (replicated) during synthesis, then tightly packaged, or condensed, during early mitosis.

Q1: Why is it important for a chromosome to be copied before mitosis?

Q2: Are sister chromatids attached at the centromere considered to be one or two chromosomes?

Q3: Why is the chromosome's DNA tightly packed for mitosis and cytokinesis? (*Hint:* Think about what would happen if it were unpackaged, as during interphase.)

next (**Figure 6.6**). For example, upon receiving the appropriate signals, cell cycle regulatory proteins advance a cell from the G_1 phase to the S phase by triggering chromosome replication and other processes associated with it.

Cell cycle regulatory proteins also respond to negative internal or external control signals. Internal signals will pause a cell in the G_1 phase under any of the following conditions: the cell is too small, the nutrient supply is inadequate, or the cell's DNA is damaged. G_2 pauses in the same circumstances, as well as when chromosome duplication in the S phase is incomplete for any reason.

Nonylphenol interferes with the G_0 and G_1 checkpoints. In essence, it gives the cell a green light to enter the cell cycle at a time when the cell would not normally divide. So, when Soto and Sonnenschein realized that nonylphenol enables human breast cells and rat uterine cells to divide

inappropriately, they became concerned. If nonylphenol was being used in everyday plastics, it was possible that healthy human cells were exposed to it on a regular basis.

In a 1991 paper detailing their discovery, Soto and Sonnenschein wrote that nonylphenol might be interfering with science experiments like theirs and that, even more important, it could be harmful to humans. "From the very beginning, we realized that this could be a health problem," says Sonnenschein.

Unequal Division

In Ohio, Patricia Hunt was having no success determining why her control mouse eggs divided abnormally, with either too many or too few chromosomes. After searching in vain for months, one day she noticed that something was wrong with

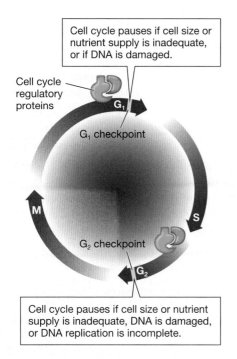

Cell cycle pauses if cell size or nutrient supply is inadequate, or if DNA is damaged.

Cell cycle regulatory proteins

G₁ checkpoint

M

S

G₂ checkpoint

Cell cycle pauses if cell size or nutrient supply is inadequate, DNA is damaged, or DNA replication is incomplete.

Figure 6.6

The cell cycle must pass through checkpoints to proceed

Just two of the known cell cycle checkpoints are depicted in this diagram. Checkpoints also operate in both the S phase and the M phase (mitosis).

> **Q1:** What could happen if the cell's checkpoints are disabled?
>
> **Q2:** What is the advantage of stopping the cell cycle if the cell's DNA is damaged?
>
> **Q3:** Which part of the cell cycle may have been influenced in Soto and Sonnenschein's breast tumor cell experiments?

the plastic mouse cages: the water bottles were leaking and the plastic cage walls were hazy.

She asked around and found out that, months earlier, a substitute janitor had used detergent with a high pH instead of the normal low-pH detergent to clean the cages and bottles. With some lab work, she identified the chemical oozing from the corroded plastic. "It took essentially one washing with the wrong detergent," says Hunt, "and once it was damaged, [the plastic] started to leach bisphenol A."

Bisphenol A, or BPA, was first synthesized in 1891. A synthetic hormone that, like nonylphenol, mimics estrogen, BPA was tested in the 1930s by clinicians seeking a hormone replacement therapy for women who needed estrogen, but it wasn't as effective as other substitutes. In the 1940s and '50s, the chemical industry found another use for BPA, as a chemical component of a clear, strong plastic. Manufacturers began to incorporate it into eyeglass lenses, water and baby bottles, the linings of food and beverage cans, and more. Unfortunately, as Hunt found out, BPA doesn't necessarily stay in those products. Not all the BPA used to make plastics gets locked into chemical bonds, and what doesn't get locked into bonds can work free—especially if the plastic is heated, such as when a baby bottle is warmed, or if the plastic is exposed to a harsh chemical, as Hunt's mouse cages were. Because of its prevalence in products and its ability to leach out of them, BPA is one of the most common chemicals we are exposed to in everyday life.

To confirm the hypothesis that BPA caused the mouse egg abnormalities, Hunt's team re-created the original event. They intentionally damaged a set of new cages and put healthy female mice in them. They also had the mice drink from damaged water bottles. Later, when they examined the eggs of these mice, they saw the same toxic effects as before: 40 percent of the eggs had abnormal chromosomes. The eggs showed errors in meiosis.

Meiosis is a specialized type of cell division that kicks off **sexual reproduction**, the process by which genetic information from two individuals is combined to produce offspring. Sexual reproduction has two steps: cell division through meiosis, and fertilization. BPA was affecting the first of these steps—meiosis.

We learned earlier that mitosis produces daughter cells with the same number of chromosomes as the parent cell. These non–sex cells are called **somatic cells**. In contrast, meiosis produces **gametes**, daughter cells that have half the chromosome count of the parent cell. The differences between mitosis and meiosis are the reason why the somatic cells of plants and animals have twice as much genetic information as their gametes have. The double set of genetic information possessed by somatic cells is known as the **diploid** set (represented by $2n$), and the single set possessed by gametes is called the **haploid** set (represented by n).

Fertilization, the fusion of two gametes, results in a single cell called the **zygote**. The

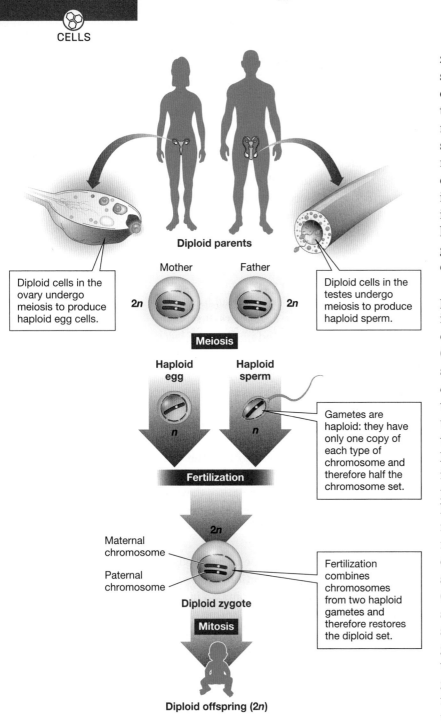

Diploid parents

Mother   Father

Diploid cells in the ovary undergo meiosis to produce haploid egg cells.

Diploid cells in the testes undergo meiosis to produce haploid sperm.

$2n$ $2n$

Meiosis

Haploid egg   Haploid sperm

n n

Gametes are haploid: they have only one copy of each type of chromosome and therefore half the chromosome set.

Fertilization

$2n$

Maternal chromosome

Paternal chromosome

Fertilization combines chromosomes from two haploid gametes and therefore restores the diploid set.

Diploid zygote

Mitosis

Diploid offspring ($2n$)

Figure 6.7

Fertilization creates a zygote from the fusion of two gametes

In species with two sexes, female gametes are *eggs* and male gametes are *sperm*. This figure shows only one of the 23 homologous pairs found in human cells.

Q1: Is a zygote haploid or diploid?

Q2: Which cellular process creates a baby from a zygote?

Q3: If a mother or father was exposed to BPA prior to conceiving a child, how might that explain potential birth defects in the fetus?

zygote inherits a haploid (n) set of chromosomes from each of the gametes, restoring the complete diploid ($2n$) set of genetic information to the offspring. Each **homologous pair** of chromosomes in the zygote consists of one chromosome received from the father and one from the mother. The zygote then divides by mitosis to create a mass of cells that will eventually develop into a mature organism (**Figure 6.7**). All cells in a mature organism are diploid, containing homologous pairs of chromosomes, except for gametes, which are haploid and contain only one of each homologous pair.

Meiosis occurs in two stages after interphase—*meiosis I* and *meiosis II*—each involving one round of nuclear division followed by cytokinesis, for a total of two cell divisions (**Figure 6.8**). **Meiosis I** reduces the chromosome number to haploid by *separating one of each homologous pair* into two different daughter cells. Each homologous chromosome lines up with its partner and then separates to the two ends of the cells, and cytokinesis occurs. **Meiosis II** *separates each sister chromatid* into two different daughter cells. This time, the phases of the division cycle are almost exactly like those of mitosis: sister chromatids separate to the two ends of the cells, and cytokinesis occurs, leading to an equal segregation of chromatids into two new daughter cells. In summary, meiosis I produces two haploid cells (n) with one of each pair of duplicated chromosomes, and in meiosis II these two haploid cells separate the sister chromatids and each divide, giving rise to a total of four haploid cells (n) with unduplicated chromosomes. Each haploid gamete now has half of the chromosome set found in the original diploid cell ($2n$) that underwent meiosis.

BPA is toxic because it disrupts the process of meiosis, hindering the ability of the chromosomes to separate into four haploid cells. Hunt realized that if BPA was disrupting meiosis in mice, it could be doing the same in humans. And if a human gamete (either the sperm or the egg) does not contain the correct number of chromosomes, fertilization typically results in a miscarriage.

Hunt was nervous about publishing the results of her experiment. "We knew we were stepping into a landmine," she says. "We knew the paper would get some press, because essentially we were publishing that this chemical—used in

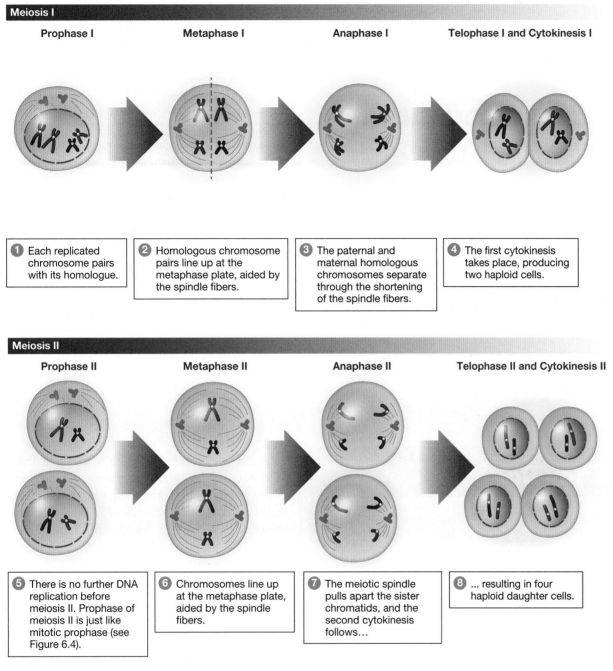

Meiosis I

Prophase I | Metaphase I | Anaphase I | Telophase I and Cytokinesis I

① Each replicated chromosome pairs with its homologue.

② Homologous chromosome pairs line up at the metaphase plate, aided by the spindle fibers.

③ The paternal and maternal homologous chromosomes separate through the shortening of the spindle fibers.

④ The first cytokinesis takes place, producing two haploid cells.

Meiosis II

Prophase II | Metaphase II | Anaphase II | Telophase II and Cytokinesis II

⑤ There is no further DNA replication before meiosis II. Prophase of meiosis II is just like mitotic prophase (see Figure 6.4).

⑥ Chromosomes line up at the metaphase plate, aided by the spindle fibers.

⑦ The meiotic spindle pulls apart the sister chromatids, and the second cytokinesis follows…

⑧ … resulting in four haploid daughter cells.

Figure 6.8

Meiosis with cytokinesis creates haploid daughter cells

The homologous chromosomes of a diploid cell after interphase occurs are paired and then separated into two haploid cells in meiosis I. Sister chromatids of these haploid cells are separated during meiosis II, each into two haploid cells. Cytokinesis occurs at the end of both meiosis I and meiosis II. ▶▤

Q1: Is a daughter cell haploid or diploid after the first meiotic division? How about after the second meiotic division?

Q2: What is the difference between homologous chromosomes and sister chromatids?

Q3: If the skin cells of house cats contain 38 homologous pairs of chromosomes, how many chromosomes are present in the egg cells they produce?

a wide variety of consumer products and that we are probably all exposed to—can cause an increased risk of miscarriage and babies with birth defects."

Shuffling the DNA

As a glance at a pair of parents and their biological children will tell you, the offspring resulting from sexual reproduction are similar to their parents, but—unlike the clones resulting from asexual reproduction—they are *not* identical. Because each half of a sexually reproducing organism's DNA comes from a different parent, meiosis and fertilization maintain the constant chromosome number of a species while allowing for genetic diversity within the population.

Meiosis generates genetic diversity in two ways: *crossing-over* between the paternal and maternal members of each homologous pair of chromosomes, as well as *independent assortment* of these paired-up chromosomes during meiosis I. **Crossing-over** is the physical exchange of identical chromosomal segments during meiosis I between the nonsister chromatids in each duplicated homologous pair. These nonsister chromatids make physical contact at random sites along their length, and each exchange the exact same segments of DNA (**Figure 6.9**). The chromatids are said to be *recombined*, and the exchange of DNA segments is known as **genetic recombination**. Without crossing-over, every chromosome inherited by a gamete would be just the way it was in the parent cell.

The **independent assortment** of chromosomes—the random distribution of the homologous chromosomes into daughter cells during meiosis I—also contributes to the genetic variety of the gametes produced. Each homologous chromosome pair in a given meiotic cell orients itself independently when it lines up at the imaginary equatorial plane known as the metaphase plate during meiosis I, leading to many possible combinations of maternal and paternal chromosomes in the daughter cells (**Figure 6.10**).

As with crossing-over, the independent assortment of chromosomes creates gametes that are likely to be genetically different from the parent, and also from each other. Then, during fertilization, the fusion of two gametes adds a tremendous amount of genetic variation because it combines a one-in-a-million egg with a one-in-a-million sperm. These three processes together—crossing-over, independent assortment, and fertilization—give each of us our genetic uniqueness.

Ten Years Later

Hunt and her team published their results in 2003. "We got a firestorm," she recalls with a grimace. The press reported the findings extensively, and many people and companies were upset over the allegations that BPA was toxic. Members of the plastics industry who did not agree with the paper's conclusions criticized Hunt's work. But there was more supporting research to come. Soto and Sonnenschein had also turned their attention to BPA because in everyday products it is a far more common chemical than nonylphenol and is therefore of greater concern.

What Can You Do?

There are things you can do to reduce your own risk of exposure to BPA. Today, the U.S. Food and Drug Administration recommends that individuals not put hot or boiling liquid intended for consumption in plastic containers made with BPA. (Some, but not all, plastics that are marked with the recycling code 3 or 7 may be made with BPA.) The organization also recommends discarding all bottles with scratches, which may harbor bacteria and, if the plastic contains BPA, may lead to greater release of the chemical.

"Get educated about your world," says Heather Patisaul, a BPA researcher at North Carolina State University. "You can either become completely reliant on the information you get from media and government, or you can educate yourself, which is vastly more useful." If you are concerned about BPA exposure, it is possible to cut it down by making lifestyle changes like not eating canned food, not drinking bottled water, and not putting plastic in the microwave. "You can be empowered," says Patisaul. "Those types of things can effect great change."

Still, BPA and similar chemicals are ubiquitous in modern life, says Ana Soto. Ultimately, the best way to avoid them will be for government regulators to take a stand and outlaw the use of these chemicals in consumer products, she notes. She encourages individuals to contact their representatives and ask them to push legislation limiting the use of BPA in manufacturing.

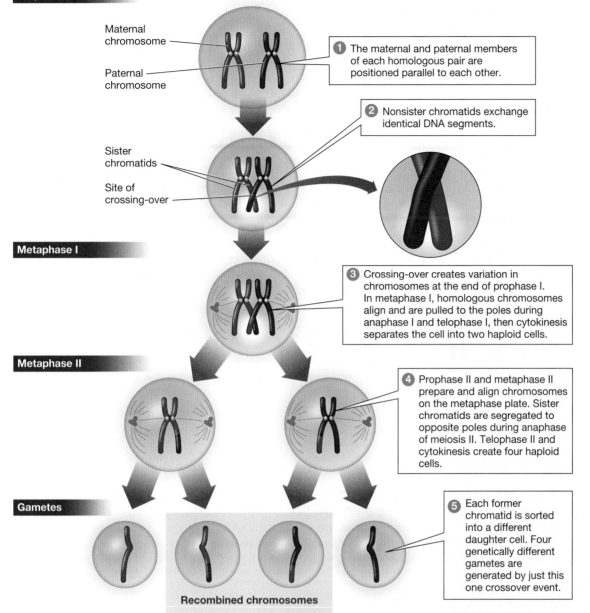

Prophase I

Maternal chromosome
Paternal chromosome

1 The maternal and paternal members of each homologous pair are positioned parallel to each other.

2 Nonsister chromatids exchange identical DNA segments.

Sister chromatids
Site of crossing-over

Metaphase I

3 Crossing-over creates variation in chromosomes at the end of prophase I. In metaphase I, homologous chromosomes align and are pulled to the poles during anaphase I and telophase I, then cytokinesis separates the cell into two haploid cells.

Metaphase II

4 Prophase II and metaphase II prepare and align chromosomes on the metaphase plate. Sister chromatids are segregated to opposite poles during anaphase of meiosis II. Telophase II and cytokinesis create four haploid cells.

Gametes

5 Each former chromatid is sorted into a different daughter cell. Four genetically different gametes are generated by just this one crossover event.

Recombined chromosomes

Figure 6.9

Crossing-over produces chromosomes with new combinations of DNA

Only one maternal and one paternal chromosome are depicted here, rather than the 23 pairs of homologous chromosomes found in humans. Only the resulting cells of prophase I, metaphase I, metaphase II, and the resulting gametes of the process of meiosis are depicted here.

Q1: Why is the term "crossing-over" appropriate for the exchange of DNA segments between homologous chromosomes?

Q2: At what stage of meiosis (I or II) does crossing-over occur?

Q3: What would be the effect of crossing-over between two sister chromatids?

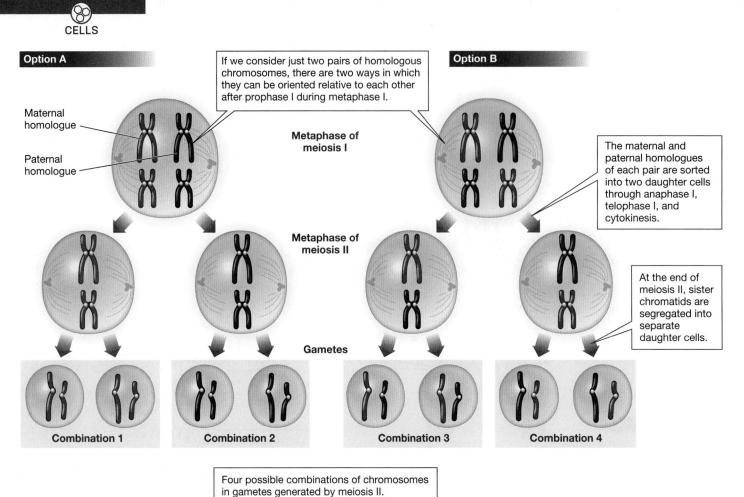

Option A

Option B

Maternal homologue

Paternal homologue

If we consider just two pairs of homologous chromosomes, there are two ways in which they can be oriented relative to each other after prophase I during metaphase I.

Metaphase of meiosis I

The maternal and paternal homologues of each pair are sorted into two daughter cells through anaphase I, telophase I, and cytokinesis.

Metaphase of meiosis II

At the end of meiosis II, sister chromatids are segregated into separate daughter cells.

Gametes

Combination 1

Combination 2

Combination 3

Combination 4

Four possible combinations of chromosomes in gametes generated by meiosis II.

Figure 6.10

The independent assortment of homologous chromosomes generates chromosomal diversity among gametes

Only two pairs of homologous chromosomes are shown here, rather than the 23 homologous pairs in human cells. Each gamete will receive either a maternal or a paternal homologue of each chromosome. Only the resulting cells of metaphase I, metaphase II, and the resulting gametes of the process of meiosis are depicted here.

Q1: During meiosis, does random assortment occur before or after crossing-over?

Q2: What would be the effect on genetic diversity if homologous chromosomes did not randomly separate into the daughter cells during meiosis?

Q3: With two pairs of homologous chromosomes, four kinds of gametes can be produced. How many kinds of gametes can be produced with three pairs of homologous chromosomes? What does this suggest for the 23 homologous pairs of chromosomes in human cells?

At the same time that Soto, Sonnenschein, and Hunt were doing their work, Frederick vom Saal at the University of Missouri found that male mice that had been exposed to BPA in utero—even at very low doses—had dramatically enlarged prostates in adulthood that were hypersensitive to hormones. This study suggests that men are also at risk of health effects from BPA.

In 2007, Hunt followed up her original work with a study that she says made the first paper look like "child's play." Her team exposed pregnant mice to BPA just as their female fetuses were producing a supply of eggs in their ovaries. When that second generation of females became adults, their eggs were also damaged, Hunt found, demonstrating that BPA exposure affects not just adult females, but two generations of their offspring.

Exposure aside, not everyone agrees that BPA is toxic. Numerous companies that manufacture plastics have conducted studies whose results do not match Hunt's and vom Saal's results. To reach a scientific consensus, on November 28, 2006, Soto, Sonnenschein, Hunt, vom Saal, and 34 other researchers from across the United States gathered at the University of North Carolina in Chapel Hill to summarize the research on BPA. The result of their two-day meeting was the "Chapel Hill Bisphenol A Consensus Statement," summarizing hundreds of studies done in vitro and in vivo over the previous 10 years. Completing this analysis, the group concluded firmly that BPA exposure at current levels in our environment presents a risk to human health (see "What Can You Do?" on page 108). "It was quite clear that there is a serious problem," says Soto.

Over time, many baby-bottle manufacturers took BPA out of their bottles, even as government regulators were slower to respond. "As scientists, our role is to call attention to what is wrong, but it is the role of the politicians to act on it and try to straighten it out," says Sonnenschein. Then, in July 2012, the FDA banned BPA from baby bottles and children's drinking cups, though the prohibition does not apply to the use of BPA in other types of containers (**Figure 6.11**). There is still concern, however, from many scientists about the chemicals that have replaced BPA, some of which are also estrogen mimics.

In the last several years, scientists all over the world have shown that BPA disrupts meiosis and mitosis and causes a plethora of health problems in mice and rats, including breast and prostate cancer, miscarriage and birth defects, diabetes and obesity, and even behavioral problems such as attention deficit hyperactivity disorder. Whether BPA is causing similar diseases in humans remains unknown; such hypotheses are hard to test because most people already have BPA in their bodies, making experimental control groups difficult to set up. BPA has been found in human blood, urine, breast milk, and amniotic fluid. In 2016, Hunt and her colleagues found "near-universal exposure" to BPA in a group of pregnant women in the United States. A major source of that exposure, they found, was cash register receipts, which can be hard to avoid touching.

Hunt, Soto, and Sonnenschein continue to explore the effects of BPA; they are studying how exposure to low doses of BPA affects monkeys, a model animal that more accurately represents the human system. "We're slowly raising awareness," says Hunt, "and slowly changing things."

Figure 6.11

BPA-free bottles and cans are now widely available

If you are concerned about being exposed to BPA, check labels before you buy.

Cancer's Big 10

When the cell cycle spirals out of control, cancer emerges: abnormal cells divide in a frenzy and can invade other tissues. There are more than 100 types of cancer, but some are more prevalent than others. And some are more deadly than others, because of their location in the body or how quickly the cells divide. New treatments, screening procedures, and vaccines can reduce these rates.

Assessment available in smartwork5

Top 10 cancer sites by rate of incidence

Incidence rates per 100,000

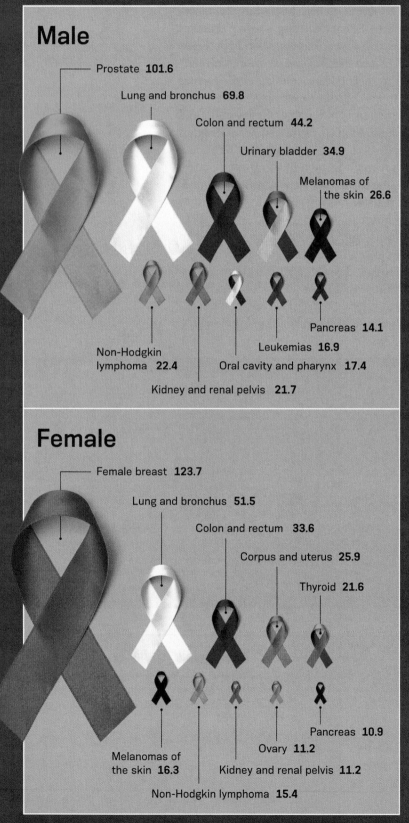

Male

- Prostate **101.6**
- Lung and bronchus **69.8**
- Colon and rectum **44.2**
- Urinary bladder **34.9**
- Melanomas of the skin **26.6**
- Non-Hodgkin lymphoma **22.4**
- Kidney and renal pelvis **21.7**
- Oral cavity and pharynx **17.4**
- Leukemias **16.9**
- Pancreas **14.1**

Female

- Female breast **123.7**
- Lung and bronchus **51.5**
- Colon and rectum **33.6**
- Corpus and uterus **25.9**
- Thyroid **21.6**
- Melanomas of the skin **16.3**
- Non-Hodgkin lymphoma **15.4**
- Kidney and renal pelvis **11.2**
- Ovary **11.2**
- Pancreas **10.9**

Top 10 cancer deaths by rate of incidence

Incidence rates per 100,000

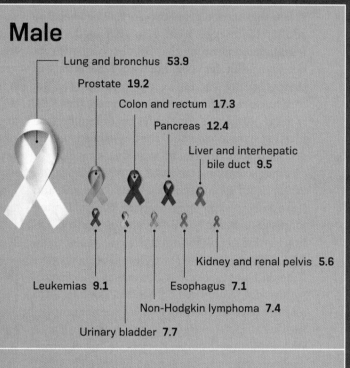

Male

- Lung and bronchus **53.9**
- Prostate **19.2**
- Colon and rectum **17.3**
- Pancreas **12.4**
- Liver and interhepatic bile duct **9.5**
- Leukemias **9.1**
- Urinary bladder **7.7**
- Non-Hodgkin lymphoma **7.4**
- Esophagus **7.1**
- Kidney and renal pelvis **5.6**

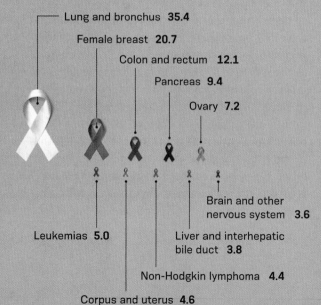

Female

- Lung and bronchus **35.4**
- Female breast **20.7**
- Colon and rectum **12.1**
- Pancreas **9.4**
- Ovary **7.2**
- Leukemias **5.0**
- Corpus and uterus **4.6**
- Non-Hodgkin lymphoma **4.4**
- Liver and interhepatic bile duct **3.8**
- Brain and other nervous system **3.6**

REVIEWING THE SCIENCE

- The **cell cycle** is the set sequence of events over the life span of a eukaryotic cell that will divide. **Interphase** and **cell division** are the two main stages of the cell cycle. Interphase is longer, and consists of the G_1, **S**, and G_2 **phases**. DNA is replicated in the S phase. Cells that will not divide exit the cell cycle and enter a G_0 **phase**.

- Cell division is necessary for growth and repair in multicellular organisms, and for **asexual** and **sexual reproduction** in all types of organisms. Many prokaryotes divide through **binary fission**, a form of asexual reproduction.

- Each **chromosome** in a cell contains a single DNA molecule compacted by packaging proteins. The **somatic cells** of eukaryotes have two of each type of chromosome matched together in **homologous pairs**. One chromosome in each pair is inherited from the mother; the other, from the father. Chromosomal replication produces two identical **sister chromatids** that are held together firmly at the **centromere**.

- Eukaryotes perform cell division through **mitosis** followed by **cytokinesis**, producing daughter cells that are genetically identical to each other and to the parent cell. The four main phases of mitosis are prophase, metaphase, anaphase, and telophase. Through these phases, the chromosomes of a parent cell are condensed and positioned appropriately, and the sister chromatids are separated to opposite ends of the cell. During cytokinesis, the cytoplasm of the parent cell is physically divided to create two daughter cells.

- The cell cycle is carefully regulated. Checkpoints ensure that the cycle does not proceed if conditions are not right. Unregulated cell growth and cancer occur when checkpoints fail.

- **Meiosis** is critical for sexual reproduction. In animals, the products of meiosis are sex cells, called **gametes**, that fuse during **fertilization** to give rise to a **zygote**. Meiosis—consisting of two rounds of nuclear and cytoplasmic divisions—produces **haploid** gametes containing only one chromosome from each homologous pair. When two gametes fuse during fertilization, a **diploid** zygote is formed, having one of each homologous chromosome from both parents.

- During **meiosis I**, the maternal and paternal members of each homologous pair are sorted into two daughter cells. **Meiosis II** is similar to mitosis in that sister chromatids are segregated into separate daughter cells at the end of cytokinesis.

- Meiosis produces genetically diverse gametes through **crossing-over** of homologous chromosomes, leading to **genetic recombination** and then the **independent assortment** of homologous chromosomes. Meiosis and fertilization together introduce genetic variation into populations.

THE QUESTIONS

The Basics

1 Homologous chromosomes are

(a) the same thing as sister chromatids.

(b) a pair of chromosomes of the same kind.

(c) identical copies of the same chromosome.

(d) always haploid.

2 Which of the following is *not* a contributor to genetic variation?

(a) binary fission

(b) crossing-over of homologous chromosomes

(c) random assortment of homologous chromosomes

(d) fertilization

3 Link each cell phase to the event that occurs within it.

CYTOKINESIS	1. Each of the chromosomes in a human cell contains two sister chromatids by the end of this phase.
S PHASE	2. Most cell growth occurs during this phase.
G_1 PHASE	3. Cells that will never replicate leave the cell cycle and enter this phase.
G_0 PHASE	4. Two separate daughter cells are produced at the end of this phase.

4 Select the correct terms in the following sentences: (**Mitosis / Meiosis**) produces daughter cells with half the number of chromosomes that the parent cell has. Cell division in prokaryotes is called (**mitosis / binary fission**). Meiosis I separates (**sister chromatids / homologous chromosomes**); meiosis II separates (**sister chromatids / homologous chromosomes**) into separate daughter cells.

5 Place the following events of sexual reproduction in the correct order by numbering them from 1 to 5.

_____a. Separation of homologous chromosomes

_____b. Separation of sister chromatids

_____c. Mitosis within the zygote, leading to a multicellular organism

_____d. Cytokinesis, leading to four haploid daughter cells

_____e. Fusion of two gametes

6 Which of the following is not a contributor to genetic variation?

(a) replication of sister chromatids

(b) crossing-over of homologous chromosomes

(c) random assortment of homologous chromosomes

(d) fertilization

7 Loss of cell cycle control may lead to

(a) pregnancy.

(b) cancer.

(c) fertilization.

(d) crossing-over of homologous chromosomes.

Challenge Yourself

8 Suppose a scientist has recently identified a new protein that prevents a cell from entering mitosis if there are any signs of DNA damage. This protein would be classified as a type of _____ protein.

(a) chromatid

(b) cell cycle checkpoint

(c) benign

(d) malignant

(e) angiogenesis

9 You've been reduced in size by a misfire from a "shrink-inator" gun! You realize that you are inside a cell during prophase I of meiosis. You see two linear molecules of DNA compacted and attached to one another by a centromere. What exactly are you looking at?

(a) a homologous pair of chromosomes

(b) sister chromatids

(c) the metaphase plate

(d) cell cycle checkpoints

(e) gametes

10 Scientists are able to isolate cells in various phases of the cell cycle. During an experiment in which Dr. Patrick McGroyn is causing cultured tumor cells to go through mitotic cell division, he isolates a group of cells that contain one and a half times more DNA than cells isolated in the G_1 phase have. What phase must these cells be in? How do you know?

Try Something New

11 Domesticated sheep cells have a total of 54 chromosomes (versus 46 chromosomes in human cells).

(a) How many separate DNA molecules are present in a sheep liver cell at the end of the G_1 phase?

(b) How many separate DNA molecules are present in the daughter cells after meiosis I in the ovary of the sheep?

(c) How many separate DNA molecules are present in the daughter cells after meiosis II in the ovary of the sheep?

12 Biopsies from aggressive cancers often have cells that contain several nuclei per cell when viewed through a microscope. Which scenario could explain how such a multinucleated cell might have come to be?

(a) The cell underwent repeated mitosis with simultaneous cytokinesis.

(b) The cell had multiple S phases before it entered mitosis.

(c) The cell underwent repeated mitosis, but cytokinesis did not occur.

(d) The cell underwent repeated cytokinesis but no mitosis.

(e) The cell actually went through meiosis and not mitosis.

13 Describe the likely consequences of bypassing the G_1 and G_2 checkpoints in the cell cycle. Why do compounds like nonylphenol lead to the multiplication of abnormal cells?

Leveling Up

14 **What do _you_ think?** Cancer begins with a single cell that breaks loose of normal restraints on cell division and starts dividing rapidly to establish a colony of rogue cells. As cancer cells spread through the body, they disrupt the normal functions of tissues and organs; unchecked, cancer can cause death through failure of multiple organ systems. Many cancers could be prevented by not smoking or chewing tobacco, eating less meat and processed foods, eating more fruits and vegetables, drinking alcohol only moderately if at all, exercising regularly, and maintaining a healthy weight. Only 5–10 percent of cancers are directly attributable to genetic causes.

A "sin tax" is a tax on a product or activity that has negative effects on others, as a way to offset some of those effects. Common targets of sin taxes are tobacco and alcohol because of their public health costs. Proponents of a sin tax on tobacco argue that such a tax would decrease the amount that people smoke (because of the increased cost) and could also partially fund the costs of medical care necessitated by increased rates of cancer and other diseases caused by smoking. Critics point out that sin taxes have historically triggered smuggling and black markets, and have a disproportionate effect on poor people because the wealthy can more easily afford to pay the higher prices.

What do you think? Should we institute taxes on tobacco? Would fewer people smoke, or would they smoke less, if tobacco were more expensive? What about higher health care premiums for smokers, based on their higher risk for cancer? Would increasing the premiums cause more people to stop smoking? Would you support such a policy?

15 *Write Now* **biology: BPA effects** The studies of BPA described in this chapter used an inbred strain of mice that is known to be especially susceptible to estrogen and estrogen-like chemicals, such as BPA. The plastics industry maintains that the susceptibility of this strain of mice to estrogen renders these studies invalid as a basis for estimating the effects of BPA on humans. BPA researchers respond that the current situation, exposing millions of people to unknown levels of BPA, constitutes a massive uncontrolled experiment. They maintain that even a small risk of harm is too great to be allowed when so many people are exposed. Bills banning the use of BPA in food and beverage containers were introduced in Congress in 2009, 2011, 2013,

2014, and 2015. All failed to pass. Another bill was introduced to the House of Representatives on September 28, 2016.

a. Go to the Embryo Project Encyclopedia (http://embryo.asu.edu), an NSF-funded online repository of information about embryo research. Search for the term "BPA" to find the most up-to-date information on BPA research.

b. Go to the website of the American Chemistry Council (http://www.americanchemistry.com), a trade organization that advocates for the chemical industry. Search for the term "BPA" to find its most up-to-date policy statement on the safety of BPA.

c. Draft a letter urging your congressional representative to support or oppose the most recent BPA bill, explaining why you believe the weight of scientific evidence makes your position a prudent response to the situation.

For more, visit digital.wwnorton.com/bionow2 for access to:

Dog Days
of Science

Two canine-loving researchers unravel the genetic secrets of man's best friend.

After reading this chapter you should be able to:

- Distinguish between the genotype and phenotype of a given genetic trait.
- Describe the importance of Gregor Mendel's experiments to our understanding of inheritance.
- Illustrate Mendel's laws of segregation and independent assortment.
- Create a Punnett square to predict the phenotype of offspring from parents with a known genotype—both for single genes and for two independent genes.
- Give examples of Mendelian traits and of traits with complex inheritance.
- Explain how an individual's phenotype may be determined by multiple genes that interact with one another and with the environment.

CHAPTER
07

PATTERNS OF
INHERITANCE

Gordon Lark's best friend was dying. Soft and shaggy, with tousled black hair, Georgie hadn't left Lark's side in 10 years, since his daughter had first purchased Georgie as a puppy from two kids by the side of the road. But as she aged, Georgie had become ill with Addison's disease, a disorder in which her body's immune system began to attack and destroy her own tissues. Georgie passed away in 1996.

Lark, a scientist at the University of Utah in Salt Lake City, was heartbroken. To help heal the wound, he decided to adopt another dog of the same breed—a Portuguese water dog (PWD), named for its history of helping Portuguese fishermen with their work (**Figure 7.1**). He contacted Karen Miller, a PWD breeder on a farm in rural New York. As part of the owner screening process, Miller asked Lark about his profession. "I said I was a soybean geneticist," Lark recalls, "but all she heard was 'blah, blah, genetics, blah, blah.' And she got really excited."

As a breeder, Miller was keenly interested in how dogs inherit characteristics from their parents, so Miller and Lark began talking by phone each week about genetics. When it came time to pick up his new puppy, Mopsa, Lark requested the bill. But Miller didn't want Lark's money. She had something else in mind: she gave him Mopsa free of charge, in the hope that Lark would start researching dog genetics. "That's silly," Lark told her. He wasn't a dog researcher. Lark had spent a career studying the genetic traits of bacteria and soybeans.

A **genetic trait** is any inherited characteristic of an organism that can be observed or detected in some manner. Some genetic traits are **invariant**, meaning they are the same in all individuals of the species. All soybeans, for example, have pods that contain seeds. Other genetic traits are **variable**; for example, soybean seeds occur in various sizes and colors, including black, brown, and green.

Apart from his love for canines, Lark wasn't intimately familiar with dogs' physical and biochemical traits. **Physical traits**, such as the shape of a dog's face, are easy to observe. **Biochemical traits**, on the other hand, such as a dog's susceptibility to Addison's disease, are often more difficult to observe. It is easy to collect physical and biochemical information from a field of soybeans but far more difficult to collect it from domesticated animals in homes all over the country. And to study dogs, Lark would also need data on **behavioral traits**, such as shyness and extroversion—factors he didn't have to take into account with soybeans. All of these traits—physical, biochemical, and behavioral—are influenced by genes.

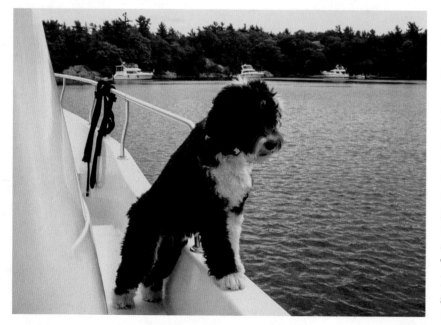

Figure 7.1

A Portuguese Water Dog

Getting to the Genes

A **gene** is the basic unit of information affecting a genetic trait. At the molecular level, a gene consists of a stretch of DNA on a **chromosome**—a threadlike molecule, made of DNA and proteins, that is found in the nucleus of a eukaryotic cell (**Figure 7.2**). A gene contains the information, or "code," for a specific protein; the protein then causes or contributes to a particular genetic trait. To study PWD traits, Lark would need dog DNA, which can be obtained from blood or saliva. Once he had that DNA, he could begin to search for **alleles**—different versions of a given gene—and link them to genetic traits. Alleles of a gene arise by **mutation**, which is any change in the DNA that makes up a gene (see Chapter 9 for more on mutation). Genetic diversity in a species, whether it is soybeans or dogs or humans, comes about because the species as a whole contains many different alleles of its genes.

Dogs are the champions of genetic variation. All dogs are the same species, yet a Pekingese weighs only a couple of pounds, while a Saint Bernard can weigh over 180 pounds. Dogs, in fact, are reported to have more variation in the size and shape of their species than any other living land mammal on Earth, with the possible exception of humans.

On a whim, Lark agreed to dabble in dog genetics, but to do so, he would need to compare the **genotypes** of individual dogs (their genetic makeup, controlled by their combinations of alleles) with their **phenotypes** (the physical expression of their genetic makeup). The genotype of a given trait is the pair of alleles that codes for its phenotype. To identify genes responsible for dog traits, Lark would need both types of information for each of many dogs.

Miller was already on the case. Three months after Lark got Mopsa, Miller sent him 5,000 PWD pedigrees—detailed health and breeding records for individual dogs. Lark was astonished. It was the first of many times that the enthusiasm and generosity of dog owners would contribute to his research.

"That was literally how this started," says Lark. Today, Lark and Miller's unlikely partnership has blossomed into a national research project producing valuable knowledge about the genetic basis of health and disease in both man and man's best friend. What's more, their effort has demonstrated how tiny genetic changes can create huge variation in a single species.

Pet Project

The Georgie Project, as Lark fondly named it, officially began in 1996. Lark's first task was to collect genotypes and phenotypes from PWDs. To his pleasant surprise, PWD owners were enthusiastic and began flooding him with pedigrees, blood samples, and X-rays taken by their veterinarians. In short order, Lark had DNA from more than 1,000 dogs and detailed body measurements for over 500. Then the hard work began.

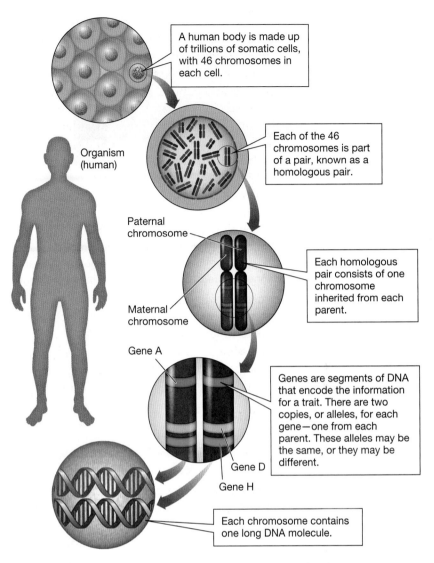

Organism (human)

A human body is made up of trillions of somatic cells, with 46 chromosomes in each cell.

Each of the 46 chromosomes is part of a pair, known as a homologous pair.

Paternal chromosome

Each homologous pair consists of one chromosome inherited from each parent.

Maternal chromosome

Gene A

Genes are segments of DNA that encode the information for a trait. There are two copies, or alleles, for each gene—one from each parent. These alleles may be the same, or they may be different.

Gene D

Gene H

Each chromosome contains one long DNA molecule.

Figure 7.2

Genes are segments of DNA that confer a genetic trait
Somatic cells (cells of the body) have two copies of most genes.

Q1: What is the physical structure of a gene?

Q2: How many copies of each gene are found in the diploid cells in a woman's body?

Q3: With 46 chromosomes in a human diploid cell, how many chromosomes come from the person's mother and how many from the father?

Using the dogs' genotypes and phenotypes, Lark set out to pinpoint the alleles for particular traits. Some genes have alleles that are **dominant** when paired with another allele; that is, one allele prevents a second allele from affecting the phenotype when the two alleles are paired together. The

Phenotype:

Genotype: *bb*   *BB* or *Bb*

Figure 7.3

Poodles illustrate variation in the coat color gene
These poodles, close cousins of the Portuguese water dog, may have a black coat (dominant allele *B*) or a brown coat (recessive allele *b*). Other coat colors, with different inheritance patterns, are found in poodles and other dog breeds.

> **Q1:** Which might you observe directly: the genotype or the phenotype?
>
> **Q2:** Which poodle could be heterozygous: the one with the black coat or the one with the brown coat?
>
> **Q3:** Can you identify with certainty the genotype of a black poodle? A brown poodle?

black-fur allele (*B*), for example, is dominant in dogs. An allele that has no effect on the phenotype when paired with a dominant allele is said to be **recessive**. In dogs, the brown-fur allele (*b*) is recessive. (When a gene has one dominant and one recessive allele, we generally use an uppercase letter for the dominant allele and a lowercase letter for the recessive allele.)

An individual that carries two copies of the same allele (such as *BB* or *bb*) is **homozygous** for that gene. An individual whose genotype consists of two different alleles for a given phenotype (*Bb*) is **heterozygous** for that gene. Having one dominant allele and one recessive allele, a heterozygous individual will show the dominant phenotype;

GORDON LARK

A geneticist at the University of Utah in Salt Lake City, Gordon Lark initiated the Georgie Project in 1996 to study the genetics of Portuguese water dogs. The national research project has led to valuable knowledge about the genetic basis of health and disease in humans and dogs.

a dog that is heterozygous for fur color (*Bb*), for example, will be black (**Figure 7.3**).

The first dog trait that Lark decided to investigate was size. What makes a Great Dane large and a Chihuahua small? To find out, Lark asked for help from the "mother of all dog projects," as Lark calls her—a researcher named Elaine Ostrander, whose entry into dog research was almost as strange as Lark's.

Crisscrossing Plants

In 1990, Ostrander was a young, enthusiastic researcher who had just completed her postdoctoral studies in molecular biology at Harvard University and was ready to start her own laboratory in California. But first she had to decide which organism to study. Typical choices included fruit flies, worms, or plants—organisms that are easy to grow and manipulate. Ostrander picked plants, just as Gregor Johann Mendel, an Austrian monk who later became known as the "father of modern genetics," had done in the mid-1800s.

Mendel famously bred pea plants in a garden at his monastery. Through his work with pea plants, Mendel discovered patterns of inheritance that today form the foundation of genetics for scientists like Ostrander. "Mendel's laws," as they are now called, describe how genes are passed from parents to offspring. These laws allow us to use parental genotypes to predict offspring genotypes and phenotypes.

Each time Mendel bred two pea plants together, he was performing a **genetic cross**, or just "cross" for short. A genetic cross is a controlled mating experiment performed to examine how a particular trait is inherited. In a series of genetic crosses, the organisms involved in the first cross are called the **P generation** ("P" for "parental").

For example, Mendel investigated the inheritance of flower color by crossing pea plants that had different flower colors (**Figure 7.4**). He had noticed that some plants always "bred true" for flower color; that is, the offspring always produced flowers that had the same color as the parents and were therefore homozygous. He performed a genetic cross with a P generation

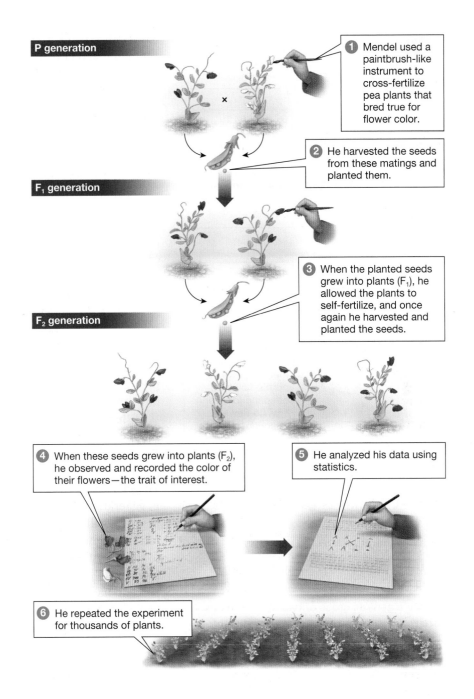

P generation

① Mendel used a paintbrush-like instrument to cross-fertilize pea plants that bred true for flower color.

② He harvested the seeds from these matings and planted them.

F₁ generation

③ When the planted seeds grew into plants (F₁), he allowed the plants to self-fertilize, and once again he harvested and planted the seeds.

F₂ generation

④ When these seeds grew into plants (F₂), he observed and recorded the color of their flowers—the trait of interest.

⑤ He analyzed his data using statistics.

⑥ He repeated the experiment for thousands of plants.

Figure 7.4

Mendel's careful experiments

Mendel was meticulous in conducting his research and making observations, following a very careful protocol. Controlled breeding is possible in flowering plants because they have both male and female reproductive structures. Mendel manipulated his P-generation plants by removing all the female structures from one plant and all the male structures from the other, thus preventing self-fertilization. He could then perform the initial cross by transferring pollen from the "male" plant to the flower of the "female" plant. ▶

Q1: What would you predict about the color of the F₁ plants' flowers?

Q2: Why was it important that Mendel begin with pea plants that he knew bred true for flower color? Why couldn't he simply cross a purple-flowered plant and a white-flowered plant?

Q3: Over the years, Mendel experimented with more than 30,000 pea plants. Why did Mendel collect data on so many plants? Why didn't he study just one cross? *Hint:* Read "What Are the Odds?" on page 124 before answering.

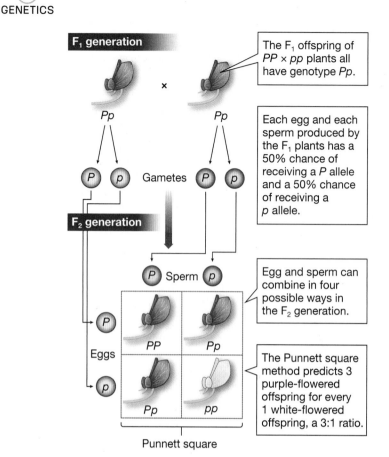

F₁ generation

The F₁ offspring of *PP* × *pp* plants all have genotype *Pp*.

Pp × *Pp*

Gametes *P* *p*   *P* *p*

Each egg and each sperm produced by the F₁ plants has a 50% chance of receiving a *P* allele and a 50% chance of receiving a *p* allele.

F₂ generation

P Sperm *p*

Egg and sperm can combine in four possible ways in the F₂ generation.

Eggs

P   PP   Pp

p   Pp   pp

The Punnett square method predicts 3 purple-flowered offspring for every 1 white-flowered offspring, a 3:1 ratio.

Punnett square

Figure 7.5

Punnett squares predict the offspring of genetic crosses

Q1: Why did Mendel's entire F₁ generation look the same?

Q2: The phenotype ratio in the F₂ generation is 3:1 purple-to-white flowers. What is the genotype ratio?

Q3: Draw a Punnett square for a genetic cross of two heterozygous, black-coated dogs. What are the phenotype and genotype ratios of their offspring?

in which one parent bred true for purple flowers (*PP*) and the other bred true for white flowers (*pp*). The first generation of offspring of a genetic cross is called the **F₁ generation** ("F" is for "filial," a word that refers to a son or daughter). When the individuals of the F₁ generation are crossed with each other, the resulting offspring are said to belong to the **F₂ generation**. Mendel allowed the F₁-generation pea plants to self-fertilize to produce the F₂ generation.

We can predict the results of an experimental cross by using a grid-like diagram called a **Punnett square** (**Figure 7.5**). A Punnett square

shows all possible ways that two alleles can be brought together through fertilization. To create a Punnett square showing how a trait is inherited, list the alleles of the male genotype across the top of the grid, writing each unique allele just once. List the alleles of the female genotype along the left edge of the grid, again writing each unique allele only once. In the case of Mendel's cross of the F₁ generation, the male genotype (*Pp*) is mated with the female genotype (also *Pp*).

Next, fill in each box (or "cell") in the grid by combining the male allele at the top of each column with the female allele listed at the beginning of each row. The Punnett square shows all four ways in which the two alleles in the sperm can combine with the two alleles found in the eggs. The four genotypes shown within the Punnett square are all equally likely outcomes of this cross.

Using the Punnett square method, we can predict that ¼ of the F₂ generation is likely to have genotype *PP*, ½ to have genotype *Pp*, and ¼ to have genotype *pp*. Because the allele for purple flowers (*P*) is dominant, plants with *PP* or *Pp* genotypes have purple flowers, while plants with *pp* genotypes have white flowers. Therefore, we predict that ¾ (75 percent) of the F₂ generation will have purple flowers and ¼ (25 percent) will have white flowers—a 3:1 (¾:¼) ratio of phenotypes. This prediction is very close to the actual results that Mendel obtained. Of a total 929 F₂ plants that Mendel raised, 705 (76 percent) had purple flowers and 224 (24 percent) had white flowers.

Results like these supported Mendel's first law, the **law of segregation**, which, in modern terms (Mendel did not know about DNA), states that the two alleles of a gene are separated during meiosis, the specialized type of cell division during sexual reproduction that was discussed in Chapter 6, and end up in different gametes—egg or sperm cells. One of the two alleles is found on one of the chromosomes in a homologous pair, and the other allele is found on the other chromosome in the pair (see **Figure 7.2**). Remember, homologous chromosome pairs are partitioned into separate daughter cells during meiosis I. Mendel's law of segregation can be used to predict how a single trait will be inherited.

You can try this on your own by making a Punnett square to predict the ratio of black and brown offspring that would result if two heterozygous (*Bb*) black-coated dogs were mated. It is important to understand that the predicted ratios simply give the *probability* that a particular offspring will have a certain phenotype or genotype; the actual ratio will vary (see "What Are the Odds?" on page 124).

Peas in a Pod

Mendel's research on pea seeds led to his second law, the **law of independent assortment**. This law states that when gametes form, the two alleles of any given gene segregate during meiosis independently of any two alleles of other genes. For example, pea seeds can have a round or wrinkled shape, and they can be yellow or green. Two different genes control the two different traits—the *R* gene, with alleles *R* (round) and *r* (wrinkled), controls seed shape, while the *Y* gene, with alleles *Y* (yellow) and *y* (green), controls the color of the seed—but neither gene affects the inheritance of the other.

Mendel tested the idea of independent assortment in the set of experiments illustrated in **Figure 7.6.** He tracked seed shape, the trait controlled by the *R* and *r* alleles, and seed color, controlled by the *Y* and *y* alleles. The test of his hypothesis came when he examined the phenotypes of the offspring produced by crossing the heterozygous F_1 plants (*RrYy*). As predicted by the hypothesis, two new phenotypic combinations were found among the F_2 offspring: plants with round, green seeds (*RRyy* or *Rryy*) and plants with wrinkled, yellow seeds (*rrYY* or *rrYy*). Figure 7.6 summarizes the ratios of the two parental phenotypes and the two novel, nonparental phenotypes.

Traits controlled by a single gene and unaffected by environmental conditions are called **Mendelian traits**. But when Mendel described his laws of inheritance, he had no idea what genes were made of, where they were located within a cell, or how they segregated and independently assorted. Today we know that genes are located on chromosomes and that these chromosomes are the basis for all inheritance. We call these assertions the **chromosome theory of inheritance**, which explains the mechanism underlying

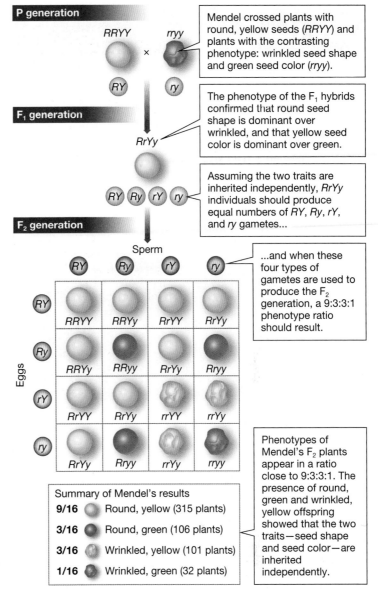

Figure 7.6

Independent assortment of pea color and shape

Mendel used two-trait breeding experiments, called **dihybrid crosses**, to test the hypothesis that the alleles of *two different genes* are inherited independently from each other. ▶️

Q1: List all the possible offspring genotypes and phenotypes.

Q2: What is the offspring phenotype ratio?

Q3: Complete a Punnett square for a genetic cross of two true-breeding Portuguese water dogs—one with a black, wavy coat (homozygous dominant, *BBWW*) and one with a brown, curly coat (homozygous recessive, *bbww*). What is the phenotype ratio of their offspring (F_1)? Now fill out another Punnett square, crossing two of the offspring. What is the phenotype ratio of the F_2 generation?

What Are the Odds?

The *probability* of an event is the chance that the event will occur. For example, there is a probability of 0.5 that a coin will turn up "heads" when it is tossed. A probability of 0.5 is the same thing as a 50 percent chance, or ½ odds, or a ratio of one heads to one tails (1:1). If you toss the coin only a few times, the observed percentage of heads may differ greatly from 50 percent. But if you toss it many, many times, that observed percentage will be very close to 50 percent. Each toss of a coin is an independent event, in the sense that the outcome of one toss does not affect the outcome of the next toss. The probability of getting two heads in a row is a product of the separate probabilities of each individual toss: 0.5 × 0.5, which is 0.25. In our cross, the probability of getting a brown puppy is ¼, or 0.25. To go back to a Punnett square to predict the ratio of puppies from a genetic cross of two heterozygous (*Bb*) black-coated dogs, the probability of getting a black puppy is ³⁄₄, or 0.75.

We cannot know with certainty what the actual phenotype or genotype of a particular offspring is going to be, except when true-breeding individuals are crossed. For example, two brown dogs, both of whom have a *bb* genotype, will have only *bb*-genotype, brown-phenotype offspring. Moreover, the probability that a particular offspring will display a specific phenotype is completely unaffected by how many offspring there are. The likelihood that we will see the 3:1 black-to-brown outcome, however, increases when we analyze a larger number of offspring, just as Mendel analyzed thousands of pea plants.

Mendel's laws by identifying chromosomes as the paired factors (where each homologous chromosome in a pair has one allele for a gene) that are shuffled and recombined, and then separated randomly into sperm and egg cells during meiosis (see Figures 6.7 and 6.8 in Chapter 6). Then, during fertilization, a one-in-a-million sperm fuses with a one-in-a-million egg to create a unique individual. That is how offspring can have genotypes and phenotypes that were not present in either parent, such as a brown puppy born to two black dogs.

Going to the Dogs

Like Mendel, Elaine Ostrander planned to study plants to unravel the secrets of genetics and inheritance. But when she arrived at UC Berkeley to open her lab, the space was not yet available. So she wandered down the hall and into the office of Jasper Rine, a geneticist who normally studied yeast but was looking for someone to start a mammalian genome research project. Ostrander volunteered.

But which mammal to study? "I was allergic to cats, and I didn't know enough about cows or pigs or horses," she recalls, so she picked dogs. Not only was Ostrander a dog lover, but the American Kennel Club had just begun offering funding to researchers trying to identify genes associated with dog diseases.

In 1993, Ostrander began identifying all the genes unique to dogs—that is, making a map of the dog **genome**. Some colleagues said she was nuts, that no one would give her money to support the research. But Ostrander is nothing if not persistent, and she knew the potential value of the research: dogs have more than 350 inherited diseases, and up to 300 of those are similar to conditions in people, including cancer, epilepsy, heart disease, and Addison's disease, the illness that killed Lark's dog Georgie. The genetics of bladder cancer is difficult to study in humans, for example, but the disease is quite common in Scottish terriers and would be easier to study in a dog species. By cracking the genetic code of dogs, Ostrander hoped to uncover causes and potential treatments for human diseases.

In 2005, she published the first full dog genome sequence, for a female boxer named Tasha. The achievement gained her scientific fame and raised awareness among scientists of the importance of dog genetics to human health. "Of the more than 5,500 mammals living today, dogs are arguably the most remarkable," Ostrander's coauthor, Eric Lander, a professor of biology at the Massachusetts Institute of Technology, said when the first dog genome sequence was published. "The incredible physical and behavioral diversity of dogs—from Chihuahuas to Great Danes—is encoded in their genomes. It can uniquely help us understand embryonic development, neurobiology, human disease, and the basis of evolution" (**Figure 7.7**).

Shadow, a standard poodle, was the first dog to have its genome partially (about 80%) sequenced.

Tasha, a boxer, was the first dog to have its complete genome sequenced. Boxers are vulnerable to hip, thyroid, and heart problems. Scientists identified a gene for cardiomyopathy in boxers, a heart disorder also found in humans.

Pembroke Welsh corgis may develop a fatal neurodegenerative condition similar to amyotrophic lateral sclerosis (ALS) in humans. The human gene mutation associated with ALS was also found in corgis with the condition.

Psychiatric disorders often have a genetic component. Doberman pinschers are susceptible to canine compulsive disorder, similar to obsessive-compulsive disorder in humans. The responsible gene in Dobermans has been linked to autism disorders in humans.

Golden retrievers are prone to cancers of the bone marrow. Ostrander's research group is analyzing the genomes of hundreds of goldens with and without cancer, hoping to identify the genes responsible.

Figure 7.7

Man's best friend

The Dog Genome Project has identified the genetic basis of several diseases and conditions in dogs, and in some cases it has been able to link the gene to a similar gene in humans.

Q1: Boxers are far more inbred than poodles. Why does that inbreeding make the former a better target for genetic studies of disease than the latter?

Q2: Explain why a geneticist interested in finding a gene linked to cancer would want to look at the DNA of senior golden retrievers with *and* without cancer?

Q3: Obsessive-compulsive disorder (OCD) in humans is characterized by obsessive thoughts and compulsive behavior, such as pacing. Canine compulsive disorder (CCD) is characterized by compulsive behavior such as "flank sucking," sometimes seen in Doberman pinschers. Would you predict that the medications given to humans with OCD would decrease compulsive behaviors in CCD dogs? Why or why not?

But years before Ostrander completed the dog genome, she had begun a different pet project. In 2001, Ostrander received a call from a scientist in Utah who wanted to talk about dogs. It was Gordon Lark, who told her he was collecting genetic trait information about PWDs. "The day I met Gordon was the best day of my life," says Ostrander. "I knew it was golden."

In 2002, the duo published a paper pinpointing genes that control dog body shape, from the tall, lanky look of a greyhound to the short, stocky frame of a pit bull. In the acknowledgments of the paper, they thanked Karen Miller and all the PWD owners who had contributed pedigree information.

In the spring of 2006, Lark and Ostrander began their second collaboration, this time to identify the genetic basis of dog size. Lark collected skeletal measurements of 92 PWDs and DNA samples from each dog. Ostrander used that genotype and phenotype information to identify a key gene for body size—*IGF1*, which controls the activity of a growth factor and is known to influence body size in mice and humans. This gene's two alleles are called *I* and *B*. Lark and Ostrander discovered that PWDs homozygous for allele *I* (*II*) were usually large dogs, and those homozygous for allele *B* (*BB*) were always small dogs. That single gene accounted for whether a PWD was large or small.

Interestingly, neither *IGF1* allele is dominant or recessive. Instead, heterozygous dogs, with an *IB* genotype, are medium-sized dogs. This is an example of a trait inherited by **incomplete dominance**—in which neither allele is able to exert its full effect, so a heterozygote displays an intermediate phenotype. Dogs with an *IB* genotype aren't large or small, but rather medium-sized (**Figure 7.8**).

Early in the twentieth century, geneticists identified yet another type of interaction among alleles—codominance—that Mendel had not observed among his pea plants. A pair of alleles shows **codominance** when the effect of the two alleles is equally visible in the phenotype of the heterozygote. In dogs, gum color is codominant. A dog's gums can be pink, black, or pink with black spots; in the latter case, both alleles are fully on display, and neither is diminished or diluted by the presence of the other allele (as in incomplete dominance) or suppressed by a dominant allele (as in the case of dominant and recessive alleles). In humans, the blood type AB is a codominant trait.

Genotype: *II*
Phenotype: Large

Genotype: *BB*
Phenotype: Small

Genotype: *IB*
Phenotype: Medium

Figure 7.8

Incomplete dominance of body size alleles

Great Danes and Chihuahuas illustrate the extreme size variation found in domestic dogs (see the chapter-opening photo). Unlike the case with Mendelian traits, dogs heterozygous for the main body size gene show an intermediate size like the Cocker Spaniel pictured here on the right.

Q1: What are the genotypes of a large and a small dog?

Q2: Is it possible to have a heterozygous large dog? Explain why or why not.

Q3: Crossing a Great Dane and a Chihuahua is likely to be unsuccessful, even though they are members of the same species (and thus have compatible sperm and egg). Why is that? What are some potential risks of such a cross?

It's Complicated

Many of the traits people tend to be curious about—body weight, intelligence, athleticism, and musical talent, to name a few—are yet more complicated. A **complex trait** is a genetic trait whose pattern of inheritance cannot be predicted by Mendel's laws of inheritance. Complex traits do not fit the straightforward single-gene, single-phenotype pattern discussed so far.

Sometimes a *single gene* influences a number of *different traits*. Such cases are examples of

Most Chronic Diseases Are Complex Traits

A disease is a condition that impairs health. It may be caused by external factors, such as infection by viruses, bacteria, or parasites, or injury from harmful chemicals or high-energy radiation. Nutrient deficiency can also lead to disease. Inadequate vitamin C consumption, for example, produces scurvy, once common among sailors and pirates. Disease may also be caused by the malfunction of one or more genes. Diseases caused exclusively by gene malfunction are described as genetic disorders, distinguishing them from infections and other types of diseases.

But many of the diseases that are most common in industrialized countries—heart disease, cancer, stroke, diabetes, asthma, and arthritis, for example—are caused by multiple genes interacting in complex ways with each other and with external factors. They are complex traits: malfunctions in key genes make a person susceptible to developing these diseases, but environmental factors affect whether the disease will actually appear and how severe the symptoms will be. A large percentage of the estimated risk of developing chronic diseases is preventable by lifestyle choices such as maintaining good nutrition, exercising regularly (see graph), and avoiding tobacco. (The word "chronic" means "unceasing"—a reference to the fact a person who develops one of these diseases will have it for the rest of their lives.)

A major goal of modern genetics is to identify genes that contribute to human disease. Researchers have identified alleles associated with increased risk of a number of common ailments, including high blood pressure, heart disease, diabetes, Alzheimer disease, several types of cancer, and schizophrenia. The hope is that one day soon, genetic tests will tell us whether we are predisposed to a disease before we become ill with it. Then, a person carrying a risky allele might take preventive measures to reduce the chance of developing the condition, and treatment could be customized to fit the particular allele involved. This tailored approach to treatment, called "personalized medicine," is already being used to treat breast cancer and other chronic diseases.

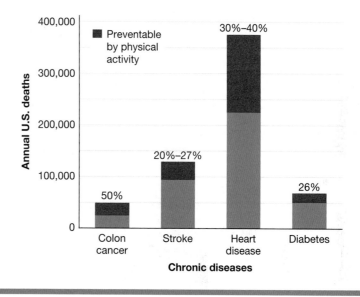

pleiotropy (*pleio*, "many"; *tropy*, "change"). In PWDs, Lark found that single genes can control multiple related skeletal traits. The shape of a dog's head and the shape of its limb bones are controlled by a single gene. That connection makes sense, says Lark, since a small head and long legs are advantageous for a fast dog, while a strong dog, like a pit bull, uses both its massive jaw and its short, thick legs for power.

Another good example of pleiotropy comes from a long-term breeding experiment to tame Russian silver foxes. Researchers found that as foxes became tamer and tamer, they also developed floppy ears instead of straight ones and had shorter legs and curlier tails than ordinary foxes have (see "The New Family Pet?," page 129).

Patterns of inheritance can get even more complicated. *Single traits* governed by the action of *more than one gene* are called **polygenic traits**. In humans, polygenic traits include eye and skin color, running speed, blood pressure, body size, and more. Of the thousands of human genetic traits, governed by an estimated 24,000 genes, fewer than 4,000 are known or suspected to be controlled by a single gene with a dominant and a recessive allele. The rest are polygenic traits.

Another twist on inheritance is **epistasis**, which occurs when the phenotypic effect of a gene's alleles depends on the presence of alleles for another, independently inherited gene. Labrador coat color, for example, is affected by epistasis (**Figure 7.9**). Dog fur, as mentioned earlier, has a dominant allele (*B*) that leads to black fur and a recessive allele (*b*) that produces brown fur. But the effects of these alleles (*B* and *b*) can be eliminated completely,

Genotype:	B-, E-	bb, E-	--, ee
Phenotype:	Black	Brown	Yellow

Figure 7.9

Epistasis in coat color

These Labrador retrievers show complex inheritance of coat color. The yellow dog carries two alleles that interfere with the deposition of melanin in hair. Both the brown and the black dogs must carry at least one allele that allows melanin deposition. A dash indicates that the allele is unknown, based on phenotype.

Q1: What are the possible genotypes (at both genes) of the black dog? The yellow dog? The brown dog?

Q2: Draw a Punnett square showing possible matings between the black dog (assuming it is heterozygous at both genes) and the yellow dog (assuming it is heterozygous at the *B* gene). List all the possible phenotypes of their offspring. (See Figure 7.6 for an example of a Punnett square made with two traits.)

Q3: If you wanted the most variable litter possible, what colors of Labrador retrievers would you cross?

ELAINE OSTRANDER

The "mother of all dog projects," Elaine Ostrander is chief and NIH distinguished investigator of the Cancer Genetics and Comparative Genomics Branch, as well as head of the Comparative Genetics Section at the National Institutes of Health. She studies genes important to growth, size variation, and cancer in dogs.

depending on which allele of the expression gene (*E* or *e*) is present. Dogs with a dominant *E* allele deposit a pigment called melanin in their fur and are therefore able to express whatever fur color genotype is present. But a recessive *ee* genotype blocks the deposition of melanin in fur, so the dog is yellow, regardless of the genotype at the *B/b* gene (*BB*, *Bb*, or *bb*).

If the environment affects the phenotype, it becomes nearly impossible to predict the phenotype when given only the genotype of an individual or its parents. The effects of many genes depend on internal and external environmental conditions, such as body temperature, carbon dioxide levels in the blood, external temperature, and amount of sunlight.

For example, cats have a gene that codes for an enzyme called tyrosinase, which is involved in melanin production. Siamese cats have a special C^t allele of the gene. The C^t allele codes for a tyrosinase that works well at colder temperatures ($\leq 35°C$) but does not function at warmer temperatures ($\geq 37°C$), so the production of melanin depends on the temperature of the surroundings (**Figure 7.10**). Because a cat's extremities tend to be colder than the rest of its body, melanin is produced there, and hence the paws, nose, ears, and tail of a Siamese cat tend to be dark. If a patch of light hair is shaved from the body of a Siamese cat and the skin is covered with an ice pack, the hair that grows back will be dark. Similarly, if dark hair is shaved from the tail and allowed to grow back under warm conditions, it will be light-colored.

Man's Best Friend

After describing the inheritance of size in Portuguese water dogs, Lark and Ostrander looked at the *IGF1* gene in over 350 dogs representing 14 small breeds and 9 giant breeds. The genotype *BB* was common in small dogs and virtually nonexistent in large dogs. "All small dog breeds had them. It didn't matter when they were bred or how; they all had the exact same pattern," says Ostrander.

"It was amazing," adds Lark. Breeders have, over time, been selecting for these alleles to create smaller and smaller dogs. "What mankind can do, without any genetic tools but just knowledge of heritability, is just extraordinary," he says.

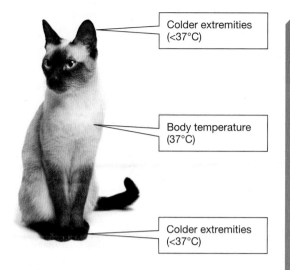

Colder extremities
(<37°C)

Body temperature
(37°C)

Colder extremities
(<37°C)

Figure 7.10

The environment can alter the effects of genes

Coat color in Siamese cats is controlled by a temperature-sensitive allele.

Q1: The gene that brings about the pale Siamese body fur is also responsible in part for the typical blue eyes of the species. What is the term for this type of inheritance?

Q2: Siamese kittens that weigh more tend to have darker fur on their bodies. Why might this be?

Q3: The Siamese cat pictured is called a "seal point" because it has seal-colored (dark brown) extremities. Some Siamese cats show the same color pattern, but the dark areas are of a lighter color or even a different shade—for example, lilac point, red point, blue point. What results would you predict if the experiments described in the text (shaving the cat and then increasing or decreasing temperature) were conducted on cats with these color patterns?

After their success with dog size, Lark and Ostrander identified genes responsible for other traits—fur color, leg length, skull shape, and more. They also identified genes related to cancer and other complex traits that might tell us something about human disease (see "Most Chronic Diseases Are Complex Traits," page 127). In border collies, Ostrander's team identified a gene

The New Family Pet?

The silver fox is the same species as the more familiar red fox. Because of its soft, silver coat, it has been bred in captivity for over 100 years to provide fur coats, stoles, and hats for the wealthy.

In 1959, a Russian geneticist, Dmitry Belyaev, began to conduct breeding experiments on silver foxes he had purchased from a fur breeder, pairing only the tamest individuals of each generation. He determined how tame a fox was by observing its response when approached and offered food. As the foxes became tamer in each generation, they did not show a "fear response" until they were older—9 weeks instead of 6 weeks. (Domestic dogs develop a fear response at about 8–12 weeks.) In addition, in the tame foxes, the hormones associated with a fear response did not increase until later. These traits were all clearly influenced by the same gene or genes—an example of pleiotropy.

Another surprising result was that the foxes' appearance began to change along with the changes in behavior. They began to develop shorter tails, wider faces, and floppier ears. All of these made the adult

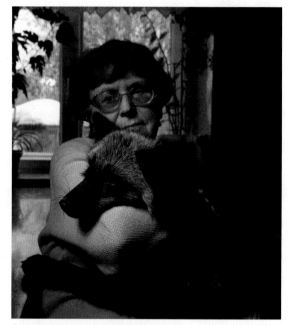

Only 50 years and 35 generations separate this tame silver fox from its wild relative

foxes look more puppy-like, and they are similar to the differences that can be observed when comparing the domestic dog to its ancestor, the wolf. Scientists conjecture that in both cases, tameness and associated changes in development, physiology, and anatomy were brought about by breeding for juvenile features.

GEORGIE

Georgie was Gordon Lark's first Portuguese water dog, a fiercely loyal and playful friend.

Does Bigger Mean Better?

Genome size is the total amount of DNA in one copy of an organism's genome, typically measured in millions of base pairs (Mb). There is a huge range in genome sizes across plants and animals, from living organisms of 150,000 Mb to the *E. coli* genome of 4.6 Mb. Dogs and humans have smaller genomes than many species, making them easier to study genetically than, say, a lungfish.

Assessment available in smartwork**5**

Lungfish
130,000 Mb

Red-Spotted Newt
39,100 Mb

White Spruce
23,600 Mb

Human
3,200 Mb

Dog
2,400 Mb

Chicken
1,200 Mb

Legend

▬▬ = 100 megabases

A megabase (Mb) is a unit of length for DNA equal to 1 million base pairs of nucleotides.

involved in an eye disease that causes blindness in both humans and dogs. Her lab also identified a gene involved in kidney cancer in dogs that causes a similar syndrome in humans.

Today, Lark and Ostrander both continue their canine research, and the Georgie Project lives on. In one investigation, the team is collecting data from tissues of deceased PWDs to determine the dogs' state of health at time of death. From data gleaned, they hope to learn how the health of aging dogs (and maybe people) is associated with their genetic makeup. "I'm really happy with the growth of the field," says Ostrander, "and there's a lot of research coming down the pike."

Mopsa, the puppy that Karen Miller gave Lark in 1996 in return for studying PWDs, died in April 2012, just a week short of her sixteenth birthday (**Figure 7.11**). But Lark has a new best friend, a PWD puppy he named Chou (pronounced "shoo"), for the French *petit chou*, meaning "little cabbage."

"We often make a mistake and call Chou 'Mopsa,' because Chou looks so similar," says Lark. After all, he adds, PWDs share similar genotypes and thus similar phenotypes. And these genetic traits make them the cuddly, devoted pets that they are.

Figure 7.11

Gordon Lark and Mopsa

REVIEWING THE SCIENCE

- A **genetic trait** is any characteristic that is inherited and may be **physical**, **biochemical**, or **behavioral**. All of these types of traits are either **invariant** (the same in all members of a species) or **variable** (different in different members of a species).

- A **gene** is a stretch of DNA that affects one or more genetic traits. Genes are formed on **chromosomes**, threadlike molecules made of DNA and proteins.

- The **genotype** is an individual's genetic makeup or, more specifically, the pair of different versions of a given gene (**alleles**) that determine a given trait. The **phenotype** is the physical expression of an individual's genetic makeup or, more specifically, the expression of a version of the given trait.

- An allele is **dominant** when it prevents a second allele from affecting the phenotype. This second allele is said to be **recessive** because it has no effect on the phenotype when paired with a dominant allele.

- When a genotype consists of two copies of the same allele, it is **homozygous** for that gene. A genotype that consists of two different alleles is **heterozygous** for that gene.

- A grid-like diagram called a **Punnett square** helps predict the probability of genotypes and phenotypes resulting from a **genetic cross**.

- Mendel's experiments enabled him to deduce two laws of inheritance: The **law of segregation** states that two alleles of a gene are separated during meiosis and end up in different gametes. The **law of independent assortment** states that during meiosis, the two alleles of any given gene segregate independently of any two alleles of any other gene.

- **Mendelian traits** are genetic traits controlled by a single gene and unaffected by environmental conditions.

- The **chromosome theory of inheritance** explains how Mendel's laws arise: genes occupy specific locations on chromosomes, and those chromosomes are randomly shuffled and recombined during meiosis.

- When a heterozygote displays an intermediate phenotype, neither allele is dominant—a condition known as **incomplete dominance**, in which neither allele is able to exert its full effect. When the effect of the two alleles is equally visible in the phenotype of the heterozygote, the pair of alleles shows **codominance**.

- **Complex traits** have patterns of inheritance that Mendel's laws cannot predict. In **pleiotropy**, for example, a single gene influences a number of different traits. **Polygenic traits**, by contrast, are governed by the action of more than one gene. And in **epistasis**, the phenotypic effect of the alleles of one gene depends on the presence of alleles for another, independently inherited gene.

THE QUESTIONS

The Basics

1 Link each term with the correct definition.

GENOTYPE

PHENOTYPE

HETEROZYGOTE

HOMOZYGOTE

DOMINANT

RECESSIVE

1. An individual that carries one copy each of two different alleles (for example, an *Aa* individual or an *IB* individual).
2. An individual that carries two copies of the same allele (for example, an *AA*, *aa*, or *II* individual).
3. An allele that does not affect the phenotype when paired with a dominant allele in a heterozygote.
4. The genetic makeup of an individual; more specifically, the two alleles of a given gene that affect a specific genetic trait in a given individual.
5. The specific version of a genetic trait that is displayed by a given individual.
6. The allele that controls the phenotype when paired with a different allele in a heterozygote individual.

2 Select the correct terms:
The (**gene / allele**) for coat color has two (**genes / alleles**)—one for brown coloring and one for black.

3 Select the correct terms:
Cells undergo (**mitosis / meiosis**) to become gametes. This process sorts the alleles of a gene into separate gametes, which is the basis for Mendel's law of (**segregation / independent assortment**). Genes on different chromosomes also sort into separate gametes during this process, which is the basis for Mendel's law of (**segregation / independent assortment**).

4 For each of the following cases, identify whether the described trait is an example of Mendelian inheritance (M) or a more complex form of inheritance (C).

_____ a. brown versus black coat color in dogs

_____ b. body size in dogs

_____ c. coat color in Siamese cats

_____ d. skin color in humans

_____ e. flower color in pea plants

5 A single phenotype that results from a combination of two different genes in which one gene interferes with the expression of another gene is known as

(a) pleiotropy.

(b) complete dominance.

(c) incomplete dominance.

(d) codominance.

(e) epistasis.

6 Before Mendel conducted his experiments with pea plants, people believed that offspring were a "blend" of their parents and would show intermediate levels of their parents' traits. What would Mendel's F_1 pea flowers have looked like if this were true?

(a) white

(b) purple

(c) red

(d) yellow

(e) light purple

Challenge Yourself

7 A riddle: In my type of inheritance, the F_1 offspring of a true-breeding black parent and a true-breeding white parent are all gray. What type of inheritance am I?

8 Some time ago, you noticed that the sunflowers in your garden were either tall or short—nothing in between. You bred the tall sunflowers for many generations, until you felt confident that they "bred true," and you did the same for the short sunflowers. You then set up a parental cross (P), and all of the resulting F_1 offspring were short sunflowers. From this experiment, you conclude that the short phenotype is

(a) pleiotropic.

(b) recessive.

(c) true-breeding in the F_1 generation.

(d) dominant.

(e) incompletely dominant.

9 In chickens, a mutant gene called *frizzle* causes not only feathers that curl outward like a Labradoodle's fur, but also an abnormal body temperature, an increased metabolism, and fewer eggs laid than by a normal chicken. From this information, you can conclude that the *frizzle* gene is _____.

Try Something New

10 The silver fox (see "The New Family Pet?" on page 127) belongs to the same species as the red fox: *Vulpes vulpes*. Two silver foxes always breed true for silver offspring. A silver fox bred to a red fox will produce either all red offspring or, occasionally, half red and half silver offspring. Red foxes bred together usually produce all red offspring, but they occasionally produce silver offspring in the ratio of 3 red to 1 silver. (*Hint:* Draw Punnett squares showing these predicted results.) Which of the following statements is/are consistent with the information provided here about inheritance of coat color in *Vulpes vulpes*? (Select all that apply.)

(a) Red foxes are all homozygous.

(b) Silver foxes are all homozygous.

(c) Red is dominant to silver.

(d) Some silver foxes are homozygous and some are heterozygous.

(e) Some red foxes are homozygous and some are heterozygous.

11 In your garden you grow Big Boy (round) and Roma (oval) tomatoes. You love the taste of Big Boys, but you think it's easier to slice Roma tomatoes. You decide to cross-pollinate a Big Boy and a Roma to see whether you can create a new strain of "Long Boys." In the first generation, all of the tomatoes are round. How would you explain this result? What would your next cross be? Write out the cross in a Punnett square, using parental genotypes. What proportion of the next generation, if any, would be oval?

12 For several hundred years, goldfish have been selectively bred in China and Japan for body color and shape, tail shape, bulging eyes, and even fleshy head growths.

Wild goldfish   Pet-shop goldfish   Black moor goldfish

Imagine that you have a tank of pet-shop goldfish and have just added a couple of black moor goldfish, hoping that they will breed. When the eggs laid by the black moor female (P generation) hatch and the young fish (F_1 generation) begin to develop, you are shocked to see that they are orange. How would you explain this result in terms of the inheritance of body color in goldfish? What breeding experiment could you conduct to test your hypothesis?

13 In 2009, a large team of researchers including Elaine Ostrander and Gordon Lark published the results of its research on coat inheritance in dogs. The study began by focusing on dachshunds and Portuguese water dogs, but then widened to more than 80 breeds. The scientists were able to explain 95 percent of the variation in dog coat types with just two alleles at each of three genes, each inherited independently of the other. These genes coded for hair length (L/l), wave or curl in the coat (W/w), and the presence of "furnishings" (F/f), which are the moustache and eyebrows often seen in wire-haired dogs (see photo). Long-haired dogs carry two copies of the long-hair allele, which is

recessive to the short-hair allele. Dogs with furnishings can be either homozygous or heterozygous for the furnishings allele; dogs without furnishings are homozygous for the no-furnishings allele.

a. At the hair length and furnishings genes, what is the genotype of a long-haired dog without furnishings?

b. At the hair length and furnishings genes, what are all the possible genotypes of a short-haired dog with furnishings?

c. Create a Punnett square of two dogs heterozygous for hair length and furnishings. What is the offspring phenotype ratio for those two traits?

Leveling Up

14 **Doing science** Do you want to get involved in dog research? If you have a purebred as a pet, you can. Find out whether the Dog Genome Project at the National Institutes of Health is doing research on your pet's breed. If they are, you can send in a swab of your dog's saliva and contribute to science. Visit the NIH website (http://research.nhgri.nih.gov/dog_genome) for more information.

15 **Is it science?** The November 18, 2003, issue of *Weekly World News* printed a story about a woman who, after repeatedly watching the movie *Shrek* while taking fertility drugs, gave birth to a baby who looked like the main character, an ogre named Shrek. Like Shrek, the newborn had dull green skin, a large flat nose, and ears protruding from stems. From what you know about genetics, do you think it's possible for a developing fetus to change so drastically (from a normal-looking baby to a "Shrek" baby) because its mother was obsessed with a movie? Why or why not? How would you explain your answer to someone who believed this news report?

16 **What do *you* think?** Many people are critical of those who breed or purchase purebred dogs, arguing that there are many mixed-breed dogs waiting to be adopted from shelters. They also point out that mixed-breed dogs are less likely than purebred dogs to suffer from genetic diseases. Those who prefer a particular breed argue that there is a strong genetic influence on dog personality and behavior, and that they don't want any surprises when they add a new member to their family. What do you think?

For more, visit digital.wwnorton.com/bionow2 for access to:

A Deadly Inheritance

How researchers identified a mysterious genetic disorder, and their risky effort to develop a cure.

After reading this chapter you should be able to:

- Interpret a human pedigree to determine whether a given condition is recessive, dominant, or sex-linked.

- Review a human karyotype to identify the sex chromosomes and any abnormalities in chromosome number.

- Diagram a chromosome, identifying genes, alleles, and loci.

- Explain how sex is genetically determined in humans, and how sex determination relates to the inheritance of sex-linked traits.

- Describe the genotype and phenotype of a "genetic carrier" and use this term appropriately for sex-linked traits only.

- Compare and contrast the inheritance of recessive, dominant, and sex-linked genetic disorders.

- Calculate the probability of inheriting a particular genetic disorder by using a Punnett square.

CHAPTER

08

CHROMOSOMES
AND HUMAN
GENETICS

F elix clutches his mother's side. Her arm is wrapped tightly around him. Lying in a rumpled white hospital bed, Felix looks away from the two red tubes protruding from his body. Blood pumps out one side of his body into a humming machine next to the bed and is pumped back into the other side. Dr. Christoph Klein steps into the room, clothed head to toe in blue scrubs, and offers a smile and a reassuring word (**Figure 8.1**).

Shortly after his birth in 2005, Felix began to bleed. His parents rushed him to an intensive care unit. The bleeding eventually stopped, but the hospital visits did not. Three years later, Felix was diagnosed with a rare and deadly disease: Wiskott-Aldrich syndrome (WAS). "Wiskott-Aldrich was the very last diagnosis I would ever want to receive," Felix's mother later said. "Every day I prayed, 'Please, Lord, let this chalice pass from us.'"

Felix's pediatrician feared the worst. WAS patients suffer from recurring infections, pneumonia, bleeding, and rashes; they often develop leukemia or lymphoma and die of complications due to infections. Some patients can be treated with a bone marrow transplant. However, if the donor is not a matched sibling or a close unrelated match, survival rates are low.

So, in 2009, Felix's family appeared in Klein's office at the Hannover Medical School in Germany, holding tightly to the hope that he might be able to save Felix's life. Klein, a pediatrician who has made a career of studying rare diseases, was running a clinical trial testing a new treatment for WAS. The treatment was risky and the results unknown, but it was Felix's only hope to be cured of his deadly disease—one that had perplexed scientists for decades.

A Mysterious Malady

More than 70 years earlier, in 1937, three young brothers had come to see Dr. Alfred Wiskott, also a German pediatrician. At first, Wiskott had no idea what was causing their symptoms. The boys bled abnormally: their blood was unable to clot, and they had bloody diarrhea. They also had recurring ear infections and blistering, weeping rashes on their skin. Wiskott recorded their symptoms, but he could not help the boys. All three died at a young age.

What had killed the brothers? Their parents also had four daughters, all of whom were healthy, so it was unlikely that an infection, a toxin, or an environmental factor had caused the illness. Instead, Wiskott suspected, the boys might have inherited a disease from their parents.

Thanks to the chromosome theory of inheritance, Wiskott knew that the boys had inherited hereditary material, in the form of chromosomes, from their parents. Recall from Chapter 7 that offspring inherit one chromosome from the mother and one from the father. Wiskott suspected that a gene on one of the inherited chromosomes was causing the mysterious illness.

A disease caused by an inherited mutation, passed down from a parent to a child, is a **genetic disorder**. Wiskott recognized the importance of researching genetic disorders, since such studies could lead to the prevention of or cure for a disease. But daunting problems have long plagued the study of human genetic disorders. From a biological point of view, humans have a long generation time, select their own mates, and decide whether and when to have children. In addition, human families tend to be much smaller than would be ideal for a scientific study. From an ethical point of view, geneticists and physicians cannot intervene and perform experiments directly on humans to determine how genetic disorders are inherited.

Figure 8.1

Felix and Dr. Klein
Dr. Christoph Klein examines Felix, who suffers from a rare genetic disease.

Painful Pedigree

Twenty years after Wiskott described the first cases of the mysterious disease and determined that it was inherited, an American pediatrician, Dr. Robert Anderson Aldrich, solved the next piece of the puzzle. Aldrich met a six-month-old boy with anemia, bloody diarrhea, and general weakness. After several emergency room visits, the baby died.

Aldrich sat down with the boy's mother to review possible causes. After an hour of asking questions, he still had no idea what might have brought on the illness. Finally, he asked about other relatives who might have had a similar illness. The boy's grandmother, who had tagged along to the meeting, exclaimed sadly, "Just like all the rest of them." Other male infants in the family, it turned out, had died under similar circumstances.

Given that information, Aldrich worked with the mother and grandmother to trace the family's history back six generations by drawing a **pedigree**, a chart similar to a family tree that shows genetic relationships among family

CHRISTOPH KLEIN

A medical doctor and cancer researcher, Christoph Klein is now chair of pediatrics at the University of Munich. In 2010, Klein began testing a new gene therapy to treat young children with Wiskott-Aldrich syndrome, a rare and life-threatening disease. The therapy, though still experimental, has been very successful.

members over two or more generations of a family's medical history (**Figure 8.2**). Pedigrees provide scientists with a way to analyze information in order to learn about the inheritance of a particular genetic trait or disorder. Aldrich found that 16 male infants in the family, but no females, had died of the syndrome (**Figure 8.3**).

Because of the "remarkable family history," as he described it, Aldrich concluded, like Wiskott, that the illness was a genetic disorder, caused by a mutation passed down from parent to child. Genetic disorders can be caused by mutations in

Generation

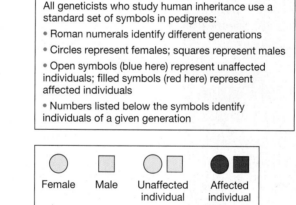

All geneticists who study human inheritance use a standard set of symbols in pedigrees:

- Roman numerals identify different generations
- Circles represent females; squares represent males
- Open symbols (blue here) represent unaffected individuals; filled symbols (red here) represent affected individuals
- Numbers listed below the symbols identify individuals of a given generation

Female   Male   Unaffected individual   Affected individual

Figure 8.2

Patterns of inheritance can be analyzed in family pedigrees

This cystic fibrosis pedigree shows six children (generation III), two of whom are affected with the disease.

Q1: Which two children in this pedigree have cystic fibrosis? How do you know?

Q2: Does either parent of these two children have cystic fibrosis? If so, which one(s)? How do you know?

Q3: Do any of the grandparents of these two children have cystic fibrosis? If so, which one(s)? How do you know?

Generation

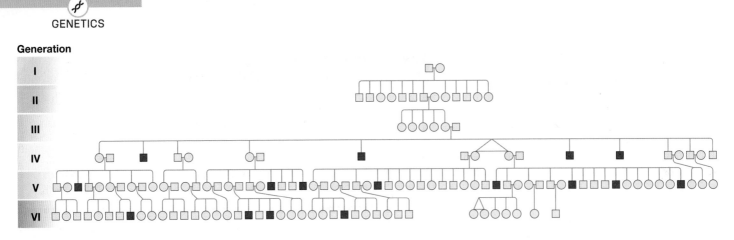

Figure 8.3

Pedigree of a family with a history of Wiskott-Aldrich syndrome

As in Figure 8.2, circles represent females and squares represent males in this pedigree. Individuals affected by WAS are shaded red.

Q1: How many male and how many female descendants (individuals that did not join the family by marriage) does generation IV of Aldrich's pedigree contain?

Q2: What proportions of the male and female descendants in generation IV were affected by the disorder?

Q3: Why did Aldrich hypothesize that the disease was X-linked? (You will need to read ahead to answer this question.)

Figure 8.4

A human karyotype

To research chromosomal abnormalities, a scientist will take a photograph of a cell's chromosomes during mitosis and then pair up each set of homologous chromosomes to create a **karyotype**. In humans, the autosomes are numbered 1 through 22, and the sex chromosomes are designated X or Y.

Q1: Is this the karyotype of a male or a female?

Q2: How would the karyotype of a person with Down syndrome differ from this karyotype?

Q3: The size of a chromosome correlates roughly with the number of genes residing on it. Why are an extra copy of chromosome 21 and a missing Y chromosome two of the least damaging chromosomal abnormalities?

individual genes or by abnormalities in chromosome number or structure.

Every species has a characteristic number of chromosomes: humans have 23 pairs of homologous chromosomes, for a total of 46, while mosquitoes, for example, have only 3 pairs, or 6 in total. One of the 23 pairs of human chromosomes consists of the **sex chromosomes**; they determine whether a person is male or female. All other chromosomes are called **autosomes**.

Autosomes are homologous chromosomes exactly alike in terms of length, shape, and the genes they carry (**Figure 8.4**). Human autosomes are labeled with the numbers 1 through 22 (for example, chromosome 4). Sex chromosomes are assigned letter names; in humans, males have one X chromosome and one Y

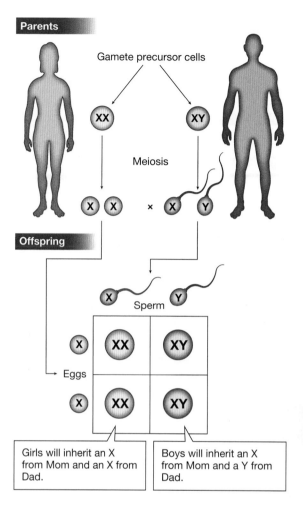

Parents

Gamete precursor cells

XX × XY

Meiosis

X X × X Y

Offspring

X Sperm Y

Eggs

	X	Y
X	XX	XY
X	XX	XY

Girls will inherit an X from Mom and an X from Dad.

Boys will inherit an X from Mom and a Y from Dad.

Figure 8.5

Dad's chromosomes determine baby's sex

Q1: What are the odds that a given egg cell will contain an X chromosome? A Y chromosome? What are those odds for a sperm cell?

Q2: If a couple has two daughters, does that mean their next two children are more likely to be sons? Explain your reasoning. (*Hint:* Refer back to "What Are the Odds?" on page 124.)

Q3: Sisters share the same X chromosome inherited from their father, but they may inherit different X chromosomes from their mother. What is the probability that brothers share the same Y chromosome? What is the probability that brothers share the same X chromosome?

ALFRED WISKOTT

A German pediatrician who lived from 1898 to 1978, Alfred Wiskott described the cases of three young brothers with a serious bleeding disease—what would later be called Wiskott-Aldrich syndrome. Wiskott deduced that the boys had inherited a gene from their parents that was causing the illness.

chromosome, whereas females have two X chromosomes. The Y chromosome in humans is much smaller than the X chromosome.

Because human females have two copies of the X chromosome, all the gametes (eggs) they produce contain one X chromosome, passed on to their offspring. Males, however, have one X chromosome and one Y chromosome, so half of their gametes (sperm) will contain an X chromosome and half will contain a Y chromosome (**Figure 8.5**).

Each chromosome has a particular structure, with genes arranged on it in a precise sequence. Any change in the chromosome number or structure, compared to what is typical for a species, is considered a **chromosomal abnormality**. The two most common types of chromosomal abnormalities in humans are changes in the overall number of chromosomes and changes in chromosome structure, such as a change in the length of an individual chromosome (**Figure 8.6**).

Changes in the number of chromosomes in humans are usually lethal, although Turner syndrome, in which females receive only one X chromosome, is an exception to this rule. Turner syndrome individuals tend to live long, healthy lives with mild to moderate reproductive issues. Similarly, Down syndrome individuals receive three copies of chromosome 21 and can live relatively long lives, but they still suffer from mild to moderate intellectual and developmental disabilities. Changes in chromosome structure can also have dramatic effects. For example, cri du chat syndrome, caused by a deletion on chromosome 5, results in slowed growth, a small head, and developmental delays.

Looking for Loci

In his 1954 paper describing the disease, Aldrich suggested that since primarily males inherit the syndrome, it is caused by a mutation on a sex chromosome. Aldrich recognized that the family's disease was tightly linked to gender and concluded that the disease was caused by

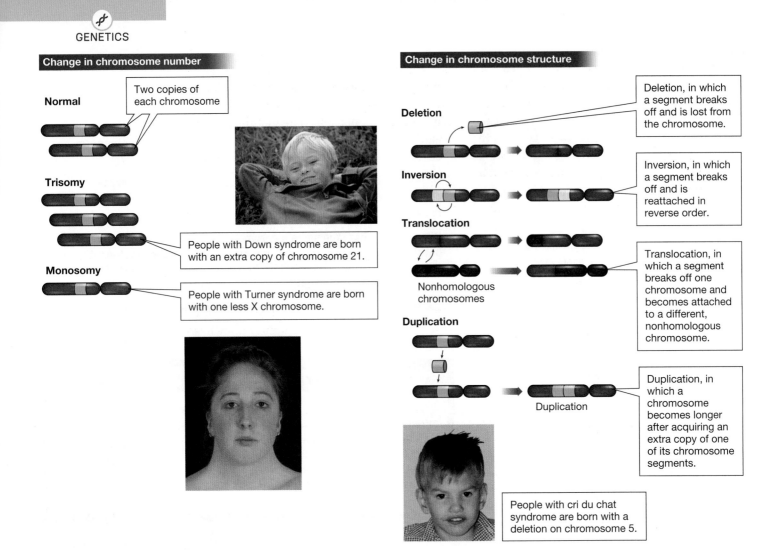

Change in chromosome number

Normal — Two copies of each chromosome

Trisomy — People with Down syndrome are born with an extra copy of chromosome 21.

Monosomy — People with Turner syndrome are born with one less X chromosome.

Change in chromosome structure

Deletion — Deletion, in which a segment breaks off and is lost from the chromosome.

Inversion — Inversion, in which a segment breaks off and is reattached in reverse order.

Translocation — Nonhomologous chromosomes — Translocation, in which a segment breaks off one chromosome and becomes attached to a different, nonhomologous chromosome.

Duplication — Duplication, in which a chromosome becomes longer after acquiring an extra copy of one of its chromosome segments.

People with cri du chat syndrome are born with a deletion on chromosome 5.

Figure 8.6

Chromosomal abnormalities can cause serious genetic disorders

Any increase or decrease in the number of chromosomes almost invariably results in spontaneous abortion of the fetus, which is estimated to occur in up to 20 percent of all pregnancies. Down syndrome and a missing or additional sex chromosome are exceptions. Changes in chromosome structure may have relatively minor or more severe effects, depending on the size and location of the change.

Q1: Why are changes in chromosome *number* almost always more severe than changes in chromosome structure?

Q2: In which part of meiosis would you predict that chromosomal abnormalities are produced? (Refer back to Chapter 6 if necessary.)

Q3: Create a mnemonic to help remember the four kinds of structural changes (for example, Doctors Improve Treatment Daily).

sex-linked inheritance—an inherited mutation on the X or Y chromosome. But where exactly was the mutation located?

The physical location of a gene on a chromosome is called its **locus** (plural "loci"). Because a gene can occur in different versions, or alleles, a diploid cell can have two different alleles at a given locus on a pair of homologous chromosomes. If the two alleles at a locus are different, the cell is heterozygous for the gene. If the two alleles at a locus are identical, the cell is homozygous for the gene (**Figure 8.7**).

The sex chromosomes, however, are different. Roughly 1,240 of the estimated 20,000 human genes are found on the X and Y chromosomes. Approximately 1,180 of those 1,240 genes are located on the X chromosome, while only about 60 are located on the much smaller Y chromosome. These 1,240 genes are said to be **sex-linked**. Sex-linked genes on the X chromosome are **X-linked**. Sex-linked genes on the Y chromosome are **Y-linked**. One of these Y-linked genes is the *SRY* gene (short for "sex-determining region of Y"). *SRY* functions as the "master sex switch," committing the sex of the developing embryo to male. In the absence of this gene, a human embryo develops as a female.

The *SRY* gene does not act alone; in both males and females, other genes on the autosomes and sex chromosomes directly influence the development of the sexual characteristics that distinguish men and women. Still, the *SRY* gene plays a crucial role because when present, it causes other genes to produce male sexual characteristics, but when absent, those genes produce female sexual characteristics. Individuals that have XY chromosomal makeup but a nonfunctional SRY gene are considered intersex; they are not reproductively functional males or females, although they appear phenotypically female (**Figure 8.8**).

Interestingly, there are no well-documented cases of disease-causing Y-linked genes. X chromosomes, however, contain genes known to be involved in many human genetic traits and

Figure 8.7

Genetic loci on homologous chromosomes

The genes shown here take up a larger portion of the chromosome than they would if they were drawn to scale. The average human chromosome has more than a thousand different genes interspersed with large stretches of noncoding DNA.

Q1: How do we know whether two chromosomes are homologous?

Q2: In one sentence, explain how the terms "gene," "locus," and "chromosome" are related.

Q3: If hair color were determined by a single gene, what would be an example of the gene's alleles?

Figure 8.8

Advocates for intersex youth: interACT

Founded by attorney Anne Tamar-Mattis in 2006 under the name "Advocates for Informed Choice," interACT was created with a focused mission of ending harmful medical interventions on intersex children. Its mission statement reads "interACT uses innovative legal and other strategies, to advocate for the human rights of children born with intersex traits."

disorders. Aldrich correctly concluded that the gene leading to the family's bleeding disease was located on the X chromosome, so it was X-linked.

X Marks the Spot

Thanks to advances in molecular biology tools, in 1994, researchers determined that the gene causing WAS is on the X chromosome and called it, unsurprisingly, *WAS* (gene names are typically italicized, while disorder names, in this case "WAS," are not). *WAS*, they discovered, is the genetic code for a protein crucial for the formation and function of blood cells and immune system cells. Without a healthy copy of this gene, individuals acquire blood and immune system disorders, are susceptible to infections, and have increased risk of lymphoma, cancer of the lymph nodes.

Felix inherited a mutated version of *WAS* from his mother and no allele from his father. We can use a Punnett square to illustrate how the X-linked recessive mutation for WAS is inherited. We label the recessive mutated *WAS* allele *a*, and in the Punnett square we write this allele as X^a to emphasize the fact that it is on the X chromosome. Then we label the dominant, healthy allele *A* and write this allele as X^A in the Punnett square (**Figure 8.9**). Individuals like Felix's mother, who have only one copy of a recessive allele, are said to be **genetic carriers** of the disorder. Carriers can pass on the disorder allele, but they do not have the disease.

If a carrier female like Felix's mother, with genotype X^AX^a, has children with a normal male (with genotype X^AY), each of their sons will have a 50 percent chance of getting the disorder. Felix had a 50 percent chance of getting WAS, and he did.

Males of genotype X^aY suffer from the condition because the Y chromosome does not have a

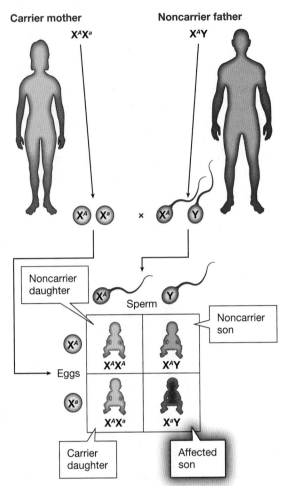

Figure 8.9

X-linked recessive conditions are more common in males

The recessive disorder allele (*a*) is located on the X chromosome and is denoted by X^a. The dominant normal allele (*A*) on the X chromosome is denoted by X^A.

Q1: Which of the children specified in this Punnett square represents Felix? What is his genotype?

Q2: Explain why Felix is neither homozygous nor heterozygous for the *WAS* gene.

Q3: Create a Punnett square to illustrate the offspring that could result if Felix had children with a noncarrier woman. What is the probability that a son would have WAS? What is the probability that a daughter would be a carrier of WAS?

ROBERT ALDRICH

Robert Aldrich was an American pediatrician who lived from 1917 to 1998. By creating a family pedigree, Aldrich demonstrated that the mysterious bleeding syndrome afflicting infant boys was a sex-linked, recessive disorder.

Genetic Diseases Affecting Americans

Wiskott-Aldrich syndrome is only one of many genetic conditions seen in humans. Here are some of the most common genetic diseases in the United States, most of which can be identified in newborns through genetic testing.

U.S. births per year: 4,000,000

• = 1 birth

6,037 births
Down Syndrome

1,140 births
Cystic Fibrosis

800 births
Marfan Syndrome

Male U.S. births per year: 2,050,000

• = 1 male birth

590 births
Duchenne Muscular Dystrophy

400 births
Hemophilia

400 births
Fragile X Syndrome

Gaucher disease
Chronic enzyme deficiency; fairly common among Ashkenazi Jews

Familial colon cancer
One in 200 people has this allele; of those who have it, 65% are likely to develop the disease

Retinitis pigmentosa
Progressive degeneration of the retina

ALD (adrenoleukodystrophy)
Nerve disease portrayed in movie *Lorenzo's Oil*

Neurofibromatosis, type 2
Tumors of the auditory nerve and tissues surrounding the brain

Amyotrophic lateral sclerosis (ALS) (Lou Gehrig disease)
Fatal degenerative nerve ailment

Adenosine deaminase (ADA) immune deficiency
Metabolic disorder that damages the immune system

Familial hypercholesterolemia
Extremely high cholesterol

Amyloidosis
Accumulation in the tissues of an insoluble fibrillar protein

Breast cancer
Roughly 5% of cases are caused by this allele

Polycystic kidney disease
Cysts resulting in enlarged kidneys and renal failure

Tay-Sachs disease
Fatal hereditary disorder involving lipid metabolism; most common in Ashkenazi Jews and French Canadians

Burkitt lymphoma
A translocation between chromosomes 14 and 18 results in cancer of the white blood cells; most common in children and young adults

Huntington disease
Neurodegenerative disorder tending to strike people in their forties and fifties

Familial polyposis of the colon
Abnormal tissue growths frequently leading to cancer

Spinocerebellar ataxia
Destroys nerves in the brain and spinal cord, resulting in loss of muscle control

Cystic fibrosis
Mucus fills up the lungs, interfering with breathing; one of the most prevalent genetic diseases in the U.S.

Multiple exostoses
A disorder of cartilage and bone

Malignant melanoma
Tumors originating in the skin

Multiple endocrine neoplasia, type 2
Tumors in endocrine glands and other tissues

Sickle-cell disease
Chronic inherited anemia in which red blood cells sickle (form crescents), plugging small blood vessels

Retinoblastoma
Relatively common eye tumor, accounting for 2% of childhood malignancies

PKU (phenylketonuria)
An inborn error of metabolism that results in mental retardation if untreated

Figure 8.10

Single-gene disorders

Mutations of single genes that lead to genetic disorders are found on the X chromosome and on each of the 22 autosomes in humans. In each of these mutations, the healthy allele at that locus codes for an important function; for example, the sickle-cell allele is a mutation in the gene that codes for the hemoglobin protein, critical for carrying oxygen in the blood. For clarity, only one such genetic disorder per chromosome is shown.

Q1: Which chromosome contains the gene for cystic fibrosis? For Tay-Sachs disease? For sickle-cell disease?

Q2: No known genetic disorders are encoded on the Y chromosome. Why do you think this is?

Q3: In your own words, explain why most single-gene disorders are recessive rather than dominant.

copy of that gene. In other words, because males cannot be heterozygous for any X-linked genes, the effects of an *a* allele cannot be masked. In general, males are more likely than females to get recessive X-linked disorders, because they need to inherit only a single copy of the disorder allele to exhibit the disorder. Females, on the other hand, must inherit two copies to be affected. X-linked recessive inheritance explains why boys are more likely to get WAS than girls are.

Other X-linked genetic disorders in humans include red-green color blindness, hemophilia, and Duchenne muscular dystrophy—a lethal disorder that causes muscles to waste away, often leading to death at a young age. All of these X-linked disorders are caused by recessive alleles.

More Common, but No Less Deadly: Zoe's Story

X-linked disorders like WAS are rare compared to autosomal disorders. Both sexes are equally likely to be affected by **recessive genetic disorders** on autosomes, since both males and females have two copies of autosomal chromosomes and therefore identical odds of being homozygous or heterozygous for a disorder allele. Several thousand human genetic disorders are inherited as recessive traits on autosomes. These include sickle-cell disease, Tay-Sachs disease, and the most common fatal genetic disease in the United States, cystic fibrosis (**Figure 8.10**).

Scott and Jada first began to suspect something was wrong when their newborn daughter, Zoe, didn't put on any weight. Every time Zoe ate, her belly became hard and bloated. She screamed in pain. "I was beside myself," recalls Jada. "I knew something wasn't right."

Like Felix, Zoe spent the first year of her life in and out of the pediatrician's office, until the eve of her first birthday, April 6, 2005. On that day, Scott and Jada sat with Zoe in a children's hospital in Florida, waiting for a second opinion. The doctor came in and asked them to sit down. The diagnosis was cystic fibrosis (**Figure 8.11**).

Cystic fibrosis (CF) is a lethal recessive genetic disorder caused by one or more mutations in the

Figure 8.11

Cystic fibrosis damages lung tissue
The panel on the left shows a chest x-ray from a cystic fibrosis patient and the panel on the right shows a normal chest x-ray. Notice the damaged fibrotic lung tissue in the cystic fibrosis x-ray.

cystic fibrosis transmembrane regulator gene (*CFTR*). A *CFTR* mutation causes the body to produce abnormally thick, sticky mucus, which clogs the airways and leads to lung infections. The thick mucus also obstructs the pancreas, preventing enzymes from reaching the intestines, where they are needed to break down and digest food. The average life span of people with CF who live to adulthood is approximately 35 years. There is no cure.

Recessive genetic disorders vary in severity; some, like cystic fibrosis, are lethal, whereas others have relatively mild effects. Adult-onset lactose intolerance, for example, is caused by a single recessive allele that leads to a shutdown in the production of lactase, the enzyme that digests milk sugar.

The only individuals who get a disorder caused by an autosomal recessive allele (*a*) are those who have two copies of that allele (*aa*). Usually, when a child inherits a recessive genetic disorder, both parents are heterozygous; that is, they both have the genotype *Aa* (**Figure 8.12**). It is also possible for one or both parents to have the genotype *aa* and thus the disease. Because the *A* allele is dominant and does not cause the disorder, heterozygous individuals (*Aa*) like Scott and Jada, Zoe's

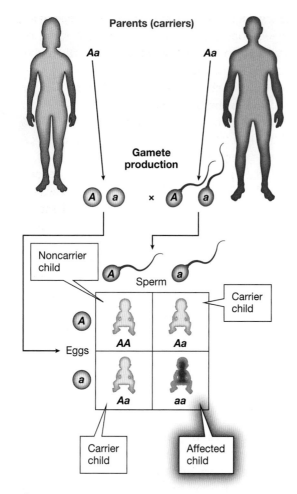

Parents (carriers)

Aa   *Aa*

Gamete production

A a × A a

Noncarrier child

A   a
Sperm

Carrier child

Eggs

A | AA | Aa
a | Aa | aa

Carrier child

Affected child

Figure 8.12

Inheritance of cystic fibrosis, an autosomal recessive disorder

The patterns of inheritance for a human autosomal recessive genetic disorder are the same as for any recessive trait (compare this figure with the pattern shown by Mendel's pea plants in Figure 7.5). Recessive disorder alleles are denoted *a*. Dominant, normal alleles are denoted *A*. Here, the parents are a carrier female (genotype *Aa*) and a carrier male (genotype *Aa*). ▶️

Q1: Which of the children in this Punnett square represents Zoe? What is her genotype?

Q2: If Zoe's parents had another child, what is the probability that the child would have cystic fibrosis? That the child would be a CF carrier?

Q3: If Zoe is able to have a child of her own someday, and the other parent is not a carrier of cystic fibrosis (he would likely be tested before they chose to have children), what is the probability that the child would have cystic fibrosis? That the child would be a carrier?

parents, are genetic carriers of the disorder; they carry the disorder allele (*a*) but do not have the disease.

If two carriers of a recessive genetic disorder have children, the patterns of inheritance are the same as for any other recessive trait. Each child, male or female, has a 25 percent chance of not carrying the disorder allele (genotype *AA*), a 50 percent chance of being a carrier (genotype *Aa*), and a 25 percent chance of inheriting the disorder (genotype *aa*). Zoe did not beat the odds.

These percentages reveal one way in which lethal recessive disorders such as cystic fibrosis can persist in the human population. Although homozygous recessive individuals (with genotype *aa*) often die before they are old enough to have children, carriers (with genotype *Aa*) are not harmed by the disorder. In a sense, the *a* alleles can hide in heterozygous carriers, and those carriers are likely to pass the disorder allele to half of their children. An estimated one in 29 European Americans has a mutated *CFTR* gene. Recessive genetic disorders can also arise in the human population because new mutations can produce new copies of the recessive alleles.

Deadly with One Allele

Cystic fibrosis is an example of a recessive genetic disorder, in which a child, like Zoe, inherits two recessive copies of a disorder allele. A more rare type of inherited disease is a **dominant genetic disorder**, caused by an autosomal dominant allele (*A*). In this case, the allele that causes a disorder cannot "hide" in the same way that a recessive allele can: *AA* and *Aa* individuals get the disorder; only *aa* individuals are symptom-free (**Figure 8.13**). Dominant genetic disorders are more rare than recessive disorders because a dominant disorder often produces serious negative effects immediately upon birth, and individuals with the *A* allele may not live long enough to reproduce. Hence, few people with a dominant genetic disorder pass the allele on to their children.

For this reason, most cases of a dominant genetic disorder are produced by a new mutation in a generation. For example, achondroplasia, a form of dwarfism, is caused by a mutation

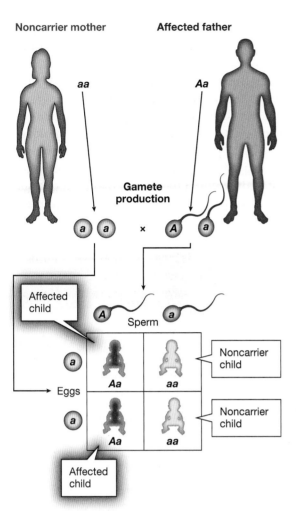

Noncarrier mother

aa

Affected father

Aa

Gamete production

a a × A a

Affected child

A Sperm a

Eggs

a — Aa (Affected child), aa → **Noncarrier child**

a — Aa (Affected child), aa → **Noncarrier child**

Figure 8.13

Inheritance of an autosomal dominant disorder

The pattern of inheritance for a human autosomal dominant genetic disorder is the same as for any other dominant trait. This Punnett square shows the possible children of a normal female (genotype *aa*) and an affected male (genotype *Aa*).

Q1: What is the probability that a child with one parent who has an autosomal dominant disorder will inherit the disease?

Q2: Why are there no carriers with a dominant genetic disorder?

Q3: Because dominant genetic disorders are rare, it is extremely rare for both parents to have the condition (genotype *Aa*). Draw a Punnett square with two *Aa* parents. What proportion of the offspring would have the disorder? What proportion would be normal?

in a gene involved in bone growth. People with achondroplasia have a decreased life span, so few of them live long enough to pass the mutation on to their offspring. Instead, infants with achondroplasia are born to unaffected parents at a rate of between one in 10,000 and one in 100,000. Almost all of these births are to fathers older than 35 years, who produce this mutation during sperm production.

Huntington disease, a dominant genetic disorder, is an exception to that rule because symptoms of the disease—uncontrolled movements and loss of intellectual faculties caused by dying brain cells—arise later in life, in one's forties, after the person carrying that allele has had the opportunity to reproduce. In this way the allele is readily passed from one generation to the next. This is why some couples whose families have a history of Huntington's may choose to screen their developing fetus for the gene that causes the disorder.

Replacing Deadly Genes: A Work in Progress

Most inherited genetic disorders, including cystic fibrosis and Huntington disease, have no cure. Patients and their families do everything they can simply to manage the disease. At the age of 7, Zoe was taking 25 pills each day just to digest food. At the age of 10, she underwent four extended hospitalizations, and almost her entire summer was spent on drip IVs. Zoe continues to take antibiotics and mucus thinners, uses nasal sprays, and spends 2 hours a day doing breathing therapy. "We don't talk about the future," says Jada, Zoe's mother. Her voice cracks. "But she is aware that she's not like everybody else."

Scientists have not given up the race to find effective treatments, even cures, for genetic disorders. Thanks to Christoph Klein, WAS is one of the few such disorders for which an effective treatment has been identified.

It was in 2003 that Klein, then at Boston Children's Hospital, saw the first glimmer of hope that there might be a therapy for boys afflicted with WAS. He and colleagues believed that they might be able to treat WAS using **gene therapy**, a technique for correcting defective genes responsible for disease development. Gene therapy is a type of **genetic engineering**, the permanent introduction of one or more genes into a cell, tissue, or organism. Klein hoped that by correcting the defective *WAS* gene in young affected boys, he could offer a short-term treatment, and potentially even a permanent cure, for the disorder.

First, Klein and his team tested their plan in mice. They bred mice that lacked the entire *WAS* gene and thus could not produce the important WAS protein. These mice had some of the same symptoms that Felix had, including a reduction in the number of blood and immune system cells. Then the researchers used a virus to insert a healthy copy of the *WAS* gene into the mice's blood cells, where they hoped it would produce the WAS protein. Months later, Klein was thrilled to find that the mouse cells were expressing healthy WAS protein, and that mature blood and stem cells were being produced in normal quantities.

Prenatal Genetic Screening

How is the baby? This is one of the first questions we ask after a child is born. Usually everything is fine, but sometimes, as with Felix and Zoe, the answer can be devastating. Today, some parents choose to have prenatal genetic screening tests performed to check their baby's health before birth.

In **amniocentesis**, a needle is inserted through the abdomen into the uterus to extract a small amount of amniotic fluid from the pregnancy sac that surrounds the fetus. This fluid contains fetal cells (often sloughed-off skin cells) that can be tested for genetic disorders. Another method is **chorionic** (kohr-ee-AH-nik) **villus sampling** (**CVS**), in which a physician uses ultrasound to guide a narrow, flexible tube through a woman's vagina and into her uterus, where the tip of the tube is placed next to the villi, a cluster of cells that attaches the pregnancy sac to the wall of the uterus. Cells are removed from the villi by gentle suction and then tested for genetic disorders.

Risks associated with amniocentesis and CVS—including vaginal cramping, miscarriage, and premature birth—have declined dramatically in recent years because of advances in technology and more extensive training. Recent studies suggest that the risk of miscarriage after CVS and amniocentesis is essentially the same: about 0.06 percent. The tests are widely used by parents who know they face an increased chance of giving birth to a baby with a genetic disorder. Older parents, for example, might want to test for Down syndrome, since the risk of that condition increases with the age of the mother and perhaps the father too. A couple in which one parent carries an allele for a dominant genetic disorder (such as Huntington disease), or both parents are carriers for a recessive genetic disorder (such as cystic fibrosis), might also choose prenatal genetic screening.

Couples who elect to have such tests performed have only two choices if their fears are confirmed: they can abort the fetus, or they can give birth to a child with a genetic disorder. Prior to conception, however, couples at risk of having a child with a genetic disorder have options to minimize that risk.

- If they are willing and can afford the procedure, a couple can choose to have a child by **in vitro fertilization** (**IVF**), in which an egg is fertilized by a sperm in a petri dish, after which one or more embryos are implanted into the mother's uterus.

- During **preimplantation genetic diagnosis** (**PGD**), one or two cells are removed from the developing embryo in the dish, usually 3 days after fertilization. The cell or cells removed from the embryo are tested for genetic disorders. Finally, one or more embryos that are free of disorders are implanted into the mother's uterus, and the rest of the embryos, including those with genetic disorders, are frozen. Typically, parents who opt for PGD either have a serious genetic disorder or carry alleles for one.

Like all other genetic screening methods, the use of PGD raises ethical issues. People who support PGD feel that amniocentesis and CVS provide parents with a bleak set of moral choices: if the fetus has a serious genetic disorder, the parents can either abort the fetus or allow the birth of a child who will live a life that may be short and full of suffering. In their view, discarding an embryo at the 4- to 12-cell stage is morally preferable to aborting a well-developed fetus, or to giving birth to a child that will suffer the devastating effects of a serious genetic disorder. Those opposed to PGD agree that the moral choices are bleak, but they argue that once fertilization has occurred, a new life has formed and it is immoral to end that life, even at the 4- to 12-cell stage. What do you think?

Bolstered by these results, the scientists tested the technique in human cells and finally, in 2005, began a human clinical trial in Germany. The first two boys, both 3 years old, were admitted in 2006. Between 2006 and 2009, 10 boys were admitted to the clinical trial, with Felix coming on board in 2009.

During the first phase of the trial, Felix had to lie still for 9 hours as a machine pumped blood out of his body in order to extract cells. These cells were then taken to a laboratory where scientists used a genetically modified virus to insert a healthy copy of *WAS* into them (**Figure 8.14**). The doctors then pumped the

genetically engineered cells back into Felix's body. And the waiting began.

Klein published the early results in 2010. After gene therapy, nine of the boys improved with regard to bleeding and infections. They were able to play soccer like their healthy peers, they responded to vaccinations, and they did not develop severe infectious diseases. Klein and his colleagues showed that gene therapy for the fatal WAS diagnosis is feasible, and that WAS can be corrected.

A new therapy does not come without risks, though. A few years after receiving the gene therapy, seven of the patients developed leukemia

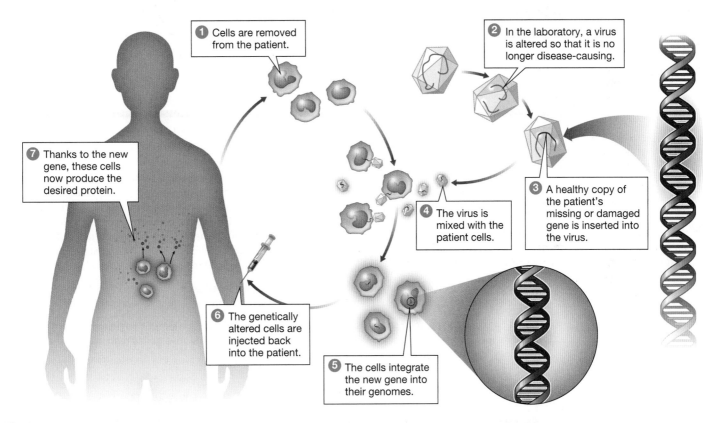

Figure 8.14

Gene therapy

In gene therapy, genetic information is transferred into cells to achieve a desired effect. Gene therapy may be used to compensate for a genetic mutation in a cell that causes the cell to malfunction. ▶▤

Q1: Which gene was missing or damaged in Felix's case? From what chromosome would a healthy copy be taken?

Q2: Why did Dr. Klein's group first conduct gene therapy on mice rather than on humans? What are the advantages and limitations of this approach?

Q3: If Felix has children of his own someday, will they run the risk of inheriting his disorder, or has gene therapy removed that possibility? Explain your reasoning.

from the type of virus used to insert the healthy gene into their cells. Three of the boys died, either from leukemia or from complications after bone marrow transplants. For this reason the trial was halted. Together with other physicians and researchers, Klein is investigating the causes of leukemia after gene therapy, and he is looking for safer ways to insert the healthy gene into cells.

A Happy Ending for Felix

Numerous gene therapy trials continue around the world for other genetic diseases, using other methods to insert new genes as a way to avoid risks like leukemia. In 2012, European health officials green-lit gene therapy for a recessive genetic disorder called lipoprotein lipase deficiency (LPLD), the first commercially approved gene therapy treatment in either Europe or the United States. Many others nip at its heels, including treatments for hemophilia and ADA-SCID, also known as the "bubble boy disease."

As for Felix, a year after the gene therapy, he returned to the hospital for tests. The treatment had worked. Today, his body is producing healthy, functional blood cells, and most of his symptoms are completely resolved. Felix now enjoys a normal life. His inherited genetic disorder was cured by gene therapy. "The most beautiful moment is when my son smiles, hugs me, and says, 'Mum, I love you so much,'" said Felix's mother after the treatment. "We are grateful that there was a treatment that helped our son."

REVIEWING THE SCIENCE

- A **genetic disorder** is a disease caused by an inherited mutation in a gene, passed down from a parent to a child. Genetic disorders can be caused by mutations in individual genes or by **chromosomal abnormalities** (changes in chromosome number or structure).

- Every person has two **sex chromosomes**: males have one X and one Y chromosome, and females have two X chromosomes. The *SRY* gene on the Y chromosome is required for human embryos to develop as males.

- The physical location of a gene on a chromosome is called its **locus**. Genes found solely on the X or Y chromosome are said to be **sex-linked**.

- A **dominant genetic disorder** is caused by a dominant allele on an **autosome** (non–sex chromosome). Dominant genetic disorders are more rare than **recessive genetic disorders**, which are caused by two autosomal recessive alleles. Several thousand human genetic disorders are inherited as recessive traits on autosomes.

- A family **pedigree** can be used to determine whether a given condition is recessive, dominant, or sex-linked.

- **Gene therapy** is a **genetic engineering** technique for correcting defective genes responsible for disease development.

THE QUESTIONS

The Basics

1 Use these terms correctly in the following sentence: **alleles, chromosomes, genes, loci.** Two homologous _____ contain the same _____, found at the same _____, but may have the same or different copies of _____.

2 A particular person is said to be a carrier of a genetic trait. What does this tell you about their phenotype?

(a) They physically show the trait.

(b) They physically show the trait more than a noncarrier would show the trait.

(c) They are almost normal but show an intermediate phenotype for the trait.

(d) They are completely normal and do not physically show the trait.

3 Match each of the following terms with the correct definition.

GENE THERAPY

IN VITRO FERTILIZATION

PREIMPLANTATION GENETIC DIAGNOSIS (PGD)

CHORIONIC VILLUS SAMPLING (CVS)

AMNIOCENTESIS

1. A procedure in which cells are gently suctioned from a pregnant woman's uterus to test for genetic disorders in the fetus.

2. A procedure in which a small amount of fluid (and fetal cells within it) is carefully extracted from a pregnant woman's uterus to test for genetic disorders in the fetus.

3. A treatment approach that seeks to correct a genetic disorder by inserting healthy copies of the mutated genes responsible for the disorder.

4. A procedure in which one or two cells are removed from a developing embryo and tested for genetic disorders; embryos that are free of genetic disorders may then be implanted into a woman's uterus.

5. A procedure in which an egg is fertilized in a petri dish, after which one or more embryos are implanted into a woman's uterus.

4 In the karyotype shown here, identify the sex chromosomes. Is this individual a male or a female?

Challenge Yourself

5 Sickle-cell disease is inherited as a recessive genetic disorder in humans; the normal hemoglobin allele (H) is dominant to the sickle-cell allele (h). For two parents of genotype Hh (carriers), construct a Punnett square to show the possible genotypes of their children.

6 Which of the following correctly predicts the possible genotypes of the children from question 5?

(a) ¼ HH, ¼ Hh, ½ hh

(b) ½ HH, ½ Hh

(c) ½ HH, ½ hh

(d) ¼ HH, ½ Hh, ¼ hh

(e) ¼ HH, ¾ Hh

7 Which of the following correctly predicts the possible phenotypes of the children from question 5?

(a) ¼ normal, ¾ sickle-cell disease

(b) all normal

(c) all sickle-cell disease

(d) ½ normal, ½ sickle-cell disease

(e) ¾ normal, ¼ sickle-cell disease

8 Given that the normal hemoglobin allele (H) is dominant to the sickle-cell allele (h), each time two Hh individuals have a child together, what is the chance that the child will have sickle-cell disease?

(a) 0%

(b) 75%

(c) 25%

(d) 50%

(e) 100%

9 Which of the following karyotypic sex chromosome abnormalities result(s) in a male phenotype? (Select all that apply.)

(a) XO

(b) XXY

(c) XXX

(d) XYY

(e) XXXY

(f) XY with a complete deletion of the *SRY* gene

Try Something New

10 On the Indonesian island of Bali, about ¾ of the feral (stray) cats have a stumpy tail, while only ¼ of the cats have a long tail. Experiments in which a bunch of random stumpy-tailed cats were mated with each other yielded some stumpy-tailed kittens and some long-tailed kittens. The same experiments with long-tailed cats produced only long-tailed kittens. What do these results show about the genotype of the long-tailed cats? (*Hint:* Draw Punnett squares using the letter "t" for the tail trait to help you determine which allele is dominant and which is recessive.)

(a) They must all be *tt*.

(b) They are both *Tt* and *tt*.

(c) They are both *Tt* and *TT*.

(d) They must all be *Tt*.

11 Recall that human females have two X chromosomes, while human males have one X chromosome and one Y chromosome.

a. Do males inherit their X chromosome from their mother or from their father?

b. If a female has one copy of an X-linked recessive allele for a genetic disorder, does she have the disorder?

c. If a male has one copy of an X-linked recessive allele for a genetic disorder, does he have the disorder?

d. Assume that a female is a carrier of an X-linked recessive disorder. With respect to the disorder allele, how many types of gametes can she produce?

e. Assume that a male with an X-linked recessive genetic disorder has children with a female who does not carry the disorder allele. Could any of their sons have the genetic disorder? How about their daughters? Could any of their children be carriers for the disorder? If so, which sex(es) could they be?

12 For the pedigree shown here, the disorder is caused by a recessive (*g*) allele on the X chromosome. Label each of the following individuals with the correct genotype (X^GY, X^gY, X^GX^G, X^GX^g, X^gX^g).

Generation

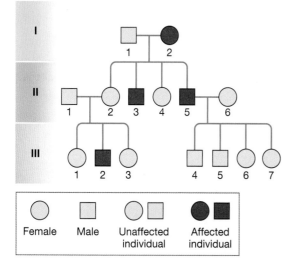

a. Generation I, number 2

b. Generation II, number 2

c. Generation II, number 5

d. Generation III, number 2

e. Generation III, number 6

13 For the pedigree shown here, the disorder is caused by a recessive (*g*) autosomal allele. Label each of the following individuals with the correct genotype (*GG*, *Gg*, *gg*).

Generation

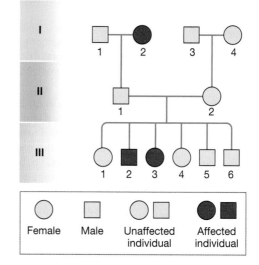

a. Generation I, number 2

b. Generation II, number 1

c. Generation II, number 2

d. Generation III, number 3

Leveling Up

14 *Write Now* biology: debating preimplantation genetic diagnosis (PGD) This assignment is designed to give you the experience of applying your knowledge of biology to a current controversy or topic of interest. You will apply the same sort of reasoning that you should be able to use as an informed citizen and consumer when making decisions that involve biology.

The scenario: The U.S. Senate Committee on Commerce, Science and Transportation (CCST) is considering proposing legislation on preimplantation genetic diagnosis (PGD; see "Prenatal Genetic Screening" on page 148). The chair of the committee has invited special-interest groups to present testimony on the pros and cons of PGD. You will contribute a position paper defining PGD, describing your group's position on the technology, and making a recommendation for legislation.

Choose one of the following special-interest groups to represent (or your instructor will assign one to you):

(a) **Reproductive Specialist Group (medical doctors).** You will describe how PGD and in vitro fertilization (IVF) can be used to screen for genetic disorders or for sex selection and "family balancing," and to increase fertility in women of advanced maternal age. You will argue that legislation is not necessary, since medical association guidelines are already in place, and that it is not the place of government to judge individuals' reasons for undergoing PGD.

(b) **Genetics and Public Policy Group.** You will present data on the beliefs of the American public in this matter. You will also present the status of current legislation in the United States and how it compares to PGD legislation internationally. You will propose limited legislation based on these findings.

(c) **Parents of Down Syndrome Children.** You will argue that people with disabilities already suffer from prejudice, and that widespread use of such testing will cause even more prejudice. You will propose legislation prohibiting PGD for almost all conditions, with the exception of deadly, infant-onset disease.

(d) **Americans with Cystic Fibrosis.** You will present the case of a couple, both of whom are carriers for CF, and discuss their options. You will propose limited legislation that disallows PGD for sex selection and nondisease conditions.

15 **What do *you* think?** *Gattaca*, a science fiction movie from 1997, is based in a future in which preimplantation genetic diagnosis (PGD) is used to create genetically superior children, and social status is determined mainly by one's DNA. Watch the movie and take careful notes as you observe how PGD has been misused by society. Do you think these outcomes are likely, if we continue to develop PGD? Are the benefits worth this risk?

16 **Life choices** While you cannot control the genes that you inherited from your parents, you can control your exposure to external factors that increase your chances of developing a disease that is only partially genetically determined. For this assignment, you will estimate your life expectancy, as determined mainly from your lifestyle choices. Go to the "Living to 100" website (http://www.livingto100.com) and complete the Life Expectancy Calculator. Answer all questions honestly (no one will see your answers), and go to the "feedback" page at the end of the questionnaire. Take notes so that you can answer the following questions.

a. What is your calculated life expectancy, given the answers you gave?

b. List the recommendations to increase your life expectancy, and how much they could increase your life span.

c. What change would increase your life span the most?

d. Identify one way that you could feasibly increase your life span. Will you do that? Why or why not?

For more, visit digital.wwnorton.com/bionow2 for access to:

Pigs to the Rescue

Using CRISPR, a hot new genome-editing tool, scientists hope to create a steady stream of transplant organs—from pigs.

After reading this chapter you should be able to:

- Describe the structure of DNA, using appropriate terminology.
- Use the base-pairing rules to determine a complementary strand of DNA based on a given template strand.
- Describe how the genome editing tool CRISPR-Cas9 works.
- Label a diagram of DNA replication, identifying the location of each step in the process.
- Identify when PCR and gene sequencing technology should be used in an experiment.
- Explain the cause of DNA replication errors, and describe how they are repaired.
- Give an example of a mutation and its potential effects on an organism.

CHAPTER

09

WHAT GENES
ARE

"Pigs," thought Marc Güell. "What if we could modify pigs?"

Güell, a young biochemist at Harvard Medical School in Massachusetts, had been working with colleagues on techniques to manipulate **DNA** (**d**eoxyribo**nucleic a**cid), the genetic code of life. In 2012, Güell and his colleague Luhan Yang, along with their boss, the geneticist George Church, had begun adding and deleting genes from organisms, a process known as genome editing, using a new technological breakthrough called CRISPR-Cas9—CRISPR for short.

Before the discovery of CRISPR (pronounced "crisper"), editing DNA was difficult and expensive, and it was typically done only in model organisms such as mice and fruit flies—species for which biologists had developed a solid tool kit for genetic manipulation. CRISPR-Cas9, a simple but creative combination of a specific protein and two single-stranded **RNA** (**ribo**-**nucleic a**cid) molecules, made genome editing inexpensive and available to all, not just to well-funded genetics labs. It also enabled researchers to quickly and efficiently change the DNA of nearly any organism—like a "molecular scalpel," as one of the CRISPR inventors called it—making it possible to edit genes in fungi, plants, humans, you name it. "It was the biggest change in my career," says Güell. "With CRISPR, everything is easier, faster, and cheaper."

CRISPR also provided the new opportunity to edit more than one gene at a time, and the team picked a unique initial specimen: *Sus scrofa domesticus*, the domestic pig. "We started to think about what would be a good application to use this technology," says Güell. "One of the things that seemed like a very interesting, very difficult problem was the lack of organs for transplantation. We thought, what if we could modify pigs to make them compatible enough to be an unlimited source of organs?" (**Figure 9.1**).

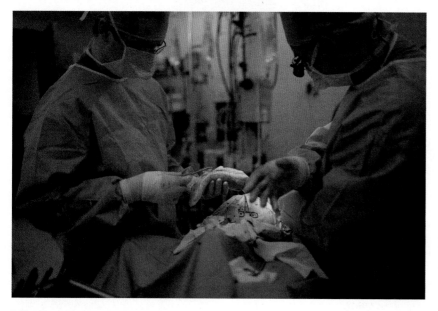

Figure 9.1

Experimental pig-to-baboon lung transplant
A lung from a pig engineered by CRISPR to prevent rejection is tested for safety and efficacy in primates by being transplanted into a baboon.

MARC GÜELL AND GEORGE CHURCH

Marc Güell (right) is a research fellow at the Wyss Institute at Harvard Medical School, where he has been working to engineer the pig genome for human compatibility. He works in the lab of geneticist George Church (left), a pioneer in genome engineering and synthetic biology.

Deep in the DNA

Each day in the United States, an average of 22 people die waiting for an organ transplant, and 119,000 men, women, and children are currently on the national transplant waiting list, each hoping their name is called before it's too late.

Researchers have explored many ways to grow and store organs for transplantation—from freezing them to building them from scratch—but one of the most promising, if you can look past the mud and flies, is pigs. Our porcine friends have long been considered an excellent potential source of organs because their organs are relatively close in size to human organs—including the heart, liver, and kidneys—and because pigs and humans have similar organ anatomies (**Figure 9.2**). In addition, pigs are an easier sell to the public: in general, people tend to prefer the idea of transplants from pigs over transplants from mammals more closely related to us, such as baboons. If we were able to transplant organs from animals into humans, a process called *xenotransplantation*, healthy organs could be available on demand and in essentially limitless supply.

Yet there has been a barrier to harvesting pig organs for humans: the pig genome is dotted with DNA from a family of viruses called p̲orcine e̲ndogenous r̲etrovi̲ruses, or PERVs. DNA is built from two parallel strands of repeating units called **nucleotides**. Each nucleotide is composed of the sugar deoxyribose, a phosphate group, and one of four **bases**: adenine, cytosine, guanine, or thymine. We identify nucleotides by their bases, using "adenine nucleotide" as shorthand for "nucleotide with an adenine base."

The nucleotides of a single strand are connected by covalent bonds between the phosphate group of one nucleotide and the sugar of the next nucleotide. The two DNA strands are connected by hydrogen bonds linking the bases on one strand to the bases on the other, like the rungs that connect the two sides of a ladder (**Figure 9.3**). The term **base pair** (or nucleotide pair) refers to two nucleotides held together by one of these bonds between their bases; that is, a base pair corresponds to one rung of the DNA ladder. The ladder twists into a spiral called a **double helix** (**Figure 9.4**). Within the long, winding double helix of the pig genome, short sections of DNA from PERVs are scattered about—sections made up of the same four nucleotides but encoding information for viral proteins rather than pig proteins.

Nucleotides do not form base pairs willy-nilly. The adenine (A) nucleotide on one strand can pair only with thymine (T) on the other strand (see **Figure 9.3**); similarly, cytosine (C) on one strand can pair only with guanine (G) on the other strand. These **base-pairing rules**, which provide **complementary base-pairing** between two nucleic acid strands, have an important consequence: when the sequence of nucleotides on one strand of the DNA molecule is known, the sequence of nucleotides on the other, complementary strand of the molecule is automatically known as well. The fact that A can pair only with T and that C can pair only with G allows the original strands

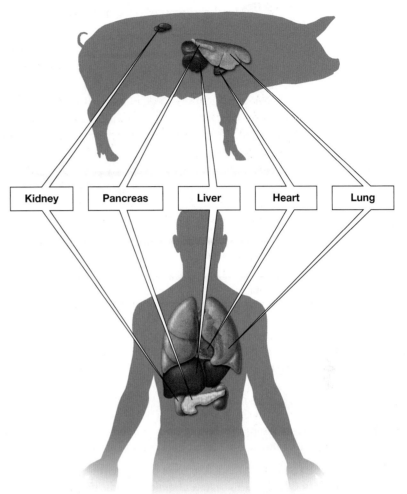

Figure 9.2

Pig and human organs are remarkably similar in size

Q1: Name one reason why, for a potential transplant, matching the size of organs shown would be important.

Q2: Name a tissue transplant for which size matching would not be as important.

Q3: Name a tissue transplant for which size matching would not be important at all.

to serve as "template strands" on which new strands can be built through complementary base-pairing. (We will see more in Chapter 10 about building new DNA strands, including how RNA can pair with DNA, which CRISPR takes advantage of.)

Still, the four nucleotides can be arranged in any order along a single strand of DNA, and each DNA strand is composed of millions of these nucleotides, so a tremendous amount of

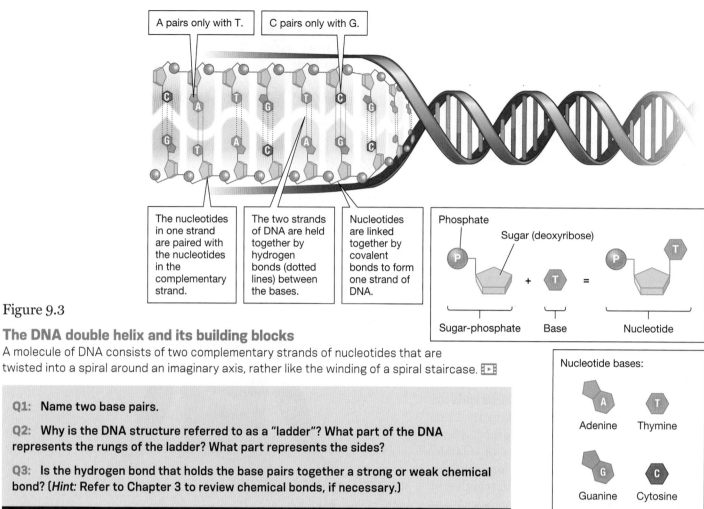

A pairs only with T.

C pairs only with G.

The nucleotides in one strand are paired with the nucleotides in the complementary strand.

The two strands of DNA are held together by hydrogen bonds (dotted lines) between the bases.

Nucleotides are linked together by covalent bonds to form one strand of DNA.

Phosphate

Sugar (deoxyribose)

Sugar-phosphate   Base   Nucleotide

Nucleotide bases:

A — Adenine   T — Thymine

G — Guanine   C — Cytosine

Figure 9.3

The DNA double helix and its building blocks

A molecule of DNA consists of two complementary strands of nucleotides that are twisted into a spiral around an imaginary axis, rather like the winding of a spiral staircase.

Q1: Name two base pairs.

Q2: Why is the DNA structure referred to as a "ladder"? What part of the DNA represents the rungs of the ladder? What part represents the sides?

Q3: Is the hydrogen bond that holds the base pairs together a strong or weak chemical bond? (*Hint:* Refer to Chapter 3 to review chemical bonds, if necessary.)

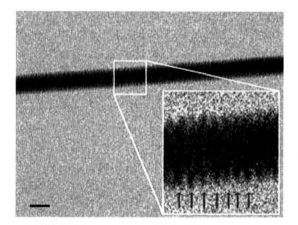

Figure 9.4

What DNA actually looks like

In November 2012, Italian researchers used an electron microscope to directly visualize DNA for the first time. This is the single thread of double-stranded DNA that they saw.

information can be stored in a DNA sequence and in a genome. The genome of the domestic pig, for example, has about 3 billion base pairs, the human genome has about 3.2 billion base pairs, a tomato has only about 900 *million* base pairs, and the bacterium *Escherichia coli* has a measly 4.6 million. The sequence of nucleotides in DNA differs among species and among individuals within a species, and these differences in genotype can result in different phenotypes (**Figure 9.5**).

In the mid-1990s, scientists became very excited about the idea of using pig organs in humans, but testing stalled because of the fear that humans would become infected with PERVs. Just breeding pigs in sterile conditions can't get rid of the virus; it's integrated right there in the double helix. The Harvard team believed CRISPR might be able to solve that

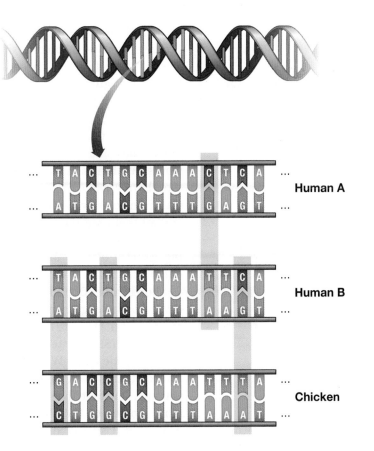

Figure 9.5

The sequence of bases in DNA differs among species and among individuals within a species

The sequence of bases in a hypothetical gene is compared for two humans (A and B) and a chicken. Base pairs highlighted in blue are variant; that is, they differ between the genes of persons A and B, and between the same genes in humans and chickens.

Q1: If all genes are composed of just four nucleotides, how can different genes carry different types of information?

Q2: Would you expect to see more variation in the sequence of DNA bases between two members of the same species (such as humans) or between two individuals of different species (for example, humans and chickens)? Explain your reasoning.

Q3: Do different alleles of a gene have the same DNA sequence or different DNA sequences?

problem by destroying the PERV DNA in pig cells once and for all.

Precise Cuts

The DNA sequences called **CRISPR**, a blessedly easy acronym for *clustered regularly interspaced short palindromic repeats*, are actually part of a defense system used by bacteria. Bacteria are constantly under bombardment from viruses that try to sneak into and take over their genomes, so bacteria evolved a set of defensive measures, including a tool to recognize and cut foreign, interloping DNA. Pioneered by microbiologists Jennifer Doudna of UC Berkeley and Emmanuelle Charpentier at Umeå University in Sweden, the CRISPR-Cas9 system is composed

① The CRISPR-Cas9 system is made up of guide RNA molecules and Cas9 proteins.

② The *pol* gene within a double-stranded PERV DNA sequence is targeted within the pig genome. The guide RNAs bring Cas9 to their target DNA sequences via complementary base-pairing.

③ Cas9 cuts through both strands of the target DNA in the pig chromosome, and the CRISPR-Cas9 complex is released.

④ During normal repair processes, the DNA sequences between the cut sites are removed and the remaining DNA is relinked, "knocking out" most of the *pol* gene. What remains of *pol* is not functional.

Figure 9.6

Genome editing with CRISPR-Cas9, an efficient and cost-effective tool

Q1: What common mechanism is employed by the guide RNA to find its target DNA sequence?

Q2: How many strands of DNA must Cas9 cut to be effective?

Q3: Does Cas9 also cause the deletion of DNA from the genome?

of two RNA molecules that guide two Cas9 proteins to chosen, precise sites in a genome. There, Cas9 efficiently cleaves both strands of the DNA (**Figure 9.6**).

The CRISPR system is like a "Delete" key or, in certain experiments, the cut-and-paste tool of a word processing program—except with nucleotides and genes instead of letters and words. While most new laboratory tools can take months or years to become widely used, CRISPR was immediately a lab favorite, with genetic studies that used the tool being published right and left: editing the zebra fish genome, correcting genetic disease mutations in adult mice, enhancing pest resistance in wheat, and more.

Shortly after Doudna and Charpentier's discovery, both Feng Zhang (at the Broad Institute of MIT and Harvard) and Church's lab engineered CRISPR systems for genome editing in eukaryotic cells. With that innovation, the Church team was ready to start editing the pig genome, but first they needed to identify how many copies of the PERV DNA the genome contained. Using laboratory tools to sequence the pig DNA, they identified 62 copies of PERVs scattered throughout the genome.

Sixty-two may sound like a lot, but it's a drop in the bucket compared to the number of genes in a mammalian genome. Pigs have over 1,600 genes for their sense of smell alone. An estimated 19,000 protein-coding genes are packed into the human genome, surrounded by even more DNA that does not code for proteins. Cells are adept at stuffing an enormous amount of DNA into a small space; they use a variety of packaging proteins to wind, fold, and compress the DNA double helix, going through several levels of packing to create the DNA-protein complex that we call a chromosome (**Figure 9.7**).

Short lengths of double-stranded DNA are wound around "spools" of proteins, known as **histone proteins**, to create a beads-on-a-string structure consisting of many histone beads, called **nucleosomes**, connected by strings of DNA. This beads-on-a-string structure is compressed and coiled into a more compact form, known as the **chromatin fiber**, by yet other types of packaging proteins. The chromatin fiber is then looped back and forth to further condense and coil again to form the resulting chromosome in the interior of the nucleus.

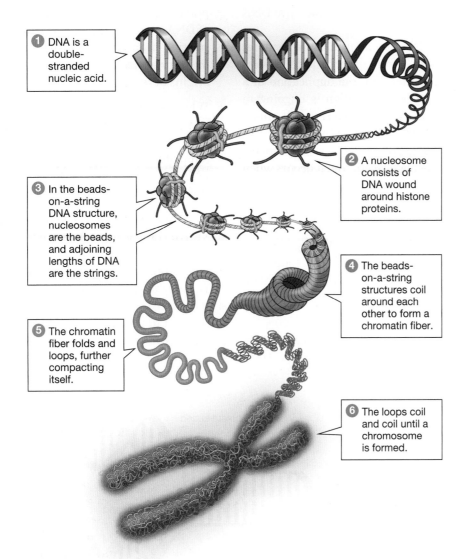

1 DNA is a double-stranded nucleic acid.

2 A nucleosome consists of DNA wound around histone proteins.

3 In the beads-on-a-string DNA structure, nucleosomes are the beads, and adjoining lengths of DNA are the strings.

4 The beads-on-a-string structures coil around each other to form a chromatin fiber.

5 The chromatin fiber folds and loops, further compacting itself.

6 The loops coil and coil until a chromosome is formed.

Figure 9.7

Chromosomes are meticulously organized DNA-protein complexes
The DNA double helix is continuously coiled and packaged around proteins until it is compacted into a chromosome.

Q1: What are the structures that result from the first level of coiling around proteins called?

Q2: What makes up a "bead" and what makes up the "string" in the beads-on-a-string structure of DNA?

Q3: What is the name for the structure that is more compact than the beads-on-a-string structure but less compact than an actual chromosome?

Double or Nothing

DNA from viruses is actually present in all mammalian genomes that we know of, including the human genome, yet not all these viral instructions behave the same. In humans, viral DNA was integrated into the genome a long, long time ago, and although it remains in the human genome, it is no longer active. In pigs, viruses entered and wove their DNA into the pig genome much more recently, and that viral DNA is still active.

All this viral DNA remains in these genomes because it is passed from generation to generation via **DNA replication**, the duplication of a DNA molecule. DNA replication is ongoing in our bodies: It occurs right before a cell enters mitosis, so that there is a copy of the DNA to pass along to the new cell. It occurs feverishly when a new embryo is being formed. And it occurs when viruses hijack our cell machinery and copy their own DNA.

Cells replicate DNA in three steps:

1. The DNA molecule unwinds through special proteins that bind the DNA at sequences known as **origins of replication**, which then break the hydrogen bonds connecting the two strands of DNA.

2. Each strand is then used as a template for the construction of a new strand of DNA. **DNA polymerase**—a key enzyme in the replication of DNA—builds the two new strands of DNA, starting from **primers** that pair complementary nucleotides at specific sites near the origins of replication.

3. When construction is completed, there are two identical copies of the original DNA molecule. Each copy is composed of a template strand of DNA (from the original DNA molecule) and a newly synthesized strand of DNA.

This mode of replication is known as **semiconservative replication** because one "old" strand (the template strand) is retained, or "conserved," in each new double helix (**Figure 9.8**).

The mechanics of copying DNA are far from simple. More than a dozen enzymes and proteins are needed to unwind the DNA, to stabilize the separated strands, to start the replication process, to attach nucleotides that are complementary

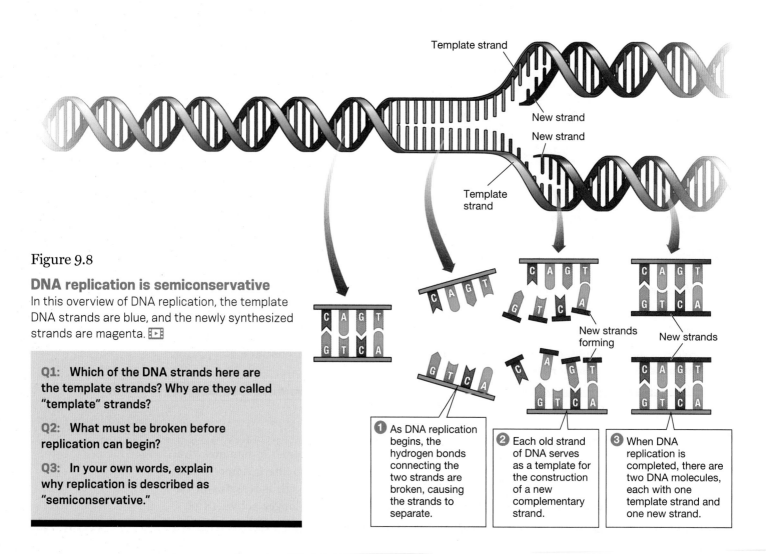

Figure 9.8

DNA replication is semiconservative

In this overview of DNA replication, the template DNA strands are blue, and the newly synthesized strands are magenta. ▶️

Q1: Which of the DNA strands here are the template strands? Why are they called "template" strands?

Q2: What must be broken before replication can begin?

Q3: In your own words, explain why replication is described as "semiconservative."

① As DNA replication begins, the hydrogen bonds connecting the two strands are broken, causing the strands to separate.

② Each old strand of DNA serves as a template for the construction of a new complementary strand.

③ When DNA replication is completed, there are two DNA molecules, each with one template strand and one new strand.

to the template strand, to "proofread" the results, and to join partly replicated fragments of DNA to one another, ultimately forming a completely replicated chromosome.

Despite the complexity of this task, cells can copy DNA molecules containing billions of nucleotides in a matter of hours—about 8 hours in humans (over 100,000 nucleotides per second). This speed is achieved in part by starting the replication of the DNA molecule at thousands of different origins of replication at once.

Before CRISPR was developed, the greatest genetic technology was a tool for replicating DNA quickly in a test tube. In the 1980s, scientists developed a way to mimic natural DNA replication in a laboratory—fast. The **polymerase chain reaction** (**PCR**) is a technique that makes it possible to produce millions of copies of a targeted DNA sequence—to "amplify" the DNA— in just a few hours, even with an extremely small initial amount of DNA (**Figure 9.9**).

PCR relies on heat, not special proteins, to cause the DNA to unwind and its strands to separate. PCR also uses targeted primers to start the replication process at points determined by the scientist, rather than at DNA's natural origins of replication. The researcher

1 Heat separates the double strands of the original DNA into two single strands.

2 As the mixture cools, the primers pair with the original DNA.

The same three-step cycle can be repeated many times, yielding billions of copies of the original DNA.

Original DNA sequence

Primer   New DNA

New DNA

New DNA

New DNA

3 DNA polymerase fills in the missing nucleotides, producing new copies of the original DNA.

Cycle 1   Cycle 2   Cycle 3

Figure 9.9

PCR can amplify small amounts of DNA more than a millionfold

Short primers consisting of synthetic DNA segments are mixed in a test tube with a sample of the target DNA, the enzyme DNA polymerase, and all four nucleotides (A, C, G, and T). The primers form base pairs with the two ends of a gene of interest. A machine then processes the mixture and doubles the number of double-stranded versions of the template sequence. The doubling process can be repeated many times (only three cycles are shown here).

Q1: PCR replicates DNA many times to increase the amount available for analysis. Why is this process called "amplification"?

Q2: During the PCR cycle, what causes the DNA strands to separate?

Q3: Identify a difference between how PCR and DNA replication are accomplished.

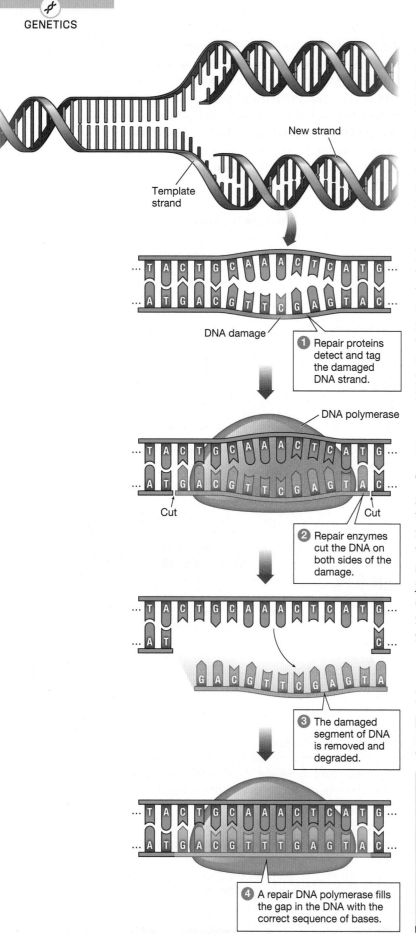

New strand

Template strand

DNA damage

1 Repair proteins detect and tag the damaged DNA strand.

DNA polymerase

Cut   Cut

2 Repair enzymes cut the DNA on both sides of the damage.

3 The damaged segment of DNA is removed and degraded.

4 A repair DNA polymerase fills the gap in the DNA with the correct sequence of bases.

creates a solution containing DNA polymerase, primers, and loose nucleotides, and the solution goes into a PCR machine, a device developed for this specific process. The machine cycles between heating, which separates the strands, and cooling, which enables the added enzymes to build two new strands of DNA through complementary base-pairing. These heating and cooling steps are repeated 25–40 times to replicate millions of exact copies of the targeted region of DNA.

PCR remains a central tool in most biology labs. The Harvard team, for example, used PCR to replicate pig DNA for their sequencing reactions, which require many copies of the same sequence in order to succeed. DNA sequencing machines, which produce a data file listing the complete sequence of nucleotides in a given strand of DNA, were used to identify all the places in the pig genome where PERV had been integrated.

Making Mutations

To inactivate all 62 copies of PERV in the pig genome at once, the Church team decided to try to prevent DNA polymerase from successfully replicating the PERV DNA. To that end, the team planned to use CRISPR to *mutate* a small region of the PERV DNA: a gene named *pol* (see **Figure 9.6**). A change to the sequence of nucleotides in an organism's DNA is called a **mutation**. The extent of a mutation can range from the change of one nucleotide in a single base

Figure 9.10

Repair proteins fix DNA damage
Large complexes of DNA repair proteins work together to fix damaged DNA.

Q1: Summarize how DNA repair works and why the repair mechanisms are essential for the normal function of cells and whole organisms.

Q2: Is DNA repair 100 percent effective?

Q3: What would happen to an organism if its DNA repair became less effective?

pair (known as a **point mutation**) to the addition or deletion of one or more whole chromosomes (a *chromosomal abnormality*, described in Chapter 8).

Güell was developing techniques to artificially mutate genes, yet point mutations occur naturally and randomly often, especially during DNA replication. When DNA is copied right before mitosis occurs in a cell, there are many opportunities for mistakes to be made. The enzymes that copy DNA sometimes insert an incorrect nucleotide in the newly synthesized strand. In addition, DNA in cells is constantly being damaged by chemical, physical, and biological agents, including energy from radiation or heat, collisions with other molecules in the cell, attacks by viruses (like PERVs), and random chemical accidents (some of which are caused by environmental pollutants, but most of which result from normal metabolic processes).

Replication errors and damage to DNA—especially to essential genes—disrupt normal cell functions. If not repaired, DNA damage leads to malfunctioning proteins, such as Felix's WAS protein in Chapter 8. DNA damage can also cause the death of cells and, ultimately, the death of an organism. Thankfully, cells have a way to recover: DNA polymerase immediately corrects almost all mistakes during DNA replication, "proofreading" complementary base pairs as they form.

DNA polymerase is not infallible. When an incorrect nucleotide is added but escapes proofreading by DNA polymerase, a mismatch error has occurred. This happens about once in every 10 million nucleotides. But cells have another backup safety program: repair proteins that correct 99 percent of mismatch errors, reducing the overall chance of an error to one mistake in every *billion* nucleotides (**Figure 9.10**).

On the rare occasions when a mismatch error is not corrected by repair proteins, the DNA sequence is changed, and the new sequence is reproduced the next time the DNA is replicated. If the mutation occurs within a gene, it will result in the formation of a new allele. Most new alleles are either neutral or harmful, but occasionally a mutation may be beneficial.

Three types of point mutations can alter a gene's DNA sequence: substitutions, insertions, and deletions. In a **substitution** point mutation, one nucleotide is substituted for another in

the DNA sequence of the gene. An **insertion** or **deletion** point mutation occurs when a nucleotide is, respectively, inserted into or deleted from a DNA sequence. Sickle-cell disease, a human genetic blood disorder, is caused by a substitution point mutation (**Figure 9.11**). Sometimes,

Normal hemoglobin DNA

Sickle-cell hemoglobin DNA

Normal hemoglobin

Sickle-cell hemoglobin

Normal red blood cells

A sickled red blood cell

Figure 9.11

A point mutation in the hemoglobin gene leads to sickle-cell disease

In people with the genetic disorder sickle-cell disease, a single base in the gene that makes hemoglobin, an important protein involved in oxygen transport in red blood cells, is altered. The red blood cells of people with sickle-cell disease become curved and distorted under low-oxygen conditions and can clog blood vessels, leading to serious effects, including heart and kidney failure.

Q1: What are the three types of point mutations?

Q2: Sickle-cell disease is an autosomal recessive genetic disorder. How many mutated hemoglobin alleles do people with sickle-cell disease have?

Q3: Because of improved treatments, individuals with sickle-cell disease are now living into their forties, fifties, or longer. How might this extension of life span affect the prevalence of sickle-cell disease in the population?

however, changing a few nucleotides in a gene's DNA sequence has little or no effect. In such cases, a mutation is said to be "silent" because it produces no change in the function of the protein, and therefore no change in the phenotype of the organism. (We further explore mutations and their effect on the resulting proteins in Chapter 10.)

Insertions and deletions can be point mutations, but they can also involve more than one nucleotide; sometimes thousands may be added or deleted. Large insertions and deletions almost always result in the synthesis of a protein that cannot function properly. Güell's goal was to use CRISPR in the pig genome to delete bits of DNA in the *pol* gene so that the PERVs could no longer function. The *pol* gene is essential for PERVs to replicate, so a large mutation should halt the formation of viral particles. In addition to being essential to the virus, *pol* is present in all 62 copies of PERVs but not elsewhere in the pig genome, so targeting CRISPR to mutate *pol* would spare damage to other parts of the pig DNA. "My biggest concern is always not to destroy the genome," says Güell. So, he and the team designed two CRISPR RNA guides to target the complex right to *pol*, and to mutate the heck out of it.

After much trial and error, the team inserted its CRISPR-Cas9 construct into pig embryos, and the system began its work. The CRISPR system made 455 different cuts in the PERV *pol* genes throughout the genome, resulting in deletions ranging from 1 to 148 base pairs. About 80 percent of the engineered mutations were small deletions of fewer than nine base pairs—tiny, but enough to disrupt the gene.

Pigs Are People Too?

In October 2015, Güell, Church, and the team presented the results of their initial work: Using CRISPR, they had inactivated 62 PERVs in pig embryos. Recall that in previous non-CRISPR-edited experiments, PERVs from the pig cells infected human cells when put in close proximity. This time, when the team placed the CRISPR-edited pig cells next to human cells, there was zero transmission of the virus. "The gene editing completely disrupted the ability of the virus DNA to replicate and form viral particles," says Güell.

Although the team has successfully destroyed the PERV genes in the pig genome, there are still a few more steps before pig organs will be ready and safe for human transplantation. Other genes need to be manipulated, says Güell, such as immune system proteins so that the human immune system doesn't reject the organs as foreign. However, the next immediate step, says Güell, is to edit an embryo, implant it back into a female pig, and raise a live, genetically engineered animal. "We're trying hard," says Güell. "It could be in the next year."

Other researchers are pursuing a different path: growing actual human organs in pigs. Again, CRISPR is the workhorse for the experiment. In 2016, a team at UC Davis used CRISPR to remove the genes that encode for the pig pancreas in a pig embryo. The procedure created a void in the embryo, which they then filled with human stem cells. Stem cells have the potential to develop into almost all human cells and organs—including, in this case, a fully functional human pancreas.

Some research groups are trying to use this method to grow other organs. **Figure 9.12** shows the process by which scientists could use CRISPR and human stem cells to grow a pig with human kidneys, which would then be used to donate a kidney to a patient in need of a transplant. No researchers have yet to complete this process, and the work remains controversial, especially because of concerns that the human cells might migrate to the developing pig's brain and make the pig, even in some small way, more human.

Overall, CRISPR has revitalized the idea of a safe, clean, limitless source of healthy organs. "It opens up the possibility of not just transplantation from pigs to humans but the whole idea that a pig organ is perfectible," Church told the BBC in 2016. "Gene editing could ensure the organs are very clean, available on demand, and healthy, so they could be superior to human donor organs."

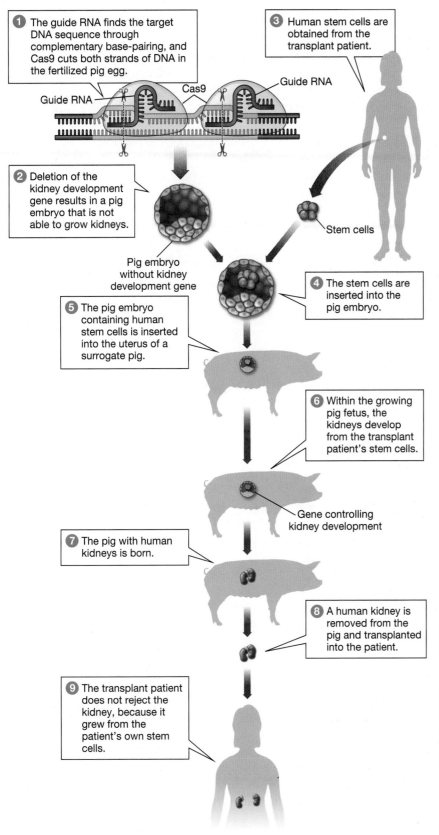

1 The guide RNA finds the target DNA sequence through complementary base-pairing, and Cas9 cuts both strands of DNA in the fertilized pig egg.

Guide RNA

Cas9

Guide RNA

3 Human stem cells are obtained from the transplant patient.

2 Deletion of the kidney development gene results in a pig embryo that is not able to grow kidneys.

Pig embryo without kidney development gene

Stem cells

4 The stem cells are inserted into the pig embryo.

5 The pig embryo containing human stem cells is inserted into the uterus of a surrogate pig.

6 Within the growing pig fetus, the kidneys develop from the transplant patient's stem cells.

Gene controlling kidney development

7 The pig with human kidneys is born.

8 A human kidney is removed from the pig and transplanted into the patient.

9 The transplant patient does not reject the kidney, because it grew from the patient's own stem cells.

Figure 9.12

Human organs could grow up in pigs
Growing a human organ (here, a kidney) in pigs modified by CRISPR to lack that organ could help meet transplant needs.

Q1: Name a step in this process that is similar to the original CRISPR method that removed the PERVs from the pig genome. (*Hint:* Review Figure 9.6.)

Q2: Name a step in this process that is not included in the original CRISPR method that removed the PERVs from the pig genome.

Q3: What parts of this process would scientists need to change in order to develop several different human organs in a single pig?

The Meteoric Rise of CRISPR

The genome editing tool CRISPR, short for "clustered regularly interspaced short palindromic repeats," has taken molecular biology laboratories by storm over the past 5 years. It has been used to edit the genomes of crops and livestock to improve breeding and production, to control populations of disease-carrying insects, to silence genetic disorders in animal models, and more. Here are a few highlights from the short but shining history of CRISPR.

Assessment available in smartwork5

Search results article count

PubMed search results for "CRISPR" by year

CRISPR

2002	2003	2004	2005	2006	2007	2008	2009	2010	2011	2012	2013	2014	2015	2016
1	1	0	5	6	12	21	32	45	79	126	282	607	1,258	2,143

CRISPR repeats are first observed in bacterial genomes. Their significance is not yet known.
1987

Researchers propose that CRISPR functions in nature as part of a bacterial adaptive immune system.
2006

The CRISPR-Cas9 system is used to edit targeted genes in both human and mouse cells, and later plant cells.
2013

The first human trial to use CRISPR genome editing gets approval from the National Institutes of Health, in a cancer therapy trial to edit a patient's own immune system cells.
2016

2002
The term "CRISPR" is coined by researchers in Spain and the Netherlands.

2011
The final necessary piece for the genome editing system is identified: a second small RNA needed to guide Cas9 to its targets.

2015
In China, scientists use CRISPR-Cas9 to edit preimplantation human embryos, repairing a mutated gene that would cause a blood disorder. Subsequently, an international ban prohibits the use of genome editing to make changes to the human genome.

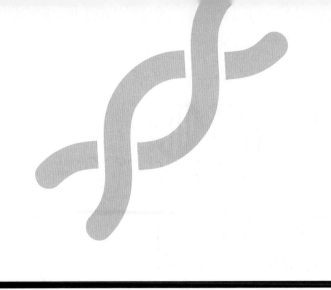

REVIEWING THE SCIENCE

- Genes are composed of **DNA**, which consists of two parallel strands of repeating units called **nucleotides** twisted into a **double helix**.

- The four nucleotides of DNA contain the **bases** adenine (A), cytosine (C), guanine (G), and thymine (T). The nucleotides exhibit **complementary base-pairing** according to **base-pairing rules**: A can pair only with T, and C can pair only with G.

- DNA is wrapped around **histone proteins**, forming **nucleosomes**. The nucleosome structures can further compact the DNA by coiling around themselves to form a **chromatin fiber**. Chromatin fibers further coil around themselves to form chromosomes.

- The **CRISPR**-Cas9 editing system is composed of two pieces of **RNA** designed to form base pairs at precise locations in a gene. This DNA-RNA interaction guides the Cas9 proteins to the sites where they efficiently cut the DNA, resulting in a gene deletion after normal repair processes take place. Additional genetic manipulations are required to generate a gene insertion.

- **DNA replication** occurs in all living organisms prior to mitosis. The double helix unwinds, and the two strands break apart. Each strand of DNA serves as a template from which a new strand is copied. **DNA polymerase** builds each new strand of DNA using **primers** located near the **origins of replication**.

- The **polymerase chain reaction**, or **PCR**, is a laboratory technique to amplify the DNA from a small initial amount to millions of copies. Amplified DNA can then be sequenced to examine specific genes or mutations.

- DNA is subject to damage by physical, chemical, and biological agents, and errors in DNA replication are common. DNA polymerase "proofreads" the DNA during replication and corrects most mistakes. Repair proteins are a backup repair mechanism and correct any errors that DNA polymerase misses.

- A change to the sequence of bases in an organism's DNA is called a **mutation**. Three types of mismatch mutations can alter a gene's DNA sequence: **substitutions**, **insertions**, and **deletions**. If only a single base is altered, it is a **point mutation**.

THE QUESTIONS

The Basics

1 DNA replication results in

(a) two DNA molecules—one with two old strands, and one with two new strands.

(b) two DNA molecules, each of which has two new strands.

(c) two DNA molecules, each of which has one old strand and one new strand.

(d) none of the above

2 The DNA of cells is damaged

(a) thousands of times per day.

(b) by collisions with other molecules, chemical accidents, and radiation.

(c) not very often and only by radiation.

(d) both a and b

3 The DNA of different species differs in the

(a) sequence of bases.

(b) complementary base-pairing.

(c) number of nucleotide strands.

(d) location of the sugar-phosphate portion of the DNA molecule.

4 Mutation

(a) can produce new alleles.

(b) can be harmful, beneficial, or neutral.

(c) is a change in an organism's DNA sequence.

(d) all of the above

5 Link each term with the correct definition.

NUCLEOTIDE	1. Two complementary bases joined by hydrogen bonds.
BASE PAIR	2. The nitrogen-containing component of a nucleotide; there are four variants of this component.
DNA MOLECULE	3. A strand of nucleotides linked together by covalent bonds between a sugar and a phosphate; two strands are linked by hydrogen bonds between complementary bases.
BASE	4. A phosphate, a sugar, and a nitrogen-containing base.

6 In the diagram of replication shown here, fill in the blanks with the appropriate terms: (a) base pair, (b) base, (c) nucleotide, (d) template strand, (e) newly synthesized strand, (f) separating strands.

7 Select the correct terms:
To work on removing PERV DNA from the pig genome, researchers first replicate the DNA many times, using (**PCR / CRISPR**). This process increases the amount of DNA so that they are able to run experiments to edit the genome, using (**PCR / CRISPR**).

Challenge Yourself

8 If a strand of DNA has the sequence CGGTATATC, then the complementary strand of DNA has the sequence

(a) ATTCGCGCA.

(b) GCCCGCGCT.

(c) GCCATATAG.

(d) TAACGCGCT.

9 Place the following steps of DNA replication and repair in the correct order by numbering them from 1 to 5.

_____ a. A template strand begins to be replicated.

_____ b. If the incorrect base is not identified and replaced, it remains as a point mutation in the DNA.

_____ c. DNA polymerase identifies and replaces most incorrect bases with the correct base, complementary to the base on the template strand.

_____ d. An incorrect base is added to the growing strand of DNA.

_____ e. Proteins identify and replace any incorrect bases missed by DNA polymerase.

10 Given the template DNA sequence GCAGCATGTT, identify each of the following mutations to the complementary strand as an insertion (I), a deletion (D), or a substitution (S).

_____ a. CGTCGTACA

_____ b. CGTGGTACAA

_____ c. CGTCGTACTAA

11 If the target sequence for PCR on one strand of DNA is ATGCAAATCCTGG, what is the sequence of each of the two strands of the double-stranded DNA produced?

Try Something New

12 Using base-pairing rules to guide you, for a DNA double helix that contains 20 percent adenine (A), specify the percentage of (a) thymine, T; (b) guanine, G; and (c) cytosine, C.

13 Why do scientists need to use CRISPR to remove PERVs from pig embryos, rather than just editing the genes in the organs of adult pigs?

14 Scientists estimate that genes encoding proteins make up less than 1.5 percent of the human genome. Other genes in our cells encode different types of nonprotein RNA molecules. The rest of our genome consists of various types of **noncoding DNA**, defined as DNA that does not code for any kind of functional RNA. Some of the remaining DNA has regulatory functions—for example, controlling gene expression. Some of it has architectural functions, such as giving structure to chromosomes or positioning them at precise locations within the nucleus. Given that these sections of DNA are noncoding, does it matter if they contain replication errors? Explain your reasoning.

Leveling Up

15 **What do *you* think?** In another line of research directed at solving the shortage of organs available for transplantation, scientists have been working feverishly on generating human tissues and organs using human induced pluripotent stem cells (iPSCs). These cells, which have the ability to transform into many different cell types, start as a patient's own skin cells and are treated to attain an undifferentiated, or embryonic-like, state. Scientists then direct them with growth factors to become kidney cells, liver cells, or any other cell type needed. For more information, see https://www.cirm.ca.gov/patients/creating-new-types-stem-cells.

Read the article "Eight Tiny Organs Grown by Scientists," about the potential of iPSCs to fill the transplantable organ void (http://www.insight.mrc.ac.uk/2015/07/20/eight-tiny-organs-grown-by-scientists). Using the information in this article, make a list of pros and cons for both pig organs and iPSC organs. Reflect on which option you think is the better course of action.

16 ***Write Now* biology: next-generation GMO** Read the article "CRISPR Brings an Early Harvest" (http://www.genengnews.com/gen-news-highlights/crispr-brings-an-early-harvest/81253507), which describes a new way to produce GMO tomatoes without using standard gene transfer techniques. Write a letter to your state or federal legislator expressing your opinion either for or against this new technology for creating GMO foods. Cover the following points in your letter.

a. Explain how CRISPR-Cas9 technology works, comparing and contrasting it to traditional GMO technology.

b. Describe how the CRISPR system was applied to the plants in the article.

c. In your final paragraph, try to persuade your government representative to either endorse or condemn this technology for improving crop yield and quality, using specific examples.

For more, visit digital.wwnorton.com/bionow2 for access to:

Tobacco's New Leaf

One company's quest to produce tomorrow's drugs in today's plants.

After reading this chapter you should be able to:

- Explain how an organism uses gene expression to create a phenotype from a genotype.

- Caption a diagram of transcription, identifying each step in the process and the relevant molecules.

- Caption a diagram of translation, identifying each step in the process and the role of each type of RNA.

- Determine the correct amino acid from a given codon.

- Justify how the genetic code fits the definitions of "redundant" and "universal."

- Describe how a cell can increase or decrease its expression of particular genes, and why this ability is important for an organism.

CHAPTER

10

HOW GENES
WORK

The greenhouse is vast, the size of half a football field. A loud, steady thrum reverberates in the room as massive metal fans push air through the hot, humid space. Mike Wanner, a tall, serious man with gray hair and piercing brown eyes, walks through rows of leafy plants, each a foot and a half tall. He stoops to rub a leaf between his fingers, then raises his hand to his nose and sniffs. "That's what tobacco smells like," Wanner shouts, straining to be heard over the noise of the fans.

Wanner is the North Carolina site manager for operations at Medicago, a Canadian biotechnology company growing tobacco at a facility outside Durham, once home to Lucky Strike cigarettes (**Figure 10.1**). But these plants are not being grown to smoke, chew, or dip. Instead, they serve a dramatically different purpose: this tobacco will be used to make flu vaccines.

Influenza vaccines are normally grown in chicken eggs—a process that takes months. As an alternative, companies and researchers have begun experimenting with "biopharming"—manufacturing vaccine proteins in plants. Since

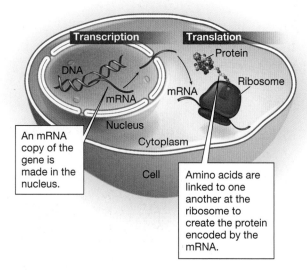

Figure 10.2

An overview of gene expression

Genetic information flows from DNA to RNA to protein during gene expression, which occurs in two steps: first *transcription* and then *translation*. The transcription of a protein-coding gene produces an mRNA molecule, which is then transported to the cytoplasm, where translation occurs and the protein is made with the help of ribosomes.

most genes contain instructions for building proteins, scientists insert a gene that codes for the protein that interests them. This production of a protein from a gene occurs via **gene expression**—the process by which genes are transcribed into RNA and then translated to make proteins (**Figure 10.2**).

Proteins participate in virtually every process inside and between cells, so gene expression is the fundamental way by which genes influence the structure and function of a cell or organism. It is the process through which an organism's genotype gives rise to its phenotype (see Chapter 7). All prokaryotes, eukaryotes, and viruses utilize gene expression.

In addition to being involved in every process in a cell, proteins are the key components of some vaccines. Injected pieces of viral proteins activate the human immune system to defend against that virus in the future (see Figure 2.1). To make these vaccines, scientists must produce large quantities of viral proteins. The traditional method is to inject chicken eggs with a virus, let the virus multiply in the chicken cells, and then

Figure 10.1

Growing tobacco to treat the flu

Tobacco plants grow in a Medicago greenhouse in Quebec City, Canada.

Figure 10.3

Tobacco or egg?

Using tobacco plants to produce proteins for influenza vaccines has several advantages over the traditional approach using chicken eggs.

	Tobacco	Egg
Speed from outbreak to vaccine production:	1 month	6 months
Cost to produce 50 million flu vaccines:	$36 million	$400 million
Allergy risks:	Minimal	Individuals with egg allergies cannot receive the vaccine
Availability:	Not yet approved by the Food and Drug Administration	Approved and currently the most widely used technology to produce flu vaccines

Q1: Which is the faster way to produce vaccines: biopharming with plants or creating vaccines in eggs? Why is this important?

Q2: How much cheaper is biopharming with plants than creating vaccines in eggs? Why is this important?

Q3: Why must tobacco-derived vaccines, or any new medications for that matter, be approved by the FDA?

extract the virus, remove its genetic material, and prepare a vaccine from the leftover viral proteins. It's a long, cumbersome process.

Unlike chicken eggs, however, plants can be grown in vast quantities, and they grow rapidly—often in just days or weeks. "The big advantage of plant systems is that they can produce massive amounts of proteins very inexpensively," says James Roth, director of the Center for Food Security and Public Health at Iowa State University, who studies biopharming in plants and animals (and is not associated with Medicago). In the event of a pandemic flu outbreak, Medicago could produce vaccines 6 times faster and 12 times cheaper than traditional egg manufacturing could, the company claims (**Figure 10.3**).

In April 2012, Medicago put its tobacco plants to the test, running the North Carolina manufacturing facility at full tilt for 30 days. The U.S. Department of Defense had given the company millions of dollars to test whether it could quickly produce enough pandemic flu vaccine from the tobacco to stem an outbreak. The pressure is on, said Wanner, standing inside the facility. He looked out over his crop. "We'll see what happens."

Fighting the Flu with Tobacco

In 1997, Louis-Philippe Vézina, then a research scientist at Agriculture and Agri-Food Canada and an associate professor at Université Laval in Quebec City, Canada, decided to start a company. A plant biotechnologist, Vézina wanted to explore the possibility of manufacturing proteins in alfalfa plants, and thus he called his company Medicago, after the genus to which alfalfa belongs. Later, as the company grew, Vézina and his team discovered that tobacco produces higher yields of proteins in a shorter time frame than alfalfa does, so the company switched plants.

MICHAEL WANNER

Michael Wanner is chief financial officer and U.S. site manager for Medicago, an innovative biotechnology company that develops vaccines in the leaves of tobacco plants.

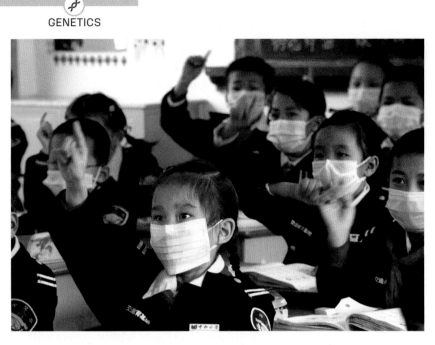

Figure 10.4

Swine flu

In 2009, people around the world donned masks as a precaution against influenza virus H1N1, nicknamed "swine flu" because it contained DNA from bird, swine, and human flu viruses. The virus caused a pandemic, killing an estimated 284,500 people.

By 2005, Medicago had begun to receive calls asking about its product and when it would begin clinical trials, the first step toward getting a drug approved by the U.S. Food and Drug Administration (FDA). "I was surprised. Even the FDA was asking, 'When are we going to see your vaccine?'" recalls Vézina. "They were keen on technologies like this because they know how much vaccines cost, and they're interested in anything that can decrease that cost."

One of the first vaccines that Medicago produced was a vaccine for influenza virus H1N1, or swine flu, the most common cause of the flu in 2009 (**Figure 10.4**). In the previous H1N1 pandemic, it had taken months for vaccines grown in chicken eggs to reach the market. By

contrast, Medicago produced its vaccine, ready for testing, in just 19 days.

To make a flu vaccine in plants, Vézina and his team use genetic engineering in much the same way that it was used to create a gene therapy for Wiskott-Aldrich syndrome, as described in Chapter 8. They identify and synthesize a single viral gene that codes for hemagglutinin, a protein found on the surface of flu viruses. To make large quantities of that viral protein, the scientists at Medicago first insert the hemagglutinin gene into small, rod-shaped bacteria called "agrobacteria," which infect plants. Loaded with the hemagglutinin gene, these agrobacteria are then exposed to the tobacco plants: a robotic arm lifts a tray containing 5-week-old tobacco plants secured to the surface, flips the tray upside down, and dips the plants into a liquid solution swimming with the bacteria. Once the plants are immersed in the liquid, the technicians turn on a vacuum, sucking air out of the leaves and pulling bacteria into them—like dipping a sponge into water, squeezing it, and then releasing it to soak up the liquid.

When the plants are returned to the greenhouse, their leaves are floppy and almost translucent, like wet tissue paper. Now things are cooking. Once the bacteria are inside the leaves, the bacterial cells release the viral hemagglutinin gene into the plant cells, where it is transported into plant cell nuclei to begin the process of gene expression: making a protein from DNA via the two steps of transcription and translation.

"When you get the *Agrobacterium* solution into the leaf, it will invade the cells and there will be a burst of viral gene expression in the plant cell," said Vézina. "The plant takes over and uses its machinery to produce the protein" (**Figure 10.5**).

Two-Step Dance, Transcription: DNA to RNA

The first step in gene expression—in this case, for a flu gene in a plant cell—is **transcription**, the synthesis of RNA based on a DNA template. In the nucleus, an enzyme called

LOUIS-PHILIPPE VÉZINA

Plant researcher Louis-Philippe Vézina cofounded Medicago in 1997 and served as the company's chief scientific officer until 2014, overseeing all scientific research and development.

RNA polymerase binds to a segment of DNA near the beginning of the gene that is called a **promoter**. The promoter contains a specific sequence of DNA bases that the RNA polymerase recognizes and binds. At Medicago, scientists attach specific promoters to the hemagglutinin gene so that the plant cell's RNA polymerase can identify the gene and actively transcribe it, maximizing the rate of transcription, says Vézina.

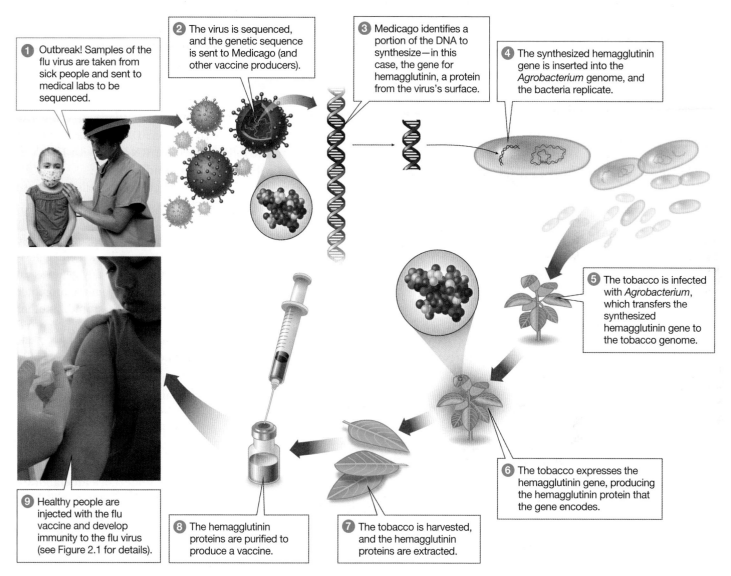

Figure 10.5

From outbreak to vaccine

During flu season or any other viral outbreak, a large network of medical professionals and scientists is activated to create a vaccine quickly and effectively. ▶

Q1: In which of the step(s) illustrated here does DNA replication occur? In which step(s) does gene expression occur?

Q2: Why do vaccine producers not simply replicate the entire viral genome, instead isolating the gene for one protein and replicating only that gene?

Q3: What role do the bacteria play in this process? Why are they needed?

Tobacco cell

1 Transcription begins when RNA polymerase binds to the promoter.

RNA polymerase

Hemagglutinin gene

Terminator (in pink)

Promoter (in pink)

2 An mRNA molecule is produced as RNA polymerase moves down the template strand of DNA.

Direction of transcripton

RNA polymerase

Template strand of DNA

RNA nucleotides

New RNA strand

Direction of transcripton

3 Transcription ends when RNA polymerase reads through the terminator.

Figure 10.6

Plants making proteins, I: Transcription

RNA polymerase transcribes the hemagglutinin gene into a molecule of RNA.

Q1: Why is only one strand of DNA used as a template?

Q2: If a mutation occurred within the promoter or terminator region, do you think it would affect the mRNA transcribed? Why or why not?

Q3: The template strand of part of a gene has the base sequence TGAGAAGACCAGGGTTGT. What is the sequence of RNA transcribed from this DNA, assuming that RNA polymerase travels from left to right on this strand?

Once bound to the promoter, the RNA polymerase unzips the DNA double helix at the beginning of the gene, separating a short portion of the two strands. Only one of the two DNA strands is used as a template, and thus it is called the **template strand**. The RNA polymerase begins to move down the DNA template strand, constructing a **messenger RNA (mRNA)** molecule, a strand of nucleotides complementary to the DNA, from free nucleotides floating around in the nucleus (**Figure 10.6**). RNA does not have all the same bases as DNA: its four bases are adenine (A), cytosine (C), guanine (G), and uracil (U). Those bases pair with the four DNA bases according to the following rules: RNA's A with DNA's T, C with G, G with C, and U with A.

Part of the reason that tobacco cells produce so much hemagglutinin so quickly is that the hemagglutinin gene inserted into the cells contains a special DNA sequence that triggers multiple RNA polymerases to transcribe a hemagglutinin gene at a single time. As for any gene, as an RNA polymerase moves away from the promoter and travels down the template strand, another RNA polymerase can bind at the promoter and start synthesizing a second mRNA on the heels of the first. At any given time, therefore, many RNA polymerases can be traveling down a DNA template simultaneously, each synthesizing an mRNA.

Transcription stops when the RNA polymerase reads through a special sequence of bases called a **terminator**. In eukaryotic cells, the mRNA then undergoes an elaborate sequence of modifications that prepare it to leave the nucleus. These steps include chemical modification of both ends of the mRNA, as well as a process called RNA splicing.

Most eukaryotic genes (and many viral genes) are embedded with stretches of sequences that don't code for anything, called **introns**. The stretches of DNA in a gene that carry instructions for building the protein are called **exons**. Because of this patchwork construction, with genes made of introns and exons interspersed, newly transcribed mRNA (pre-mRNA) is also a patchwork of coding sequences intermixed within noncoding sequences. During **RNA splicing**, the introns are snipped out of a pre-mRNA and the remaining pieces of mRNA—the exons—are joined to generate the

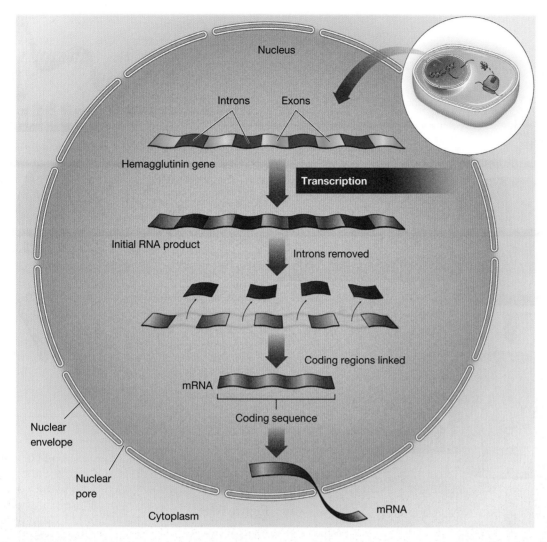

Figure 10.7

Processing mRNA for export to the cytoplasm

In eukaryotes, introns must be removed before an mRNA leaves the nucleus.

Q1: In your own words, define RNA splicing. When during gene expression does it occur?

Q2: What do you predict would happen if the introns were not removed from RNA before translation? Why would it be a problem if the introns were not removed?

Q3: Where is the mRNA destined to go once it has been transported out of the nucleus?

---

mature mRNA (**Figure 10.7**). This mRNA is then ready to leave the nucleus.

To review, transcription occurs when RNA polymerase binds to a promoter, unzips the DNA helix, and constructs a strand of mRNA based on the DNA template strand. Transcription ends at the terminator sequence, and the mRNA is then processed, at which time noncoding introns are spliced out of the sequence. Voilà! mRNA is created from a gene.

Two-Step Dance, Translation: RNA to Protein

The microscopic molecular dance inside the tobacco cells continues with translation. Once the hemagglutinin gene has been transcribed into mRNA in the nucleus, it is time to make

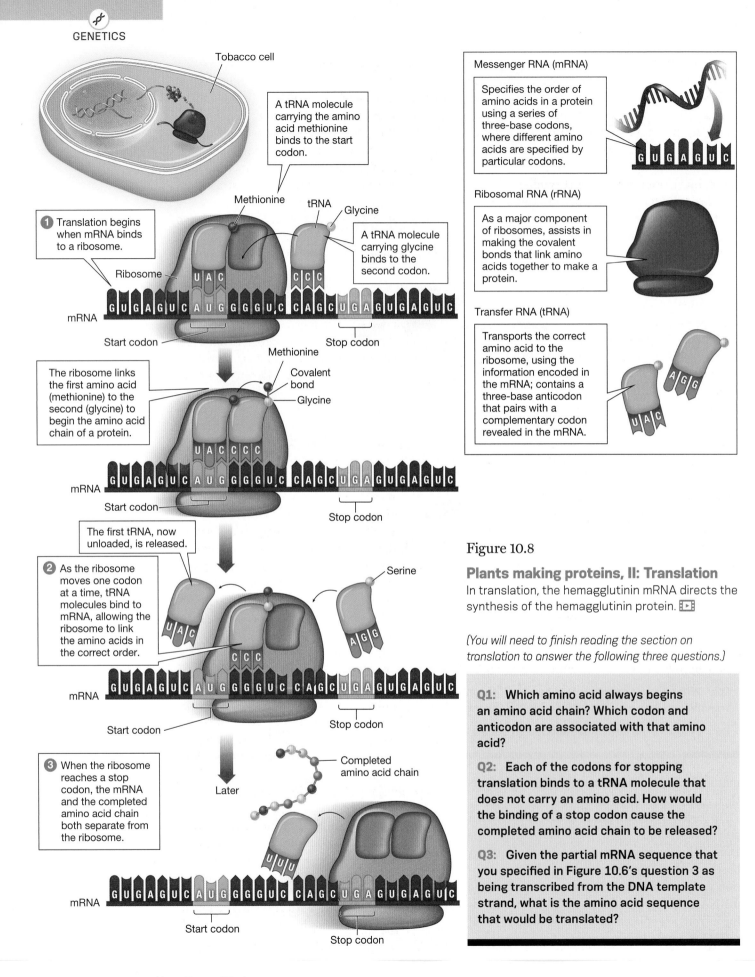

Tobacco cell

A tRNA molecule carrying the amino acid methionine binds to the start codon.

Messenger RNA (mRNA)

Specifies the order of amino acids in a protein using a series of three-base codons, where different amino acids are specified by particular codons.

Ribosomal RNA (rRNA)

As a major component of ribosomes, assists in making the covalent bonds that link amino acids together to make a protein.

Transfer RNA (tRNA)

Transports the correct amino acid to the ribosome, using the information encoded in the mRNA; contains a three-base anticodon that pairs with a complementary codon revealed in the mRNA.

Methionine — tRNA — Glycine

A tRNA molecule carrying glycine binds to the second codon.

① Translation begins when mRNA binds to a ribosome.

Ribosome — UAC — CCC

mRNA — GUGAGUCAUGGGGUCCAGCUGAGUGAGUC

Start codon — Stop codon

The ribosome links the first amino acid (methionine) to the second (glycine) to begin the amino acid chain of a protein.

Methionine — Covalent bond — Glycine

UACCCC

mRNA — GUGAGUCAUGGGGUCCAGCUGAGUGAGUC

Start codon — Stop codon

The first tRNA, now unloaded, is released.

② As the ribosome moves one codon at a time, tRNA molecules bind to mRNA, allowing the ribosome to link the amino acids in the correct order.

Serine

UAC — AGG — CCC

mRNA — GUGAGUCAUGGGGUCCAGCUGAGUGAGUC

Start codon — Stop codon

③ When the ribosome reaches a stop codon, the mRNA and the completed amino acid chain both separate from the ribosome.

Later

Completed amino acid chain

UUU

mRNA — GUGAGUCAUGGGGUCCAGCUGAGUGAGUC

Start codon — Stop codon

Figure 10.8

Plants making proteins, II: Translation
In translation, the hemagglutinin mRNA directs the synthesis of the hemagglutinin protein. ▶️

(You will need to finish reading the section on translation to answer the following three questions.)

Q1: Which amino acid always begins an amino acid chain? Which codon and anticodon are associated with that amino acid?

Q2: Each of the codons for stopping translation binds to a tRNA molecule that does not carry an amino acid. How would the binding of a stop codon cause the completed amino acid chain to be released?

Q3: Given the partial mRNA sequence that you specified in Figure 10.6's question 3 as being transcribed from the DNA template strand, what is the amino acid sequence that would be translated?

the protein, the actual product that will be extracted from the tobacco leaves. First the mRNA is transported from the nucleus, where it was made, to the sites of protein synthesis: the cell structures called **ribosomes** in the cytoplasm. To escape the nucleus, the long strand of mRNA passes through a nuclear pore, like a noodle slipping through the hole of a colander. Once the mRNA molecule arrives in the cytoplasm, the information it contains must be translated, with the help of ribosomes, from the language of mRNA (nitrogenous bases) to the language of proteins (amino acids). **Translation** is the process by which ribosomes convert the information in mRNA into proteins.

During translation, ribosomes "read" the mRNA code like a grocery list (bread, milk, etc.), and collect the corresponding amino acids, linking them in the precise sequence dictated by mRNA (**Figure 10.8**). Ribosomes read the mRNA information in sets of three bases at a time, and each unique sequence of three mRNA bases is called a **codon**. The hemagglutinin gene has about 1,770 bases, of which 1,695 code for the protein. That makes 565 codons (1,695 divided by 3), and therefore the hemagglutinin protein is composed of 565 amino acids.

The four bases of mRNA (A, C, G, U) can be arranged to create a three-base sequence in 64 different ways (because $4^3 = 64$). Therefore, there are 64 possible codons (**Figure 10.9**). Most of the 64 codons specify a particular amino acid. A couple of amino acids are specified by only one codon, while other amino acids are specified by anywhere from two to six different codons. Some codons do not code for any amino acid and instead act as signposts that communicate to the ribosomes where they should start or stop reading the mRNA. The **start codon** (AUG) is the ribosome's starting point on the mRNA strand, and there are three possible **stop codons** (UAA, UAG, and UGA). By beginning and ending at fixed points, the cell ensures that the mRNA message is read in precisely the same way every time.

The information specified by all 64 possible codons is the **genetic code** (see **Figure 10.9**). The genetic code has several significant characteristics. First, it is *unambiguous*: each codon specifies only one amino acid. It is also *redundant*: since there are a total of 64 codons but only 20 amino acids, several different codons call for the same amino acid, as already mentioned. Finally, the genetic code is virtually *universal*: nearly every organism on Earth uses the same code, from agrobacteria to tobacco cells to human cells—a feature that illustrates the common descent of all organisms.

Making a protein from an mRNA strand requires two additional types of RNA: The first is **ribosomal RNA** (**rRNA**), an important component of ribosomes. The second is **transfer RNA** (**tRNA**), which is the caddy for the process, delivering specific amino acids to the ribosomes as the codons are read off the mRNA "list."

Each tRNA specializes in binding to a specific amino acid and recognizes and pairs with a specific codon in the mRNA, like a puzzle piece that fits one amino acid on one end and one codon on the other. At one end of a tRNA molecule, a special sequence of three nitrogenous bases, called an **anticodon**, binds the correct codon on the mRNA. At the other end, the specific amino acid attaches (see **Figure 10.8**).

Let's recap. For translation to occur, an mRNA molecule must first bind to a ribosome. The ribosomal machinery then "scans" through the mRNA until it finds a start codon (AUG). Next the ribosome recruits the appropriate tRNAs one by one, as determined by the codons read in the mRNA sequence. A special site on the ribosome facilitates the linking of one amino acid to another, like beads on a string. Finally, the ribosome reaches a stop codon. The amino acid chain cannot be extended further, because none of the tRNAs will recognize and pair with any of the three stop codons. At this point the mRNA molecule and the completed amino acid chain separate from the ribosome. The new protein then folds into its compact, specific three-dimensional shape and is ready to go to work in the cell.

But this process does not always go as planned. In Chapter 9 we learned that a mutation is a change in the sequence of DNA bases. Mutations affect an organism by disrupting or preventing the healthy formation of a protein. For example, a mutation can cause a DNA sequence not to be translated or transcribed, prompt the amino acid chain to end prematurely, or make the final protein fold incorrectly,

Figure 10.9

The genetic code

(Top) The genetic code is composed of the 64 possible codons found in the mRNA. Each codon specifies an amino acid or is a signal that starts or stops translation. (Bottom) The genetic code is used during the translation of mRNA to protein.

Q1: How many codons code for isoleucine? For tryptophan? For leucine?

Q2: What codons are associated with asparagine? With serine?

Q3: From the partial mRNA sequence that you specified in Figure 10.6's question 3 as being transcribed from the DNA template strand, remove only the first A. What amino acid sequence would be translated as a result of this change? How does that sequence compare to the amino acid sequence you translated from the original mRNA sequence? *Bonus:* What kind of mutation is this? (*Hint:* See Chapter 9.)

among other possibilities. While single-base substitutions are not always a problem, single-base insertions and deletions cause a genetic "frameshift," shifting all subsequent codons "downstream" by one base (**Figure 10.10**). This shift scrambles the entire downstream DNA message, in turn scrambling the entire RNA message and causing the ribosomes to assemble a very different sequence of amino acids from the mutation point onward—as if every letter in this phrase was shifted to the left one space while the word length and spaces between words were retained (a si fever ylette ri nthi sphras ewa sshifte dt oth erigh ton espac ewhil eth ewor dlengt han dspace sbetwee nword swer eretaine d).

Tweaking Gene Expression

Inside a tobacco cell, as in most living cells, the expression of many genes can be turned on or off, slowed down (**down-regulated**), or sped up (**up-regulated**). This **gene regulation** enables organisms to change which genes they express in response to internal signals (from inside the body) or external cues in the environment. In this way, by producing different proteins as needed, organisms can adapt to their surroundings.

All cells in a multicellular individual have essentially the same DNA, yet different cells express different sets of genes, and within a given cell the pattern of gene expression can change over time. Single-celled organisms, such as bacteria, face a more difficult challenge: they are directly exposed to their environment, and they have no specialized cells to help them deal with changes in that environment. One way they meet this challenge is to express different genes at different times.

The expression of most genes in prokaryotes and eukaryotes is regulated by both internal and external signals. Many genes are also developmentally regulated, meaning that their expression can change, sometimes dramatically, as an organism grows and develops. Gene expression is regulated at many different points in the cell, including DNA packing (the way DNA is compressed or unwound in the genome), transcription, mRNA processing, and several

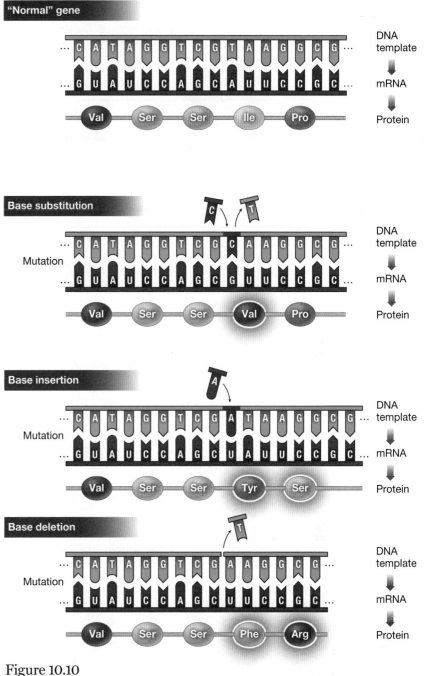

Figure 10.10

Effects of point mutations

Q1: Why is an insertion or a deletion in a gene more likely to alter the protein product than a substitution, such as A for C, would?

Q2: Which would you expect to have more impact on an organism: a point mutation as shown here, or the insertion or deletion of a whole chromosome (discussed in Chapter 8)?

Q3: Which mechanisms in a cell prevent mutations? (*Hint:* Refer back to Chapter 6 if needed.)

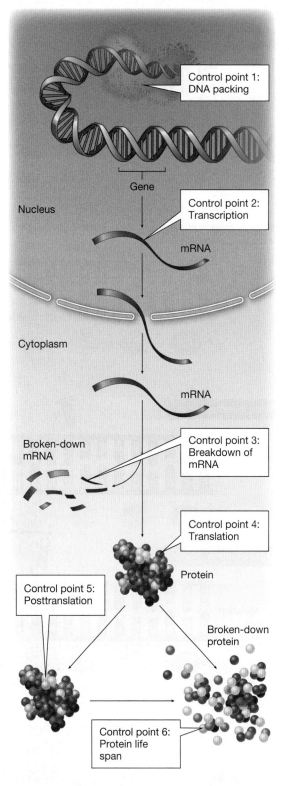

Control point 1:
DNA packing

Gene

Nucleus

Control point 2:
Transcription

mRNA

Cytoplasm

mRNA

Broken-down
mRNA

Control point 3:
Breakdown of
mRNA

Control point 4:
Translation

Protein

Control point 5:
Posttranslation

Broken-down
protein

Control point 6:
Protein life
span

points during translation (**Figure 10.11**). But for all living cells, a few genes are always expressed at a low level; their transcription is not regulated, because these genes are needed at all times.

At Medicago, the company takes advantage of gene regulation in tobacco cells to produce as much hemagglutinin protein as possible. After the agrobacteria are vacuum-sucked into the leaves and transcription and translation begin to actively occur in the tobacco cells, the plants are sent to grow in an incubation room, where technicians can alter the humidity, temperature, and level of light to maximize the amount of protein expressed by the plants.

"We've been able to tweak environmental conditions of the plants to boost gene expression," says Vézina. "You name it, we've tried it." It's an important step in the process, adds Wanner. "We determined what the best conditions for protein expression are."

To the Market

It's the end of April, and the final steps are being taken to isolate and purify the hemagglutinin protein from the tobacco plants, to see how much vaccine can be made in a month. After the plants have incubated for several days, the leaves are stripped and diced into green confetti, then digested with enzymes to

Figure 10.11

How gene expression is regulated

In eukaryotes, gene expression can be controlled at several points along the pathway from gene to protein to phenotype: before transcription, at transcription, during RNA processing, or at translation. Gene expression can also be regulated after translation, by control of the activity or life span of the protein.

Q1: As illustrated here, at what control point is transcription regulated?

Q2: What is a possible advantage of regulating gene expression before transcription, versus after?

Q3: If you wanted to up-regulate production of the hemagglutinin protein in a tobacco plant carrying the hemagglutinin gene, at which control point(s) would that be possible? Justify your reasoning.

The Deadly Price of a Pandemic

The Spanish flu of 1918 devastated populations across Europe, killing not only the young and old, but also healthy adults. Since then, new strains of flu have reared their ugly heads, and doctors and vaccine manufacturers struggle to anticipate the next outbreak.

1918
H1N1

50M	**650**K
dead worldwide	dead in U.S.

1957
H2N2

1.5M	**70**K
dead worldwide	dead in U.S.

1968
H3N2

750K	**34**K
dead worldwide	dead in U.S.

2009
H1N1

285K	**12**K
dead worldwide	dead in U.S.

break up the leaf material so that the desired proteins are released into solution. The resulting solution, which resembles green-pea soup, is filtered several times to isolate clusters of hemagglutinin, which will then be processed into a vaccine product that is safe to inject into people.

Medicago is not the first company to produce a human drug using a plant. The first human-like enzyme was produced from tobacco back in 1992 at Virginia Polytechnic Institute. Numerous other plant biopharming companies have sprung up since then, experimenting with various plant species, including corn, soybean, duckweed, and more. But the field is not without risks and controversy. "The main concern and risk is spread—that the gene will get out into nature and spread," says Roth. "But there are techniques to make sure that doesn't happen, and it is closely regulated."

One of those techniques is to grow the plants in contained environments, where there is no risk of contaminating food crops. Israel-based biotech company Protalix Biotherapeutics, for example, grows carrot cells inside in large hanging bags of fluid and cells. In May 2012, the FDA approved the first biopharmed drug for humans (produced by Protalix)—a therapy for Gaucher disease, a rare genetic disorder—grown in the company's carrot cells. "This approval demonstrates a proof of concept for the power of this technology," the CEO of Protalix told the journal *Nature*.

Today, Medicago has completed safety trials for its pandemic flu vaccine and has had positive results in the first of two large clinical trials in humans that tested its seasonal flu vaccine's efficacy. When the April manufacturing test was completed, the company had produced an astounding 10 million doses of flu vaccine in a single month. It would have taken 5–6 months to produce the same amount using the traditional method of growing vaccines in chicken eggs. With that success under its belt, in 2015 the company began building a second production complex. Based in Quebec City, the manufacturing facility will have the capacity to deliver up to 50 million doses of seasonal flu vaccine.

Medicago is also developing novel vaccines against rotavirus and rabies virus, and the company recently received a U.S. government contract to manufacture antibodies to treat Ebola virus infection. "It might have taken a bit longer than we thought for biopharming to be accepted, but it deserves the visibility and attraction it has now," says Vézina. "Biopharming is here to stay. I'm convinced of this."

REVIEWING THE SCIENCE

- Most genes contain instructions for building proteins. **Gene expression** is the process by which genes are transcribed into RNA and then translated into a protein.

- During **transcription**, which occurs in the nucleus, RNA polymerase binds the **promoter** of a gene and produces a **messenger RNA (mRNA)** version of the gene sequence from free nucleotides.

- Next, during **RNA splicing**, **introns** are snipped out of the pre-mRNA sequence, and the remaining **exons** are joined. The mRNA is transported out of the nucleus.

- During **translation**, which occurs in the cytoplasm, ribosomes convert the sequences of bases in an mRNA molecule to the sequence of amino acids in a protein, with the help

of **ribosomes** composed of **ribosomal RNA (rRNA)** and proteins, and associated with **transfer RNA (tRNA)**.

- Ribosomes read the mRNA information in sets of three bases at a time, called **codons**. There are 64 possible codons, including a **start codon** (AUG) and three possible **stop codons** (UAA, UAG, and UGA).

- The **genetic code** is redundant because there are 64 codons and only 20 amino acids, and universal because nearly all organisms on Earth use the same code.

- Gene expression is regulated at many different points in the pathway from gene to protein. Organisms rely on **gene regulation** to respond to signals inside the body and to external cues in the environment.

THE QUESTIONS

The Basics

1 Link each term with the correct definition.

GENE EXPRESSION

1. The production of RNA, using the information in the DNA sequence of a gene.

GENE REGULATION

2. The flow of information from gene to protein.

TRANSCRIPTION

3. The control of gene expression in response to environmental or developmental needs.

TRANSLATION

4. The linking of amino acids in the precise sequence dictated by an mRNA base sequence.

2 For each of the following, identify the type of RNA involved (mRNA, rRNA, or tRNA).

_____ a. Transports the correct amino acid to the ribosome, using the information encoded in the mRNA.

_____ b. Is a major component of ribosomes.

_____ c. Specifies the order of amino acids in a protein, using a series of three-base codons, where different amino acids are specified by particular codons.

_____ d. Contains a three-base anticodon that pairs with a complementary codon revealed in the mRNA.

_____ e. Assists in making the bonds that link amino acids together to make a protein.

3 Select the correct terms:
The genetic code demonstrates (**ambiguity / redundancy**) because some amino acids are coded by more than one codon. The lack of (**ambiguity / redundancy**) in the genetic code is evidenced by the fact that each codon codes for one, and only one, amino acid.

4 In the diagram of transcription shown here, fill in the blanks with the appropriate terms: (a) gene; (b) promoter; (c) terminator; (d) RNA polymerase; (e) mRNA.

5 Place the following steps of translation in the correct order by numbering them from 1 to 9.

_____ a. A tRNA molecule carrying the amino acid methionine binds at its anticodon site to the mRNA start codon.

_____ b. The ribosome links the first amino acid to the second amino acid to begin the amino acid chain.

_____ c. The ribosome continues to link each amino acid to the growing amino acid chain.

_____ d. The ribosome reaches a stop codon.

_____ e. An mRNA binds to a ribosome.

_____ f. The mRNA and the completed amino acid chain separate from the ribosome.

_____ g. The first tRNA, separated from its amino acid, releases from the mRNA.

_____ h. A tRNA molecule carrying the second amino acid binds to the second mRNA codon.

_____ i. Each tRNA releases from the mRNA after it is separated from its amino acid.

6 Which of the following are possible reasons that a cell would regulate its expression of a gene? (Select all that apply.)

(a) an increased need for a particular enzyme

(b) a decreased need for a particular enzyme

(c) increasing temperature in the external environment

(d) changing needs as an organism ages

(e) death

Challenge Yourself

7 Using the genetic code shown in Figure 10.9, find the amino acid coded by each of the following codons.

(a) AAU

(b) UAA

(c) AUA

(d) GGG

(e) CCC

8 Using the genetic code shown in Figure 10.9, find a codon that codes for each of the following amino acids.

(a) arginine

(b) alanine

(c) methionine

(d) glycine

9 During transcription, what RNA molecule will be made from the DNA template CGTTACG?

(a) CGTTAGC

(b) GCAAUGC

(c) GCATTGC

(d) CGUUAGC

10 Which amino acid sequence will be generated during translation from the following small mRNA: ...CCC-AUG-UCU-UCG-UUA-UGA-UUG...? (*Hint:* Remember where translation starts and stops.)

(a) Met-Glu-Arg-Arg-Glu-Leu

(b) Met-Ser-Ser-Leu-Leu

(c) Pro-Met-Ser-Ser-Leu-Leu

(d) Pro-Met-Ser-Ser-Leu

(e) Met-Ser-Ser-Leu

Try Something New

11 Some diseases (for example, Huntington's and Parkinson's) appear to be related to increasing protein levels in brain cells, which lead eventually to cell death. At which of the control points shown in Figure 10.11 might a gene regulation error be occurring with these diseases? Identify one control point at which the error would result in up-regulation of gene expression, and one control point at which the error would result in down-regulation of gene expression.

12 The following nucleic acid is an entire primary transcript (pre-mRNA not yet processed): ACGCAUGCGaugaugccccucagGUCUuguuuccgugaUGCCGUUGACCUGA. The nucleotides in capitals are exons; and the nucleotides in lowercase type are introns. Appropriately splice this primary transcript.

(a) ACGCAUGCGGUCUUGCCGUUGACCUGA

(b) augccuuucagguuuccguga

(c) augccuuucagGUCUuguuuccguga

(d) ACGCAUGCGaugGUCUUGCCGUUGACCUGA

(e) ACGCAUGCGaugccuagGUCUuguuuccgugaUGCCGUUGACC UGA

13 How is gene expression similar to DNA replication, and how is it different? Give at least one similarity and one difference.

Leveling Up

14 *Write Now* **biology: genotype to phenotype** Your roommate, who is also taking a biology class, has become a little confused. He informs you that the genetic code is known to be ambiguous because a given genotype may give rise to a variety of phenotypes during gene expression (for example, his twin brother is an inch taller and more tan than he is). You like your roommate and would like him to pass his next biology exam, so you decide to help him. Write him a brief note explaining (a) why the genetic code is not, in fact, ambiguous and (b) how gene expression derives a phenotype from a genotype.

15 **What do *you* think?** Most people carry two copies of a normal gene that codes for an enzyme, glucosylceramidase, that is involved in breaking down lipids no longer needed in cells. (Enzymes are proteins that cause specific chemical changes; they are biological catalysts.) One in 100 people in the United

States carries a recessive mutation that codes for a defective glucosylceramidase enzyme. And about one in 40,000 people carries two copies of the mutation and displays the symptoms of Gaucher disease. These symptoms, caused by the accumulation of lipids in cells, include anemia, enlarged organs, swollen glands and joints, and, in severe cases, neurological problems and early death.

Enzyme replacement therapy is effective but very expensive—about $200,000 annually—and must be continued, every 2 weeks, for life. The Israeli biotech company Protalix Biotherapeutics, working with the U.S.-based Pfizer Pharmaceuticals, has developed a process to genetically modify carrots to produce a replacement enzyme. The biopharmed enzyme will cost about 25 percent less than the standard enzyme therapies, which are grown in mammalian cell lines. Protalix is now working on treatments for other enzyme deficiency diseases.

The FDA's May 2012 approval of the drug developed by Protalix alarmed some environmental activists and health advocates, who fear that the company's genetically modified carrot is just the beginning of a wedge that will lead to an underregulated and potentially dangerous industry. There is some legitimacy to their concerns: the U.S. Department of Agriculture (USDA) does not require an environmental impact assessment for biopharmed crops; nor does it require biotech companies to share the location of their test fields or the identity of the biopharmed molecules being produced. Furthermore, the USDA is not sufficiently staffed to effectively monitor companies involved in biopharming.

What do you think? Should biopharming be allowed in the United States? If so, under what conditions and with what limits? For example, should it be allowed to produce drugs for only life-threatening illnesses, or only under highly controlled conditions? Be prepared to discuss your observations and reflections in class.

16 **Life choices** Go to the Centers for Disease Control and Prevention (CDC) influenza website (http://www.cdc.gov/flu) and read the pages "Key Facts About Influenza (Flu)" (under "Flu Basics") and "Key Facts About Seasonal Flu Vaccine" (under "Prevention—Flu Vaccine"). You can also go to the Mayo Clinic's influenza website (http://www.mayoclinic.org/diseases-conditions/flu/home/ovc-20248057). Then answer the following questions.

a. What is the flu? How is it passed on?

b. What are the possible symptoms and complications of the flu?

c. How can you decrease your chance of getting the flu, and what treatments are available if you become infected?

d. What are the benefits and risks of the flu vaccine?

e. Why is there a new flu vaccine every year?

f. Why is the flu vaccine more effective in some years than in others?

g. Who would you recommend should get the flu vaccine? Explain your reasoning.

h. Do you get a flu vaccine every year? Why or why not?

For more, visit digital.wwnorton.com/bionow2 for access to:

Whale Hunting

Fossil hunters discover Moby Dick's earliest ancestor—a furry, four-legged land lover.

After reading this chapter you should be able to:

- Define evolution and list the six types of evidence for evolution.
- Compare and contrast artificial selection and natural selection.
- Summarize the argument that the fossil record provides evidence in support of evolution.
- Give an example of a homologous or vestigial trait, and explain how such traits support the theory of common descent.
- Explain why even distantly related species have similar DNA.
- Use your knowledge of evolution and continental drift to make a prediction about the geographic location of a given set of fossils.
- Relate similarities in embryonic development among species to their shared evolutionary past.

CHAPTER

11

EVIDENCE FOR
EVOLUTION

Fossils break all the time. This time, the 50-million-year-old ear bone of a small, deerlike mammal called *Indohyus* snapped clean off the skull. Sheepishly, the young laboratory technician cleaning the fossil handed the broken piece to his boss, paleontologist and embryologist J. G. M. "Hans" Thewissen at Northeast Ohio Medical University. Thewissen tenderly turned the preserved animal remains over in his hand. Then, as the tech reached for the fossil to glue it back onto the animal's skull, Thewissen went rigid.

"Wow, that is weird," said Thewissen. The *Indohyus* ear bone, which should have looked like the ear bone of every other land-living mammal—like half a hollow walnut shell, but smaller—was instead razor thin on one side and very thick on the other (**Figure 11.1**). "Wow," repeated Thewissen. This wasn't the ear of a deer, or any other land mammal. Thewissen squinted closer. "It looks just like a whale," he said.

Although they live in the ocean like fish, whales are mammals like us. So are dolphins and porpoises (**Figure 11.2**). Like all mammals, whales are warm-blooded, have backbones, breathe air, and nurse their young from mammary glands. Numerous fossils have been found documenting whales' unique transition from land-living mammals to the mammoths of the sea, during which whale populations developed longer tails and shorter and shorter legs. But one crucial link in the fossil record was missing: the closest land-living relatives of whales. What did the ancestors of whales look like before they entered the water? Staring at the strange fossil in his hand, Thewissen realized he could be holding the ear of that missing link.

Whales are but one of the many organisms that share our planet. Every species is exquisitely fit for life in its particular environment: whales in the open ocean, hawks streaking through the sky, tree frogs camouflaged in the green leaves of a rainforest. There is a great diversity of life on Earth—animals, plants, fungi, and more—with each species well matched to its surroundings. This diversity of life is due to evolution.

"Evolution," in everyday language, means "change over time." In science, biological **evolution** is a change in the inherited characteristics of a

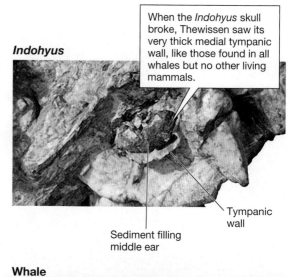

Indohyus

When the *Indohyus* skull broke, Thewissen saw its very thick medial tympanic wall, like those found in all whales but no other living mammals.

Tympanic wall

Sediment filling middle ear

Whale

Thick medial tympanic wall

White-tailed deer

Thin medial tympanic wall

Figure 11.1

The mysterious ear bone

The *Indohyus* fossil ear bone (top) looks more like the ear bone of whales (middle) than that of any modern land mammal (bottom). (Source: *Indohyus* and whale photos courtesy of Thewissen Lab, NEOMED.)

J. G. M. "HANS" THEWISSEN

Paleontologist and embryologist J. G. M. "Hans" Thewissen is a professor and whale expert at Northeast Ohio Medical University in the Department of Anatomy and Neurobiology. He and his lab study ancestral whale fossils and modern whale species.

Figure 11.2

A montage of mammals
Earth boasts an amazing diversity of life, as evidenced here. These are only a few of the almost 5,500 species of mammals currently living on the planet.

group of organisms over generations. Whales, for example, evolved from four-legged land-living animals into sleek emperors of the ocean, slowly becoming suited for the water over tens of millions of years. Whales changed as a population. Populations evolve; individuals do not.

You may wonder how we can be sure of evolutionary change, especially for a transformation as extreme as that of a furry, four-legged beast into a whale. There is strong evidence for evolution not only from fossils like those that Thewissen studies, but also from features of existing organisms, common patterns of how embryos develop, DNA evidence, geographic evidence, and even direct observation of organisms evolving today—including man's best friend.

Artificial to Natural

As we saw in Chapter 7, all dogs are a single species. That species is *Canis lupus familiaris*, a subspecies of the gray wolf. Scientists estimate that domestication of the gray wolf began about 16,000 years ago as humans habituated the

animals to civilization. From there, people bred the wolf for desired qualities, such as decreased agression and the ability to follow commands. Millennia later, dogs were selectively bred for specific traits like the long legs of a greyhound or the short snout of a bulldog. This selective process has led to incredible variation in the size and shape of dog breeds, from 6-pound Chihuahuas to 200-pound Great Danes (**Figure 11.3**). Dogs are an example of evolution that we can directly observe, and they evolved via artificial selection. **Artificial selection** is brought about by **selective breeding**, in which humans allow only individuals with certain inherited characteristics to mate.

Through selective breeding, humans have crafted enormous evolutionary changes within not only dogs, but many other domesticated organisms, including ornamental flowers, pet birds, and food crops. We can observe evolution happening through artificial selection via selective breeding.

Artificial selection happens when humans choose which individuals of a particular species are allowed to breed. Without the intervention of humans, can the environment itself

Figure 11.3

Selective breeding of dogs produces myriad traits

Dogs were domesticated only a few times, and always from gray wolves. Thus, the remarkable diversity of dogs represents the effects of selective breeding on a small number of lineages of domesticated wolves.

Q1: What is selective breeding, and how does it work?

Q2: Explain how selective breeding leads to artificial selection.

Q3: Name as many organisms as you can whose current characteristics are due to artificial selection.

"choose" who survives and breeds? It does. In nature, evolution occurs mainly via *natural selection*, as well as through other mechanisms we will discuss later. **Natural selection** is the process by which individuals with advantageous inherited characteristics for a particular environment survive and reproduce at a higher rate than do individuals with other, less useful characteristics. In other words, whoever has the most kids wins!

After the environment "chooses" the winners—those that successfully breed the most—the characteristics of those individuals become more common in successive generations because they have produced more offspring. For example, in 1977 a terrible drought struck the Galápagos Islands off the coast of Ecuador. One species of small ground finches—petite birds with sharp, pointy beaks—starved as the small, tender seeds they ate became scarce. But some heat-loving, drought-resistant plants still produced large, hard seeds. The finches with larger beaks could eat those seeds. They survived and reproduced, and by 1978, in just one generation, the average beak size in the population had increased (**Figure 11.4**).

Birds with smaller beaks were less likely to find seeds to eat and died at higher rates, thus failing to have sex, produce offspring, and pass on their genetic traits.

Birds with large beaks were more likely to survive, and passed on their large beak size to their offspring.

Figure 11.4

Natural selection results in larger beak size in finches

After a drought, only birds with larger beaks were able to eat the available food: large, hard seeds. In the span of just one generation, the average size of the species' beak was visibly larger.

Q1: What is natural selection?

Q2: If heavy rains caused an abundance of small, tender seeds and fewer large seeds, what do you predict would happen to the average beak size of the finches?

Q3: Compare and contrast artificial selection and natural selection. Name two ways in which they are similar. How are they different?

This is one of many examples of how a population can evolve via natural selection so that more and more individuals have beneficial traits, and fewer and fewer have disadvantageous traits. This is called **adaptation**—an evolutionary process by which a population becomes better matched to its environment over time. The finch population quickly adapted to its new, drier environment. Over time, the small-beaked finches died off and the large-beaked birds survived and reproduced; the finch population adapted to its environment in just a few years. Other adaptations take millions of years, such as whale ancestors adapting to aquatic life.

It is important to realize that biological evolution includes human evolution. Surveys taken since 2013 reveal that about 40 percent of adults in the United States do not believe that humans evolved from earlier species of animals. This statistic is startling because evolution has been a settled issue in science for nearly 150 years. Scientists like Thewissen go to work every day and see evolution in action. In fact, the vast majority of scientists of all nations agree that the evidence for evolution is overwhelming.

Six lines of evidence provide compelling support for biological evolution:

1. Direct observation of evolution through artificial selection

2. Fossil evidence

3. Shared characteristics among living organisms

4. Similarities and differences in DNA

5. Biogeographic evidence

6. Common patterns of embryo development

Nowhere is all this evidence more present and intriguing than in one of the most dramatic transitions to occur on Earth: the evolution of small, land-living mammals into dolphins, porpoises, and mighty whales.

Fossil Secrets

Fossils are the mineralized remains of formerly living organisms or the impressions of formerly living organisms (**Figure 11.5**). For Thewissen, getting his hands on the

Soft-bodied animals such as this one dominated life on Earth 600 million years ago (mya).

A fossil of a trilobite that lived between 410 and 355 mya.

Fossilized leaf of a 300-million-year-old seed fern.

This 20-million-year-old termite is preserved in amber, the fossilized resin of a tree.

A fossil of a *Velociraptor* entangled with a *Protoceratops*, which bit down on the predator's claw, locking both in a death grip.

Once solid wood has fossilized into solid rock, it is known as petrified wood.

Figure 11.5

Fossils through the ages

Myriad fossils exist, ranging from imprints of organisms, to preserved organisms, to completely mineralized bone and wood. Each fossil can be dated, and when the results are compiled, they can tell the life history of Earth.

Indohyus fossils was no easy task. Beginning in 2003, he made an annual pilgrimage to Dehradun, India, a city nestled in the foothills of the Himalayas. There he visited the widow of A. Ranga Rao, an Indian geologist who had hoarded piles of fossils excavated from Kashmir, a disputed border area between India and Pakistan. Most early whale fossils have come from the India-Pakistan region, where whales first evolved. But because of political tensions, in recent decades it has been too dangerous to travel to Kashmir, much less dig for fossils. Thewissen was frustrated by his inability to collect fossils in Kashmir.

The fossil record enables biologists to reconstruct the history of life on Earth, and it provides some of the strongest evidence that species have evolved over time. The relative depth or distance from the surface of Earth at which a fossil is found indicates its *order* in the fossil record. The ages of fossils correspond to their order: older fossils are found in deeper, older rock layers.

The fossil record contains excellent examples of how major new groups of organisms arose from previously existing organisms. The record includes numerous **transitional fossils**, evidence of species with some similarities to the ancestral group (land-living mammals) and some similarities to the descendant species (whales). Thewissen spent decades studying these intermediates—from the first known whale, the wolflike *Pakicetus* that waded in shallow freshwater; to the larger crocodile-like *Ambulocetus* that stalked its prey underwater; to the fully aquatic *Dorudon*, with its blowhole, flippers, and tail (**Figure 11.6**). Yet Thewissen and others had long been searching for the

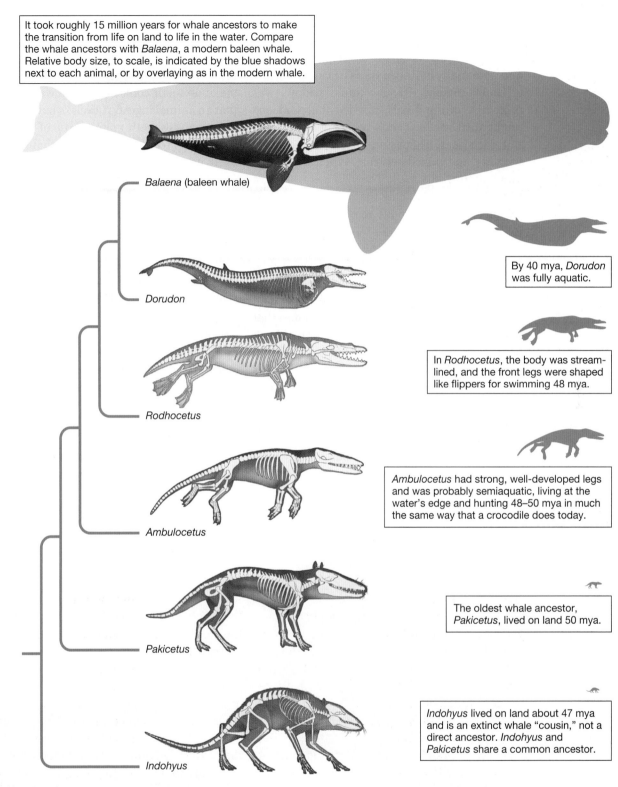

It took roughly 15 million years for whale ancestors to make the transition from life on land to life in the water. Compare the whale ancestors with *Balaena*, a modern baleen whale. Relative body size, to scale, is indicated by the blue shadows next to each animal, or by overlaying as in the modern whale.

Balaena (baleen whale)

Dorudon

By 40 mya, *Dorudon* was fully aquatic.

Rodhocetus

In *Rodhocetus*, the body was stream-lined, and the front legs were shaped like flippers for swimming 48 mya.

Ambulocetus

Ambulocetus had strong, well-developed legs and was probably semiaquatic, living at the water's edge and hunting 48–50 mya in much the same way that a crocodile does today.

Pakicetus

The oldest whale ancestor, *Pakicetus*, lived on land 50 mya.

Indohyus

Indohyus lived on land about 47 mya and is an extinct whale "cousin," not a direct ancestor. *Indohyus* and *Pakicetus* share a common ancestor.

Figure 11.6

Skeletons and body sizes of modern whales and fossil ancestors

These reconstructed skeletons from modern whales (top) and various ancestors are in chronological order.

Q1: What is the general definition of a fossil?

Q2: How are the ancestors of modern whales different from their present form?

Q3: What is meant by the term "intermediate fossil" when referring to the fossil record?

animal that preceded them all—the ancestor that lived on land. If Thewissen had to guess where those fossils might be, it was Kashmir, and potentially in Ranga Rao's basement.

Unfortunately, Ranga Rao's widow was protective of the fossils, worried that someone might steal her husband's property and legacy. But each year, Thewissen visited her, chatted with her, and gained her trust. When she passed away in 2007, she made Thewissen cotrustee of her estate, and suddenly the fossils, which had sat in dusty piles for 30 years, were available for study.

"I focused on taking the rocks back to the U.S., and having my fossil preparers remove the fossils from the rocks, which is very difficult," says Thewissen. From Ranga Rao's collection, Thewissen identified more than 400 bones that belonged to *Indohyus*. By collecting a thighbone here and a jawbone there, his team compiled a Frankenstein-like skeleton of a single *Indohyus* individual (**Figure 11.7**). After the discovery of the whalelike ear, the researchers looked even more carefully at the features of the fossils and found additional evidence that *Indohyus* was a relative of whales. This unassuming little animal, with a pointy snout and slender legs tipped with hooves, lived close to and had an affinity for the water.

The first clues about *Indohyus*'s lifestyle came from its teeth. Oxygen in the molecules that make up teeth comes from the water and food that an animal ingests. Levels of oxygen isotopes in *Indohyus*'s teeth match those of water-going mammals today, suggesting that *Indohyus* lived near and potentially spent a significant amount of time in the water. It also had large, crushing molars with levels of carbon isotopes that suggest it grazed on plants, as do hippopotamuses and muskrats that graze near and in water (**Figure 11.8**).

Lisa Cooper, a graduate student in Thewissen's lab at the time, identified another adaptation to the water: *Indohyus*'s leg bones. From the outside, the limbs of *Indohyus* look like those of any other mammal walking around on land. But on the inside, it's another story. Cooper cut out a section of bone from a limb, ground it down until she could see light through it, and then looked at the bone under a microscope. She saw that a thick layer of bone was wrapped around the bone marrow.

"Hans already had lots of bones of the earliest whales, and they all had extraordinarily

Figure 11.7

Fossilized skeleton of *Indohyus*, oldest cousin of the whales
A reconstructed fossilized skeleton of *Indohyus* was compiled from multiple sources and locations. The illustration is an artist's depiction of the living animal about 47 million years ago. (Source: Photo courtesy of Thewissen Lab, NEOMED.)

Crushing basins

Crushing basins

Crushing basins

The molars of *Indohyus* (top left) are similar to the shape of molars in contemporary aquatic plant-eating animals like hippos (top right and bottom left). These molars have crushing basins for grinding up tough plant fibers.

Figure 11.8

Comparing the skulls and jaws of fossilized *Indohyus* and a modern hippopotamus
These organisms' teeth indicate their ability to eat plant material.

thick bones," says Cooper, now an assistant professor at Northeast Ohio Medical University. Modern animals that live in shallow water, such as manatees and hippos, also have thick bones, which help prevent them from floating and enable them to dive quickly (**Figure 11.9**). "It isn't just isolated to whales," says Cooper. "Bones have thickened again and again as different groups of vertebrates entered the water. When you trace back through the fossil record, there is a pretty good correlation between thickness of bone and whether something was living in the water."

Indohyus's thick bones are an example of an **adaptive trait**, a feature that gives an individual improved function in a competitive environment. By being able to easily wade and dive in water, *Indohyus* had an advantage over other organisms in escaping predators and accessing plants to eat on the river floor. Adaptive traits take many forms, from an anatomical feature

LISA COOPER

Lisa Cooper is an assistant professor at Northeast Ohio Medical University in the Department of Anatomy and Neurobiology. She earned her PhD in Thewissen's lab.

Figure 11.9

Cross sections of the femurs of *Indohyus*, water-dwelling animals, and land-dwelling animals

Aquatic organisms have thick, dense bone around a narrow marrow space, while terrestrial organisms have thin bone and a large marrow space. *Indohyus* bone structure is a trait that is shared with other water-dwelling animals.

Q1: Why do water-dwelling animals have thicker bones than land-dwelling animals?

Q2: Why does this thick-bone adaptation suggest a water-dwelling lifestyle?

Q3: How did this adaptation likely increase survival or reproduction in *Indohyus*?

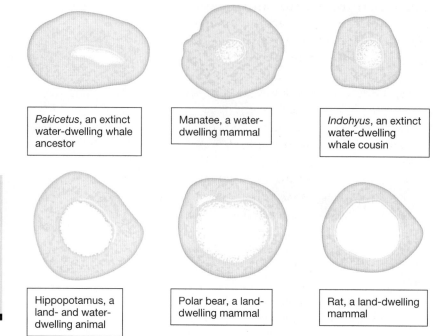

Pakicetus, an extinct water-dwelling whale ancestor

Manatee, a water-dwelling mammal

Indohyus, an extinct water-dwelling whale cousin

Hippopotamus, a land- and water-dwelling animal

Polar bear, a land-dwelling mammal

Rat, a land-dwelling mammal

like *Indohyus*'s bones, to behaviors, to the functions of individual proteins. Echolocation in bats, for example, is an adaptation for catching insects in the dark. Stick insects physically and behaviorally mimic the plants they live on—an adaptation that helps them avoid detection by predators (**Figure 11.10**).

Figure 11.10

Stick insects avoid detection by predators

Stick insects are well adapted to their environment; they move slowly and look just like the branches they live on.

The Ultimate Family Tree

Thick bones are not restricted to just the water-loving ancestors of whales, as we noted, but can be seen in many other animals, such as hippos. This similarity across organisms is another type of evidence for evolution—shared characteristics among species. Many shared characteristics—such as thick bones for animals that take to shallow water, or sexual reproduction via egg and sperm, or eukaryotic cells—result from organisms sharing traits that evolved from a **common ancestor**, an organism from which many species have evolved. A group of organisms have **common descent** if they share a common ancestor.

When one species splits into two, the two resulting species share similar features, or **homologous traits**, because they have common descent—though these features may begin to look different from one another over time

(**Figure 11.11**). For example, whales are so different from humans that it can be difficult to identify similarities, but we evolved from a common mammalian ancestor and do share homologous traits: humans and whales both nurse their young and have a single lower jawbone because our common ancestor had those traits.

Vestigial traits are another type of trait that many organisms have because of a common ancestor. These features are a piece of the evolutionary past, inherited from a common ancestor but no longer used. Vestigial traits may appear as reduced or degenerated parts whose function is hard to discern (**Figure 11.12**). For example, many modern whales have vestiges of thighbones, also called femurs, embedded in the skin next to the pelvis. In land mammals, birds, and other tetrapod vertebrates, these bones are critical for walking, running, and jumping. Aquatic whales have no need of this bone, yet its traces remain.

Whales also have small muscles devoted to nonexistent external ears, likely from a time when they were able to move their ears as land animals such as dogs do for directional hearing. Vestigial traits are not adaptations. In fact, they can be detrimental. Most humans no longer need wisdom teeth to replace lost teeth during adolescence, yet most people still have them. They tend to erupt around the twentieth year of life, often causing severe pain and displacing other teeth, and they usually require removal.

Clues in the Code

Within the cells of every organism is one of the strongest pieces of evidence for evolution: DNA. Living things universally use DNA as hereditary, or genetic, material (see Chapter 8 for review). The fact that all organisms on Earth—even those as different as bacteria, redwood trees, and humans—use the same genetic code is further evidence that the great diversity of living things evolved from a common ancestor.

Researchers have analyzed the DNA sequences of whales and other animals and shown that, of all animals, whales are most

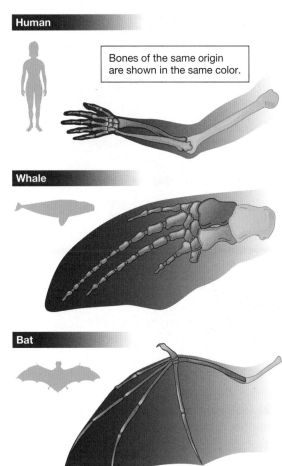

Bones of the same origin are shown in the same color.

Figure 11.11

Homologous traits are shared characteristics inherited from a common ancestor

The human arm, the whale flipper, and the bat wing are homologous structures due to common descent. All three structures have a matching set of five digits and a matching set of arm bones that have been altered by evolution for different functions.

Q1: What is meant by the term "common ancestor"? Give an example.

Q2: Why are homologous structures among organisms evidence for evolution?

Q3: Aside from skeletal structural similarities, what other commonalities among organisms might be considered homologous?

External surface

Skeletal structure

Reduced hind legs

Figure 11.12

Vestigial traits are reduced or degenerated remnants with no apparent function

Snakes are limbless reptiles with no apparent use for the degenerated remnants of hind legs that they still have. The python shown here has extremely reduced hind legs that are only barely visible externally.

Q1: Why are vestigial structures among organisms evidence for evolution? Give an example of another vestigial structure.

Q2: Are vestigial structures also homologous structures? Explain.

Q3: Why do vestigial structures still exist if they are no longer useful?

---

closely related to even-toed ungulates—a group of hoofed mammals that includes modern deer, giraffes, camels, pigs, and hippos. *Indohyus* is an example of an extinct even-toed ungulate. Molecular studies therefore confirm the prediction that whales and *Indohyus* share a common ancestry.

According to DNA sequence similarity, hippos are whales' closest living relatives. Whale DNA is more similar to hippo DNA than to the DNA of other marine mammals, such as seals and sea lions. **DNA sequence similarity** is a measure of how closely related two DNA molecules are to each other. For example, in the DNA sequences of the same insulin gene in humans and mice, 83 percent of the nucleotides are identical at corresponding positions (**Figure 11.13**). We share a more recent common ancestor with mice than we do

with chickens, for whom the DNA sequence of the same gene is only 72 percent identical to ours. In fact, the insulin gene of our closest living relative on Earth, the chimpanzee, has 98 percent similarity to the human insulin gene. That comparability implies that humans and chimpanzees share a very recent common ancestor.

The fact that these separate lines of evidence—anatomical features and DNA—yield the same result over and over again for diverse groups of organisms is strong evidence for evolution.

Whale evolution is "one of the best case studies documenting how a vertebrate can go from a terrestrial to an aquatic environment," says Lisa Cooper. Another type of evidence supporting whale evolution comes from the locations where whale fossils have been found.

Figure 11.13

DNA sequence similarities of the insulin gene

The complete coding sequence of the human insulin gene contains 333 nucleotides. Only the first 50 nucleotides are shown here. Unshaded paired sequences are identical nucleotides at that position; those shaded in yellow are different.

Q1: If a sequence from another species showed a 96 percent sequence similarity to humans, would that species be more or less closely related to humans than chimpanzees are?

Q2: Should similarities in the DNA sequences of genes be considered evolutionary homology? Explain.

Q3: How is the increased similarity in the DNA sequences of genes between more closely related organisms—and the decreased similarity between less closely related organisms—evidence for evolution? Use the example in this figure to support your answer.

Birthplace of Whales

Earth's continents are on massive tectonic plates, which slowly move over time in a process called continental drift or plate tectonics. About 250 million years ago, South America, Africa, and all of the other landmasses of Earth had drifted together to form one giant continent called Pangaea. About 200 million years ago, Pangaea slowly began to split up, separating to ultimately form the continents as we know them. That separation continues today; for example, each year South America and Africa drift farther apart by about an inch.

We can use knowledge of evolution and continental drift to make predictions about the **biogeography** of a species—the geographic locations where its fossils will be found. For example, today the lungfish *Neoceratodus fosteri* is found only in northeastern Australia, but its ancestors lived during the time of Pangaea. As predicted, therefore, fossils of those ancestors are found on all continents except Antarctica (**Figure 11.14**).

The biogeography of whale fossils matches the pattern predicted by evolution: all early

Portions of the supercontinent Pangaea began to drift apart about 200 million years ago.

Pangaea

Red dots indicate places where *N. fosteri* fossils have been found.

This species is currently found in only the orange-shaded region of Australia.

Figure 11.14

Biogeography can reveal a species' evolutionary past

Ancestors of the freshwater lungfish *Neoceratodus fosteri* lived during the time of Pangaea. *N. fosteri* fossils have been found on all continents except Antarctica.

Q1: Why should we expect to find *N. fosteri* fossils all over the world, given that it first evolved in Pangaea?

Q2: Can we use biogeographic evidence to support evolution without using fossil evidence? Give examples.

Q3: How might we use DNA sequence similarities together with biogeography as evidence for evolution?

species of whales, such as the crocodile-like *Pakicetus*, that lived in rivers and lakes but did not swim in the ocean are found near India and Pakistan. "It makes sense," says Thewissen. "You don't have crocodiles crossing the Atlantic." But fossils of fully aquatic protocetids, which emerged about 40 million years ago, are geographically much more widespread; they have been found as far away from Pakistan as Canada. "Protocetids are good swimmers, so we find their fossils all around the world," says Thewissen.

Growing Together

Though Thewissen has built a career on finding and describing whale fossils, he has recently become enamored with another vein of evolutionary evidence: embryology. A major prediction of evolution is that organisms should carry within themselves evidence of their evolutionary past, and they do. Evidence of evolution can be observed in shared patterns of **embryonic development**.

Once again, these common patterns are caused by descent from a common ancestor. Rather than evolving new organs "from scratch," new species inherit structures that may have been modified in form and sometimes even in function.

Upon fusion of sperm and egg, an animal embryo begins to grow and develop. The manner in which an embryo develops, especially at the early stages, may mirror early developmental stages of ancestral forms. For example, anteaters and some baleen whales do not have teeth as adults, but as fetuses they do. And the embryos of fishes, amphibians, reptiles, birds, and mammals (including humans) all develop pharyngeal pouches or gill slits (**Figure 11.15**). In fish, the pouches develop into gills that adult fish use to absorb oxygen underwater. In human embryos, these same features become parts of the ear and throat.

"I was interested to get embryos to look at some of these processes that we see happen in evolution, to see if they happen in development," says Thewissen. The first trait he examined was

☐ Gill slits ☐ Tail

Fish   Reptile   Bird   Human

Embryos all share gill slits and tails because fishes, reptiles, birds, and humans all evolved from a common ancestor that had these features.

Figure 11.15

Evolutionary history can be extrapolated from similarities in embryo development

Complex structures in descendant species are generally elaborations of structures that existed in their common ancestor.

Q1: Why are the similarities among organisms during early development evidence for evolution? Give an example.

Q2: Are the similar structures among vertebrate species during embryogenesis homologous structures? Explain.

Q3: Why do embryonic structures still exist at points during embryogenesis if they are not used after birth?

hind limbs. "We know, from fossil evidence, that early whales lose their hind limbs," says Thewissen. So he wondered if hind limbs exist in whale or dolphin embryos (dolphins are also mammals and are closely related to whales). And if so, what makes them subsequently disappear before the animal is born.

Examining spotted dolphin embryos, Thewissen saw that when the embryos are the size of a pea, they do develop hind limb buds, but by the time they grow into the size of a bean, the limb buds are gone. In 2006, he and researchers at several other universities studied the genes that are active in whale and dolphin embryos and concluded that whales' hind limbs regressed over millions of years through small changes in

Watching Evolution Happen

In the laboratory, scientists can manipulate populations of organisms to watch evolution in real time via artificial selection. In 2012, researchers at UC Irvine manipulated the growing environment of *Escherichia coli* bacteria, exposing them to far hotter temperatures than normal, to see if they would adapt. Most of the bacteria adapted via one of just two primary pathways.

Assessment available in smartwork**5**

37°C

42.2°C

1 | Scientists grow **115** separate, genetically identical populations of *E. coli* at a comfortable temperature of **37°**C, or **98.6°**F.

2 | The populations reproduce through **2,000** generations while being subjected to an increased heat of **42.2°**C, or **107.96°**F.

3 | One sample is taken from each of the **115** populations, and the genome is sequenced.

4 | **1,258** molecular changes, averaging **11** genetic mutations per clone, are detected. To help survive the heat, the *E. coli* tend to mutate along two different trajectories:

- Mutations in the RNA polymerase complex, an enzyme that transcribes RNA.

- Mutations in the *rho* gene, which encodes a protein that stops RNA transcription.

5 | The research continues. The next step is to figure out how the mutations in the RNA polymerase complex and in *rho* helped the *E. coli* survive the heat.

a number of genes relatively late in embryonic development (**Figure 11.16**). The loss of hind limbs in the embryo corresponds to the disappearance of hind limbs in the fossil record. "It was an awesome way to combine embryology with fossils," says Thewissen. In 2015, researchers at the Smithsonian Institution likewise found that the development of fetal ear bones in the womb paralleled changes observed in the whale fossil record.

With their discovery of *Indohyus*, Thewissen, Cooper, and their team bridged a 10-million-year gap in the fossil record, identifying an important transition species to whales. It is another rock in the mountain of evidence surrounding whale evolution, which was further fortified in April 2015 with the discovery of fossils from a new species of extinct pygmy sperm whale found in Panama.

Evolution is supported by mutually reinforcing, independent lines of evidence: direct observation, fossils, shared characteristics among living organisms, similarities and divergences in DNA, biogeographic evidence, and common patterns of embryo development. Just as the theory of gravity forms the foundation of physics, so evolution is the central tenet of biology.

There is no question that evolution happens. The intriguing, fascinating question is *how* does it happen? Through scientific research, we know that whales descended from a group

Forelimb Hind limb

Dolphins maintain their front limb buds, which become flippers.

Dolphins' hind limb buds do not develop into limbs or flippers.

Figure 11.16

Dolphin embryonic development
The embryos of dolphins from weeks 4–9 of development show the formation and then subsequent loss of the hind limb buds. (Source: Based on a photo by Thewissen Lab, NEOMED, permission granted.)

of land-living mammals including the petite *Indohyus*, munching on freshwater plants and splashing through the shallows. But how did a population of small, furry animals become the mammoths of the sea? Next we investigate the mechanisms of evolution—how and why it works.

REVIEWING THE SCIENCE

- Biological **evolution** is the change in characteristics of a population of organisms over inherited generations.

- **Artificial selection** results in biological evolution. Humans choose which organisms survive and reproduce—a process known as **selective breeding**.

- **Natural selection** is the process by which individuals with advantageous genetic characteristics for a particular environment survive and reproduce at a higher rate than competing individuals with other, less useful characteristics.

- **Adaptation** is an evolutionary process by which a population becomes better matched to its environment over time. An adaptation is also a trait—an **adaptive trait**—that has evolved as a result of this process.

- **Fossils** are the preserved remains (or their impressions) of formerly living organisms. The fossil record enables biologists to reconstruct the history of life on Earth, and it provides some of the strongest evidence that species have evolved over time.

- Many similarities among organisms are due to the fact that the organisms evolved via **common descent** from a **common ancestor**. When one species splits into two, the two resulting species share similar features, called **homologous traits**. If a homologous trait is no longer useful, it is called a **vestigial trait**.

- The fact that organisms as different as bacteria, redwood trees, and humans show **DNA sequence similarity** is evidence that the great diversity of living things evolved from a common ancestor.

- **Biogeography** uses knowledge about evolution and plate tectonics to make predictions about the geographic locations where fossils will be found.

- Similarities in **embryonic development** of different organisms show that the characteristics of modern organisms arose through evolutionary modifications of traits inherited from common ancestors.

THE QUESTIONS

The Basics

1 Intermediate fossils

(a) share some similarities with their ancestral group.

(b) share some similarities with their descendant groups.

(c) both of the above

(d) none of the above

2 If two different organisms are closely related evolutionarily, then they will

(a) be similar in size.

(b) share a recent common ancestor.

(c) have very different DNA sequences in their genes.

(d) be randomly located throughout the world.

3 When two organisms are very distantly related in an evolutionary sense,

(a) they should have extremely similar embryonic development.

(b) they must share a very recent common ancestor.

(c) the sequences of DNA in their genes should be less similar (more different) than those of two more closely related organisms.

(d) they should share more homologous traits than two more closely related organisms share.

4 Link each term with the correct definition.

BIOGEOGRAPHY	1. A reconstruction of the history of life on Earth.
FOSSIL RECORD	2. The similarities in the nucleotide sequences among related organisms.
DNA SEQUENCE SIMILARITY	3. The similarities among organisms that are due to the fact that the organisms evolved from a common ancestor.
EMBRYONIC SIMILARITY	4. The geographic locations where related organisms and fossils are found.
HOMOLOGOUS TRAITS	5. Specifically during development, complex structures in descendant species that are generally elaborations of structures that existed in the common ancestor.

Challenge Yourself

5 All mammals have tailbones and muscles for moving a tail. Even humans have a reduced tailbone and remnant tail-twitching muscles, though these features have no apparent usefulness. These traits in humans would best be described as

(a) convergent structures.

(b) fossil evidence.

(c) evidence from biogeography.

(d) vestigial traits.

6 Reduced tailbones and the associated remnant muscles in humans are an example of what type of evidence for common descent?

(a) artificial selection

(b) homologous traits

(c) biogeography

(d) fossil evidence

7 When a population evolves to be better fitted to its environment, this is an example of _____, which is brought about by the process of _____.

8 In one sentence each, identify the similarities and the differences between artificial selection and natural selection.

Try Something New

9 Cat DNA is much more similar to dog DNA than to tortoise DNA. Why?

(a) Cats and dogs are both carnivores and take in similar nutrients.

(b) Cats and dogs have lived together with humans for a long period of time, so they have grown more similar.

(c) Cats and dogs have more offspring during their lifetime than tortoises have, so their DNA changes less rapidly.

(d) Cats and dogs have a common ancestor that is more recent than the common ancestor of cats and tortoises.

10 DNA sequences were analyzed from humans and three other mammals: species X, Y, and Z. Which of these mammals is most closely related to humans? (*Note:* Regions identical to human DNA are shown in bold type.)

Human:
AATGCTTTGGGGGATCGCGAGCGCAGCGC

Species X:
GGGTT**TTT**ATCGCTATATATATATA

Species Y:
AATGCTTTGGGGGATCGCGAGCGCATATA

Species Z:
AATGCGGGTTTTT**ATC**TATATATATATA

11 Which of the following is *not* an example of artificial selection?

(a) Your younger sibling got a hamster as a birthday present, and it turned out to be pregnant. Several of the offspring had long, silky hair, and your sibling put them together in an enclosure to try to produce more baby hamsters with long, silky hair. Your sibling continued to breed long- and silky-haired hamsters to each other and now, several years later, your sibling's bedroom is full of cages containing long- and silky-haired hamsters.

(b) Your mother has been saving the best seeds from her lima bean plants every summer and replanting them the next year. She likes seeds that are plump and bright green and saves only these each year. Within 10 years, almost all of her lima beans have become plump and bright green.

(c) Farmer Brown has a duck that can type. He sets up an online dating profile for his duck to find a female duck that can also type. Seven years after the arranged wedding of the two typing ducks, Farmer Brown has an entire flock of ducks, most of which can type.

(d) Female fish in a natural pond have variable skill at depositing their eggs near the murky shore, where the eggs are better hidden from predators. Eggs that are not deposited near the murky shore are quickly devoured by fish. Female hatchlings from the eggs deposited near the murky shore grow up to be good at depositing eggs near the shore and therefore have a survival advantage.

12 The fossil record shows that the first mammals evolved 220 million years ago. The supercontinent Pangaea began to break apart 200 million years ago. On which continents would you predict that fossils of the first mammals will be found?

Leveling Up

13 **What do *you* think?** The prerequisites for medical school application always include courses in cell biology, genetics, and biochemistry but rarely include a formal course in evolution. Do you think medical schools should require a formal course in evolution as a prerequisite for admission? Why or why not? Research the issue and support your case using information you find from credible sources.

14 ***Write Now* biology: evidence for evolution** This assignment is designed to expand your knowledge of the evidence for evolution. View the following videos, found at http://www.pbs.org/wgbh/evolution/educators/teachstuds/svideos.html, and answer the questions accompanying each one here.

Video 1: "Isn't Evolution Just a Theory?"
Why is evolution not just a theory? Use specifics from the video to defend your answer.

Video 2: "Who Was Charles Darwin?"
Why do you think Charles Darwin's ideas and book *On the Origin of Species* were so groundbreaking and "revolutionary"? Use specifics from the video to defend your answer.

Video 3: "How Do We Know Evolution Happens?"
Describe how the video portrays whale evolution. Include specific examples of transitional fossils described in the video, and explain why the scientists at the time the video was made considered the fossils to be whale ancestors.

Video 5: "Did Humans Evolve?"
DNA sequences of different species can be used to provide evidence of common descent. Using examples from the video, explain why DNA sequence similarity is the best evidence of evolution on Earth.

Video 6: "Why Does Evolution Matter Now?"
Describe why the theory of evolution matters to the field of medicine and to individual doctors. Use the example of tuberculosis from the video to support your answer. How does this video affect your answer to question 13?

For more, visit digital.wwnorton.com/bionow2 for access to:

Battling Resistance

An antibiotic-resistant superbug is evolving to overcome our drugs of last defense. How do we beat a champion pathogen?

After reading this chapter you should be able to:

- Clarify why evolution occurs only in populations, not in individuals.
- Distinguish among directional, stabilizing, and disruptive selection.
- Explain how natural selection brings about increased reproductive success of a population in its environment.
- Compare and contrast convergent evolution and evolution by common descent, and give an example of each.
- Describe how DNA mutations can create new alleles at random.
- Illustrate how gene flow works and how it may inhibit evolution.
- Relate the process of genetic drift to genetic bottlenecks and the founder effect, using examples.

CHAPTER
12

MECHANISMS
OF EVOLUTION

Dawn Sievert vividly remembers June 14, 2002—the day disaster struck. Sievert is an epidemiologist, or "disease detective"—someone who studies the patterns and causes of human disease. At the time, she was working at the Michigan Department of Community Health, monitoring reported cases of antibiotic-resistant bacteria, a major health care concern. Sievert spent a significant amount of her time investigating outbreaks of an increasingly common and worrisome microbe called MRSA, short for "methicillin-resistant *Staphylococcus aureus.*"

S. aureus, commonly known as "staph," is a small, round bacterium that usually lives benignly in our nostrils and on our skin. But on rare occasions, staph slips beneath the surface of a burn or cut and causes an infection, which can be especially dangerous for individuals with suppressed immune systems, such as the elderly and patients on chemotherapy.

Staph is one of the most common causes of hospital infections today, and it is treated with antibiotics, drugs that kill bacteria but not human cells. Penicillin (**Figure 12.1**) was the first antibiotic used against staph, but the wily microbe evolved resistance to penicillin even before the drug became commercially available to the public in the 1940s.

When penicillin stopped working against staph, doctors switched to an antibiotic called methicillin. Methicillin worked for about 20 years, until populations of staph evolved widespread resistance to that antibiotic as well. To the chagrin of doctors and patients everywhere, bacteria are able to adapt rapidly to new antibiotics, thanks to their short generation time and ability to share resistance genes among themselves. (See Chapter 15 for more on bacteria.) These tiny microbes are the Navy SEALs of evolution—the best of the best at evolving.

As we saw in Chapter 11, biological *evolution* is a change in the frequencies of inherited traits in a population over generations. Staph adapted to the presence of methicillin: Only bacteria with genetic traits protecting the microbe from the antibiotic survived. These traits were passed among populations and down from one generation to the next until methicillin resistance was frequent across staph populations (**Figure 12.2**).

Today, MRSA is rampant in hospitals, so doctors have been forced to turn to one of medicine's last lines of defense against the superbug. Vancomycin, a strong, blunt antibiotic that was first isolated from the mud of the Borneo jungle, is considered one of the "drugs of last resort" for fighting these serious infections—which brings us back to June 14, 2002, and Dawn Sievert.

On that day, a lab technician at a dialysis center in Detroit, Michigan, took two swabs of an infected foot ulcer belonging to a 40-year-old diabetic woman (**Figure 12.3**). The patient had previously suffered from numerous foot infections, including MRSA, which had been treated with vancomycin for 6½ weeks. The swabs of this latest infection were sent to a local laboratory, where technicians grew the bacteria in a dish to test its susceptibility to various antibiotics.

When the results of the first test came in, the laboratory staff immediately picked up the phone and called Sievert's office. The bacteria, they told Sievert and the health department, appeared to be resistant to vancomycin. "First, we needed laboratory confirmation and had to

Figure 12.1

Penicillin, produced by mold, kills bacteria
The mold *Penicillium*, the fuzzy white growth with blue spores growing on this petri dish, secretes the antibiotic penicillin into the agar medium, killing the red bacteria surrounding it.

DAWN SIEVERT

Dawn Sievert is an infectious disease epidemiologist who worked for the Centers for Disease Control and Prevention until 2016. While at the Michigan Department of Community Health previously, Sievert investigated the first-ever VRSA infection.

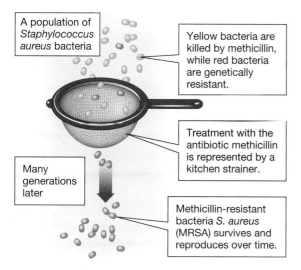

A population of *Staphylococcus aureus* bacteria

Yellow bacteria are killed by methicillin, while red bacteria are genetically resistant.

Treatment with the antibiotic methicillin is represented by a kitchen strainer.

Many generations later

Methicillin-resistant bacteria *S. aureus* (MRSA) survives and reproduces over time.

Figure 12.2

Natural selection results in resistance to antibiotics

The use of antibiotics allows any bacteria (in this case, *S. aureus*) that are resistant to the antibiotic to survive and reproduce. Over time, the frequency of resistant bacteria increases in the surviving populations.

Q1: What is natural selection selecting for here?

Q2: Why do bacteria that are not genetically resistant to antibiotics die out when exposed to antibiotics?

Q3: Why is the antibiotic represented by a kitchen strainer in this figure?

Figure 12.3

MRSA infection in the foot of a patient with diabetes

control any potential for panic," says Sievert. Her team asked the local lab to run its own test again and to send the health department a sample to independently test. Both teams waited.

The tests came back from each lab. Both were positive; the woman's foot was infected with the first reported vancomycin-resistant strain of *S. aureus*. VRSA had arrived.

Birth of a Superbug

"At that point, it was the first-ever VRSA in the world," says Sievert. Unfortunately, it was not the last.

Through evolution, staph first survived penicillin, then methicillin, then vancomycin (**Figure 12.4**). The evolution of *Staphylococcus*

MRSA (methicillin-resistant *Staphylococcus aureus*) bacteria growing on an agar plate

Paper tab infused with vancomycin. The surrounding area it has diffused into, killing off all bacteria in the area, is called the **zone of inhibition**.

VRSA (vancomycin-resistant *Staphylococcus aureus*) bacteria growing on an agar plate

Paper tab infused with vancomycin. Note that the zone of inhibition is much smaller, demonstrating that the bacteria can grow in the presence of the antibiotic.

Figure 12.4

MRSA versus VRSA in a vancomycin-resistance test

In the presence of vancomycin, MRSA (top) does not grow, while VRSA (bottom) grows relatively well.

Q1: What is the difference between MRSA and VRSA?

Q2: Why is there a clear zone (the "zone of inhibition") around the paper disk in the top dish but not in the bottom dish?

Q3: Why is the lack of a clear zone around the paper disk in the bottom dish so alarming?

aureus is a profound example of a species changing over time.

In the mid-1800s, two English biologists, Charles Darwin and Alfred Russel Wallace, studied the diversity of life and concluded that species were not, as was generally thought at that time, the unchanging result of separate acts of creation. Instead, both men came to the bold new conclusion that species "descend with modification" from ancestor species; that is, new species arise from previous species.

Descent with modification, we know today, occurs not only in populations of large organisms like whales and finches, but also in the tiniest single-celled bacteria and viruses. We generally think of evolution happening over millions of years, but some evolutionary changes, such as adaptations for antibiotic resistance, take place over very short time spans as particular alleles spread rapidly through a population. Recall from Chapter 7 that *alleles* are different versions of the same gene (sequences of DNA) produced by random mutation, and therefore *allele frequencies* are the proportions of specific alleles in a population (**Figure 12.5**). Evolution corresponds to changes in the relative proportions or frequencies of alleles in a population over time.

When allele frequencies in a population change, becoming more or less common, the attributes or phenotypes of the population change as well; that is, the population evolves. As more and more staph containing the allele for methicillin resistance survived and reproduced, the whole population of staph evolved, becoming new, more powerful bacteria. But how exactly does this happen? Where do new alleles come from, and how do the frequencies of alleles in a population change?

Sievert and a host of other researchers and doctors experienced the emergence of a new allele firsthand. When the results of the foot ulcer test came in positive for VRSA, Sievert's team immediately called the Centers for Disease Control and Prevention (CDC), a government agency that investigates disease outbreaks and makes public health recommendations. The Michigan Health Department and the CDC converged on the dialysis center where VRSA had been found. They pored over the medical history of the patient, examined her wound, took swabs from the nostrils and wounds of anyone

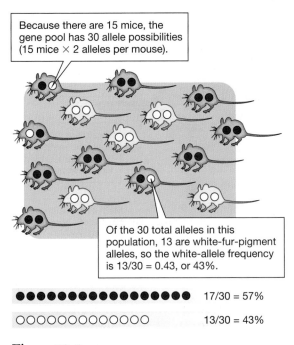

Because there are 15 mice, the gene pool has 30 allele possibilities (15 mice × 2 alleles per mouse).

Of the 30 total alleles in this population, 13 are white-fur-pigment alleles, so the white-allele frequency is 13/30 = 0.43, or 43%.

17/30 = 57%

13/30 = 43%

Figure 12.5

Allele frequencies are calculated as percentages in a population

These mice have two white-fur-pigment alleles and appear white, have two black-fur-pigment alleles and appear gray, or have one black and one white allele and appear gray. To calculate the white-fur-pigment allele frequency in the population, the number of white alleles is counted and divided by the total number of alleles.

Q1: What would the white-fur-pigment allele frequency be if three of the homozygous black allele mice (having two black alleles) were heterozygous (having one white and one black allele) instead?

Q2: What would the white-fur-pigment allele frequency be if all of the white mice died and were therefore removed from the population? Would the black-fur-pigment allele frequency be affected? If so, how?

Q3: What would the white-fur-pigment allele frequency be if all of the gray mice died and were therefore removed from the population?

who had come in contact with her, and then waited anxiously to see whether the dangerous microbe had spread.

Thankfully, they found that the vancomycin-resistant bug had not yet spread. "We're lucky.

If that highly resistant organism had the ability to spread rapidly, we'd have some very sick people at risk in hospitals and other health care settings," says Sievert.

Yet it was not the end of VRSA. Since 2002, there have been 13 additional cases of vancomycin-resistant staph infections in the United States: in the urine of a woman in New York with multiple sclerosis, in the toe wound of a diabetic man in Michigan, in the triceps wound of a woman in Michigan, and more. So far, each infection has been isolated; the microbe has never been transmitted from person to person, as MRSA has. VRSA, while dangerous, doesn't appear to spread through the human population.

But that's not to say it won't evolve to do so. When MRSA first emerged, it seemed restricted to hospital settings. Today, however, clinicians have been horrified to see cases of MRSA pop up from simple scrapes on a playground, suggesting that the microbe is out in the community, where it can do widespread damage. In theory, the same is possible for VRSA, raising the bone-chilling specter of the superbug evolving into an "apocalyptic bug," as one reporter called it.

How did staph acquire vancomycin resistance 14 separate times? How likely is vancomycin resistance to become more widespread? We can find answers to these questions by understanding four mechanisms of evolution:

1. Natural selection

2. Mutation

3. Gene flow

4. Genetic drift

Bacteria are ideal organisms for examining these evolutionary mechanisms, for the same reason that they are so dangerous: because they evolve incredibly fast.

Rising Resistance

Harvard Medical School microbiologist Michael Gilmore has long tracked the ways that bacteria evolve antibiotic resistance. After staph evolved widespread resistance to methicillin in the 1980s and vancomycin began to be used in hospitals, "we waited and waited and waited" for vancomycin resistance

to emerge, says Gilmore. He knew that once vancomycin was widely used to kill staph, the microbe population would evolve resistance to the poison. The process by which a population gains one or more alleles that enable it to survive better than other populations is called *natural selection*, as we saw in Chapter 11. Darwin and Wallace were the first to propose natural selection, and today we know that it is the central driver of evolution.

During natural selection, individuals with particular inherited characteristics survive and reproduce at a higher rate than other individuals in a population. Natural selection acts by favoring some phenotypes over others (**Figure 12.6**). For example, in an environment where bacteria are exposed to vancomycin, the bacteria that can resist the antibiotic will continue to live and reproduce, while those that cannot will perish. Although natural selection acts directly on the phenotype, not on the genotype, of a population, the alleles that code for a phenotype favored by natural selection tend to become increasingly common in future generations. Bacteria that survive an antibiotic attack, for example, pass on alleles that confer that resistance to their offspring.

Natural selection acts on adaptive traits, and therefore a population can become better suited to survive and reproduce in its environment. And unfortunately for us, natural selection is the mechanism by which staph is adapting to our use of vancomycin.

Because staph's vancomycin-resistance adaptation is very recent, scientists are studying the patterns of how natural selection gives rise to VRSA. In nature we observe three common patterns of natural selection: *directional selection*, *stabilizing selection*, and *disruptive selection*. All types of natural selection operate by the same principle: individuals with certain forms of an inherited trait have better survival rates

MICHAEL GILMORE

Michael Gilmore is a microbiologist at Harvard Medical School. He and his laboratory uncovered the genetic basis for the recent emergence of VRSA.

A population of *Staphylococcus aureus* bacteria.

After treatment with methicillin, shown as a red kitchen strainer, the resistant bacteria (MRSA) survive and reproduce.

Additional treatments with methicillin do not reduce the MRSA population.

After treatment with vancomycin, shown as a purple kitchen strainer, the resistant bacteria (VRSA) survive.

The frequency of the VRSA bacteria containing the resistance allele increases dramatically. This is evolution.

Figure 12.6

Evolution happens

Imagine there is a population of *Staphylococcus aureus* bacteria living on your skin. Most of them are susceptible to the antibiotic methicillin (red strainer). A few, however, are randomly resistant to methicillin (red), like the ones in Figure 12.2. A few of these bacteria, in turn, could also be resistant to vancomycin (purple).

Q1: Why does the population of *S. aureus* bacteria *not* pose a life-or-death health threat outright?

Q2: Why do vancomycin-resistant bacteria have a higher frequency in the population after treatment with vancomycin?

Q3: If this figure used the mouse example of allele frequency from Figure 12.5, and the white mice increased in numbers like the vancomycin-resistant bacteria here did, what would happen to the frequencies of the white-fur-pigment and black-fur-pigment alleles?

and produce more offspring than do individuals with other forms of that trait.

Directional selection is the most common pattern of natural selection, in which individuals at *one extreme* of an inherited phenotypic trait have an advantage over other individuals in the population. The peppered moth provides a vivid example: Before 1959, the number of dark-colored moths in both England and the United States increased after industrial pollution blackened the bark of trees, causing dark-colored moths to be harder for bird predators to find than light-colored moths. A reduction in air pollution following clean-air legislation, enacted in 1956 in England and in 1963 in the United States, caused the bark of trees to become lighter, and suddenly the reverse occurred: light-colored moths became harder for predators to find than dark-colored moths. As a result, the proportion of dark-colored moths plummeted because they were easily seen and eaten by predators (**Figure 12.7**). Similarly, when methicillin became widely used to fight staph, MRSA evolved via directional selection: the bacteria in hospital settings that were resistant to the antibiotic survived, while those that were not perished.

In cases of **stabilizing selection**, individuals with *intermediate values* of an inherited phenotypic trait have an advantage over other individuals in the population. Birth weight in humans provides a classic example of this pattern of natural selection (**Figure 12.8**). Historically, light or heavy babies did not survive as well as babies of average weight, and as a result there was stabilizing selection for intermediate birth weights. Today, however, this stabilizing trend is not as strong, because advances in the care of low-birthweight babies and an increase in the use of cesarean deliveries for large babies have allowed babies of all weights to thrive.

Finally, **disruptive selection** occurs when individuals with *either extreme of an inherited trait* have an advantage over individuals with an intermediate phenotype. This type of selection is the least commonly observed in nature, but one example of a trait affected by disruptive selection is the beak size within a population of birds called African seed crackers (**Figure 12.9**). During one dry season, birds with large beaks survived on hard seeds and

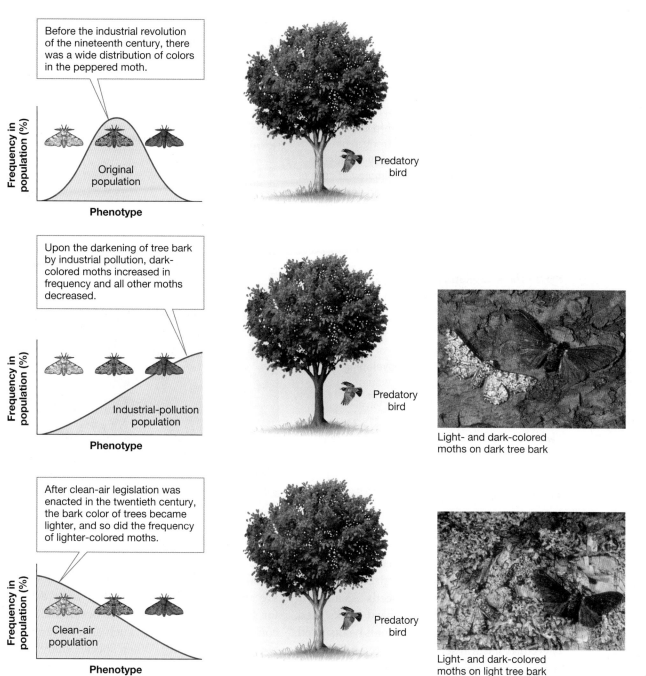

Figure 12.7

The peppered moth has undergone directional selection two different times in the last 200 years
The industrial revolution of the nineteenth century caused extreme air pollution in both England and the United States, due mostly to soot from the mass burning of coal.

Q1: If one extreme phenotype makes up most of a population after directional selection, what happened to the individuals with the other phenotypes?

Q2: What do you think would happen to the phenotypes of the peppered moth if the tree bark was significantly darkened again by disease or pollution?

Q3: What do you think would happen to the phenotypes of the peppered moth if the tree bark became a medium color, neither light nor dark? (You will need to read the next paragraph to answer this question.)

Figure 12.8

Stabilizing selection and human birth weight from 1935 to 1946

This graph is based on data for 13,700 babies born between 1935 and 1946 in a hospital in London. In countries that can afford intensive medical care for newborns, the strength of stabilizing selection has been greatly reduced in recent years. However, even with improved care of low-birthweight babies and cesarean deliveries of large babies, babies at the extremes of newborn weight still survive at a lower rate than those closer to the median weight.

Q1: Think of another example of stabilizing selection in human biology. Has modern technology or medicine changed its impact on the resulting phenotypes?

Q2: How do you think a graph of birth weight versus survival for a developing country with little health care would compare to the graph shown here?

Q3: How do you think a graph of birth weight versus survival for an affluent area of the United States today would compare to the graph shown here?

Babies that weighed 8 pounds at birth survived at a higher rate than did lighter or heavier babies.

birds with small beaks survived on soft seeds, but birds with intermediate-sized beaks fed inefficiently on both types of seeds. As a result, the birds with intermediate-sized beaks died, and the birds with large and small beaks lived. Therefore, natural selection favored both large-beaked and small-beaked birds over birds with intermediate beak sizes.

Any of these patterns of natural selection can cause distantly related organisms to evolve similar structures because they survive and reproduce under similar environmental pressures. This type of evolution, called **convergent evolution**, results in organisms that appear very much alike despite vastly dissimilar genetics. Cacti found in North American deserts and distantly related plants (euphorbias) found in African and Asian deserts offer an excellent example of convergent evolution. These two types of desert plants have very different genetics, but they look similar and function in a similar manner. In

another example, sharks and dolphins are only distantly related (sharks are fish, and dolphins are marine mammals), yet they both evolved for success as predators in the ocean and share common characteristics, such as a streamlined body (**Figure 12.10**). When species share characteristics due to convergent evolution and not because of modification by descent from a recent common ancestor, those characteristics are called **analogous traits** (instead of homologous traits).

Enter *Enterococcus*

Intent on determining how VRSA was evolving after the first Michigan infection, Gilmore began to track the appearance of the bug. He and others wanted to know where the allele for vancomycin resistance came from, and how staph had acquired it 14 different times.

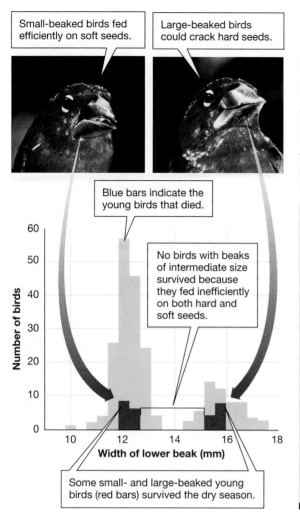

Small-beaked birds fed efficiently on soft seeds.

Large-beaked birds could crack hard seeds.

Blue bars indicate the young birds that died.

No birds with beaks of intermediate size survived because they fed inefficiently on both hard and soft seeds.

Some small- and large-beaked young birds (red bars) survived the dry season.

Number of birds (y-axis: 0, 10, 20, 30, 40, 50, 60)

Width of lower beak (mm) (x-axis: 10, 12, 14, 16, 18)

Figure 12.9

Disruptive selection for beak size

Among a group of young African seed crackers hatched in one year, only those with small or large beaks survived the dry season, when seeds were scarce. Although many of the small- and large-beaked birds did not survive, none of the intermediate-beaked birds survived.

Q1: Almost all birds starved during the dry season depicted here. What type of selection would have been present if only the intermediate-beaked birds had survived (instead of the small- and large-beaked birds)?

Q2: Describe a scenario in which African seed crackers would experience directional selection for either smaller- or larger-beaked birds. What kind of environmental conditions might bring about such a situation?

Q3: Of the three patterns of natural selection presented in this discussion, which one always results in two different phenotypes in the following generations?

As Gilmore searched for patterns among the cases of VRSA infection, he noticed some peculiar things. First, most of the infections were turning up in people with diabetes, typically in bad foot wounds. Second, in most cases, when scientists looked closely at the samples, they saw not only staph but also a small spherical bacterium called *Enterococcus*, which had evolved resistance to vancomycin years earlier (**Figure 12.11**). Upon close observation, Gilmore and others noted that the enterococci, growing cozily side by side with staph, contained a vancomycin-resistance gene identical to the one in the staph. The presence of this gene suggested that staph had acquired vancomycin resistance directly from *Enterococcus*, rather than via random mutations in its own genome.

New alleles in a species emerge via mutation. A **mutation** is a change in the sequence of any segment of DNA in an organism, and it is the only means by which new alleles are generated. DNA mutations create new alleles at random, thereby providing the raw material for evolution. In this sense, all evolutionary change depends ultimately on mutation. Mutations can stimulate the rapid evolution of populations by providing new genetic variation—differences in genotypes between individuals within a population. Then, natural selection and other mechanisms of evolution act on the resulting phenotypes.

In sexually reproducing species, genetic mutations that occur in an organism's germ line cells—the cell lineage that produces gametes such as eggs and sperm—can contribute to evolution. Mutations in other cells of the body, such as skin or blood cells, can affect the individual by causing cancer or other problems, but those mutations are not passed to that individual's offspring. If mutations are not passed to offspring, they cannot contribute to evolution.

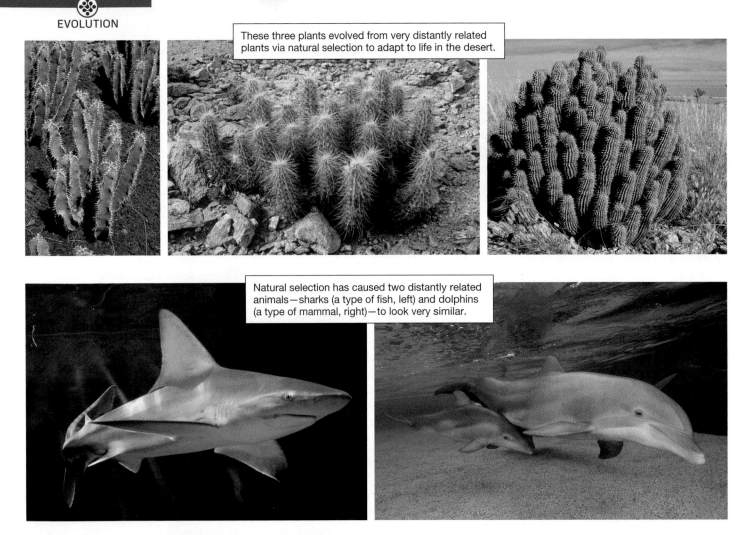

These three plants evolved from very distantly related plants via natural selection to adapt to life in the desert.

Natural selection has caused two distantly related animals—sharks (a type of fish, left) and dolphins (a type of mammal, right)—to look very similar.

Figure 12.10

Natural selection can result in convergent evolution

Q1: How is convergent evolution different from evolution by common descent?

Q2: What is the main difference between a homologous trait (see Figure 11.11) and an analogous trait?

Q3: Why are convergent traits considered evidence for evolution (see Chapter 11)?

Figure 12.11

Staphylococcus aureus (left) and ***Enterococcus*** (right)

The same does not hold true for bacteria, which are single cells and reproduce via asexual reproduction. All genetic mutations in a bacterial cell are passed on to the offspring. (See Chapter 6 for a review of cellular replication.) If a mutation passed to an offspring increases the individual's ability to reproduce, that mutation is favored by natural selection. These favored mutated genes are passed from parent to offspring, spreading through future generations in a way that alters the population as a whole.

But bacteria don't always wait for the right random mutations to pop up in their genomes. Sometimes they simply borrow new alleles from other organisms. That appeared to be the case with VRSA, researchers found. "In most cases, VRSA has developed in a perfect storm," says Sievert. It emerges within "a very bad wound that's not healing and is a soup mix of organisms coming together and sharing genes."

Horizontal gene transfer is the process by which bacteria pass genes to one another (**Figure 12.12**). Bacteria store these genes on small, circular pieces of DNA called *plasmids*. Some bacteria send these plasmids to each other through small tunnels, like dropping a package down a trash chute.

Horizontal gene transfer is one example of another mechanism by which evolution occurs: gene flow. **Gene flow** is the exchange of alleles between populations. Gene flow can occur between two different species—in this case, between a population of staph and a population of *Enterococcus*—or between two populations of the same species, such as strains of staph passing methicillin resistance among themselves in the community. An individual that migrates between two otherwise isolated populations of a species may facilitate gene flow as well (**Figure 12.13**). Gene flow can also occur when only gametes move from one population to another, as happens when wind or pollinators transport pollen from one population of plants to another.

The introduction of new alleles via gene flow can have dramatic effects. Two-way gene flow consists of an exchange of alleles between one population and another, so it tends to make the genetic composition of different populations more similar. If one strain of staph shares the methicillin-resistance allele with another strain of staph, for example, the two populations

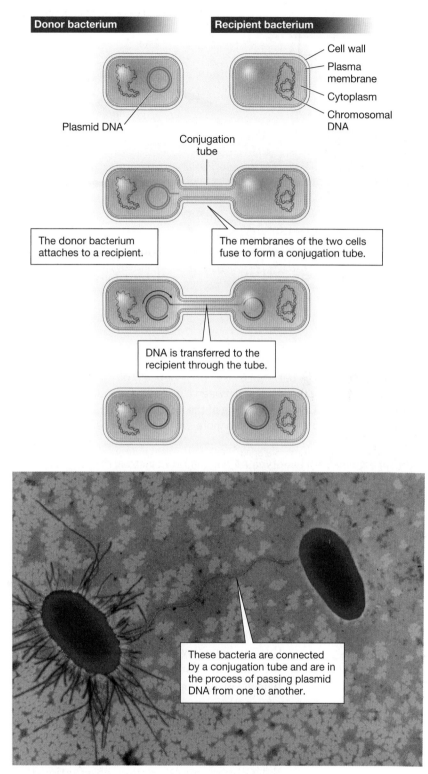

Figure 12.12

Horizontal gene transfer accelerates the rate of evolution in prokaryotes

The diagram depicts the horizontal gene transfer of plasmid DNA through **conjugation**, the physical process of transferring genetic material through direct contact. In the case of VRSA, *Staphylococcus aureus* acquired the vancomycin-resistance gene through horizontal gene transfer from *Enterococcus*.

Population 1

Population 1 is a large population containing birds of genotypes *AA*, *Aa*, and *aa*.

Population 2

Population 2 is located far from population 1 and initially has only birds of genotype *AA*.

A bird of genotype *aa* migrates from population 1 to population 2, thereby introducing the *a* allele to population 2.

Figure 12.13

Migrants can move alleles from one population to another

As a result of gene flow between these two genetically disparate populations of geese, the populations will eventually become more similar to each other. Mating between a migrant *aa* genotype goose and a population 2 goose of *AA* genotype will result in *Aa* genotype offspring. Continued mating of the *Aa* genotypes will result in all three possible genotypes (*AA*, *Aa*, and *aa*). The resulting population 2 gene pool will look much more similar to the gene pool of population 1.

Q1: If a goose with genotype *AA* had migrated instead of the goose with genotype *aa*, would the scenario described here still be considered gene flow? Why or why not?

Q2: If a goose with genotype *Aa* had migrated instead of the goose with genotype *aa*, would the scenario still be considered gene flow? Why or why not?

Q3: If the goose with genotype *aa* had migrated to population 2 as shown but had failed to mate with any of the *AA* individuals, would the scenario still be considered gene flow? Why or why not?

become more alike. A mutual exchange of alleles through gene flow can *counteract* the effects of the other mechanisms, such as mutation, that tend to make populations more different from one another.

Gene flow appeared responsible for the emergence of VRSA; staph picked up an allele for vancomycin resistance from *Enterococcus*. But Gilmore made another observation that suggested why this horizontal gene transfer was happening now, and not back in the 1980s when vancomycin had first begun to be widely used.

Gilmore noted that all the VRSA samples taken from patients were clonal cluster 5 (CC5) strains of *S. aureus*. "There were implications that there was something special about clonal cluster 5 strains that was leading to this vancomycin resistance bubbling up in these strains," says Gilmore. Compared to other strains of staph, CC5 strains appeared to have evolved the ability to readily take up and use vancomycin-resistance alleles from *Enterococcus*.

Primed for Pickup

In 2012, Gilmore and his team at Harvard analyzed the DNA of 11 of the 12 known cases of VRSA at the time (samples from one of the 12 weren't available, and the thirteenth and fourteenth cases had yet to occur). In the genome of every sample of VRSA, they found three traits that demonstrate how the CC5 strains of staph evolved to effectively pick up the allele for vancomycin resistance.

First, all the vancomycin-resistant CC5 staph bacteria have the same mutation in a gene called *DprA*. *DprA* appears to be involved in preventing horizontal gene transfer. A mutation in this gene might make it easier for the staph to take up DNA from other bacteria, such as *Enterococcus*.

Second, the CC5 strains lack a set of genes that encode an antibiotic that kills other bacteria. Perhaps this antibiotic normally kills *Enterococcus* near the staph, which would explain why horizontal transfer between the two species does not typically occur.

Finally, Gilmore and his team found that in place of that missing set of antibiotic genes, the vancomycin-resistant CC5 staph have a unique cluster of genes encoding proteins that confuse the human immune system. These proteins could make it easier for staph and other bacteria to grow in a wound because the host immune system would be less able to fight them off.

The lack of the antibiotic genes and the presence of new genes create the ideal conditions for a mixed infection, in which different species of pathogens mingle in a festering soup of contamination. Mixed infections are breeding grounds for antibiotic resistance because they are sites of gene flow among different organisms. In this explanation, the CC5 staph evolved via the three usual mechanisms—mutation, natural selection, and gene flow (via horizontal transfer)—to be more susceptible to take up the vancomycin-resistance allele from *Enterococcus*.

In addition to natural selection (and its intriguing permutation called *sexual selection*; see "Sex and Selection"), mutation, and gene flow,

Sex and Selection

Staphylococcus and other bacteria replicate asexually by copying their DNA and dividing in two. But adding sex to the equation complicates reproduction. Another mechanism by which species evolve is called sexual selection. In **sexual selection**, nature selects a trait that increases an individual's chance of mating—even if that trait decreases the individual's chance of survival.

Sexual selection favors individuals that are good at getting mates, and it often helps explain differences between males and females in size, courtship behavior, and other traits. Species whose males and females are distinctly different in appearance, such as peacocks, lions, and ducks, are said to exhibit **sexual dimorphism**. In many species, the members of one sex—often females—are choosy about whether to mate. In birds, for example, brightly colored males may perform elaborate displays in their attempts to woo a mate. In other species, males may attract attention by other means, such as calling vigorously. Females then select as their mates the males with the loudest calls.

Yet, some characteristics that increase an individual's chance of mating can *decrease* its chance of survival. For example, male túngara frogs perform a complex mating call that may or may not end in one or more "chucks." Females prefer to mate with males that emit chucks, but frog-eating bats use that same sound to help them locate their prey. As a result, a frog's attempt to locate a mate can end in disaster.

A male peacock performs an elaborate mating display to a potential mate.

Male túngara frogs face an ecological trade-off: the same type of call that is most successful at attracting females for mating also attracts predatory bats to dinner.

In this small population of upland burrowing crayfish, similar numbers of blue, orange, and green crayfish survive under normal environmental conditions.

In this small population of upland burrowing crayfish, several more blue crayfish than orange or green crayfish survive after a random catastrophe kills off most of the population.

Time

After generations of reproduction, the population contains frequencies of blue, orange, and green crayfish similar to those of the original population.

After generations of reproduction, the rebounded population contains a higher frequency of blue crayfish than orange or green crayfish.

Figure 12.14

A genetic bottleneck is a type of genetic drift

Two small populations of upland burrowing crayfish are compared. The one on the right experiences a genetic bottleneck event. This is genetic drift—a change in the frequency of a trait that is not associated with natural selection. In fact, the blue crayfish could be less well adapted to the environment than the other crayfish are.

Q1: Why do you think a genetic bottleneck is more likely to occur in a small population than in a large population?

Q2: Genetic drift is often described as a "chance event." Give other examples of chance events that could cause a genetic bottleneck.

Q3: Which resulting population has more genetic diversity?

there is one other mechanism by which organisms evolve. Although alleles that code for beneficial traits like antibiotic resistance for bacteria are usually selected for and maintained in a population by the nonrandom action of natural selection, in some cases, chance events can cause allele frequencies to change from a parent generation to the next generation.

Genetic drift is a change in allele frequencies caused by random differences in survival and reproduction among the individuals in a population. Just by chance, some individuals in a generation may leave behind more descendants than other individuals do. In this case, the genes of the next generation will be the genes of the "lucky" individuals, not necessarily the "better" individuals. Genetic drift occurs in all populations, including mammals and bacteria, but it is more likely to cause evolution in a small population than in a large one.

Two ways in which strong genetic drift may occur include genetic bottlenecks and the founder effect. A **genetic bottleneck** is a drop in the size of a population, for at least one generation, that causes a loss of genetic variation (**Figure 12.14**). A genetic bottleneck can threaten the survival of a population.

In the 1970s, for example, the population of the endangered Florida panther plummeted because of hunting and habitat destruction. The species barely escaped extinction. At one point, experts believed that only six wild individuals in the whole species were still alive. This rapid population reduction created a genetic bottleneck in which an estimated half of the genetic variation within the species was lost, and severe inbreeding among the individuals that were left resulted in maladies including low sperm counts and abnormally shaped sperm in male panthers (**Figure 12.15**).

Thankfully, panther numbers have increased to about 80–100 individuals in recent years, in part because of captive breeding programs. Most

Figure 12.16

Early Dutch colonists in South Africa

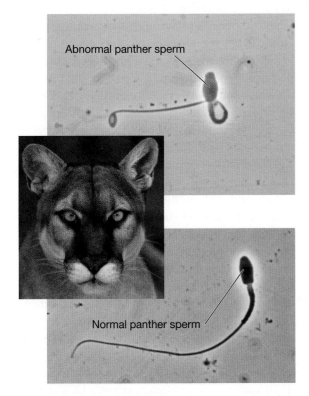

Figure 12.15

Abnormally shaped sperm in the rare Florida panther

Florida panthers have more abnormal sperm than do panthers from other populations—a possible effect of a genetic bottleneck promoting the increase of harmful alleles in the Florida panther population.

of the members of this rebounded population are incredibly similar genetically, because of the limited variety of alleles present in the original six individuals. In an effort to restore gene flow to the population, in 1995 researchers released eight female Texas cougars, close genetic cousins to the Florida panthers, into the panthers' habitat. Five of those females produced eight litters of kittens—more genetically diverse, stronger, and longer-lived than their fathers.

The **founder effect** occurs when a small group of individuals establishes a new population isolated from its original, larger population. For example, in South Africa a population of people called Afrikaners descended primarily from a few Dutch colonists (**Figure 12.16**). Today, the Afrikaner population has an unusually high frequency of the allele that causes Huntington disease because those original colonists by chance carried the allele when they settled in the area.

After Vancomycin

When the initial case of VRSA popped up back in 2002, the patient's foot was carefully treated with an old topical drug that is rarely used today. The only other options were one or two new drugs that hadn't yet been released on the market. "There are still some antibiotics we can use for treatment, but we recommend those be used very responsibly in order to maintain their effectiveness," says Sievert. "You don't want to give them out when not necessary because you don't want to see resistance to them. You save them as the end of the line, so something is available if the commonly used drugs end up failing because of resistance." After 9 months of careful treatment, the patient healed and was able to keep her foot. Other cases of VRSA have similarly been resolved by the use of the newer antibiotics and the delicate care of doctors.

But what happens when even these drugs no longer work against staph? Pharmaceutical companies are developing few new antibiotics, says Frank DeLeo, acting chief of the

How Can *You* Make a Difference? Help Prevent Antibiotic Resistance!

The U.S. Centers for Disease Control and Prevention (CDC) recommends the following guidelines:

- Take antibiotics exactly as the doctor prescribes. Do not skip doses. Complete the prescribed course of treatment, even when you start feeling better.
- Only take antibiotics prescribed for you; do not share or use leftover antibiotics. Antibiotics treat specific types of infections. Taking the wrong medicine may delay treatment and allow bacteria to multiply.
- Do not save antibiotics for the next illness. Discard any leftover medication once the prescribed course of treatment is completed.
- Do not ask for antibiotics when your doctor thinks you do not need them. Remember that antibiotics have side effects. When your doctor says you don't need an antibiotic, taking one might do more harm than good.
- Prevent infections by practicing good hand hygiene and getting recommended vaccines.

Source: http://www.cdc.gov/Features/AntibioticResistance.

Laboratory of Human Bacterial Pathogenesis at the National Institute of Allergy and Infectious Diseases, and bacteria evolve more quickly than we can develop new antibiotics. It's a vicious cycle, he says. "As we continue to use antibiotics, populations of microbes will develop resistance to antibiotics." Recently, microbes appear to be winning the struggle. In July 2016, researchers reported the second U.S. appearance of a superbug resistant to colistin, an antibiotic of last resort. "The golden age of antibiotics appears to be coming to an end," the *Los Angeles Times* wrote.

VRSA has not yet become as widespread as MRSA. Scientists believe the reason may be that staph does not handle the resistance allele for vancomycin very well; the gene from *Enterococcus* is bulky and difficult to manage, so the staph individuals that receive the gene are often outcompeted by other staph individuals in the environment. "Though it allows them to survive, they don't do well with it. There's a [survival] cost to having this resistance," says Gilmore. In the absence of vancomycin, staph is better off without the resistance allele, so perhaps staph individuals lose it or don't pick it up when not in the presence of the drug. VRSA therefore remains "a rare, unstable organism in the environment that isn't very hardy, so it doesn't last and spread," says Sievert. "That's the good news."

But there's still trepidation that staph may adapt and get comfortable with this resistance gene, just as staph adapted to methicillin resistance, says Gilmore. If staph survives and reproduces more easily with the resistance gene, the gene will likely spread to staph throughout the host/human population. "We don't know if that's a possibility," says Gilmore. "My suspicion is that it is." (See "How Can *You* Make a Difference? Help Prevent Antibiotic Resistance!")

In fact, the thirteenth VRSA case, from a man in Delaware in July 2012, was not a CC5 strain, says Gilmore—suggesting that vancomycin resistance is spreading to other strains of staph. "It's disconcerting," he adds. It seems that other populations of staph may now also be evolving vancomycin resistance. "The main worry is that this would move into a strain that was highly transmissible in the community," adds DeLeo. "The potential is there."

Race against Resistance

"Each year in the United States, more than 2 million people become infected with antibiotic-resistant bacteria, including methicillin-resistant *Staphylococcus aureus* (MRSA), resulting in at least 23,000 deaths. To combat antibiotic resistance, doctors need novel antibiotics, yet fewer and fewer new antibiotics come to pharmacies each year. In 2011, Congress passed the Generating Antibiotic Incentives Now (GAIN) Act to stimulate the development of new antibiotics, and it's working. So far, six new antibiotics have been approved through that program.

Assessment available in smartwork**5**

Antibacterial drugs approved by the FDA

19	11	11	11	4	3	8
1980–84	1985–89	1990–94	1995–99	2000–04	2005–09	2010–15

Time line of antibiotic resistance

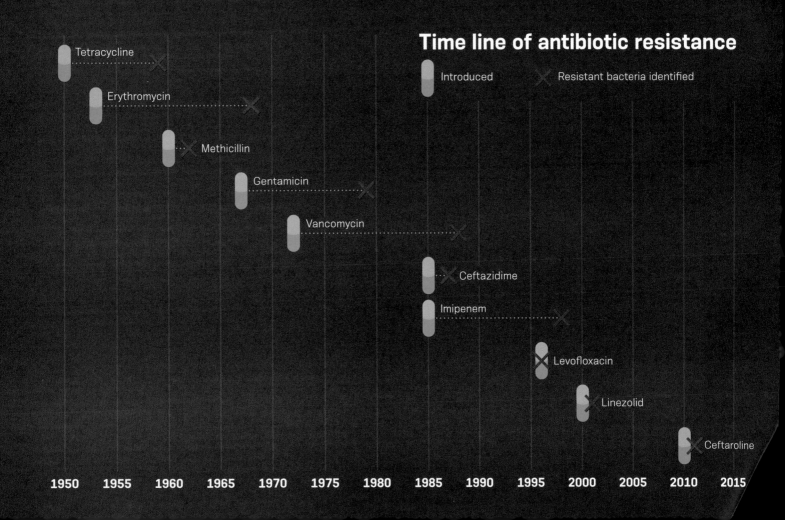

Introduced   ✕ Resistant bacteria identified

Tetracycline

Erythromycin

Methicillin

Gentamicin

Vancomycin

Ceftazidime

Imipenem

Levofloxacin

Linezolid

Ceftaroline

| 1950 | 1955 | 1960 | 1965 | 1970 | 1975 | 1980 | 1985 | 1990 | 1995 | 2000 | 2005 | 2010 | 2015 |

REVIEWING THE SCIENCE

- Natural selection for inherited traits occurs in three common patterns: directional, stabilizing, and disruptive.

- In **directional selection**, individuals at one phenotypic extreme of a given genetic trait have an advantage over all others in the population.

- In **stabilizing selection**, individuals with intermediate phenotypes have an advantage over all others in the population.

- During **disruptive selection**, individuals with either extreme phenotype have an advantage over those with an intermediate phenotype.

- In **convergent evolution**, distantly related organisms (those without a recent common ancestor) evolve similar structures in response to similar environmental challenges.

- All mechanisms of evolution depend on the genetic variation provided by new alleles created by **mutation**.

- **Sexual selection** occurs when a trait increases an individual's chance of mating even if it *decreases* that individual's chance of survival.

- **Gene flow** is the exchange of alleles between separate populations.

- **Genetic drift** is a change in allele frequencies produced by *random* differences in survival and reproduction in a small population, and most dramatically occurs through one of two processes: a genetic bottleneck or the founder effect.

- A **genetic bottleneck** occurs when a drop in the size of a population leads to a loss of genetic variation in the new, rebounded population.

- The **founder effect** occurs when a few individuals from a large population establish a new population, leading to a loss of genetic variation in the new, isolated population.

THE QUESTIONS

The Basics

1 The founder effect is a type of (**genetic drift / gene flow**) in which individuals in one small group of a large population (**establish a new distant population / are the only survivors**) and then reproduce.

2 Unlike natural selection, _____ is not related to an individual's ability to survive and may result in offspring that are less well adapted to survive in a particular environment.

(a) genetic drift

(b) sexual selection

(c) directional selection

(d) convergent evolution

3 Which of the following statements about convergent evolution is true?

(a) It demonstrates how similar environments can lead to different physical structures.

(b) It demonstrates how similar environments can lead to the same physical structures.

(c) It demonstrates that similarity of structures is due to descent from a common ancestor.

(d) It demonstrates that similarity of structures is due to random chance.

4 Evolution is most accurately described as a change in allele frequencies in _____ over time.

(a) an individual

(b) a species

(c) a population

(d) a community

Challenge Yourself

5 In a population, which individuals are most likely to survive and reproduce?

(a) The individuals that are the most different from the others in the population.

(b) The individuals that are best adapted to the environment.

(c) The largest individuals in the group.

(d) The individuals that can catch the most prey.

6 A study of a population of the goldenrod wildflower finds that large individuals consistently survive and reproduce at a higher rate than small or medium-sized individuals. Assuming size is an inherited trait, the most likely evolutionary mechanism at work here is

(a) disruptive selection.

(b) directional selection.

(c) stabilizing selection.

(d) natural selection, but it is not possible to tell whether it is disruptive, directional, or stabilizing.

7 Explain how, because of sexual selection, an individual might be very successful at surviving (natural selection), but not pass on genes to the next generation.

Try Something New

8 Two large populations of the same species found in neighboring locations that have very different environments are observed to become genetically more similar over time. Which of the four main evolutionary mechanisms is the most likely cause of this trend? Justify your answer.

9 The Tasmanian devil, a marsupial indigenous to the island of Tasmania (and formerly mainland Australia as well), experienced a population bottleneck in the late 1800s when farmers did their best to eradicate it. After it became a protected species, the population rebounded, but it is now experiencing a health crisis putting it at risk for disappearing again. Many current Tasmanian devil populations are plagued by a type of cancer called devil facial tumor disease, which occurs inside individual animals' mouths. Afflicted Tasmanian devils can actually pass their cancer cells from one animal to another during mating rituals that include vicious biting around the mouth.

Unlike the immune systems of other species, including humans, the Tasmanian devil's immune system does not reject the passed cells as foreign or nonself (as we reject a liver transplant from an unmatched donor), but accepts them as if they were their own cells. Why would a population bottleneck result in the inability of one devil's immune system to recognize another devil's cells as foreign?

10 Global warming is causing more and more ice to melt each year at far-northern latitudes, exposing more bare ground than ever before. These vast areas of brown ground coloration make polar bears (which are white) much more conspicuous to their prey. Recently, an infant polar bear was born with brown fur. This polar bear survived to adulthood and has sired several offspring with brown fur. Which of the following is a plausible explanation of how the brown fur trait appeared in these polar bears?

(a) A polar bear realized it would be better to be brown in order to hide more effectively. It induced mutations to occur in its fur pigment gene, which resulted in a change in pigment from white to brown fur.

(b) One or more random mutations occurred in the fur pigment gene in an individual polar bear embryo, which resulted in a change in pigment from white to brown fur.

(c) Increased temperatures due to global warming caused targeted mutations in the fur pigment gene in an individual polar bear embryo, which resulted in a change in pigment from white to brown fur.

(d) A female polar bear realized it would be better for her offspring to be brown and therefore mated with a grizzly bear to achieve this result.

11 In the garden shed belonging to one of this text's authors, stabilizing selection has occurred over the past 10 years in the house mouse, *Mus musculus*. Which of the following scenarios is an example of stabilizing selection?

(a) Small and medium-sized mice cannot reach the seed shelf in the shed and therefore are at a disadvantage for finding food, so they do not survive and reproduce as well as large mice do.

(b) Small mice cannot reach the seed shelf, and large mice are easily seen by hawks circling above. Medium-sized mice therefore survive and reproduce better than both small and large mice.

(c) Small mice can easily cross the yard to the vegetable garden, and large mice can easily reach the seed shelf. Medium-sized mice have trouble with the seed shelf and are seen by hawks in the yard. Small and large mice survive and reproduce much better than medium-sized mice.

(d) All of these are examples of stabilizing selection.

(e) None of these are examples of stabilizing selection.

Leveling Up

12 **What do *you* think?** One way to prevent a small population of a plant or animal species from going extinct is to deliberately introduce some individuals from a large population of the same species into the smaller population. In terms of the evolutionary mechanisms discussed in this chapter, what are the potential benefits and drawbacks of transferring individuals from one population to another? Do you think biologists and concerned citizens should take such actions?

13 *Write Now* **biology: mechanisms of evolution** This assignment explores the mechanisms of evolution through five selected short stories from *Welcome to the Monkey House* by Kurt Vonnegut Jr. Answer the questions associated with each story.

"Harrison Bergeron"
What message is this story trying to send? Cite examples from the story and relate them to the mechanisms of evolution from this chapter.

"Welcome to the Monkey House"
Is this story an example of sexual selection? Why or why not? Cite examples from the story and from this chapter to support your thinking.

"The Euphio Question"
If technology could produce such an instrument, how would it affect the evolution of humans? What about the evolution of other species on Earth?

"Unready to Wear"
Relate this story to as many of the mechanisms of evolution from this chapter as you can. Cite examples from the story and the chapter to support your thinking.

"Tomorrow and Tomorrow and Tomorrow"
Do you think these types of drugs are a good or bad thing? Where would you draw the line on technology's ability to extend life? How would drugs like these affect the natural selection and evolution of humans? What about the evolution of other species on Earth?

For more, visit digital.wwnorton.com/bionow2 for access to:

Fast Lizards, Slow Corals

The rapid evolution of lizards and the slow growth of corals offer clues about how species are born.

After reading this chapter you should be able to:

- Diagram the relationships among adaptive traits, reproductive fitness, and evolution by natural selection.
- Defend the utility of the biological species concept, and discuss its shortcomings.
- Define speciation and the role of genetic divergence among populations in creating new species.
- Differentiate between allopatric and sympatric speciation.
- Describe the two kinds of reproductive isolating mechanisms and explain their role in speciation.
- Define and give an example of coevolution.

CHAPTER
13

ADAPTATION
AND SPECIES

In 1971, the United States was at war in Vietnam, a gallon of gas cost 40 cents, a computer engineer sent the first e-mail—and an Israeli biologist named Eviatar Nevo captured 10 lizards on Pod Kopište, a small, rocky island off the coast of Croatia. Each lizard was about the length of a pinky finger and as heavy as a nickel. There was nothing remarkable about them.

Nevo and his team released the captured lizards, a common Mediterranean species called *Podarcis sicula*, on a nearby island just 3 miles away, within sight of the original island but separated by a deep ocean gulf (**Figure 13.1**). Pod Mrčaru is a smaller, plant-covered island that was already inhabited by two other lizard species. Nevo was curious to find out what would happen when the three species began to compete for resources on Pod Mrčaru. But Nevo never got the chance to return to the island. Shortly after he and his team departed, unrest broke out across Croatia, and war held the region in a stranglehold throughout the 1980s and '90s.

Scientists did not return to Pod Mrčaru until 2004. While visiting the University of Antwerp in Belgium, biologist Duncan Irschick from the University of Massachusetts, Amherst, and his Belgian colleagues decided to investigate the mysterious island. One of the researchers, familiar with an obscure paper that Nevo had published in 1972, thought it would be interesting to see what, if anything, had happened to the lizards. Further, Irschick would be able to complete some research on lizard behavior that he had planned to do in Europe anyway. "We didn't know what we'd find," recalls Irschick. "We decided, 'Let's just go check.'" This time, the lizards on the island were quite remarkable.

Figure 13.1

View of Pod Kopište from Pod Mrčaru
The island Pod Kopište (top), the original home of the *Podarcis sicula* lizards (bottom), is only 3 miles from their new home on the island Pod Mrčaru.

Leaping Lizards

Back in the 1970s, Nevo's ten *Podarcis sicula* lizards had been well adapted to the rocky, sparsely vegetated environment of Pod Kopište. They had specific **adaptive traits**, inherited characteristics that enabled them to survive and reproduce successfully on the island. Adaptive traits can be structural features, biochemical traits, or behaviors. In this case, the lizards were fast, with long legs that may have helped them catch the insects that made up most of their diet on the island. They were also territorial, fighting with other lizards over space and mating partners. These adaptive traits, among others, enabled the lizards to survive and reproduce better than competitors lacking those traits on Pod Kopište.

When transplanted to Pod Mrčaru, however, the lizards faced a new environment. Pod Kopište is large and rocky with few plants, and the lizards ate primarily insects; Pod Mrčaru is smaller yet has an abundant supply of plants, including leaves and stems of local shrubs and grasses (**Figure 13.2**). Either the lizards would adapt to their new surroundings, or they would die. The term **adaptation** is commonly applied to adaptive traits *or* the process of evolution through natural selection that brings about adaptive traits, as discussed in Chapter 12. Therefore, an adaptation can be a trait that is advantageous to an individual or a population or, in broader terms, it can be the evolutionary

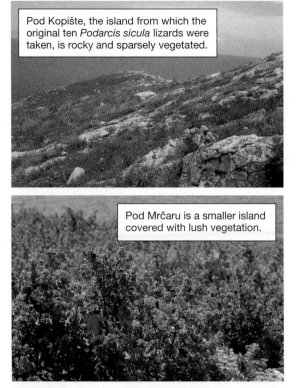

Pod Kopište, the island from which the original ten *Podarcis sicula* lizards were taken, is rocky and sparsely vegetated.

Pod Mrčaru is a smaller island covered with lush vegetation.

Figure 13.2

Pod Kopište's sparse vegetation and Pod Mrčaru's lush vegetation

Longer, wider head

Thicker body

New digestive structure

Figure 13.3

***Podarcis sicula* lizard on Pod Mrčaru, 33 years after introduction of the species**

process of natural selection that enables a good match between a population of organisms and their environment. If the population of lizards was going to survive, it would need to adapt to its new environment.

When Irschick and his team landed on Pod Mrčaru in 2004, they set to work catching lizards. It wasn't hard; the island was swarming with them. Irschick could sit on a rock and simply pick up lizards as they ran by. There were several thousand lizards on the 7-acre island. When Irschick and his team looked closely, they realized that *all* the lizards were *P. sicula*. The ten transplants had wiped out the other two species of lizards. What's more, the adapted lizards looked strange compared to *P. sicula* lizards on other islands nearby (**Figure 13.3**). "They were these really big, chunky lizards," recalls Anthony Herrel, then a postdoctoral fellow in Irschick's lab and now a researcher at the French National Centre for Scientific Research in Paris. "They were unlike lizards on any of the other islands."

Twice a year for 3 years, the group returned to Pod Mrčaru to collect, weigh, and measure the lizards. They also took small pieces of the tails (which grow back) to test the lizards' DNA and compare it to the original population of *P. sicula* on Pod Kopište. Those DNA tests confirmed that the lizards on Pod Mrčaru were indeed descended from the original ten that Nevo had transported to the island. In addition to taking size and weight measurements, the team tested the strength of the lizards' bite. They also dissected two samples of dead lizards.

What the researchers found was unprecedented. In about 33 years, from the time Nevo stepped off the island to the time Irschick stepped on, the species had evolved dramatically. The descendant lizards' heads were larger and shaped differently, making the lizard's bite much stronger than that of the original lizards brought to the island. The descendant lizards also had a unique digestive-tract structure called a cecal valve, a set of muscles between the large and small intestine that slow down food digestion, enabling the lizards to better process the cellulose of plants. It's rare for lizards to have these structures, says Herrel. Only a few plant-eating lizard species, like iguanas, have cecal valves.

DUNCAN IRSCHICK

Duncan Irschick is a biologist at the University of Massachusetts, Amherst, who studies animal function and evolution, specializing in research on animal movement and gecko adhesion.

The new traits were the result of the lizards adapting to a different food source, says Irschick: on Pod Kopište, they ate insects, but on Pod Mrčaru, they began to feast on plants. Flushing the stomachs of a few lizards revealed that the lizards' diet was made up of two-thirds plant material. Over time, the lizards that had evolved a new head shape and digestive structures survived and reproduced better than those that had not, because these traits enabled them to take advantage of the new food sources on Pod Mrčaru.

The lizards evolved not only physical adaptations, but behavioral ones too. They became less territorial and less aggressive and were mating more often, probably because more space and food were available. Adaptations, whether behavioral, physical, or biochemical, have three important characteristics. First, they closely match an organism with its environment; in this case, the lizards evolved to match the ecosystem on Pod Mrčaru. Second, adaptations are often complex, such as the lizards' new gut structure. Finally, adaptations help the organism accomplish important functions, such as feeding and mating.

The lizards on Pod Mrčaru show that evolution by natural selection can improve the adaptive fit of organisms to their environment not only over long periods of time, as with the evolution of whales described in Chapter 11, but also over surprisingly short periods of time. In just 33 years, the *P. sicula* lizard population evolved both physical and behavioral traits that helped it flourish on the new island. "It was really, truly rapid evolution across multiple facets," says Irschick.

What Makes a Species?

Despite how impressive the lizards' quick adaptation may be, natural selection does not always result in a perfect match between an organism and

its environment. In many cases, animals fail to adapt successfully. In fact, scientists estimate that 99 percent of all species that have ever lived are now extinct. Every extinct species is a silent testament to a failure to adapt in the face of adversity.

It's tempting to label the descendant lizards as a new species, but Irschick is hesitant to do so without more testing. Most commonly, the term **species** is used to refer to members of a group that can mate with one another to produce fertile offspring. According to the **biological species concept**, a species is a group of natural populations that can interbreed to produce fertile offspring and cannot breed with other such groups; that is, they are **reproductively isolated** from other populations (**Figure 13.4**). Irschick and Herrel have yet to test whether the Pod Mrčaru lizards can still mate with their cousins back on Pod Kopište. According to the

Figure 13.4

A male and a female South Pacific rattlesnake confirm that they are the same species
After performing this mating ritual, these snakes successfully mated, producing viable offspring.

Q1: How do we know that these rattlesnakes are members of the same species?

Q2: How would you design an experiment to determine whether two populations of snakes are distinct species according to the biological species concept?

Q3: For which types of populations does the biological species concept *not* work as a way of determining how they are related?

ANTHONY HERREL

Anthony Herrel worked as a research fellow in Duncan Irschick's lab before starting his own lab at the French National Centre for Scientific Research in Paris, where he studies the evolution of feeding and locomotion in vertebrates.

Figure 13.5

One species or two?
Though they look quite different, these tree frogs are genetically similar enough to be considered distinctly colored variations of the same species. Breeding these frogs with each other would determine their classification under the biological species concept.

Q1: List three kinds of information that scientists use to distinguish between species.

Q2: What differences can you observe between the individuals in the photos? Why are these differences not enough to confirm that they are from two different species?

Q3: How is genetic divergence among populations determined?

biological species concept, if individuals from the two populations of lizards can still breed and produce fertile offspring, they are not different species. This same idea holds true for all sexually reproducing organisms. (Incidentally, it's not obvious why sexual reproduction is so common; see "Why Sex?")

The definition of "species" is not a black-and-white issue; it is a frequently discussed, multi-faceted topic in evolution, and the biological species concept doesn't always apply. For example, not all species can be defined by their ability to interbreed, such as prokaryotes like bacteria, which reproduce asexually. To handle these cases and others, scientists may use biogeographic information, DNA sequence similarity, and **morphology**—the organisms' physical characteristics—to identify and distinguish species (**Figure 13.5**).

Why Sex?

Sex is ubiquitous in the animal and plant kingdoms. An estimated 99 percent of multicellular eukaryotes are capable of sexual reproduction, which involves the joining of two haploid gametes produced through meiosis. Yet sex is very costly for individuals, so scientists have struggled to explain why it is so prevalent compared to asexual reproduction. And no, it's not because sex feels good; the first eukaryotes to engage in sex were single-celled protists some 2 billion years ago, long before animals developed neurons capable of giving an individual a sense of pleasure.

Costs of Sex

1. Time and energy must be invested to find or attract a mate.

2. Parents pass on only 50 percent of their genetic material to offspring, as opposed to the 100 percent that is passed on through asexual reproduction.

3. Gene combinations that have benefited the parents may be shuffled and broken apart during meiosis and recombination.

Possible Benefits of Sex

1. The genetic diversity created by sexual reproduction is critical for adaptation to new environments.

2. Sexual reproduction can help a population eliminate detrimental alleles and generate new beneficial alleles.

3. Rapid genetic change that occurs through sexual recombination can help a population evolve resistance to parasitic infections.

The lizards that adapted to the island of Pod Mrčaru are physically different enough from their cousins on Pod Kopište to be considered a different species, says Herrel, yet their DNA sequences are close to identical, so he believes they might still be able to interbreed. **Speciation** is the process by which one species splits to form two species or more. Fundamentally, speciation occurs because of **genetic divergence**, the accumulation of differences in the DNA sequences of genes in two or more populations of organisms over time; as a result, the populations become more and more genetically dissimilar.

The two populations of lizards may continue to diverge, becoming more and more different because of their **geographic isolation** from one another, and may eventually become two unarguably distinct species. This is one of the most common ways that new species form: individuals of a single population become geographically separated from one another. This process can begin when a newly formed geographic barrier, such as a river, a canyon, or a mountain chain, isolates two populations of a single species (**Figure 13.6**). Such geographic isolation can also occur when a few members of a species colonize a region that is difficult to reach, such as an island located far outside the usual geographic range of the species.

Geographically isolated populations are disconnected genetically; there is little or no gene flow between them. Without gene flow, the other mechanisms of evolution we saw in Chapter 12—mutation, genetic drift, and natural selection—can more easily cause populations to diverge from one another. If populations remain isolated long enough, they can evolve into new species. The formation of new species from geographically isolated populations is called **allopatric speciation** (*allo*, "other"; *patric*, "country"), as shown in **Figure 13.7**.

But what happens when there isn't a physical barrier between two populations? In the ocean, for example, plants and animals can drift

The Kaibab squirrel is confined to the North Rim of the Grand Canyon.

Abert's squirrel lives on the South Rim and other southern locations, all the way into Mexico

Figure 13.6

The Grand Canyon is a geographic barrier for squirrels

The Kaibab squirrel population became isolated from the Abert's squirrel population when the Colorado River cut the Grand Canyon—which is as deep as 6,000 feet in some places. With gene flow between them blocked, probably beginning about 5 million years ago, the two populations eventually accumulated enough genetic differences to become two distinct species.

Q1: What is the definition of gene flow? How was gene flow blocked between these species?

Q2: Name as many types of geographic barriers as you can. Which do you think would be the best at blocking gene flow?

Q3: Are geographic barriers universal for all species? If not, name a geographic barrier that might block gene flow for one species but not another.

around or swim extremely long distances. There are few physical barriers yet many separate, unique species. How do new species form when populations are free to mix and mingle?

Caribbean Corals

Carlos Prada did not spend much of his time at the University of Puerto Rico in a classroom. Instead, he spent it in the ocean, studying corals, a type of marine invertebrate. On regular dives onto the Caribbean reef, the graduate student documented the morphology of a species of coral named *Eunicea flexuosa*, commonly called "sea fans." At 12 different locations along the Puerto Rican coast, Prada observed different morphologies of the coral at different depths. In shallow areas, less than 15 feet from the surface, the sea fans were wide, with a broad network of thick branches like a bush. In deeper areas, between 15 and 55 feet from the surface, the sea fans were much taller and spindly, resembling trees with networks of thin branches.

"They're supposed to be the same species, but they really looked different at different depths," says Prada. Curious about whether the morphologies were due to genetics or a result of the surrounding environment shaping the animals as they grew, Prada began

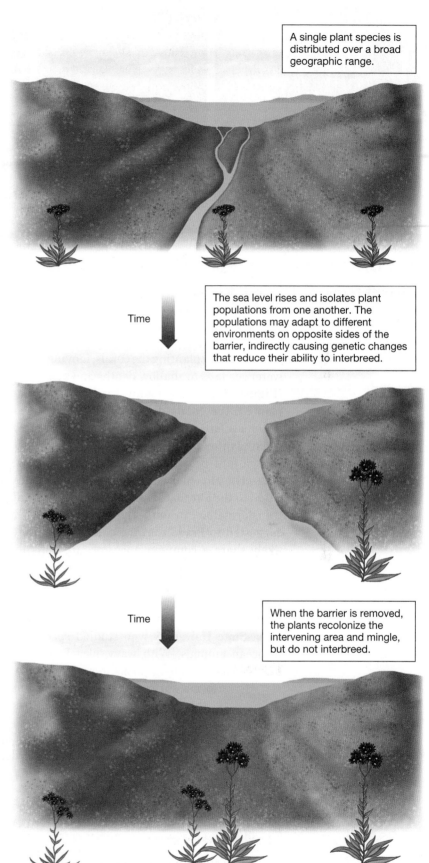

A single plant species is distributed over a broad geographic range.

The sea level rises and isolates plant populations from one another. The populations may adapt to different environments on opposite sides of the barrier, indirectly causing genetic changes that reduce their ability to interbreed.

When the barrier is removed, the plants recolonize the intervening area and mingle, but do not interbreed.

Range of overlap

Figure 13.7

Physical barriers can produce allopatric speciation by blocking gene flow
Allopatric speciation can occur when populations are separated by a geographic barrier, such as a rising sea.

Q1: What factors must be present for allopatric speciation to occur?

Q2: If a geographic barrier is removed and the two reunited populations intermingle and breed, what attributes must the offspring have in order for the two populations, according to the biological species concept, to be considered still the same species?

Q3: If the two populations in question 2 are determined to still be the same species, did allopatric speciation occur?

Shallow-water sea fan

Deep-water sea fan

Carlos Prada transplants coral to waters of different depths.

Figure 13.8

Different corals at different depths

These two corals were once considered the same species, *Eunicea flexuosa*, commonly called a "sea fan." (Source: Photos courtesy of Carlos Prada.)

carefully transplanting the corals, moving deep-water sea fans to shallow depths and vice versa (**Figure 13.8**). He found that when transplanted, the corals did change. The shallow-water sea fans became taller and more spindly when planted in deep water, and the deep-water sea fans became wider in shallow waters, but—critically—neither made a complete transition to the alternate shape. The lack of a total transformation by either form to the other suggested to Prada that the corals, while they likely share a common ancestor, are actually two species that have adapted to their respective water depths.

When Prada finished his graduate work in Puerto Rico, he e-mailed a professor at Louisiana State University who studied speciation in ocean animals. With wavy, bleached-blond hair, Michael Hellberg looks more like a California surfer than a professor, but this evolutionary biologist has long been fascinated with how one species splits to become two, especially in the ocean. "Say you have a new lake forming, and a species becomes isolated in the lake. Then it's pretty obvious there's not going to be a lot of interbreeding to fight against, and the species just adapts. To me, there's no mystery in that," says Hellberg. This would be an example of allopatric speciation. "I've always tried to target groups where species look closely related and where ranges of the species overlap. That makes things a lot harder."

Hellberg welcomed Prada into his crew, and the two set out to extend Prada's work to find out whether his idea—that coral evolve different adaptive traits at different depths in the ocean, leading to the formation of new species—was unique to coral reefs in Puerto Rico or could be observed in other areas around the Caribbean. With Hellberg's support, Prada traveled to the Bahamas, Panama, and Curaçao to observe and take samples from sea fan colonies.

As he waited for Prada to return home with the data, Hellberg remained skeptical of the idea of **ecological isolation**, that two closely related species in the same territory could be reproductively isolated by slight differences in habitat. But

MICHAEL HELLBERG AND CARLOS PRADA

Michael Hellberg (right) is an evolutionary biologist at Louisiana State University who studies how species evolve in marine environments. Carlos Prada (left) was a graduate student in Hellberg's lab, and is now a postdoctoral researcher at Penn State studying how organisms cope with changes in the environment.

as Prada sampled more and more locations, the evidence was convincing. Indeed, everywhere he looked, Prada saw the same thing he had observed in Puerto Rico: broad, leaflike coral at shallow depths, and tall, sticklike coral in deeper waters. Distance between the populations did not matter; some shallow-water and deep-water fans were close enough together that Prada could reach out and bend them to touch each other. But depth did matter. The corals were physically close enough to easily interbreed in the water, yet the species had somehow become specialized to two different depths (**Figure 13.9**).

Prada brought his coral samples back to the lab and performed tests to see how genetically similar the two groups of corals were. Prada and Hellberg found that all the shallow-water sea fans across the Caribbean were more closely related to each other than they were to any of the deep-water sea fans. The same held true for the deep-water sea fans: they were more closely related to each other than to any of the shallow-water fans. In the DNA data, Hellberg had seen some genetic exchange between the two populations in the past, yet that exchange of genetic materials had been limited across the two groups, and each species had bred almost exclusively among its own populations.

Different Depths, Different Habitats

How did a few yards of depth produce habitats so different that the corals evolved different adaptations to each depth? The scientists are still investigating the differences between the shallow and deep habitats, but they have some hypotheses. One is that the coral have adapted their morphologies, and possibly their biochemistry, to suit different symbiotic algae that grow on them at different depths.

Symbiotic algae live on coral and, through photosynthesis, use sunlight to produce energy and organic compounds that coral use to maintain and grow calcium carbonate skeletons. In turn, the coral provide the algae with a sheltered place to live and produce carbon dioxide that the algae use during photosynthesis. (This cooperative relationship between species is known as mutualism; see Chapter 20.) But different species of algae have

In ecological isolation, the two species are usually physically close enough to breed, but somehow do not.

Shallow

Deep

Figure 13.9

Depth of water provides ecological isolation for sea fans
The two depths shown here differ slightly in their quantity and quality of light, the force of the waves or current, the amount of sediment deposited on them, the number and type of predators, and the availability and type of food.

different light and nutrient requirements, and organisms 15 feet below the surface receive a lot of sunlight, while those at 30 feet receive less light, and light of different wavelengths. Prada found different types of algae living on the two coral populations (**Figure 13.10**), so it's possible that the sea fans evolved to be better hosts for the most successful algae at their depth.

In many cases in the ocean, as with coral and algae, the interaction between two species so strongly influences their survival that they have evolved in tandem—a phenomenon known as **coevolution**. The term "coevolution" encompasses a wide variety of ways in which an adaptation in one species evolves alongside a complementary adaptation in another species.

Figure 13.10

Different species of corals and their resident algae
Although all these algae (single-celled eukaryotic organisms) look the same under a microscope, they are actually different species with different light requirements and photosynthetic capacities.

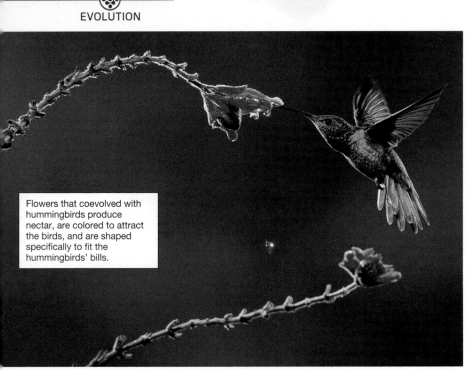

Flowers that coevolved with hummingbirds produce nectar, are colored to attract the birds, and are shaped specifically to fit the hummingbirds' bills.

Figure 13.11

Coevolution at its finest

This hummingbird's bill fits perfectly into this flower for easy access to the hummingbird's favorite food, nectar. The hummingbird feasts and then distributes the flower's pollen by carrying the pollen along to its next meal. It's a win-win situation!

Q1: Describe how coevolution, as with the hummingbird bill and hummingbird-pollinated flowers, is different from the kind of evolution described in Chapters 11 and 12.

Q2: Is coevolution the same thing as convergent evolution, described in Chapter 12? Why or why not?

Q3: Do you think one species' adapting over time to feed specifically and extremely successfully on another species is an example of coevolution? Why or why not?

So Many Chromosomes

New plant species can form in a single generation as a result of **polyploidy**, a condition in which an individual gains an extra full set or two (or three) of chromosomes. Humans and most other eukaryotes are *diploid* (having two sets of chromosomes), but some organisms are *triploid* (three sets) or *tetraploid* (four sets), or have an even higher number of chromosomes. Polyploidy is invariably fatal in people, but in many plant species it is not lethal.

| Haploid (*n*) | Diploid (2*n*) | Triploid (3*n*) | Tetraploid (4*n*) |

Polyploidy can occur when two different species hybridize and produce an offspring with an odd number of chromosomes, because of improper alignment of the chromosomes during mitosis (see Chapter 6). This increase in chromosome sets can lead to reproductive isolation because the chromosome number in the gametes of the new polyploid individual no longer matches the number in the gametes of either of its parents.

Polyploidy has had a large effect on life on Earth: more than half of all plant species alive today are descended from species that originated by polyploidy. A few animal species also appear to have originated by polyploidy, including several species of lizards, salamanders, and fish.

Another example of coevolution is the relationship between hummingbirds and certain species of flowers (**Figure 13.11**).

Alternatively, or in addition to hosting specific algae, Hellberg hypothesizes that the shallow-water and deep-water sea fans may have adapted to the type and density of sedimentation found at different ocean depths. Or perhaps predators are more abundant at certain depths. Whatever the major habitat factors driving the corals' adaptation, natural selection led to the formation of two different coral populations tailored to the ocean depths at which they grow.

Today, Hellberg considers the two populations different species. "What makes this interesting is that you have populations that could potentially interbreed, yet they maintain their differences in the face of each other," says Hellberg. "For just about any biological question you want to ask, they are independent entities." The formation of new species in the *absence* of geographic isolation is called **sympatric speciation** (from *sym*, "together"; **Figure 13.12**). Sympatric speciation is a particularly important process in plants (see "So Many Chromosomes").

Still, a question lingers: Why don't the two coral species interbreed? They appear to be capable of it, because of evidence of some genetic

Haplochromis chilotes feeds on insects.

Haplochromis macrognathus feeds on other fishes.

Macropleurodus bicolor feeds on snails and other mollusks.

Astatotilapia elegans is a generalized bottom feeder.

Figure 13.12

Sympatric speciation drives diversity among Lake Victoria cichlid species

Scientists have described some 500 species of cichlid fishes in Lake Victoria. Genetic analyses indicate that they all descended from just two ancestor species over the past 100,000 years. These four species show some of the differences in feeding behavior and morphology.

Q1: What is the main difference between allopatric and sympatric speciation?

Q2: Name two events that must happen for both allopatric speciation and sympatric speciation to occur.

Q3: Do you think all of the 500 species in Lake Victoria arose through sympatric speciation? Why or why not?

mixing in the past, yet there is no evidence that interbreeding is common; only a few rare hybrids grow on the ocean floor between them. When two species are reproductively isolated from each other, we say that **reproductive barriers** exist between those species. Reproductive barriers are often divided into two categories: *prezygotic* and *postzygotic*.

Barriers that prevent a male gamete (such as a human sperm) and a female gamete (such as a human egg) from fusing to form a zygote are **prezygotic barriers**. Prezygotic barriers act *before* the zygote exists (**Figure 13.13**). Barriers that prevent zygotes from developing into

Blue-footed boobies point their beaks, wings, and tails upward in a mating dance called "sky pointing."

Figure 13.13

The blue-footed booby courtship dance is a prezygotic, behavioral reproductive barrier

This species of booby has a unique ritual dance that must be accurately completed before mating. Other booby species do not perform exactly the same dance and therefore do not mate with the blue-footed booby.

Q1: What does "prezygotic" mean?

Q2: How is the booby's ritual dance a prezygotic reproductive barrier?

Q3: What are some other prezygotic reproductive barriers besides a mating dance?

Table 13.1

Reproductive Barriers That Isolate Two Species in the Same Geographic Region

Type of Barrier	Description	Effect
Prezygotic		
Ecological isolation	The two species breed in different portions of their habitat, in different seasons, or at different times of day.	Mating is prevented.
Behavioral isolation	The two species respond poorly to each other's courtship displays or other mating behaviors.	Mating is prevented.
Mechanical isolation	The two species are physically unable to mate.	Mating is prevented.
Gametic isolation	The gametes of the two species cannot fuse, or they survive poorly in the reproductive tract of the other species.	Fertilization is prevented.
Postzygotic		
Zygote death	Zygotes fail to develop properly, and they die before birth.	No offspring are produced.
Hybrid sterility	Hybrids survive but are unable to produce viable offspring.	No offspring are produced.
Hybrid performance	Hybrids survive poorly or reproduce poorly.	Hybrids are not successful.

healthy and fertile offspring are called **postzygotic barriers**, which act *after* the zygote is formed.

A wide variety of cellular, anatomical, physiological, and behavioral mechanisms generate pre- and postzygotic reproductive barriers, but they all have the same overall effect: little or no mating occurs, and therefore few or no alleles are exchanged between species (**Table 13.1**). Hellberg suspects that some prezygotic barriers exist between the two sea fan species—perhaps the gametes of the two species don't fuse successfully, as in some sea urchins (**Figure 13.14**)—but there may also be a postzygotic barrier such as generation time.

It seems likely that the reason hybrids between the two species do not succeed is the uniquely long generation time for these corals, says Prada. Sea fans don't reach reproductive age until they are 15–20 years old, and they

continue reproducing until they are 60 years old or more. So, while coral gametes and larvae can and do disperse far from their parents and may interbreed or take root at incorrect depths, natural selection then has 15–20 years to winnow out the less successful offspring. If there are small differences in survival rates between the two species at a particular depth—if a deep-water coral fares slightly worse in a shallow area than a shallow-water coral does—those differences will become amplified over 15 years, and by the time the corals reach reproductive age it is likely that only the shallow-water species will still be alive in the shallow area.

New species form and adapt to their environments in many different ways, from island-bound lizards to depth-dependent corals. And from those varied beginnings comes the vast diversity of life that exists on our planet today, the result of billions of years of evolution. In

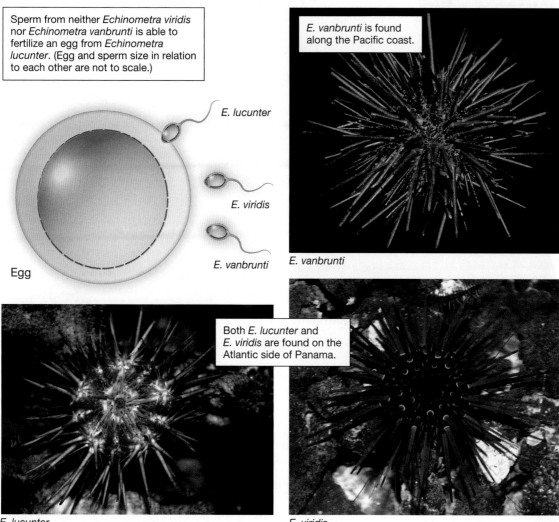

Sperm from neither *Echinometra viridis* nor *Echinometra vanbrunti* is able to fertilize an egg from *Echinometra lucunter*. (Egg and sperm size in relation to each other are not to scale.)

E. lucunter

E. viridis

E. vanbrunti

Egg

E. vanbrunti is found along the Pacific coast.

E. vanbrunti

Both *E. lucunter* and *E. viridis* are found on the Atlantic side of Panama.

E. lucunter

E. viridis

Figure 13.14

Gametic isolation among sea urchin species

The sea urchin species found along both coasts of Panama illustrate gametic isolation as a reproductive barrier. At one point long ago, the Panama landmass was not continuous and did not separate the two oceans. These three species likely evolved through both allopatric and sympatric speciation from a recent common ancestor after the oceans were separated.

Q1: Which species is/are sympatric with *Echinometra lucunter*?

Q2: Which species is/are allopatric with *E. lucunter*?

Q3: If two individuals have incompatible gametes, what will be the result of a mating event between them?

the next section of this book, we explore the dramatic transitions in life-forms that have walked, crawled, and swum around our planet, from small dinosaurs with wings to the most successful, most cunning species ever to traverse the planet: humans.

On the Diversity of Species

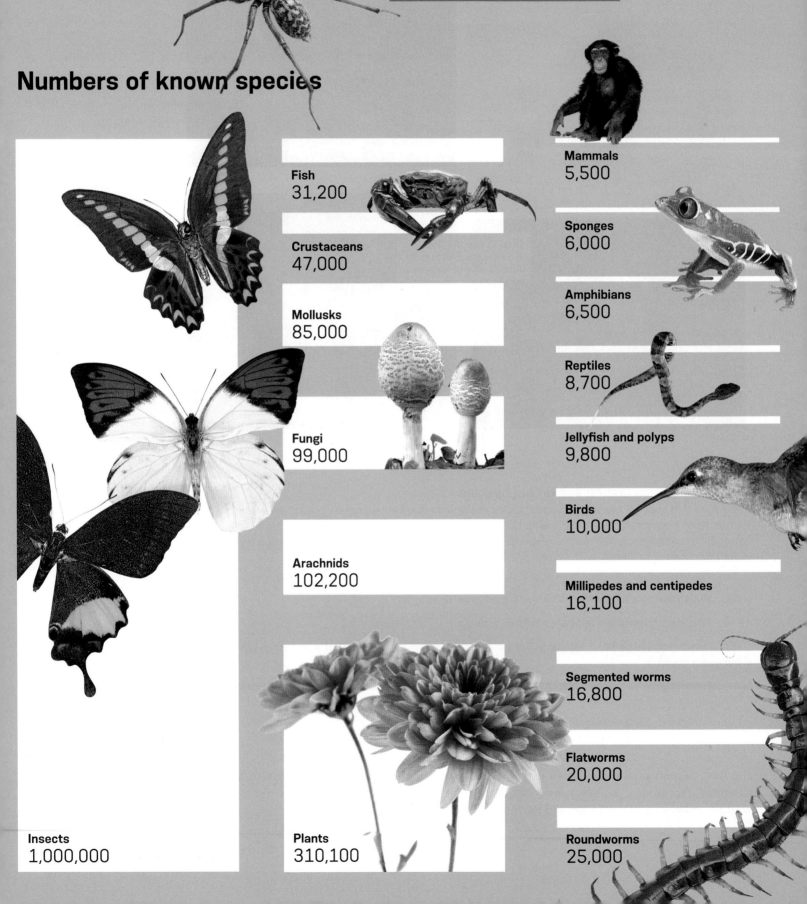

There are an estimated 8.74 million eukaryotic species on Earth, yet we are familiar with only a fraction of those. Scientists estimate that 86% of land species and 91% of aquatic species have not been discovered. Of the known species, insects top the chart with an estimated 1 million different species.

Assessment available in smartw⊙rk**5**

Numbers of known species

Fish
31,200

Crustaceans
47,000

Mollusks
85,000

Fungi
99,000

Arachnids
102,200

Insects
1,000,000

Plants
310,100

Mammals
5,500

Sponges
6,000

Amphibians
6,500

Reptiles
8,700

Jellyfish and polyps
9,800

Birds
10,000

Millipedes and centipedes
16,100

Segmented worms
16,800

Flatworms
20,000

Roundworms
25,000

REVIEWING THE SCIENCE

- **Adaptive traits,** or adaptations, are inherited characteristics that improve an individual's chances of surviving and reproducing in a specific environment.

- Natural selection leads to **adaptation**, genetic changes that improve the reproductive success of organisms in their environment over time. Adaptation does not produce "perfection."

- According to the **biological species concept**, a **species** is a group of populations that interbreed and can produce live and fertile offspring.

- The process of **speciation**, one species splitting into two or more species, is the by-product of **reproductive isolation** and **genetic divergence** between populations.

- **Allopatric speciation** occurs when populations of a species become **geographically isolated**, limiting gene flow and making genetic divergence more likely.

- Speciation that occurs between populations lacking geographic isolation is called **sympatric speciation**.

- In **ecological isolation**, two closely related species in the same area are reproductively isolated by slight differences in habitat.

- When adaptive traits of one species evolve in concert with adaptive traits in another species, **coevolution** has occurred.

- Reproductive isolating mechanisms can be **prezygotic barriers** (before zygote formation) or **postzygotic barriers** (after zygote formation).

The Basics

1 When two populations are reproductively isolated, what else must occur for speciation to happen?

(a) gene flow

(b) genetic divergence

(c) coevolution

(d) convergent evolution

(e) none of the above

2 Traits that are inherited and improve an individual's ability to survive and reproduce are called _____ traits.

(a) adaptive

(b) polymorphic

(c) biological

(d) sympatric

(e) allopatric

3 Prezygotic isolating mechanisms prevent hybrid offspring from occurring between species because

(a) the resulting offspring are not fertile and cannot reproduce.

(b) the egg and sperm fuse and form a zygote, but it does not survive.

(c) the egg and sperm do not ever meet or, if they do, cannot fuse to form a zygote.

(d) all of the above

4 Which of the following reproductive isolating mechanisms prevents mating because the two species are physically unable to mate?

(a) ecological isolation

(b) behavioral isolation

(c) mechanical isolation

(d) gametic isolation

(e) all of the above

5 Select the correct terms:
Speciation due to geographic separation and (**genetic divergence / gene flow**) of two populations is considered (**allopatric / sympatric**) speciation.

6 Select the correct phrases:
(**Prezygotic barriers / Postzygotic barriers**) are reproductive isolating mechanisms that occur after a zygote has been formed. An example would be (**an infertile hybrid / gametes that cannot fuse**).

Challenge Yourself

7 Natural selection

(a) leads to species being better adapted to their environment.

(b) may lead to speciation if there is no gene flow between populations.

(c) may lead to genetic divergence among populations in differing environments.

(d) all of the above

(e) none of the above

8 Which of the following examples describe an adaptive trait? (Select all that apply.)

(a) A rainforest tree that is vulnerable to storm damage.

(b) A male bird that is more successful than others at attracting a female mate.

(c) A rabbit that is better camouflaged in its environment.

(d) A desert plant that is able to survive drought.

(e) A frog that is more noticeable to predators.

9 Place the following events in the order in which they are most likely to occur by numbering them from 1 to 5.

_____ a. Speciation occurs.

_____ b. Genetic divergence takes place.

_____ c. Geographic barrier arises.

_____ d. Gene flow is discontinued.

_____ e. Species adapts to a new environment.

Try Something New

10 Distinct species that are able to interbreed in nature are said to "hybridize," and their offspring are called "hybrids." The gray oak and the Gambel oak can mate to produce fertile hybrids in regions where they co-occur. However, the gene flow in nature is sufficiently limited that, overall, the two species remain phenotypically distinct. If the hybrid offspring survive well and reproduce to the extent that there is a large population of hybrid individuals that breed between themselves but do not interbreed

with either of the two original parent species (the gray and the Gambel), which of the following would you say most likely led to the new hybrid species?

(a) prezygotic reproductive barriers

(b) sympatric speciation

(c) allopatric speciation

(d) postzygotic reproductive barriers

(e) none of the above

11 The four-eyed fish, *Anableps anableps*, really has only two eyes, which function as four, enabling the fish to see clearly through both air and water. *A. anableps* is a surface feeder, so the ability to see above water helps it locate prey such as insects. Its unique eyes also enable it to scan simultaneously for predators attacking from above (such as birds) or below (such as other fish). This species of fish _____ crafted by natural selection.

Anableps anableps

(a) is an example of coevolution

(b) has an evolutionary adaptation

(c) is an example of sympatric speciation

(d) is an example of allopatric speciation

12 Give an example of a barrier that might bring about allopatric speciation. In addition to this barrier, what would be necessary for speciation to occur?

13 Researchers from the Smithsonian Institution were startled to discover that the 3 species of *Starksia* blennies they had been studying in the Caribbean islands were really 10 different species. How could these researchers have thought that these 10 different reef fish species were only 3 species?

Leveling Up

14 *Write Now* **biology: adaptations** Select an organism (other than humans) that you find interesting. Research two adaptations of your organism and describe them in detail. Explain carefully why each of these features is considered an important adaptation for its species. Discuss how a change in the environment might change the usefulness of these adaptations.

15 **Doing science** The eastern and western meadowlarks are very similar in morphology. Both are grassland species with some overlap in their range in the upper Midwest, and each sings a distinctly different song. Create a hypothesis as to whether these are two different species or two populations of the same species, and describe an experiment that would enable you to test your hypothesis. What result would support your hypothesis, and what result would cause you to reject your hypothesis?

16 **What do *you* think?** According to the Defenders of Wildlife organization, there were 20–30 million American range bison in the old West. During the late 1800s the American range bison were hunted to the brink of extinction, leaving behind a bottleneck population of only 1,091 individuals. The population has since rebounded to about 500,000 bison. Unfortunately, almost all of these bison are the descendants of these few individuals crossbred with domestic cattle by ranchers. Scientists and conservationists want to genetically test bison to find those of pure bison origin to preserve the species. Only these, they argue, should be called American range bison and be allowed to roam free in the national parks as bison. They think hybrids should be confined to farms and ranches, should be called "beefalo" rather than bison, and should not be afforded the protection that pure bison currently have. What do you think? Should bison tainted with cattle genes be removed from free-range parks? Should the government spend scarce conservation monies on genetic testing and breeding efforts to preserve the pure bison population? Investigate conservation efforts and the costs of genetically testing and relocating bison to help you with your decision. Is speciation at the hands of human beings now part of evolution as we know it?

For more, visit digital.wwnorton.com/bionow2 for access to:

The First Bird

A fresh wave of fossils challenges the identity of the first dino-bird.

- -

After reading this chapter you should be able to:

- Identify critical events on a time line of life on Earth and explain their evolutionary importance.

- Interpret an evolutionary tree of a group of organisms.

- Explain how new scientific data may change an evolutionary tree without challenging our understanding of evolution.

- Classify a given species at each level of the Linnaean system.

- Give an example of an adaptation that enabled a group of organisms to transition from an aquatic existence to life on land.

- Describe the impact of prehistoric mass extinctions on Earth's biodiversity, and infer the likely impact of the current extinction rate on biodiversity.

CHAPTER

14

THE HISTORY
OF LIFE

I n 1861, quarry workers in Germany unearthed the fossil of a crow-sized bird. The 150-million-year-old preserved skeleton looked bizarre: it had feathers and clawed hands like a bird, but teeth and a long, bony tail like a reptile.

After close examination by paleontologists, the fossil was hailed as a transitional form between birds and reptiles and labeled the oldest known bird. Named *Archaeopteryx*, the feathered dinosaur became famous around the world as the first solid evidence that birds descended from dinosaurs (**Figure 14.1**). Charles Darwin, who proposed the theory of descent with modification, called the discovery "a grand case for me."

Over 100 years later, a young fossil hunter named Xu Xing bent over the dirt at a dig site in Liaoning province in northern China. It was Xu's first time in the field since finishing college. An excellent student, Xu had wanted to study economics at Peking University in Beijing, but at the time, students in China did not pick their majors, and Xu was assigned paleontology. Luckily, farmers in Liaoning had recently begun to uncover huge numbers of fossils, and research was booming.

So, as a grudging paleontologist, Xu began searching for dinosaurs in Liaoning. He had a particular interest in feathered dinosaurs. By the late twentieth century, the idea that birds descended from dinosaurs was no longer

a hypothesis, but a well-established scientific theory backed by a mountain of data. And atop that mountain of data sat *Archaeopteryx*, the earliest known bird.

Little did Xu know, as he dug through the dirt that day, that he would soon be the one to knock *Archaeopteryx* from its perch.

Dinosaurs and Domains

Although the *Archaeopteryx* fossil is old—150 million years old, in fact—much older fossils have been found.

Our solar system and Earth formed 4.6 billion years ago. Some of the oldest known rocks on Earth are 3.8 billion years old and contain carbon deposits that hint at life. Cell-like structures have been found in layered mounds of sedimentary rock called stromatolites that formed 3.7 billion years ago, and projections based on DNA analysis also support the idea that life appeared on Earth at that time.

The question of how life arose from nonlife is one of the greatest riddles in biology, but scientists have little doubt that all life on Earth is related. As noted in Chapter 1, all living organisms are united by a basic set of characteristics.

Figure 14.1

The original *Archaeopteryx* fossil

There are 11 complete *Archaeopteryx* fossils, but this one from Germany was the first to be found. Note the faint impressions of feathers on the wings and tail.

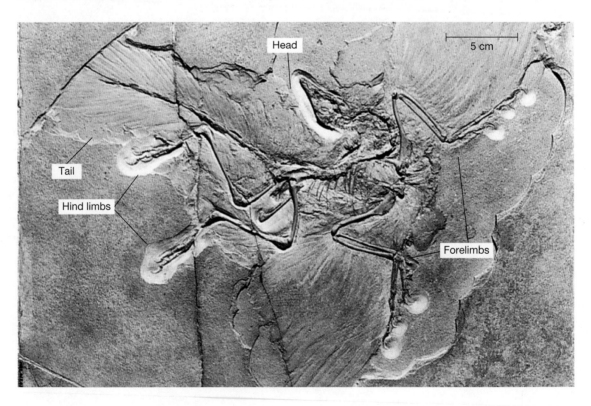

Life shares this set of common properties because all living organisms descended from a common ancestor, known as the *universal ancestor*. This hypothetical ancestral cell is placed at the base of the tree of life. From that cell, all life emerged. Biological diversity, or **biodiversity**, embraces the variety of all the world's living things, as well as their interactions with each other and the ecosystems they inhabit. Biodiversity can be described at the level of genes, or species, or entire ecosystems.

In spite of intense worldwide interest, scientists do not know the exact number of species alive today. Most traditional estimates have fallen in the range of 3–30 million species, but in 2016, scientists combined microbial, plant, and animal data sets from around the world, scaled up, and estimated that Earth could contain nearly a trillion species. So far, about 1.5 million species have been collected, named, and placed into an evolutionary tree, meaning it is possible that only one ten-thousandth of 1 percent of species are known to us. In other words, 99.9999 percent of species are yet undiscovered.

Life throughout history and today is so diverse that biologists created a classification system to organize it into categories. The **domains** form the highest, most inclusive hierarchical level in the organization of life, describing the most basic and ancient divisions among living organisms. There are three domains of life (**Figure 14.2**):

- **Bacteria**, which includes familiar disease-causing bacteria such as *E. coli*

Domain Bacteria's key evolutionary innovations include diverse lifestyles and communication and coordination within populations. Bacteria are microscopic and single-celled, and they reproduce asexually by splitting in two.

Domain Archaea's key evolutionary innovations include adaptation to extreme environments, which has allowed them to become widespread, diverse, and abundant. Like the bacteria, archaeans are microscopic and single-celled, and they reproduce asexually by splitting in two.

Domain Eukarya's key evolutionary innovations include a larger cell size, the presence of a nucleus and other membrane-enclosed internal compartments, and multicellularity in some groups. Most eukaryotes have the ability to reproduce sexually, through meiosis and the combination of gametes into a zygote.

Figure 14.2

Three domains of life
This "tree of life" shows the relationships of the three domains of life.

Q1: Why is there a shared line from the universal ancestor for Archaea and Eukarya?

Q2: Where would birds be found within this figure? What about humans?

Q3: To which domain would you expect a disease-causing organism to belong? What if the organism was multicellular?

Billions of years ago (bya)		Millions of years ago (mya)					
4.6 3.8		540	490	445	415	360	

Precambrian		Paleozoic					

Geologic period

Precambrian	Cambrian	Ordovician	Silurian	Devonian	Carboniferous

Major events

Origin of life; photosynthesis causes oxygen content of Earth's atmosphere to increase; first eukaryotes; first multicellular organisms	Large and relatively sudden increase in the diversity of animal life; increase in diversity of algae; first vertebrates	Further increases in diversity of marine invertebrates and vertebrates; plants and fungi begin to colonize land; mass extinction at end of period	Increase in diversity of fishes; first hints of colonization of land by insects and other invertebrates	Increase in diversity of land plants; first amphibians colonize land; mass extinction late in period	Extensive forests; amphibians dominate life on land; increase in diversity of insects; first reptiles
Life begins	**Invertebrates fill the seas**	**Plants begin to colonize land**	**Fish diversity increases**	**Amphibians appear**	**Earth is covered with forests**

Figure 14.3

The geologic timescale and major events in the history of life

The history of life can be divided into 12 major geologic time periods, beginning with the Precambrian (4.6 bya to 540 mya) and extending to the Quaternary (2.6 mya to the present). This time line is not drawn to scale; to do so would require extending the diagram off the book page to the left by more than 5 feet (1.5 meters).

- **Archaea**, which consists of single-celled organisms best known for living in extremely harsh environments
- **Eukarya**, which includes all other living organisms, from amoebas to plants to fungi to animals

Humans, dinosaurs, and birds are all part of the Eukarya domain. They are *eukaryotes*. Bacteria and Archaea are two different domains—Archaea are more closely related and in some ways more similar to Eukarya than

to Bacteria—yet because neither Bacteria nor Archaea are eukaryotes, the two have traditionally been lumped under a common label: *prokaryotes*.

Prokaryotes first appear in the fossil record at about 3.7 billion years ago (**Figure 14.3**), but the first eukaryotes did not evolve until over a billion years later. Luckily for us, and all other eukaryotes, roughly 2.8 billion years ago a group of bacteria evolved a type of photosynthesis that releases oxygen as a by-product. As a result, the oxygen concentration in the atmosphere increased over time, and about 2.1 billion years ago the first single-celled eukaryotes evolved. When the oxygen concentration reached its current level, by about 650 million years ago (mya), the evolution of larger, more complex multicellular organisms became possible, including fish, then land plants, then insects, amphibians, and reptiles. One group of reptiles, which would eventually dominate most

XU XING

Xu Xing is a paleontologist at the Chinese Academy of Sciences in Beijing. He has discovered more than 60 species of dinosaurs and specializes in feathered dinosaurs and the origins of flight.

300	250	200	145	65	2.6 0.0

Mesozoic — Cenozoic

Permian	Triassic	Jurassic	Cretaceous	Paleogene/Neogene	—Quaternary

Continents come together to form Pangaea; reptiles dominate life on land by 265 mya; mass extinction at end of period	Early dinosaurs; first mammals; mass extinction at end of period	Continents begin to separate; diverse dinosaurs; first birds; first flowering plants	Flowering plants begin to dominate life on land; mass extinction at end of period, including extinction of the last non-avian dinosaurs	Continents near present positions; increased diversity of flowering plants, birds, mammals, and pollinating insects	Repeated advance and retreat of glaciers; evolution of humans; extinction of large mammals and birds
Age of reptiles begins	**Dinosaurs evolve and spread**	**Large plant-eating dinosaurs thrive**	**Flowering plants increase**	**Age of mammals begins**	**Humans evolve**

Q1: During what geologic period did life on Earth begin?

Q2: How long ago did species begin to move from water to land? What period was this?

Q3: In what period would *Archaeopteryx* have been alive?

other species, was the dinosaurs. Dinosaurs first appeared about 230 mya, during the Triassic period, and they took over the planet.

Feathered Friends

Xu Xing may not have wanted to be a paleontologist when he went to college, but by the time he graduated, he was hooked on dinosaurs. Over the next 20 years, Xu became one of the most productive researchers in his field. To date, he has discovered and named more than 60 extinct species—mostly dinosaurs, but also a reptile and a salamander. And the majority of those dinosaur fossils have feathers (**Figure 14.4**).

As scientists traced back the **lineage**, or line of descent, from birds to dinosaurs, it became clear that birds are most closely related to theropods—fast-moving dinosaurs that ran

Figure 14.4

A feathered dinosaur tail, trapped in amber

In 2016, scientists discovered a 99-million-year-old dinosaur tail, complete with its feathers, in a piece of amber.

on two legs and had hollow, thin-walled bones (as birds do). Theropods were a diverse group of dinosaurs (**Figure 14.5**). Most were carnivores, including insectivores, but a few were herbivores or omnivores. There were theropods that could swim and eat fish. Some theropod species boasted enlarged scales, and of course, many theropods had feathers.

Xu and other scientists map out lineages using a diagram called an **evolutionary tree**, a model of evolutionary relationships among groups of organisms based on similarities and differences in their DNA, physical features, biochemical characteristics, or some combination of these. An evolutionary tree maps the relationships between ancestral groups and their descendants, and it clusters the most closely related groups on neighboring branches.

In an evolutionary tree, organisms are depicted as if they were leaves at the tips of the tree branches. A given ancestor and all its descendants make up a **clade**, or branch, on the evolutionary tree. *Archaeopteryx* and all subsequent animals that evolved from it are considered a clade (**Figure 14.6**).

A **node** marks the moment in time when an ancestral group split, or diverged, into two

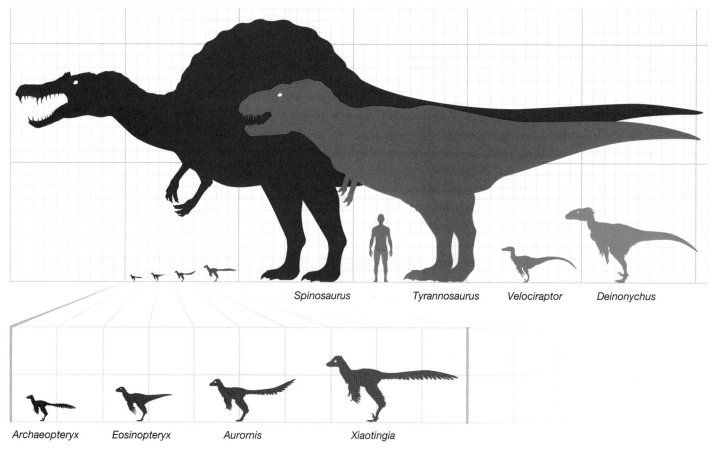

Spinosaurus   Tyrannosaurus   Velociraptor   Deinonychus

Archaeopteryx   Eosinopteryx   Aurornis   Xiaotingia

Figure 14.5

Dinosaurs large and small

Theropods ranged in size from tiny, like a chicken, to huge, like the group's most famous member, *Tyrannosaurus rex*.

Q1: In what ways were theropods the same as modern birds? Give at least two similarities.

Q2: In what ways did theropods differ from modern birds? Give at least two differences.

Q3: Birds are often referred to as "living dinosaurs." Is this accurate? Why or why not?

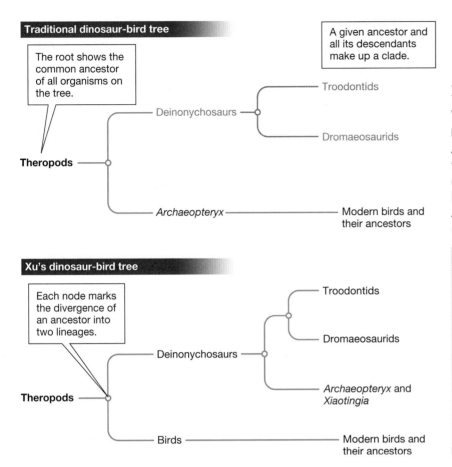

Traditional dinosaur-bird tree

The root shows the common ancestor of all organisms on the tree.

A given ancestor and all its descendants make up a clade.

Theropods — Deinonychosaurs — Troodontids

Dromaeosaurids

Archaeopteryx — Modern birds and their ancestors

Xu's dinosaur-bird tree

Each node marks the divergence of an ancestor into two lineages.

Theropods — Deinonychosaurs — Troodontids

Dromaeosaurids

Archaeopteryx and *Xiaotingia*

Birds — Modern birds and their ancestors

Figure 14.6

The evolutionary origins of birds

(*Top*) The traditional evolutionary tree, showing *Archaeopteryx* as an early bird that split off from the deinonychosaurs (birdlike, carnivorous dinosaur). (*Bottom*) The evolutionary tree proposed by Xu after the discovery of *Xiaotingia*, with *Archaeopteryx* as a deinonychosaur rather than an early bird.

Q1: In the traditional tree, identify the node showing the common ancestor for early birds and dinosaurs.

Q2: What do both the traditional tree and Xu's tree suggest about troodontids and dromaeosaurids?

Q3: In both trees, identify the node for the common ancestor of *Archaeopteryx* and other birds. In what way are the nodes different in the two trees?

separate lineages. A node represents the **most recent common ancestor** of two lineages in question—that is, the most *immediate* ancestor that *both* lineages share. For over 100 years, researchers considered *Archaeopteryx* the most recent common ancestor of both birds and dinosaurs and thus placed it at the root of the avian clade—the first bird (**Figure 14.6**, top).

The first bird, that is, until Xu stumbled across a fossil that threw the field into controversy. In 2008, Xu visited the Shandong Tianyu Museum of Nature, a dinosaur museum in eastern China. There, he happened upon a unique fossil that had been collected by a farmer in Liaoning and sold through a dealer to the museum. The fossil, entombed in yellowish rock, shows a small, birdlike dinosaur seemingly craning its neck forward and spreading its short wings (**Figure 14.7**). "I saw it and said, 'Oh, this is an important species,'" recalls Xu. He asked the museum to let him study it.

As Xu examined the fossil, which he later named *Xiaotingia zhengi*, he wondered where it belonged on the dinosaur-bird evolutionary tree.

To find out, he analyzed *shared derived traits* of the fossil and similar early birds. **Shared derived traits** are unique features common to all members of a group that originated in the group's most recent common ancestor and then were passed down in the group (but not in groups that are not direct descendants of that ancestor). In this case, the original ancestor

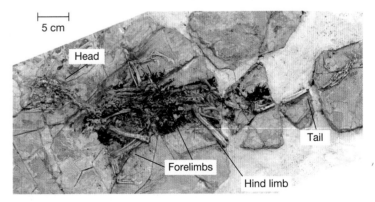

5 cm

Head

Forelimbs

Hind limb

Tail

Figure 14.7

***Xiaotingia zhengi*: a controversial new leaf on the dinosaur-bird evolutionary tree**

in question was assumed to be *Archaeopteryx*, and the shared traits included feathers, clawed hands, and a long, bony tail.

By comparing *Xiaotingia* to *Archaeopteryx* and other related species, Xu created a new evolutionary tree of early birds (**Figure 14.6**, bottom). Suddenly, *Archaeopteryx* wasn't in the avian clade. Instead, *Archaeopteryx* and *Xiaotingia* were in a different clade, grouped with deinonychosaurs—the small, birdlike, carnivorous dinosaurs commonly called raptors. (The term "raptor" refers to carnivorous modern birds like hawks and owls, but it is also used informally to describe this group of dinosaurs.) *Archaeopteryx*, Xu believed, was not the first bird, but a raptor. It had feathers, but its descendants did not evolve into birds.

"It was a big change," says Xu. Yet, he adds, it wasn't entirely unexpected. In the last 20 years, more and more fossils of early avian species have been discovered—and the more Xu compared *Archaeopteryx*, discovered in Europe, to other early birds discovered in China, the less it looked like a bird and the more it looked like a raptor. Early birds have small, thick skulls and two toes on each foot; *Archaeopteryx*, on the other hand, has a long, almost pointy skull and three toes on its feet. "*Archaeopteryx* is just so different from other early birds," says Xu.

Xu published his revised evolutionary tree in 2011. It took the scientific world by storm. Some researchers embraced the idea: "Perhaps the time has come to finally accept that *Archaeopteryx* was just another small, feathered bird-like theropod fluttering around in the Jurassic," wrote one paleontologist in the journal *Nature*. Others disagreed, arguing that Xu's analysis was not convincing. For them, *Archaeopteryx* remained the first bird.

The History of Life on Earth

Xu knew that one new fossil was but a single shred of evidence and therefore unlikely, by itself, to convince the scientific community. "The evidence is not very strong," he admits, "but it was a question we wanted to discuss and add more analyses to, including old and new fossils." The identity of the first bird is important because our understanding of how flight evolved is based on the classification of *Archaeopteryx*. If *Archaeopteryx* is not the species from which flight evolved, then our understanding of flight is wrong.

Xu's redesign of the early bird evolutionary tree raised another important question: Did flight evolve more than once? If *Archaeopteryx*, which looked capable of flight, was a bird, then flight probably evolved just once among dinosaurs, in the avian lineage. If *Archaeopteryx* was not a bird, and was instead simply a raptor that could fly, then flight evolved at least twice among dinosaurs, once in raptors and once in avians—but only avians went on to evolve into modern birds.

Biological classification helps us answer important evolutionary questions like these. In addition to recognizing three broad domains of life, biologists group the Eukarya into four distinct **kingdoms**, the second-highest level in the hierarchical classification of life (**Figure 14.8**).

Figure 14.8

The tree of life

The domains Bacteria and Archaea do not have separate kingdoms within them. The domain Eukarya encompasses four kingdoms: Protista (protists, an artificial grouping that includes organisms such as amoebas and algae), Plantae (plants), Fungi (including yeasts and mushroom-producing species), and Animalia (animals).

These kingdoms are **Protista**, a diverse group that includes amoebas and algae; **Plantae**, which encompasses all plants; **Fungi**, which includes mushrooms, molds, and yeasts; and **Animalia**, which encompasses all animals, including dinosaurs, birds, and humans. The members of each of these kingdoms share evolutionary innovations that adapted the organisms to their environments, enabling them to live and reproduce successfully (**Figure 14.9**).

Below the level of kingdom, biological classification can get even more specific using the **Linnaean hierarchy**, a system of biological classification devised in the eighteenth century by a Swedish naturalist named Carolus Linnaeus. The smallest unit of classification in the Linnaean hierarchy—the **species**—reflects individuals that are the most related to each other. The most closely related species are grouped together to form a **genus** (plural "genera"). Using these two categories in the hierarchy, every species is given a unique, two-word Latin name, called its **scientific name**. The first word of the name identifies the genus to which the organism belongs; the second word defines the species. Comparing our own species name, *Homo sapiens*, to that of the Neanderthals, *Homo neanderthalensis*, shows that the two are classified as separate species but belong to the same genus, *Homo* (**Figure 14.10**).

In the Linnaean hierarchy, each species is placed in successively larger and more inclusive categories beyond genus. Closely related genera are grouped into a **family**. Closely related families are grouped into an **order**. Closely related orders are grouped into a **class**. Closely related classes are grouped into a **phylum** (plural "phyla"). And, you guessed it, closely related phyla are grouped together into a kingdom.

The earliest forms of life evolved in water. About 650 mya (during the Precambrian period), the number of organisms appearing in the fossil record increased. At that time, much of Earth was covered by shallow seas filled with small, mostly single-celled organisms floating freely in the water.

Then, about 540 mya, the world experienced an astonishing burst of evolutionary activity, with a dramatic increase in biodiversity. Most of the major living animal groups first appear in the fossil record during this time, popularly known as the **Cambrian explosion**. The Cambrian explosion changed the face of life on Earth from a world of relatively simple, slow-moving, soft-bodied scavengers and herbivores to a world dominated by large, fast predators. The presence of predators sped up the evolution of Cambrian herbivores, judging by the variety of scales and shells and other protective body coverings typical of many Cambrian, but not Precambrian, fossils (**Figure 14.11**).

The Cambrian explosion, however, occurred primarily in the oceans. Because life first evolved in water, the colonization of land by living organisms posed enormous challenges. Indeed, many of the functions basic to life, including support, movement, reproduction, and the regulation of heat, had to be handled very differently on land than in water. About 480 mya, near the beginning of the Ordovician period, plants were the first organisms to meet these challenges. These early terrestrial colonists were single-celled or had just a few cells.

Fungi are thought to have made their way onto land next, according to fossil evidence. For example, scientists have found fossils of terrestrial fungi that are 455–460 million years old. Reconstructing the evolutionary history of eukaryotes from DNA data, scientists estimate that the common ancestor of fungi and animals diverged from all other eukaryotes about 1.5 billion years ago, and fungi diverged from their closest cousins, the animals, about 10 million years after that.

Next, land plants evolved and diversified from the original green algae that made it to land. By 360 mya, at the end of the Devonian period, Earth was covered with plants. The first land animals likely emerged about 400 mya. Many of the early animal colonists on land were carnivores; others fed on living plants or decaying plant material.

The first vertebrates to colonize land were amphibians, the earliest fossils of which date to about 365 mya (see Chapter 17 for more on vertebrates). Early amphibians descended from lobe-finned fishes. Amphibians were the most abundant large organisms on land for about 100 million years. Then, in the late Permian period, reptiles took over as the most common vertebrate group. Reptiles were the first group

Protista

The kingdom Protista is an artificial grouping defined by what members of this group are not: protists are not plants, animals, fungi, bacteria, or archaeans.

Although most protists are harmless, the best-known ones are pathogenic, like *Plasmodium vivax*, the protist that causes malaria.

Animal-like protists are consumers.

Fungus-like protists are decomposers.

Plant-like protists are photosynthetic.

Most protists are single-celled and microscopic, and they can swim with the help of one or more flagella, or by waving a carpet of tiny hairs called cilia.

Protista

Diplomonads, others | Euglenoids, others | Ciliates, others | Brown algae, diatoms, others | Forams, others | Red algae | Green algae

The evolutionary history of the protists remains unresolved.

Common ancestor of plants and the green algae

Common ancestor of the Eukarya

Plantae

Plants are multicellular autotrophs and mostly terrestrial. Because plants are producers, they form the basis of essentially all food webs on land.

Bryophytes (mosses and liverworts)

Plants have a waxy covering, known as the *cuticle*, that covers their above-ground parts. A waxy cuticle holds in moisture—an important adaptation to life on land.

Ferns

Gymnosperms

Gymnosperms were the first plants to evolve pollen, a microscopic structure that contains sperm cells, which freed them from a dependence on water for fertilization. Gymnosperms were also the first to evolve seeds, which can be disseminated so they will not compete with the mother plant for sunlight, or for water and nutrients in the soil.

Angiosperms

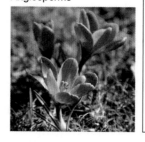

Flowering plants, or angiosperms, are dominant and the most diverse group of plants on our planet. The keys to the success of angiosperms are the flower, a structure that evolved through modification of early plant reproductive organs, and the fruit, a fleshy ovary wall that protects and helps disperse the seeds inside it.

Figure 14.9

The four kingdoms of the domain Eukarya

Q1: **What group of organisms shares the most recent common ancestor with plants?**

Zygomycetes (molds)

Fungi digest organic material outside the body and absorb the molecules released as breakdown products.

Ascomycetes (sac fungi)

Basidiomycetes (club fungi)

Fungi play several roles in terrestrial ecosystems. Many are decomposers, acting as garbage processors and recyclers by speeding the return of the nutrients in dead and dying organisms to the ecosystem.

Fungi are similar to animals in that they store surplus food energy in the form of glycogen. Like some animals, such as insects and lobsters, fungi produce a tough material called *chitin* that strengthens and protects the body. Unlike animals, fungal cells have a protective cell wall that wraps around the plasma membrane and encases the cells.

Plantae	Protista	Animalia	Fungi

Plants

Amoebas, others

Animals

Fungi

Common ancestor of animals and fungi

Animals are multicellular ingestive heterotrophs, obtaining energy and carbon by ingesting food. All animals are consumers, and some are important decomposers in the ecosystems they inhabit.

Animal cells differ from those of plants and fungi in that they lack cell walls. Instead, many cells in the animal body are enveloped in, or attached to, a felt-like layer known as the *extracellular matrix*. An important evolutionary innovation of animals is the development of true tissues. Most animals have two or three main tissue layers that give rise to a structurally complex body.

Sponges

Cnidarians

Mollusks

Arthropods

Echinoderms

Chordates

The sponges are the most ancient animal lineage. Cnidarians, a group that includes jellyfish and corals, evolved next. The remaining animal phyla fall into two groups: the *protostomes* and *deuterostomes*, distinguished by different patterns of embryonic development. Protostomes comprise more than 20 separate subgroups, including mollusks (such as snails), annelids (segmented worms), and arthropods (including spiders and insects). Deuterostomes include echinoderms (sea stars and their relatives) and the chordates. The chordates are a large group that includes all animals with backbones, such as fish, birds, and humans.

Q2: Are fungi more closely related to plants or to animals? Does the answer surprise you? Why or why not?

Q3: If you were to create an evolutionary tree in which amoebas were included within the kingdom of organisms to which they were the most closely related (rather than with protists, where they are currently placed), where would you put them?

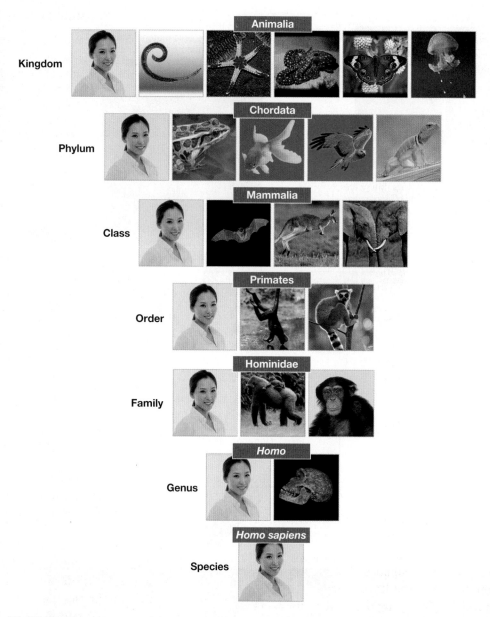

Figure 14.10

The Linnaean hierarchy of classification

Q1: Within which category are individuals most closely related to one another?

Q2: Within which category are individual species most distantly related?

Q3: Are individual species more closely related within the same order or within the same family?

of vertebrates that could reproduce without returning to open water, because they lay amniotic eggs that have a built-in food source and are protected from drying out by a hardened shell, compared to the jellylike sac that encloses the eggs of other vertebrates. The evolution of the amniotic egg was a major event in the history of life because it established a new evolutionary branch, the amniotes, which later included all reptiles, birds, and mammals.

And so, with the rise of reptiles 230 mya, the age of dinosaurs began.

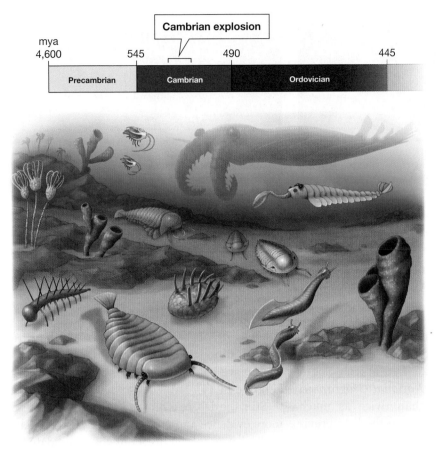

mya
4,600   545   490   445

Cambrian explosion

Precambrian   Cambrian   Ordovician

Figure 14.11

Cambrian biodiversity

During the Cambrian period, animal diversity increased dramatically. The remains of many of these species have been found in Canada (the Burgess Shale), China (the Maotianshan Shale), and fossil beds in Greenland and Sweden. Some of the fossils look familiar, resembling sponges and brachiopods, but do not appear to be related to any living groups of animals.

Tussling with Trees

In January 2013, a year and a half after Xu published his controversial paper about *Xiaotingia* and *Archaeopteryx*, his work received unexpected support. Pascal Godefroit, a paleontologist at the Royal Belgian Institute of Natural Sciences, reported the discovery of a feathered dinosaur called *Eosinopteryx brevipenna*. A commercial collector in northeastern China had dug up *Eosinopteryx* in the same area where *Xiaotingia* was discovered. The tiny 161-million-year-old dinosaur was preserved as a virtually complete skeleton, with its legs bent and arms out, as if about to jump (**Figure 14.12**).

When Godefroit and his team added *Eosinopteryx* to the evolutionary tree of feathered dinosaurs, they came to the same surprising conclusion that Xu had: *Archaeopteryx* was not a bird. Instead, *Archaeopteryx* was a raptor along with *Eosinopteryx* and *Xiaotingia*. All three species share traits that early birds did not have, such as arms longer than their legs, reduced tail plumage, and primitive feather development.

But science is a continuously changing process, and only 4 months later, Godefroit published data about another birdlike fossil that caused him to revise his hypothesis once again. This time it was a feathered dinosaur that Godefroit found collecting dust in the archives of a Chinese museum. The 18-inch-long fossil had small, sharp teeth and long forelimbs. Godefroit believed the dinosaur, which his team named *Aurornis xui* to honor Xu's work, probably couldn't fly, but instead used its wings to glide from tree to tree. But its other features, including the hip bones, were clearly shared by modern birds (**Figure 14.13**).

Using *Aurornis*, Godefroit constructed another evolutionary tree. This time he started from scratch, compiling data on almost a

PASCAL GODEFROIT

Pascal Godefroit is a paleontologist at the Royal Belgian Institute of Natural Sciences in Brussels. He has discovered numerous feathered dinosaurs.

Figure 14.12

Another feathered dinosaur, *Eosinopteryx brevipenna*, supports Xu's new tree

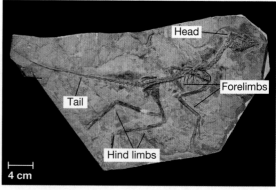

Figure 14.13

A namesake fossil, *Aurornis xui*, illuminates the dinosaur-bird transition

thousand characteristics of skeletons of 101 species of dinosaurs and birds. "It's very impressive," paleontologist Mike Lee at the South Australian Museum told *National Geographic*. "They considered more than twice as much anatomical information as even the best previous analyses."

Contrary to the tree that arose from his first study, Godefroit's second tree places *Archaeopteryx* back on its roost in the bird family, although no longer as the oldest bird. That place belongs to *Aurornis*, says Godefroit (**Figure 14.14**).

Still, the debate is not over. "Of course we need more evidence and more work," says Xu. "Many of these new species are a possible candidate for

Figure 14.14

The early birds

Godefroit's 2013 study places *Archaeopteryx* and *Xiaotingia* with birds rather than dinosaurs, as in the traditional dinosaur-bird evolutionary tree (see Figure 14.6, top). But it places *Aurornis* as the earliest known bird on the evolutionary tree.

Q1: Is *Xiaotingia* an earlier or later bird than *Archaeopteryx* in this tree?

Q2: If a future study, based on more fossils or new measurements, placed *Archaeopteryx* back with dinosaurs, would this suggest that birds are not related to dinosaurs? Why or why not?

Q3: If you were to create an evolutionary tree of modern birds, where would you expect to place the roadrunner (judging by its appearance in this figure) as compared to a house sparrow or pigeon?

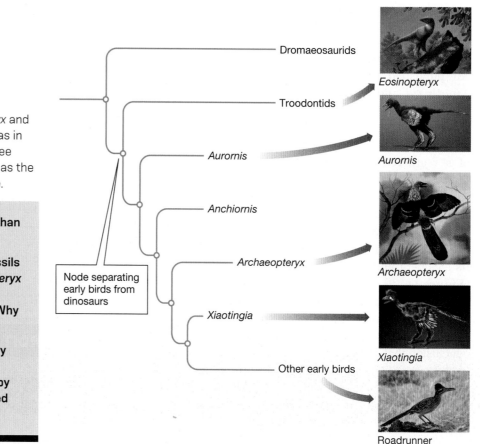

the earliest bird." There are likely to be more fossils that will shake up the bird tree, says Xu, but that's a good problem to have in science. "There are so many new species, it just makes it difficult for us all to agree."

Xu, Godefroit, and others are sure to continue to dig up dinosaur fossils for some time yet. Dinosaurs arose about 230 mya and dominated the planet from about 200 mya to about 65 mya. Then the majority of them went extinct, except for those that evolved into birds. As the fossil record shows, species have regularly gone extinct throughout the history of life. The rate at which this has happened—that is, the number of species that have gone extinct during a given period—has varied over time, from low to very high. At the upper end of this scale, the fossil record shows that there have been five **mass extinctions**, periods of time during which great numbers of species go extinct (**Figure 14.15**).

Although difficult to determine, the causes of the five mass extinctions are thought to include such factors as climate change, massive volcanic eruptions, changes in the composition of marine and atmospheric gases, and sea level changes. The Cretaceous extinction event occurred

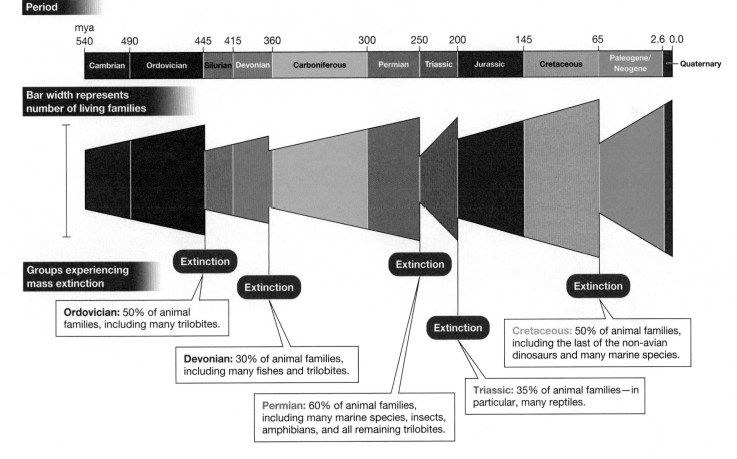

Figure 14.15

Mass extinctions and biodiversity

In addition to the marine and terrestrial animal groups shown here, plant groups were severely affected by the five mass extinctions that have occurred in Earth's history. After each extinction, life again diversified.

Q1: What extinction event occurred about 200 mya? What animal groups were most affected by this event?

Q2: Which of the mass extinctions appears to have removed the most animal groups? How long ago did this extinction occur?

Q3: The best studied of the mass extinctions is the Cretaceous extinction. Why do you think it has been better studied than the other extinctions?

The Sixth Extinction

On at least five occasions, mass extinctions have occurred across the globe, caused at different times by climate change, volcanic eruptions, and possible asteroids. Today, scientists agree we are in the midst of a sixth extinction, and this time, we, the human race, are the cause.

Assessment available in smartw⊛rk5

Ghosts of species past

Passenger Pigeon

The passenger pigeon, the most common bird in North America 200 years ago, was hunted to extinction in the nineteenth century. The last wild bird was shot in 1900, and the last captive bird died in 1914.

Xerces Blue Butterfly

The Xerces blue butterfly once lived on sand dunes around San Francisco—until its habitat was destroyed by urban development. It was last seen in 1943.

Caribbean Monk Seal

The Caribbean monk seal was the only seal native to the Caribbean Sea and Gulf of Mexico. It was overhunted for oil, and its food sources were overfished. It was last sighted in 1952.

Golden Toad

Once common in the cloud forests of Monteverde, Costa Rica, the golden toad has not been seen since 1989. Pollution and global warming likely contributed to its extinction.

Number of species known to be extinct or extinct in the wild since 1500

Arachnids	9
Crustaceans	12
Reptiles	22
● Amphibians	36
● Insects	58
Fishes	71
● Mammals	79
Plants	134
● Birds	145
Mollusks	324

about 65 mya and wiped out three-fourths of plant and animal species on Earth, including non-avian dinosaurs. Researchers suspect that a massive comet or asteroid slammed into the Gulf of Mexico, choking the skies around the planet with debris and decreasing the ability of plants to photosynthesize. As the plants died, so, too, did animals further up the food chain.

Mass extinctions affect the diversity of life in two main ways: First, entire groups of organisms perish, changing the history of life forever. Second, the extinction of one or more dominant groups of organisms can provide new opportunities for groups that previously were of relatively minor importance, thereby dramatically altering the course of evolution. When a group of organisms expands to take on new ecological roles and to form new species and higher taxonomic groups, that group is said to have undergone an **adaptive radiation**. Some of the great adaptive radiations in the history of life occurred after mass extinctions, such as when the mammals diversified after the extinction of the dinosaurs.

Today, one species of mammal, *Homo sapiens*, dominates life on land, and our impact on biodiversity is unprecedented. Because of human activities, the world is losing species at an alarming rate. In July 2016, researchers at the United Nations Environment Programme World Conservation Monitoring Centre in the United Kingdom announced that worldwide biodiversity had fallen below predetermined "safe" levels, thresholds below which ecological function is likely to be negatively affected. An estimated 58 percent of the world's land has lost more than 10 percent of its biodiversity, the researchers found, with grasslands and biodiversity hot spots such as the Amazon rainforest hit the hardest.

That is alarming news, because nonhuman organisms provide even the most basic requirements for human life. Plants and plantlike protists produce the oxygen we breathe and provide us with food. Whole ecosystems provide so-called "ecosystem services," environmental benefits that humans rely on. For example, coast redwood trees in northern California intercept fog, mist, and rain, channeling the water onto and into the ground.

Biologists today assert that we are on our way toward a new mass extinction, and the cause is clear: the activities of the ever-increasing number of humans living on, and exploiting, Earth.

REVIEWING THE SCIENCE

- The first single-celled organisms resembled bacteria and probably evolved about 3.7 billion years ago.

- The most basic and ancient branches of the tree of life define three **domains**: **Bacteria**, **Archaea**, and **Eukarya**. All life-forms fall into one of these three domains. The Eukarya are further divided into four kingdoms: **Protista**, **Fungi**, **Plantae**, and **Animalia**. The variety of these life-forms, as well as their interactions with each other and the ecosystems they inhabit, is **biodiversity**.

- Scientists use **evolutionary trees** to model ancestor-descendant relationships among different organisms. The tips of branches represent existing groups of organisms, and each **node** represents the moment when an ancestor split into two descendant groups. A **clade** is an ancestral species and all its descendants.

- Closely related groups of organisms share distinctive features that originated in their **most recent common ancestor**. These **shared derived traits** are used to identify the **lineages** of a species.

- The **Linnaean hierarchy** places each **species** in successively larger and more inclusive categories. Closely related species are grouped together into a **genus**, related genera are grouped into a **family**, related families into an **order**, related orders into a **class**, related classes into a **phylum**, and finally, related phyla into a **kingdom**.

- Organisms are identified by their genus and species names, together referred to as their **scientific name**.

- The release of oxygen by photosynthetic bacteria caused oxygen concentrations in the atmosphere to increase. Rising oxygen concentrations made possible the evolution of single-celled eukaryotes about 2.1 billion years ago. Multicellular eukaryotes followed about 650 mya.

- Life in the oceans changed dramatically during the **Cambrian explosion**, when large predators and well-defended herbivores suddenly appear in the fossil record. The Cambrian explosion is an example of **adaptive radiation**, in which a group of organisms take on new ecological roles and form new species and higher taxonomic groups.

- The land was first colonized by plants (about 480 mya), fungi (about 460 mya), and invertebrates (insects about 400 mya), which were followed later by vertebrates (about 365 mya). Each group evolved unique adaptations that enabled life to flourish on land.

- There have been five **mass extinctions** during the history of life on Earth. The extinction of a dominant group of organisms may provide new opportunities for other groups.

THE QUESTIONS

The Basics

1 The first single-celled organisms on Earth probably evolved

(a) 3,700 years ago

(b) 3.7 million years ago

(c) 3.7 billion years ago

(d) We have no idea when the first life on Earth appeared.

2 The production of _____ by prokaryotes increased its atmospheric concentration and enabled more complex forms of life to evolve.

(a) carbon dioxide

(b) oxygen

(c) nitrogen

(d) all of the above

3 The term "Cambrian explosion" refers to

(a) an increase in biodiversity.

(b) a mass extinction.

(c) an atmospheric detonation due to increasing gas concentrations.

(d) an oceanic eruption caused by underwater volcanic activity.

4 Which of the following terms most specifically describes what occurs when a group of organisms expands to take on new ecological roles, forming new species and higher taxonomic groups in the process?

(a) speciation

(b) mass extinction

(c) evolution

(d) adaptive radiation

5 Match each term with its definition.

CLADE
1. A distinctive feature that originated in two groups' most recent common ancestor.

NODE
2. An ancestor and all its descendants.

LINEAGE
3. The point at which an ancestral group splits into two separate lineages.

EVOLUTIONARY TREE
4. A diagram showing the evolutionary relationships among a related group of organisms.

SHARED DERIVED TRAIT
5. The line of descent of a group of organisms.

6 Select the correct terms:

The domains Archaea and Bacteria are referred to as (**prokaryotes / eukaryotes**). The domain (**Eukarya / Prokarya**) includes four kingdoms. The kingdom (**Plantae / Fungi**) was first to make the transition to land. The kingdom (**Animalia / Plantae**) is most closely related to the kingdom Fungi.

7 All of the following are considered possible causes of the five mass extinctions *except*

(a) climate change.

(b) change in the composition of atmospheric or marine gases.

(c) comet or asteroid strike.

(d) worldwide thunderstorms.

(e) volcanic eruptions.

Challenge Yourself

8 *Archaeopteryx*

(a) belongs to the domain Prokarya.

(b) belongs to the kingdom Protista.

(c) lived during the Precambrian period.

(d) was much larger than modern birds.

(e) is an early example of the evolution of birds.

9 Identify the domain and kingdom for each of the following organisms.

(a) *Archaeopteryx*

(b) chanterelle mushrooms

(c) palm tree

(d) green algae

(e) you

10 "King Philip came over for good soup" is a mnemonic to remember each level of the Linnaean hierarchy.

a. List each level of the Linnaean hierarchy, in the order indicated by the mnemonic.

b. What level of the Linnaean hierarchy is missing from the mnemonic?

c. Create your own mnemonic to remember the Linnaean hierarchy.

11 Place the following evolutionary events in the correct order from earliest to most recent by numbering them from 1 to 5.

_____ a. Cambrian explosion

_____ b. origin of life on Earth

_____ c. plants' transition to land

_____ d. oxygen-rich environment created by bacteria

_____ e. evolution of birds from dinosaurs

12 Fill in the tree below to show the evolutionary relationships among the domains and kingdoms of life.

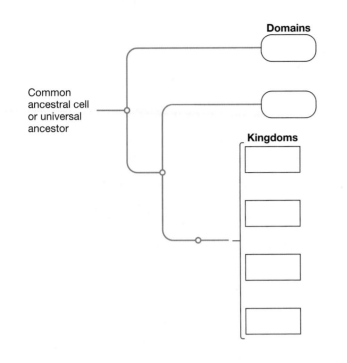

Try Something New

13 In each of the following cases, which domain(s) or kingdom(s) might contain organisms with the traits described? (You may need to review Figure 14.9 to answer.)

a. motile, with flagella or cilia

b. All organisms within the domain or kingdom are single-celled.

c. found in extreme environments (for example, high temperature, low oxygen, high salt)

d. multicellular with organ systems

e. photosynthetic

14 The traditional dinosaur-bird tree (Figure 14.6, top) can be restated as a hypothesis: "We hypothesize that *Archaeopteryx* is an early bird, and that birds split off from the closely related dinosaur groups Troodontidae and Dromaeosauridae." Restate Xu's tree (Figure 14.6, bottom) and Godefroit's tree (Figure 14.14) as hypotheses.

Leveling Up

15 *Write Now* **biology: early life on Earth** Read the *New York Times* article "World's Oldest Fossils Found in Greenland" (http://www.nytimes.com/2016/09/01/science/oldest-fossils-on-earth.html). Answer the following questions, writing one paragraph for each.

a. Summarize the findings of the study reported in the *Times* in your own words.

b. If the scientific community reached a consensus that these are *not* fossils, how would this conclusion affect our understanding of early life on Earth? Explain why such a conclusion would not be a failure of the scientific method, but rather a natural result of it.

c. Define the Late Heavy Bombardment (LHB). Describe how our understanding of the LHB has changed, and how this change has influenced our hypotheses about early life on Earth.

16 **Is it science?** Some people became quite upset when scientists agreed that the dinosaur *Brontosaurus* was in fact *Apatosaurus*, that *Triceratops* was only a juvenile *Torosaurus*, and that Pluto was not a planet. They clung to their *Brontosaurus*, or *Triceratops*, or planet Pluto, dismissing the scientists' arguments by saying, "Scientists are always changing their minds." Explain how scientists' "changing their minds" is not a problem but instead a necessary and useful feature of science, using the science described in this chapter as an example.

For more, visit digital.wwnorton.com/bionow2 for access to:

Navel Gazing

Scientists discover hundreds of new species of microbes—burrowed deep in our belly buttons.

After reading this chapter you should be able to:

- Differentiate between bacteria and archaea, and explain why they are together referred to as prokaryotes.

- Explain why prokaryotes can reproduce more rapidly than eukaryotes and why rapid reproduction can be advantageous.

- Give examples of the environments in which archaeans are found.

- Illustrate through example the breadth of metabolic diversity found in bacteria.

- Explain the function of at least one cellular structure that is unique to prokaryotes.

- Explain how citizen science contributes to scientific research.

CHAPTER

15

BACTERIA
AND ARCHAEA

The gathering crowd jostled for position around the table. The scientists behind the table smiled as they handed out long, thin cotton swabs. Holly Menninger thanked them for the swab, then stared down at her shirt. She had been told what to do: lift your shirt, insert the swab into your belly button, swirl it around, place it into a test tube, seal the top. "Oh, the things we do for science," Holly thought. Then she swirled.

After returning her tube, Menninger—an entomologist by training and, at the time, coordinator of the New York Invasive Species Research Institute at Cornell University—struck up a conversation with the researchers manning the table, including Rob Dunn of North Carolina State University. Dunn, an applied ecologist and writer with an infectious passion for science, had initiated the project to answer a simple question: What lives on our skin? The belly button, as small and strange as it is, was an excellent place to look.

If you pick away the lint, the leftover schmutz in your belly button isn't dirt; it's alive. Our bodies are ecosystems, home to an estimated 39 trillion resident **microbes** (microscopic organisms), primarily bacteria. Considering that the average human body is made up of 30 trillion human cells, that's about a 1:1 ratio of microbial cells to human cells. (The old 10:1 ratio is a miscalculation that has unfortunately become enshrined in popular culture.) The **human microbiome**—the complete collection of microbes that live in and on our cells and bodies—affects human gut health, brains, and even body odor.

Microbes live in our guts, on our eyelashes, and yes, in our belly "holes," as Dunn calls them. In fact, the "belly hole" is an ideal place for a scientific study of resident microbes because it is a protected, moist patch of skin and one of the few areas that individuals don't regularly wash. Dunn knew that his students, if they could determine which microbes were swimming around in individual volunteers' navels, could then dig into the more burning question: *Why* do we each have the microbes we do?

"We know that which microbes you have on your skin influences your risk of infection, how attractive you are to other people, and how attractive you are to mosquitoes," says Dunn. "So this question of 'What determines which microbes are on your skin?' is super intriguing."

Of the three domains of life on Earth, Bacteria was the first to split off from the shared ancestor of the Archaea and Eukarya (**Figure 15.1**). Some fossil evidence places that split at about 3.48 billion years ago, yet a 2016 discovery in Greenland suggests that bacteria existed as early as 3.7 billion years ago, close to a period of time when Earth was being bombarded by asteroids (**Figure 15.2**).

Figure 15.1

Two of the three domains of life

Bacteria and Archaea are two distinct domains. They share many characteristics that are not found in the Eukarya.

Figure 15.2

Earth's first life left its mark in Greenland rocks

In 2016, Australian scientists found stromatolites, the fossilized secretions of bacteria, in rocks exposed by receding glaciers. These rocks formed 3.7 million years ago in a part of the world that is now Greenland.

Archaea likely split from Eukarya much later, around 2.7 billion years ago. The microbes that make up Bacteria and Archaea display many small but significant differences in their DNA, plasma membrane structure, and metabolism. They also share several important characteristics and so they are traditionally lumped under a common label: *prokaryotes*.

Prokaryotes are single-celled organisms, with a single loop of DNA floating free in the cytoplasm of the cell (review Figure 4.8 for a comparison of prokaryotic and eukaryotic cells). Unlike cells found in the Eukarya, prokaryotes do not have membrane-enclosed organelles. Prokaryotic cells are not only simpler than eukaryotic cells; they are smaller—almost exclusively microscopic, and invisible to the naked eye because of their diminutive size. The simple structure and single loop of DNA enable prokaryotes to reproduce at a much more rapid rate than eukaryotes, doubling in number every 10–30 minutes (**Figure 15.3**).

Prokaryotes in both domains are widespread and extremely abundant, and they display an astonishing diversity in metabolism. The vast majority of life on Earth is single-celled and prokaryotic. Scientists estimate that the number of prokaryotes on Earth is about 5,000,000,000, 000,000,000,000,000,000,000 (5 nonillion, or 5×10^{30}). For example, prokaryotes—not fishes or algae—are the most abundant organisms in the open ocean, where they play a crucial role in the ecology of our biosphere.

The success of prokaryotes is due in part to how quickly a prokaryotic population reproduces, and also to the fact that they can live practically anywhere, including many places where few other forms of life can, such as deep, hot thermal vents or the acidic environment of the human intestine. Humans are saturated in microbes. We are each a zoo.

Menninger was so intrigued by Dunn's project that when a job became available in his lab a few months later, she immediately applied for and got the position. By then, the team had gathered about 60 swabs at local events, including the aforementioned conference, and now they had bigger ambitions: Dunn hired Menninger to spearhead a massive public outreach effort to acquire belly button swabs from all around the country. In **citizen science** projects such as this, the public participates in research by collecting

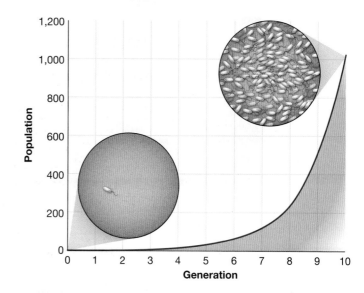

Figure 15.3

Prokaryotes are capable of extremely rapid population growth
An individual prokaryote is able to divide in two within 10–30 minutes. Those two can each divide in two in the same amount of time. This means that even a single bacterium or archaean can become a large population in a very short period of time.

Q1: If an individual prokaryote divides every 20 minutes, how many individuals will there be after an hour?

Q2: If the generation time is 20 minutes, how much time will have gone by when the final generation shown has doubled?

Q3: Many bacteria are able to reproduce more quickly in warmer conditions. What does this suggest to you about the importance of refrigerating foods?

and sometimes even analyzing data in cooperation with professional scientists (**Figure 15.4**).

"At the time, we were right at the forefront of an explosion in research, learning about how important the microorganisms that live on and in our bodies are to our health and well-being," says Menninger. "And this was a really great project to get people talking about the skin microbiome."

ROB DUNN

Rob Dunn is an applied ecologist at North Carolina State University. His fascination with species around us in our everyday lives has led him to research projects on microbes in our belly buttons, armpits, homes, and more.

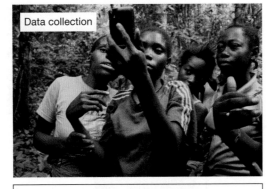

Data collection

Citizen scientists from the Republic of the Congo are learning to map the forest with GPS.

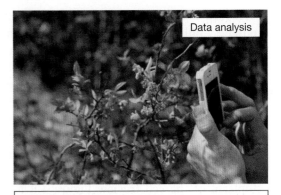

Data analysis

A citizen scientist in North America uses an app to document plant budding dates.

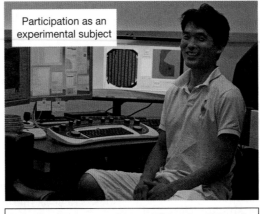

Participation as an experimental subject

Sebastian Seung, a professor at Princeton University, works with citizen scientists around the world to map the brain through a video game.

Figure 15.4

Citizen scientists work alongside professional scientists to increase our knowledge about the natural world

There are many ways you can contribute to science—by collecting data, analyzing data, or even being an experimental subject like the video game players who work with Sebastian Seung. And the advantages are not one-way: citizen scientists learn new things, often have fun, and report feeling a sense of purpose and contribution.

Q1: In which of the three ways did the navel microbiome participants contribute?

Q2: Which of the advantages listed above do you think the navel microbiome citizen scientists received?

Q3: Would you be willing to contribute to the navel microbiome project? Why or why not?

Dutch tradesman Antonie van Leeuwenhoek first opened our eyes to the microbial world back in 1668, when he used simple, handcrafted microscopes to discover bacteria swimming around in pond water. In the intervening 350 years, our knowledge of the microbial world has come a long way.

Today, we know that bacteria and archaeans make up more than two-thirds of the species found on Earth. In 2016, microbiologists at UC Berkeley published a revised evolutionary tree of life that, for the first time, included the vast diversity of bacteria and archaeans lurking in Earth's nooks and crannies. With the data used to generate that tree, they showed that life is clearly dominated by prokaryotes (**Figure 15.5**).

Merry Microbes

In 2010, as a gag, an undergraduate student in Dunn's lab decided to make a Christmas card with streaks of microbes taken from people in the lab. She asked for belly button swabs from her coworkers, spread each sample on a petri dish, and grew the bacteria into colonies. As she grew bacteria from multiple individuals, it quickly became clear that there was more variety among the microbes growing on people in the lab than had been expected.

"Then it went from being a fun lab project to a serious question," says Dunn. The Christmas gag had raised an important question: Why do one person's microbes differ from another's? From his experience studying ecology, Dunn knew that microbes are critical in our lives. But neither Dunn nor others knew which factors determine which particular skin microbes a person has, or lacks.

So Dunn and his team began to study the locally collected swabs. An early part of the experiment engaged participants by visually depicting each person's microbial menagerie. The team plated each sample onto a petri dish and then took a picture of it for the microbes' owner. Although prokaryotes are single-celled organisms, some form colonies or long chains of identical cells, produced by repeated splitting begun by one original cell. This bacterial pattern of reproduction resulted in bright, dramatic patterns on the plates (**Figure 15.6**).

Under the microscope, the individual microbes were even more diverse and elaborate. Bacteria and Archaea cells are quite variable in shape, ranging from rods to spheres to spirals (**Figure 15.7**). Still, they all have a basic structural plan. Most bacteria and many archaeans have a protective cell wall that surrounds the plasma membrane. Some have an additional wrapping around that cell wall called a **capsule**. The capsule, made of slippery biomolecules, works like an invisibility cloak: it helps disease-causing bacteria evade the immune system that protects organisms like us from foreign invaders.

Dunn's team observed many bacteria whose surface was covered in short, hairlike projections called **pili** (singular "pilus"), a common bacterial

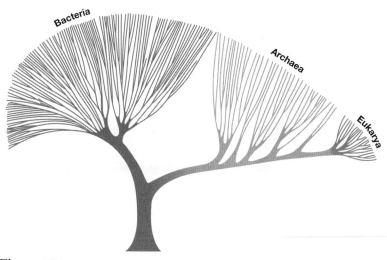

Figure 15.5

The most recent tree of life
Researchers sequenced DNA from individuals in 3,083 genera across all three domains. This information was combined by phylum, and the placement of each phylum in the tree was based on its relatedness to other phyla, as measured by DNA similarity.

> **Q1:** Where in the figure would you place the first life found on Earth?
>
> **Q2:** Identify where in the figure the Bacteria split off from the ancestor of Archaea and Eukarya.
>
> **Q3:** The figure (and thus the study) demonstrates that Archaea and Eukarya are more closely related to each other than to Bacteria. How is that illustrated?

Bacillus   *Staphylococcus*

Micrococcus   Clostridiales

Figure 15.6

Belly button bacterial biodiversity
Each petri dish contains colonies of the bacteria collected from one person's belly button sample.

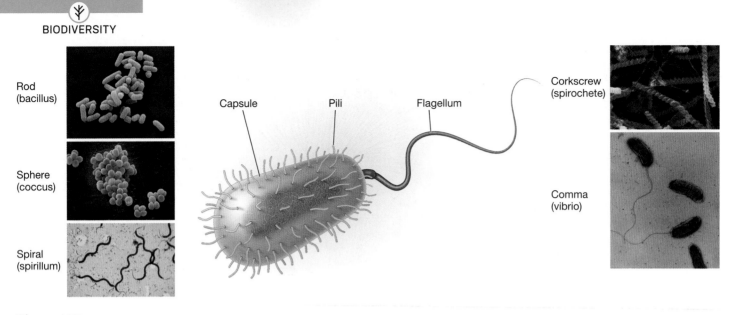

Rod
(bacillus)

Sphere
(coccus)

Spiral
(spirillum)

Capsule   Pili   Flagellum

Corkscrew
(spirochete)

Comma
(vibrio)

Figure 15.7

A simple structure, a diversity of forms

Bacteria and Archaea share a simple cell structure relative to eukaryotes, but they display a striking diversity of shapes and may have additional structures that perform special functions.

Q1: Which of these shapes do you think *Streptococcus* would take?

Q2: From the micrographs here, does it appear that all prokaryotes have a flagellum?

Q3: Which one of these shapes is most clearly capable of self-motility? Why?

feature. Bacteria use pili to link together to form bacterial mats, or to attach to surfaces in their environment, such as the cells of the human intestine. Some bacteria have one or more long, whiplike structures called **flagella** (singular "flagellum"), which spin like a propeller to push the bacterium through liquid.

Under the microscope, one obvious cellular feature is missing: prokaryotes do not have a true nucleus. They typically have much less DNA than eukaryotic cells have. Prokaryotes have less DNA because they have far fewer genes and because they contain relatively little non-coding DNA (DNA that is not used to construct proteins but may have regulatory or other functions). In contrast, eukaryotes generally have many more genes and far more non-coding DNA.

Dunn's lab partnered with Noah Fierer, an ecologist at the University of Colorado, to sequence the DNA from each belly button sample to determine the identity of the resident microbes, especially those that don't grow in a laboratory culture. To do so, lab members first had to access that DNA by mechanically crushing

the microbes and chemically dissolving their cell membranes with detergent, similar to how dish soap dissolves grease clinging to a dirty dish.

Next, they ran the sample through a silica strainer, called a column, to capture the DNA while allowing other debris to pass through. Once the DNA was isolated, the team amplified it using the polymerase chain reaction technique (see Figure 9.9) and then sequenced one short section of it, about 250 nucleotides of a well-studied gene called 16S ribosomal RNA (16S rRNA). Although 16S rRNA is found in all bacteria and archaeans, it is highly variable from one species to another, providing an identifying tag, like a fingerprint or barcode, to distinguish one species of microbe from another.

In 2012, the team published results from the initial set of swabs: among 92 belly buttons, the team identified 1,400 species of bacteria. "About 600 or so don't match up in obvious ways with known species, which is to say either they are new to science or we don't know them well enough," Dunn said in a media interview after the paper was published.

Archaea found in mineral hot springs are able to withstand extreme acidity *and* high temperatures.

Hydrogen-eating, methane-producing archaeans have been found in deep-sea thermal vents.

A new archaean species was recently found in an abandoned copper mine. Others have been found living in acidic drainage from mines.

Figure 15.8

Archaeans are everywhere

Archaeans, and bacteria to a lesser extent, thrive in environments that humans are incapable of surviving in.

Q1: Which shape in Figure 15.7 corresponds to the archaeans from deep-sea thermal vents?

Q2: Why are many archaeans referred to as "extremophiles"?

Q3: Is there anywhere you think archaeans could *not* survive? Justify your answer.

They had discovered new species of bacteria playing house in our navels. Still, Dunn was most interested in finding patterns hiding in the data. The first pattern they found was that the microbes in belly buttons are somewhat predictable. Six species showed up in about 80 percent of people, and when present, those species tended to be most widespread. "If I go to a cocktail party, I can tell you which species will be most abundant in the room," says Menninger with a laugh. That distribution pattern reminded Dunn of rainforest ecology: in any given forest, the types of plants might vary, but an ecologist can depend on a certain few tree types dominating the landscape.

Likewise, infrequent species tended to be present only in small quantities in belly buttons. Some were familiar: *Bacillus subtilis*, known to cause foot odor, also lives in some belly buttons (where it likely causes the same smell). Others were quite rare: one participant hosted a species previously found only in the soil of Japan, yet the man had never been to Japan. Dunn found a

species on himself known solely for eating pesticides. Why it was on his skin remains a mystery.

Amazingly, the team found more than just bacteria. On one particularly fragrant individual, who said he had not washed in years, they detected two species of archaeans. Although some bacteria thrive in unusual environments, the domain Archaea is well known for the extreme lifestyles of its members (**Figure 15.8**). Some are extreme **thermophiles** that live in geysers, hot springs, and hydrothermal vents—cracks in the seafloor that spew boiling water. The cells of most organisms cannot function at such high temperatures, but thermophiles have come up with evolutionary innovations that enable them to succeed where others cannot. Other archaeans, classified as **halophiles**, thrive in very salty, high-sodium environments where nothing else can live.

Yet archaeans have also been found in less exotic locations. In another citizen science project from Dunn and Fierer, students analyzed the microbial DNA swabs from home surfaces—kitchen counters, door frames, television screens, and more. Once again, they found several species of archaeans. "They are actually a lot more common than people tend to realize," says Menninger. "We're starting to have the molecular tools that allow us to find them." Dunn thinks the discovery is a bit ironic: "We spent so much time finding archaeans in the first place, in hard-to-reach, faraway places, and somebody could have just had some introspection with their own belly hole."

Yet back in the belly button, Dunn was unable to answer his main question about the bacteria: Why do individuals have the bacteria they do? The team looked at gender, ethnicity, how often participants washed their belly buttons, age, and more, yet they were unable to attribute any of the variation to biological or lifestyle factors.

So, as scientists tend to do, Dunn went looking for more samples. When Menninger joined the team, she initiated a program to have students and volunteers from around the country send in belly button swabs. Thanks to Menninger's outreach efforts, volunteers nationwide swabbed, swirled, and sealed. Soon the lab had over 600 samples to work with. That's a lot of microbes.

Talk, but No Sex

Like the plants and animals in a rainforest, microbes in an ecosystem communicate with each other, reproduce, and more. Researchers used to think prokaryotes were the ultimate loners—self-sufficient and maintaining strict single-celled lifestyles. Then they discovered that bacteria physically interact with each other in many ways.

Some bacteria (and Archaea) have a unique system of cell-to-cell communication called **quorum sensing** that enables them to sense and respond to other bacteria in the area in accordance with the density of the population. Disease-causing bacteria, for instance, may begin to multiply rapidly upon sensing that their numbers are high enough to overwhelm the host organism's immune system. Other bacteria coordinate their behavior by forming tough aggregates called *biofilms* that are made up of the same or different species (**Figure 15.9**).

It's strange to think about, but all these activities are happening on our skin: "Everything you can imagine life doing, happens in you," says Dunn, including reproduction. Prokaryotes

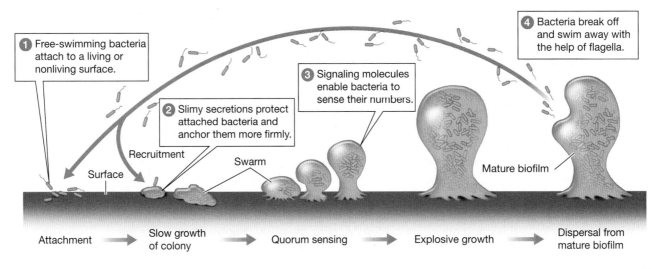

Figure 15.9

Quorum sensing enables coordinated behavior in bacteria

With quorum sensing, a bacterial population can increase its virulence, reproductive rate, or antibiotic resistance under the appropriate conditions. Biofilms are produced via quorum sensing to protect bacteria from environmental hazards.

Q1: How do individual bacteria know that they have a "quorum"?

Q2: There is a well-known biofilm found in your mouth. What is it?

Q3: Under what conditions might bacteria want to coordinate (via quorum sensing) to increase their reproductive rate?

typically reproduce by splitting in two in a process called **binary fission**, a form of asexual reproduction (see Figure 6.3). The DNA in the parent cell is copied before fission, and one copy is transferred to each of the resulting daughter cells. The genetic information in the daughter cells is virtually identical to that of the parent cell, as is invariably the case in asexual reproduction.

Although sexual reproduction has not been seen in prokaryotes, they are adept DNA pickpockets: microbes can capture bits of DNA from their environment or other bacteria and incorporate them into their own genetic material. The transfer of genetic material between microbes is known as *horizontal gene transfer* and it involves plasmids, loops of extra DNA in the cytoplasm of prokaryotes (see Figure 12.12). (Horizontal gene transfer has been seen in some eukaryotes as well. Bdelloid rotifers, for example, a type of microscopic freshwater animal, stole about 8 percent of the DNA in their genome from bacteria.) A bacterium can directly trade DNA with another bacterium through a process known as *bacterial conjugation*. Alternatively, when a bacterium dies, the cell may burst open and another bacterium may simply take up the released DNA.

Some types of bacteria—but no archaeans—can undergo **sporulation**, the formation of thick-walled dormant structures called spores. Spores are the bomb shelters of microbes: they can survive boiling and freezing, thereby allowing the microbes to hang out for a long period of time until the conditions are again favorable to reproduce. In many cases, even antibiotics can't kill bacterial spores.

All Hands on Deck

Dunn, Menninger, and the team analyzed the additional samples. Still, they were unable to explain the differences among belly button microbes. They could not identify any factors affecting the number and types of species a person might host.

So they decided to take a step that more and more scientists are taking—they made all their data freely available online for other researchers to use (with all identifying information about the participants removed). "We asked other researchers to let us know what their insights are, because we're all in this together," says Menninger. "We need all hands on deck."

The gambit worked. Mathematician Sharon Berwick at the University of Maryland took a new approach to the data. Instead of focusing on characteristics of the participants, such as gender or location, she looked at the characteristics of the microbes and found that people tend to be dominated by one of two types of belly button bacteria: *aerobic* or *anaerobic*. **Aerobes** are prokaryotes that need oxygen gas to survive. *Micrococcus* species, which Dunn found in the navel, are aerobes; they need oxygen. Therefore, species of *Micrococcus* are unlikely to do well deep inside a belly button, says Dunn, but they appear to thrive on the surface.

Another type of navel resident, species of *Clostridia*, do not use oxygen. They are **anaerobes**, prokaryotes that survive without oxygen. In fact, some anaerobes may actually be poisoned by oxygen. Among the anaerobic archaeans are several species of **methanogens**, which feed on hydrogen and produce methane gas as a by-product of their metabolism. The ability to exist in both oxygen-rich and oxygen-free environments is another reason why prokaryotes can be found in most habitats.

Some prokaryotes can even switch between the two. One of the most common bacteria Dunn's team found in belly buttons was *Staphylococcus*. Although it is best known as a pathogen, staph on your skin is typically a good guy. On the skin, it is beneficial for your health, fighting off other pathogens that want to crowd in. Some species of staph, such as *Staphylococcus epidermidis*, are typically aerobes, but when oxygen is in short supply, they switch over to a special type of anaerobic metabolism known as **fermentation**, involving the breakdown of sugars (yes, the same process that is used to ferment beer and wine; to review fermentation, see Figure 5.12). In other words, as Dunn wrote in one description of *S. epidermidis*, "Right now you might be making a teeny tiny bit of navel wine."

Like staph, *Bacillus subtilis*, a rod-shaped, spore-forming bacterium, can grow in both aerobic and anaerobic conditions. *B. subtilis* is

HOLLY MENNINGER

Holly Menninger is the director of public science at North Carolina State University, where she works to engage the public in citizen science and science education.

one of the warrior clans in our belly buttons and other areas of the skin, producing antibiotics that kill off other bacteria and even foot fungi, says Dunn. Prokaryotes are either consumers, called **heterotrophs**, or producers, called **autotrophs (Figure 15.10)**. Autotrophs make food on their own, but heterotrophs, like *B. subtilis*, obtain energy by taking it from other sources. Specifically, *B. subtilis* is a **chemoheterotroph**, an organism that consumes organic molecules to get energy (in the form of chemical bonds) and carbon (in the form of carbon-containing molecules). Some species of blue-green algae are **photoheterotrophs**, which acquire carbon from organic sources but their energy from sunlight.

Some autotrophs, called **photoautotrophs**, absorb the energy of sunlight and take in carbon dioxide to conduct photosynthesis. Others, called **chemoautotrophs**, get their energy from inorganic chemicals in their environment, including

iron ore, hydrogen sulfide, and ammonia, instead of from sunlight.

In fact, the very first photoautotrophs on the planet were prokaryotes. One type of aquatic bacteria, the cyanobacteria, is believed to be responsible for changing Earth's chemistry by evolving photosynthesis and producing oxygen gas as a by-product. Oxygen gas accumulated in the air and water, and the levels rose from next to nothing to almost 10 percent about 2.1 billion years ago. At that time in the fossil record, eukaryotes appear, suggesting that the oxygen generated by cyanobacteria may have facilitated the evolution of eukaryotes, especially multicellular forms.

Today, prokaryotes continue to aid eukaryotes. Bacteria directly help plants through a process known as **nitrogen fixation**. Plants need nitrogen in the form of ammonia or nitrate, which they cannot make themselves. Bacteria, however, can take nitrogen gas from the air—our

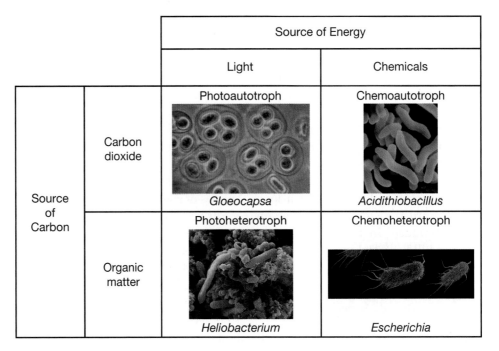

Figure 15.10

How prokaryotes feed themselves
Prokaryotes have many more ways to feed themselves than do eukaryotes. They are categorized by their energy source, and by the source of carbon they use.

Q1: What source of energy would you expect a cave-dwelling prokaryote to use?

Q2: In which of these categories would you place the bacteria responsible for nitrogen fixing? Why?

Q3: In which of these categories do decomposers belong? Explain your reasoning.

atmosphere is 78 percent nitrogen—and convert it to ammonia, making it available for plants.

Autotrophs aren't the only microbes that enable life on Earth. Many heterotrophic bacteria and archaeans are decomposers, consumers that extract nutrients from the remains of dead organisms and from waste products such as urine and feces. Decomposers play a crucial role in **nutrient cycling**: by breaking down dead organisms or waste products, decomposers release the chemical elements locked in the biological material and return them to the environment. Those released elements, such as potassium or nitrogen or phosphorus, are used then by autotrophs and eventually by heterotrophs as well.

Healthy Balance

Since van Leeuwenhoek first looked through his microscope, people tended to regard microbes with suspicion and fear (**Figure 15.11**). We vilified

Figure 15.11

The first illustrations of microbes

Antonie van Leeuwenhoek's descriptions and illustrations of microscopic life led to his being considered the "father of microbiology."

Q1: From the prokaryotic structures shown in Figure 15.7, what shape would you assign to drawing number 8 in van Leeuwenhoek's illustration?

Q2: Which of the large prokaryote drawings has the coccus shape?

Q3: Do you think all of these "animalcules" drawn by van Leeuwenhoek are prokaryotes? Why or why not?

The Bugs in Your Belly Button

There's no way around it: Your belly button is full of bugs. According to research from Rob Dunn's lab at North Carolina State University, the most abundant species in a belly button—those with the biggest populations, such as *Corynebacterium* and *Staphylococcus*—are found in most people. Yet no two people have exactly the same set of microbes. In other words, our belly button microbiomes look alike, but we each have a unique navel fingerprint.

Assessment available in smartwork**5**

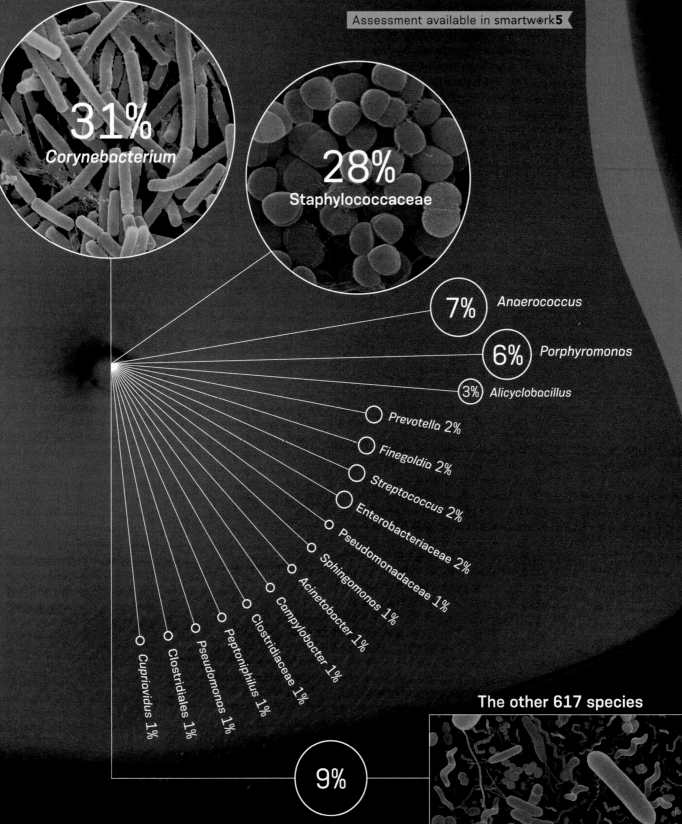

31% *Corynebacterium*

28% Staphylococcaceae

7% *Anaerococcus*

6% *Porphyromonas*

3% *Alicyclobacillus*

Prevotella 2%

Finegoldia 2%

Streptococcus 2%

Enterobacteriaceae 2%

Pseudomonadaceae 1%

Sphingomonas 1%

Acinetobacter 1%

Campylobacter 1%

Clostridiaceae 1%

Peptoniphilus 1%

Pseudomonas 1%

Clostridiales 1%

Cupriavidus 1%

The other 617 species

9%

microbes as the bad guys that our bodies were constantly at war with. But now we recognize that our ecosystems wouldn't function without microbes: we would have no oxygen to breathe, no plants to eat. And microbes in our gut and on our skin, when in healthy balance, promote human health.

We emphasize a healthy balance because it is possible for communities in the human microbiome to shift out of balance—a phenomenon called *dysbiosis*, which may cause illness. Alternatively, pathogenic bacteria can make us sick as well. Although the great majority of bacteria are harmless and many are actually beneficial to humans, some cause mild to deadly disease. Interestingly, archaeans are not known to be pathogens of any organism.

The biodiversity of microbes on our skin actually helps keep pathogens away. When a dangerous microbe lands on your skin, even before it meets the immune system it meets other microbes. If the skin has a diversity of microbes, odds are "one of them has the ability to kick the butt" of the pathogen, says Dunn.

After the overwhelming response to the belly button project, the lab kicked off a new citizen science spectacular: "Armpit-pa-looza" (**Figure 15.12**). As with the belly button, the team began collecting swabs of microbes living in people's armpits. In a preliminary study, the team found that the use of antiperspirants or deodorants dramatically inhibits bacterial growth, affecting the composition of the microbial populations on the skin more strongly than any other factor. "It now looks

Figure 15.12

Sampling armpits for science
Citizen scientists volunteer to have samples taken of the prokaryotes living under their arms.

like one of the biggest things affecting skin microbes in general is the use of antiperspirant or anti-odor products," says Dunn.

Now the team is taking a look at microbes on the skin of dogs. Skin wounds on pet pooches actually heal about four times faster than human skin wounds do, says Dunn, and he suspects that skin microbes are involved. "We're starting to figure out if we can predict wound healing rates as a function of which microbes are there to start with," says Dunn. "I'm pretty excited about that."

REVIEWING THE SCIENCE

- The non-Eukarya, or **prokaryotes**, fall into two domains: Bacteria and Archaea.

- All prokaryotes are **microbes**—microscopic, single-celled organisms—but the Bacteria and Archaea differ in significant ways, such as in their DNA, plasma membrane structure, and metabolism.

- Prokaryotes can reproduce extremely rapidly and are the most numerous life-forms on Earth. They also have the most widespread distribution.

- Many prokaryotes, particularly bacteria, have specialized structures. Bacteria with a **capsule** surrounding the cell wall can avoid detection by

organisms' immune systems. Short, hairlike **pili** help bacteria attach to surfaces and to each other. Whiplike **flagella** assist in locomotion.

- Some prokaryotes, particularly archaeans, thrive in extreme environments. **Thermophiles**, for example, live in extremely hot places, and **halophiles**, in very salty places.

- Prokaryotes typically reproduce by **binary fission**, and they may acquire DNA from each other or their environment via horizontal gene transfer.

- Some prokaryotes use **quorum sensing** to communicate with each other, and some undergo **sporulation** for protection.

- Prokaryotes exhibit unmatched diversity in methods of getting and using nutrients and energy. **Autotrophs** (producers) make their own food, whereas **heterotrophs** (consumers) obtain food from other sources. **Chemoautotrophs** and **chemoheterotrophs** use inorganic chemicals as their energy source; **photoautotrophs** and **photoheterotrophs** use sunlight.

- Prokaryotes perform key tasks in ecosystems, including photosynthesis, **nitrogen fixation** (providing nitrogen to plants), and **nutrient cycling** (decomposing dead organic matter).

- Although prokaryotes are useful to humanity in many ways, some cause deadly diseases.

contains some species that are pathogens. (**Prokaryotes / Eukaryotes**) have smaller, less complex cells and are able to reproduce at a more rapid rate. Some (**prokaryotes / eukaryotes**) are multicellular.

THE QUESTIONS

The Basics

1 Prokaryotes include

(a) archaeans.

(b) bacteria.

(c) fungi.

(d) both a and b

(e) both b and c

2 Some prokaryotes can

(a) break down chemicals for energy.

(b) use sunlight for energy.

(c) create their own energy.

(d) harvest energy from other organisms.

(e) all of the above

3 Prokaryotes are extremely abundant because

(a) they can survive in a narrow range of environments.

(b) they have a single loop of DNA.

(c) they reproduce very rapidly.

(d) they are able to form biofilms.

(e) all of the above

4 Quorum sensing

(a) is the transfer of plasmid DNA from one bacterium to another.

(b) enables bacteria to form biofilms.

(c) is the formation of thick-walled dormant structures, called spores, under conditions unfavorable for growth.

(d) enables bacteria to switch from cellular respiration to fermentation when they sense that oxygen levels are low.

5 Select the correct terms:
Of the two types of prokaryotes, (**Archaea / Bacteria**) are more closely related to eukaryotes. The domain (**Archaea / Bacteria**)

Challenge Yourself

6 Which of the following can prokaryotes *not* do?

(a) communicate with each other about environmental conditions

(b) reproduce sexually

(c) live in extreme environments, including high salt, high temperature, and high atmospheric pressure

(d) share DNA with other prokaryotes

(e) double in population size two or more times an hour

7 Which of the following descriptions is true *only* of prokaryotes?

(a) They reproduce through binary fission.

(b) Their genetic material is DNA.

(c) They are unicellular.

(d) They contain organelles within the cell.

(e) none of the above

8 Place the following steps of the belly button research described in this chapter in the correct order by numbering them from 1 to 5.

_____ a. An observation was made that belly button bacteria seemed to differ across individuals.

_____ b. The DNA was sequenced.

_____ c. DNA was isolated from the samples.

_____ d. Volunteers had their navels swabbed to collect bacterial samples.

_____ e. The bacterial samples were grown in petri dishes.

Try Something New

9 Which of the following is citizen science?

(a) reporting when the first hummingbirds arrive at your feeders in the spring

(b) completing a Facebook quiz on your horoscope

(c) filling out your medical history at the doctor's office

(d) all of the above

(e) none of the above

10 There are more genera of bacteria and archaeans in the world than of eukaryotes, and more individuals. However, the total mass of prokaryotes is thought to be approximately equal to that of eukaryotes. How can this be?

11 Why are prokaryotes able to replicate so much more quickly than eukaryotes? (*Hint:* What cellular components must be copied before a cell splits in two?) Why is this difference in replication rate an important part of our vulnerability to bacterial pathogens?

Leveling Up

12 Doing science The research discussed in this chapter would not have been possible without the many citizen scientists who shared samples of their belly button microbes with the research team. Choose a citizen science project: Perform an Internet search on the term "citizen science project," or find regularly updated lists of projects at Scitable (http://www.scitable.com) and Your Wild Life (http://www.yourwildlife.org). Write a one-paragraph summary of the goals of the project you have chosen. Would you participate in this project? Why or why not? Describe a citizen science project that you *would* want to contribute to.

13 Looking at data The table here is based on the published results of the study discussed in the chapter. It shows the number of bacterial phylotypes (kinds of bacteria) that were found on a particular percentage of people.

a. How many bacterial phylotypes were found on only 10 percent or fewer of the people sampled? How many were found on more than 90 percent of people?

b. How many phylotypes were found on more than half of the people sampled?

c. Describe in one or two sentences the frequency of phylotypes found in the study. Were most phylotypes rare or common?

Phylotypes	Percentage of Human Samples Where Found
2,188	1–10
97	11–20
31	21–30
12	31–40
17	41–50
11	51–60
4	61–70
2	71–80
2	81–90
4	91–100

For more, visit digital.wwnorton.com/bionow2 for access to:

The Dirt on Black-Market Plants

Nighttime poaching and illegal trade threaten to drive rare plants and fungi to extinction.

After reading this chapter you should be able to:

- Compare and contrast prokaryotes and eukaryotes.
- Explain what is distinctive about the kingdom Protista.
- Outline the key evolutionary innovations of plants.
- Connect structure and function in how fungi obtain energy from the environment.
- Identify key characteristics and give an example organism for each kingdom covered in this chapter.

CHAPTER
16

PLANTS, FUNGI, AND PROTISTS

Jacob Phelps spent a lot of his time in Thailand trying not to arouse suspicion. Walking among vendor stalls at wildlife markets in Bangkok, the tall, thin graduate student stopped occasionally to help a trader trim dead leaves off his plants, or to chat about the weather. Wherever he walked, plants surrounded him— hanging from the ceiling, stuffed into boxes, piled into mounds.

Phelps was careful about the questions he asked, because he was there to document illegal activity. The market traders had tens of thousands of plants for sale, most of which weren't supposed to be there. But when Phelps cautiously asked whether the plants were wild or rare, the traders were surprisingly relaxed. Concern for plant conservation, and enforcement of trading laws, is so limited that illegal sales occur openly at public markets across Southeast Asia.

Even Thailand's wildlife trade management authority at the time claimed that illegal trade in ornamental flowers was limited, "found in small case [*sic*] in some parties." But Phelps was familiar enough with plants to know that most of what he saw in the Thailand markets were wild, protected plants that were illegal to sell: the charismatic pink blossoms of *Dendrobium* orchids; the dense, fragrant plumes of *Rhynchostylis* flowers; the delicate, wavy petals of the coveted lady slipper orchid, *Paphiopedilum*. Plant enthusiasts prize such species for their beauty, fragrance, and rarity. Coveted wild orchids make up more than 80 percent of the plants traded at these unregulated markets (**Figure 16.1**).

As a graduate student at the National University of Singapore, Phelps received the blessing of his PhD supervisor, plant ecologist Ted Webb, to conduct an extensive survey of the illegal plant trade at the four largest plant markets in Thailand. It was the first time such a survey had been conducted of plant markets anywhere in Southeast Asia. In fact, the illegal trade of plants is often called the "invisible wildlife trade" because it is so rarely discussed or documented, in contrast to the widely publicized illegal trade of animals and animal products such as elephant ivory and rhinoceros horns.

Plants often take second-tier status behind animals—it's easier to get the public to care about a fuzzy baby tiger than a sprouting redwood tree— but their biology is no less wondrous. In Chapter 15 we explored the microscopic worlds of the Bacteria and Archaea domains of life. Here, we begin to meet the Eukarya, consisting of four kingdoms: Protista, Plantae, Fungi, and Animalia. Animalia will be covered in detail in Chapter 17.

The defining feature of the Eukarya is a true nucleus: instead of floating free in the cytoplasm, eukaryotic DNA is enclosed in two concentric layers of cell membranes that form a nuclear envelope. In addition to the nucleus, eukaryotes have a great variety of membrane-enclosed subcellular compartments, many of which are specialized for various tasks—such as sending messages, producing energy, or

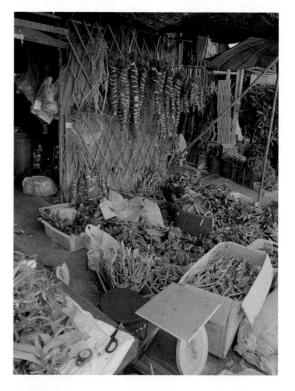

Figure 16.1

A plant market in Thailand
Jacob Phelps visited plant markets like this one to conduct research on illegal trade in endangered orchid species.

JACOB PHELPS

Jacob Phelps is an environmental scientist at Lancaster University in the United Kingdom, studying illegal wildlife trade and tropical deforestation.

cleaning up—so that the cell can function efficiently (review Figure 4.8 for a comparison of prokaryotic and eukaryotic cells, and Figure 4.9 for an overview of eukaryotic organelles). All of these organelles take up space, so the diameter of a eukaryotic cell is, on average, 10 times larger than that of a prokaryotic cell, and the eukaryotic cell volume is a thousandfold greater.

This compartmentalization of the cell interior enables eukaryotic cells to do things that most prokaryotic cells cannot do. For example, some eukaryotes engulf their prey and digest them internally. That's the way many single-celled eukaryotes, such as the blob-like amoebas, eat: they extend gooey cytoplasmic arms to engulf other cells whole, then digest their prey with an elaborate system of internal compartments for digestion, ridding the cell of waste and storing surplus food (**Figure 16.2**).

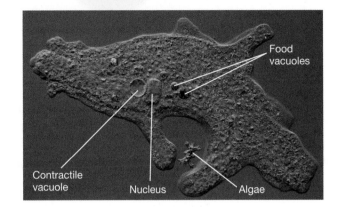

Figure 16.2

Eukaryotes have a true nucleus and compartmentalized cells
The more complex cell structure of eukaryotes enables them to perform functions that are impossible for prokaryotes. Intracellular compartments such as food vacuoles enable this amoeba to digest its food, a single-celled alga. The amoeba expels excess water with the help of its contractile vacuole, another type of intracellular organelle.

Fungi Play Well with Others

Plants probably would not have been as successful on land if they had not entered into a mutualistic relationship with fungi almost immediately on their arrival. Today, the vast majority of plants in the wild have mutualistic fungi, known as **mycorrhizal fungi**, associated with their root systems. (The mutualistic relationships between plants and mycorrhizal fungi are called "mycorrhizae.") Truffles, morels, and chanterelles, all beloved by gourmets, are the reproductive structures of mycorrhizal fungi.

Mycorrhizal fungi form thick, spongy mats of mycelium on and in the roots of their plant hosts and also extend into the surroundings, sometimes permeating several acres of the soil around the root. Mycelia are thinner, more extensively branched, and in closer contact with the soil than even the thinnest branches on a plant root. As a result, a mycelial mat plumbs far more water and mineral nutrients, such as phosphorus and nitrogen, than the plant's root system could absorb on its own. In return for sharing absorbed water and mineral nutrients, the fungus obtains sugars that the plant manufactures through photosynthesis.

Mycorrhizal fungi assist in providing nutrients to orchid seeds, which are tiny and lack stored food. The embryo within a newly sprouted orchid seed could not survive without the mycorrhizal network linking the seedling to mature photosynthesizing plants, from which the seedling draws nourishment until it can photosynthesize on its own.

Fungal mutualisms are not found only with other eukaryotes. A **lichen** is a mutualistic association between

Plants that have no mycorrhizal fungi associated with their roots do not grow as well…

…as those that host the beneficial fungi.

a photosynthetic prokaryote and a fungus. The fungus receives sugars and other carbon compounds from its photosynthetic partner, usually a green alga or a cyanobacterium. In return, the fungus produces lichen acids, a mixture of chemicals that scientists believe may function to protect both the fungus and its partner from being eaten by predators.

Lichens grow very slowly and are often pioneers in barren environments. Lichen acids wear down a rocky surface, facilitating soil formation. Soil particles build up from the slow weathering of rock, and over time, other life-forms, including plants, gain a toehold in the newly made soil.

Peculiar Protists

Amoebas are single-celled eukaryotes. Most single-celled eukaryotes are **protists**. Protista is a strange kingdom that is defined by exclusion instead of inclusion: although all protists are eukaryotes, they are grouped together in this kingdom simply because they are *not* plants, animals, or fungi (**Figure 16.3**). Researchers have proposed several classification schemes to split the protists into multiple kingdoms; however, no consensus exists on the best way to do so. For now, protists remain divided into two traditional, broad categories: the **protozoans**, which are nonphotosynthetic and motile (capable of moving); and the **algae**, which are photosynthetic and may or may not be motile.

Protists are diverse in size, shape, cellular organization, and mode of nutrition. Most protists are single-celled and microscopic, but some protists have evolved from free-living single cells into multicellular associations, such as slime molds and kelp, large seaweed such as kelp. Certain single-celled protists are bound by nothing more than a flexible plasma membrane, while others are covered in protective sheets, heavy coats, or other types of armor.

Most protists are motile and can swim with the help of one or more flagella, or by waving a carpet of tiny hairs called cilia. Others can crawl on a solid surface with the help of cellular projections called *pseudopodia* (false feet).

Many protists are heterotrophs and eat other organisms. Some of these heterotrophs function as decomposers, breaking down waste material and releasing nutrients into the environment to be taken up by producers and cycled back into the food chain. Other protists are nutritional

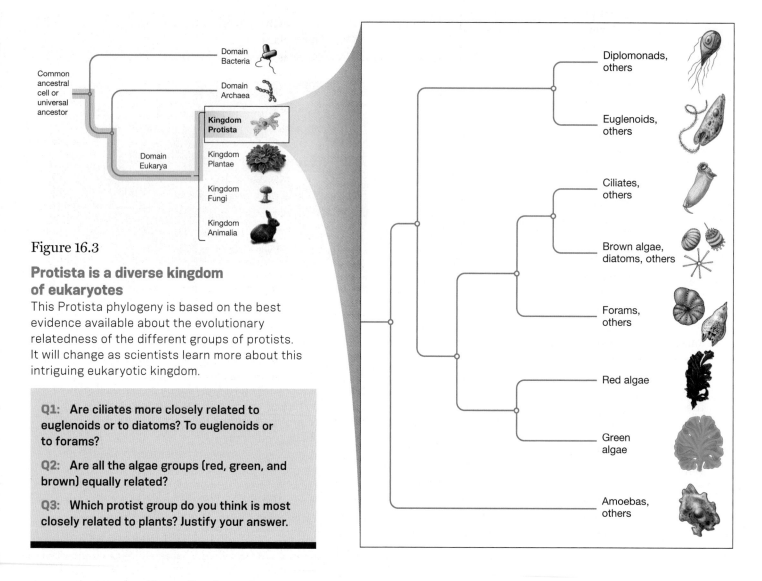

Figure 16.3

Protista is a diverse kingdom of eukaryotes

This Protista phylogeny is based on the best evidence available about the evolutionary relatedness of the different groups of protists. It will change as scientists learn more about this intriguing eukaryotic kingdom.

Q1: Are ciliates more closely related to euglenoids or to diatoms? To euglenoids or to forams?

Q2: Are all the algae groups (red, green, and brown) equally related?

Q3: Which protist group do you think is most closely related to plants? Justify your answer.

opportunists, or *mixotrophs*, organisms that use energy and carbon from a variety of sources to fuel their growth and reproduction.

Although most protists are harmless, many of the best-known protists are pathogens (disease-causing agents), such as *Plasmodium*, which is transmitted by mosquitoes and causes malaria, and *Toxoplasma gondii*, passed to humans through poorly cooked food or cat feces. Toxoplasmosis typically causes mild flu-like symptoms, but can be passed from mother to child in utero and cause more severe symptoms.

Two Cells Are Better Than One

Multicellular forms evolved several times among different lineages of the eukaryotes. A multicellular organism is a well-integrated assemblage of genetically identical cells, in which different groups of cells perform distinctly specialized functions. This functional compartmentalization makes it easier for multicellular eukaryotes to sense and respond to the external environment, through the use of complex structures such as leaves, fruit, eyes and wings.

Multicellularity also enables an individual organism to grow large, which can be advantageous for evading potential predators. A bigger individual can also gather resources from its environment more effectively than can a smaller individual. Having more resources, such as light or food, may translate into producing more surviving offspring, the ultimate measure of biological success.

To make those offspring, eukaryotes reproduce via either asexual or sexual reproduction. Asexual reproduction is common among species in this domain and generates genetically identical offspring. Protists, for example, can split into two in a process similar to binary fission in prokaryotes. Many plants reproduce asexually by fragmenting into pieces, each piece developing into a new individual.

But it is through the second type of reproduction that eukaryotes have made an indelible mark on our planet. By combining genetic information from two parents, sexual reproduction produces offspring that are genetically different from each other and from both parents. Although their life cycle is distinctly different from that of animals, plants similarly produce embryos: the fusion of egg and sperm produces a single cell, called a zygote, which then divides to produce a multicellular structure called an embryo. Sexual reproduction is one means by which natural populations become genetically diverse; it is why there are an estimated 400,000 species of plants on Earth.

Through these unique traits—a true nucleus, cellular compartmentalization, multicellularity, and sexual reproduction—eukaryotes have evolved into amazingly diverse and dynamic species. Of these species, Phelps was particularly concerned about orchids. In addition to spending time in Bangkok, the budding ecologist devoted months traveling along Thailand's borders with Laos and Myanmar, wandering hours upon hours through wildlife markets. Over time, he built local networks that allowed him to interview more than 150 plant harvesters and intermediaries, including market traders, online traders, nursery owners, and more.

Orchids are, on paper, one of the most heavily protected families, representing almost 75 percent of all species, plant or animal, for which international trade is regulated. Artificially propagated orchids can be traded legally, but wild orchid species in Southeast Asia are protected, says Phelps: "None should be collected from the wild, unless you have specialized permissions." Yet, as he came to discover, there is no enforcement of those protections in Thailand or across the region. Orchids and other ornamental and medicinal plants were not only being traded, but also poached from nearby countries, threatening the very existence of certain species and the overall biodiversity of the region.

Green-Fingered Thieves

Under the dark of night, the thieves worked swiftly. Armed with shovels, they snuck past the closed gift shop and made a beeline for their quarry. They knew exactly what to take.

The next morning, the staff at the Quail Botanical Gardens in Encinitas, California, discovered the theft. In a frantic effort, they spread details of the heist to local newspapers, television, and online community message boards. Their quick action paid off; the loot became too hot. Within days, an anonymous tip led authorities to a rural road where the thieves

Figure 16.4

Cycads: ancient, rare, and hard to move

Several workers struggle to move a cycad. In spite of their size, many cycads have been stolen from public and private gardens, and then sold on the black market.

had dumped their haul in a sorry-looking pile: 21 stocky plants, specimens with thick trunks covered in woody scales and a crown of long, green palm-like leaves, like a pineapple.

The thieves had attempted to steal a group of rare African cycads, an ancient type of plant that once lived alongside dinosaurs in the Jurassic period (**Figure 16.4**). As with the Thai orchids, horticulturalists around the world highly prize cycads as ornamental plants. Today, a rare, mature cycad can fetch $20,000 or more on the international black market. That's right—enough to fund a year of college. There are about 300 species of cycads, most threatened with extinction.

Ironically, 4 of the 21 cycads stolen from the Quail Botanical Gardens (now the San Diego Botanic Garden) were part of a rescue program; the plants had been illegally brought into the United States, where they were seized upon import by authorities, and the botanical garden had taken over their care. Some of the plants survived the ordeal and were replanted. But they weren't the only cycads stolen that year; a nursery in San Diego lost almost 40 large cycads to thieves, and many homes in the Long Beach area had plants swiped right out of their front yards.

Plants are multicellular autotrophs that are mostly terrestrial (**Figure 16.5**). Like the green

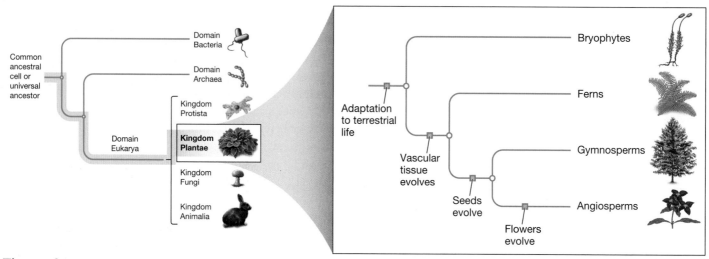

Figure 16.5

Plants are eukaryotic autotrophs

Plants are photosynthetic, making their own food from sunlight, and are almost exclusively found on land. They are at the base of almost all terrestrial food webs.

Q1: What evolutionary innovation separates all land plants from their aquatic ancestors?

Q2: How do ferns differ from bryophytes? Do they share this difference with other plant groups? (You will need to read ahead to answer this question.)

Q3: What group(s) might a plant with seeds belong to? What about a plant with flowers?

---

algae from which they evolved, plants use chloroplasts in order to photosynthesize. Most photosynthesis in plants takes place in their leaves, which typically have a broad, flat surface—a design that maximizes light interception. Because plants are producers, they form the basis of essentially all food webs on land. They may not be as cute, fuzzy, or exciting as animals, but animals wouldn't exist without them: nearly all organisms on land ultimately depend on plants for food, either directly by eating plants or indirectly by eating other organisms that eat plants.

In addition to being food, plants are valuable for many other reasons. Many organisms live on or in plants, or in soils largely made up of decomposed plants. By soaking up rainwater in their roots and other tissues, plants prevent runoff and erosion that can contaminate streams. Plants (and algae) also recycle carbon dioxide and produce the oxygen we breathe.

Many of these traits are adaptations to the challenges of living on land (**Figure 16.6**). The greatest of those challenges is how to obtain and conserve water. The first plants, the ancestors of

Algae (aquatic)

- Photosynthesis occurs and CO_2 is absorbed throughout the organism.

- Water and minerals are absorbed by the whole organism.

- Water supports the whole organism.

- Dehydration is not an issue in an aquatic environment.

Plants (terrestrial)

- Photosynthesis occurs and CO_2 is absorbed primarily in the leaves. Air enters leaf cells through minute openings in the cuticle. Many plants have more elaborate pores (*stomata*) that open and close to regulate the flow of gases into and out of the leaf.

- Roots absorb water and minerals from the soil.

- Roots anchor and support the plant within the ground. Lignin and vascular tissues help support the plant aboveground.

- The waxy cuticle holds in moisture to keep plant tissues from drying out, even when exposed to sunlight and air throughout the day.

Figure 16.6

Moving to land brought unique challenges to plants
Terrestrial plants evolved in response to challenges that their aquatic ancestors had not faced, which resulted in adaptations designed to slow dehydration, provide support and anchoring, and enable photosynthesis and nutrient uptake.

Q1: In what ways are terrestrial plants and their aquatic ancestors the same? Give at least two similarities.

Q2: In what ways do terrestrial plants and their aquatic ancestors differ? Give at least two differences.

Q3: Would you predict that aquatic plants (which have secondarily evolved to live in water) would be more like plants in a rainforest or more like desert plants? Explain your reasoning.

present-day **bryophytes** (liverworts and mosses), grew as ground-hugging carpets of greenery. These simple plants had relatively thin bodies, often just a few cells thick, and absorbed water through a wicking action. However, absorption by direct contact cannot transport fluids effectively in a plant that rises a foot or more aboveground, so more complex plants, including cycads, have a network of tissues, called the **vascular system**, that is made up of tubelike structures specialized for transporting fluids. Roots, found in most plants, also have an extensive vascular system. The first land plant with a vascular system was the ancestor of present-day **ferns**.

Like green algae from the protist kingdom, plant cells have strong but flexible cell walls composed of the substance known as **cellulose**, which gives structural strength to the cells, including those of the low-growing bryophytes. Beginning with the ferns, plants evolved yet another type of strengthening material, **lignin**, that allowed them to grow even taller. One of the strongest materials in nature, lignin links together cellulose fibers in the cell wall to create a rigid network. Lignin is the reason that some cycads in Japan can grow over 20 feet (7 meters) tall, though it takes them about 50–100 years to achieve such heights.

One reason rare cycads are so prized is that it can take a long time to raise and grow them. Cycads are **gymnosperms**, the first plants to evolve **pollen**, a microscopic structure containing sperm cells that can be lofted into the air in massive quantities. The evolution of pollen freed gymnosperms from a dependence on water for fertilization: instead of having to transport sperm cells through water, they could do it through air.

Gymnosperms were also the first plants to evolve the **seed**, which consists of the plant embryo and a supply of stored food, all encased in a protective covering. The embryo uses this stored food to grow until it is able to make its own food via photosynthesis. Seeds also provide embryos with protection from drying and from attack by predators. Unfortunately, they don't provide protection from poachers. Today, the San Diego Botanic Garden keeps not only its prized cycads, but also their valuable seeds, in a greenhouse under lock and key.

Searching for Flowers

Botanical gardens also go to extreme measures to keep flowering plants, or **angiosperms**, safe for long-term conservation. Orchids, as we have seen, are the most common illegally traded angiosperms. Others include *Galanthus*, or "snowdrop" plants, with delicate white flowers. In 2012, a single bulb of a rare variety of snowdrop fetched $945 at auction.

Compared to gymnosperms, angiosperms are a relatively recent development in the history of life. With about 250,000 species today, angiosperms are the most dominant and diverse group of plants on our planet. Nearly all agricultural crops are flowering plants. These plants also provide humans with materials such as cotton and pharmaceuticals.

Angiosperms' key evolutionary innovation is the flower, a structure that evolved through modification of the cone-like reproductive organs of gymnosperms. **Flowers** are structures that enhance sexual reproduction in angiosperms by bringing male gametes (sperm cells) to the female gametes (egg cells) in highly efficient ways, by attracting animal pollinators through scent, shape, and color (**Figure 16.7**).

Figure 16.7

A flower attracts many pollinators
Pollination in the angiosperms is more efficient than in the wind-pollinated gymnosperms. Flowers attract animal pollinators that carry pollen to and from individual plants of the same species.

Gymnosperms produce "naked" seeds that sit bare, unwrapped in any additional layers, on the modified leaf (the scale in a pine cone). In angiosperms, the modified leaf evolved into the ovary wall, which consists of tissue layers that enclose and protect the egg-bearing structures, or **ovules**. After fertilization, the ovules develop into seeds, and the ovary wall that enclosed them becomes the fruit wall.

At the Thailand plant markets, Phelps had trouble identifying orchid species until they flowered, so he went back again and again—four times per year to four different markets—looking for newly opened flowers, listing species he recognized, taking photos of others, and even occasionally asking for a flower off a plant he did not recognize, which he quickly stored in a vial of alcohol to take back for identification. In the end, Phelps gathered evidence of 348 orchid species in 93 genera, representing 13–22 percent of the area's known orchid flora, and tens of thousands of individual plants, including several new species.

Phelps's results were shockingly different from those published in preexisting government reports on plant trades among Thailand, Laos, and Myanmar. The Convention on International Trade in Endangered Species of Wild Fauna and Flora, or CITES, is an international treaty of over 175 countries that monitors and regulates international trade of plants and animals. Member countries are required to produce permits for wildlife trade of species protected under the agreement, including all wild orchids. These permits are used to guarantee that plants are legally harvested in a sustainable way that does not endanger either the species or the environment. "CITES is about conservation and sustainable use," says Anne St. John, a biologist with the U.S. Fish and Wildlife Service's Division of Management Authority, which implements CITES in the United States. "The goal is to ensure that these species are around for our grandchildren. That includes not only tigers and elephants, but also bigleaf mahogany and Brazilian rosewood."

And the orchids of Southeast Asia. Over nine years, Laos reported permits for the export of just 20 wild-collected orchids into Thailand; Myanmar reported none. Yet during just one day with a single market trader at the border between Laos and Thailand, Phelps documented that the woman sold at least 168 plants of eight different genera. In just one day, she sold eight times more plants than the government reported as sold over

9 years. "It's totally anecdotal, but incredibly illustrative of the problem," says Phelps. "This trade is completely unacknowledged. It's an open secret."

Other countries are working hard to crack down on the plant black market. American ginseng, a short leafy plant with a tan, gnarled root commonly used in Chinese herbal medicine, is the largest CITES-regulated plant export of the United States (**Figure 16.8**). A pound of quality, dried ginseng can sell for up to $900, so some people try to bypass CITES permits, harvesting plants that are too young (legally

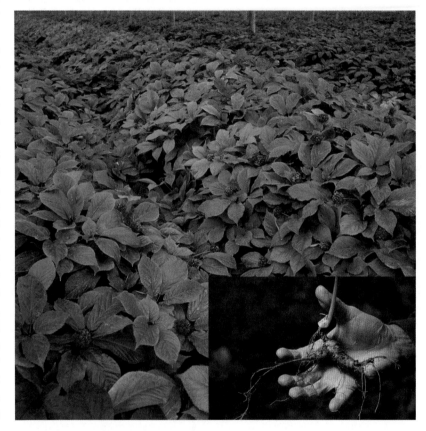

Figure 16.8

Ginseng plants carpet a forest floor
After pollination, the ginseng flowers develop a bright-red seed head that helps ginseng hunters (inset) find it in the filtered light of the forest.

Q1: What feature(s) of the ginseng plant tell you that it is not a bryophyte?

Q2: What feature(s) of the ginseng plant tell you that it is not a fern or gymnosperm?

Q3: Because of the CITES classification of ginseng, you are not allowed to sell plants younger than 5 years even if they grew on your own land. Do you agree with that law? Why or why not?

traded roots must be 5 years or older), out of season, or poached from federal lands.

"As you would expect, species that are high value and in high demand are the subject of illegal trade," says St. John. The illegal trade of ginseng has even spawned a reality show on the History television channel: *Appalachian Outlaws*. To combat the criminal activity, some states have taken to spraying ginseng plants on state and federal lands with paint, so if a marked plant shows up in a batch of ginseng to be exported to China, authorities will know where it came from. Those caught selling protected plants face fines and jail time, and violating CITES is a federal crime.

Truffle Trouble

The illegal wildlife trade is a problem not just for animals and plants, but also for **fungi**, none of which have been listed as CITES-protected species so far. Fungi are absorptive heterotrophs:

they digest organic material outside the body and absorb the breakdown products. The majority of fungal species fall into three main groups: **zygomycetes**, which contains many species of molds; **ascomycetes**, a diverse group informally known as sac fungi; and the more familiar **basidiomycetes**, or club fungi (**Figure 16.9**). Each of these groups differs in—and is named for—its unique reproductive structures.

Fungi have properties in common with both plants and animals. Like plant cells, all fungal cells have a protective cell wall that wraps around the plasma membrane and encases the cell. However, fungi are similar to animals in that they store surplus food energy in the form of glycogen.

In 2012, two bandits broke through security gates and stole an estimated $60,000 of fungi from a locked warehouse. The loot was truffles, the fruiting body of a particular group of ascomycetes. Certain white European truffles can sell for as much as $3,600 per pound, making them the most expensive food in the world. That high price also makes them a target for thieves, especially since

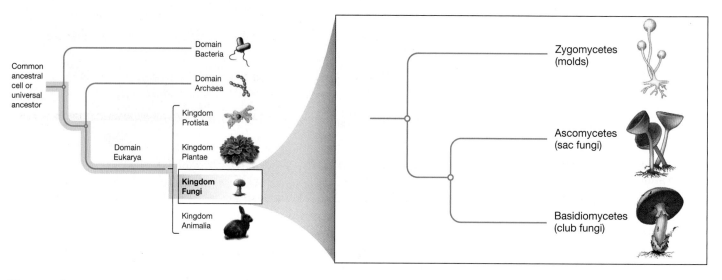

Figure 16.9

Fungi are eukaryotic absorptive heterotrophs
Fungi must take their food from other organisms, and they do this in a unique way. Instead of ingesting their food, as most heterotrophs do, fungi use chemicals to break down their food outside of their bodies and then absorb the nutrients.

Q1: What group of fungi most resembles the mushrooms you buy in a grocery store?

Q2: Are sac fungi more closely related to molds or to club fungi?

Q3: How do we know that fungi are eukaryotes rather than prokaryotes?

Figure 16.10

A truffle hunter and his dog search for a fungus beloved by gourmets

While farmers continue attempts to grow truffles commercially, their success has been limited. For now, we must rely on truffle hunters and their dogs (or pigs, which can also be trained to sniff out truffles) for this fungal delicacy.

The **fruiting body** of the fungus is what we informally call a "mushroom."

Spores are the reproductive structures of the fungus.

The **mycelium** is the bundle of hyphae that make up an individual fungus.

A **hypha** is a single fungal thread.

Figure 16.11

A fungus lives both belowground and aboveground

The main body of a fungus is unseen, belowground. To reproduce, the fungus generates a fruiting body that is usually aboveground and releases spores. Spores travel by wind and develop into new fungi.

> **Q1:** Why is it important that the fruiting body is aboveground?
>
> **Q2:** What part of a fungus is the mushroom that you can buy in the grocery store?
>
> **Q3:** Write a sentence in your own words that uses the terms "mycelium," "fruiting body," and "spore" correctly.

white truffles cannot be cultivated in greenhouses as other fungi can.

There's danger in the forests too. Truffle hunters use dogs and pigs to sniff out wild truffles (**Figure 16.10**), and some have reported competitors planting spiked traps or poisoned meatballs in forests to eliminate trained dogs. Other hounds have been stolen from their owners, never to be seen again.

That's a lot of drama for some fungi. Most fungi are multicellular, but there are some single-celled species collectively known as **yeasts**. Many of us are familiar with yeasts thanks to two important products they produce: alcohol and carbon dioxide, crucial to the rising of bread, the brewing of beer, and the fermenting of wine.

Fungi's key evolutionary innovation is their body form, which is well suited for absorptive

heterotrophy (**Figure 16.11**). They are made up of a network of fine, colorless, branching, hair-like threads called **hyphae** (singular "hypha"), which absorb nutrients from the environment. The entire bundle of hyphae, composing the main body of the fungus, is called the **mycelium** (plural "mycelia").

Because of this unique body form, many fungi are decomposers and consume nonliving organic material. As they eat, these fungi release back into the environment inorganic chemicals that had previously been trapped in the bodies of dead organisms. Once these chemicals are back in the environment, plants and algae scoop them up and use them to manufacture food. Fungi interact with plants in other important ways too: the vast majority of wild plants have mutualistic fungi in their root systems that help the plants

Food Banks

Over millennia, farmers have saved seeds from crops that were easiest to grow, process, or store, and have used those seeds and their offspring year after year. The results of this domestication process are staple crops we rely on today: corn, soybeans, wheat. Yet to feed a growing human population, we're going to have to look beyond what we're familiar with, to the smorgasbord of edible plants that pepper the planet.

Assessment available in smartwork**5**

30,000
terrestrial plants are known to be edible.

7,000
are cultivated or collected by humans for food.

30
crops provide 95% of the food energy taken in by the world's human population.

Rice   Maize   Wheat   Millet   Sorghum

5
cereal crops provide 60% of the food energy taken in by the world's human population.

absorb nutrients from the soil (see "Fungi Play Well with Others" on page 287).

Despite that assistance, fungi and plants are not always close friends. Fungi are the most significant parasite of plants: they are responsible for two-thirds of all plant diseases, causing more crop damage than bacteria, viruses, and insect pests combined.

Like many plants (and animals and protists), fungi can reproduce both asexually and sexually. Some species appear to multiply only asexually, and most multicellular fungi can reproduce asexually through fragmentation—that is, by simply breaking off from the mother colony. When fungi do reproduce sexually, they do not have distinct male and female individuals. Instead, a sexually reproducing mycelium belongs to one of two (or more!) mating types. Each mating type can mate successfully with only one of the other types. After mating, a **fruiting body** is formed that may be large enough to be readily observed.

Fungal fruiting bodies release offspring as sexual spores. A **spore** is a reproductive structure that can survive for long periods of time in a dormant state and will sprout under favorable conditions to produce the body of the organism. Spores released from a fruiting body that is raised up in the air are better able to catch a ride on wind currents or to attract animals that can carry them far and wide. Many fungi also produce asexual spores.

Fighting for the Future

The illegal sale of plants and fungi could drive species to extinction. In 2015, for example, the International Union for Conservation of Nature (IUCN), a global environmental authority, announced that 31 percent of cactus species—renowned for their unique forms and beautiful flowers—are in danger of extinction, and that the greatest threat to these plants is illegal trade (**Figure 16.12**).

During his studies, Phelps observed traders selling plants that are highly endangered in the wild. He published results from his thesis in 2015 in the journal *Biological Conservation*, pleading with scientific and policy communities for a greater focus on these plants and calling

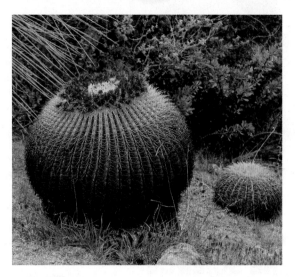

Figure 16.12

The greatest threat to endangered cacti is illegal trade
Like orchids and cycads, the main threat to most species of cacti is illegal trade. Collectors pay high prices to smugglers who are willing to risk fines and prison time to meet the demand.

the illegal plant trade "a major conservation challenge that has been almost completely overlooked."

"It's hard to get people to care about plants, and nobody can care if we don't have data to show it's a problem," says Phelps, who is now trying to engage governments and experts to raise awareness and initiate policies that will better regulate plant trade. At the CITES conference in September 2016, country delegates increased protections for a large number of plant species, including ponytail palms, which are in demand as houseplants, and all rosewoods, whose wood is used to manufacture fine musical instruments.

"All these plants are part of a larger ecosystem," says St. John. "They serve important purposes in temperature regulation, in creating oxygen, and more. We need to recognize that plants are an integral part of a healthy ecosystem and key to the diversity of life."

ANNE ST. JOHN

Anne St. John is a biologist with the Division of Management Authority for the U.S. Fish and Wildlife Service.

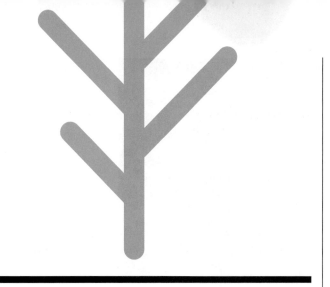

- There are at least three main groups of fungi—**zygomycetes**, **ascomycetes**, and **basidiomycetes**—and each group is characterized by distinctive reproductive structures (fruiting bodies).

- Most plant roots in natural habitats form close associations with beneficial fungi, called **mycorrhizal fungi**. A **lichen** is a mutually beneficial association between a fungus and a photosynthetic microbe, usually a green alga or a cyanobacterium.

REVIEWING THE SCIENCE

- The domain Eukarya is traditionally divided into four major kingdoms: Protista, Plantae, Fungi, and Animalia. Eukaryotes possess a true nucleus; they have complex subcellular compartments, which enable larger cell size. Sexual reproduction and multicellularity are key evolutionary innovations of the Eukarya.

- Kingdom Protista, the **protists**, lumps together many evolutionarily distinct lineages under one highly diverse grouping. Most protists are single-celled and microscopic.

- Protists are traditionally divided into two categories: **protozoans**, which are nonphotosynthetic and motile; and **algae**, which are photosynthetic and may or may not be motile.

- **Plants** are descended from green algae and have evolved numerous evolutionary innovations to adapt to life on land. All plants photosynthesize and use **cellulose** to strengthen their cell walls.

- The first plants, the ancestors of present-day **bryophytes** (liverworts and mosses), had relatively thin bodies and absorbed water through a wicking action.

- **Ferns** were the earliest plant group to grow larger and taller than the bryophytes. This extended growth required both cellulose and **lignin** to provide structure to plant bodies, as well as a network of fluid-transporting tissues called the **vascular system**.

- **Pollen** and **seeds** first evolved among the **gymnosperms**, the plant group that includes cycads.

- **Angiosperms** evolved **flowers**, and they enclose their seeds in the fruit. Many angiosperms recruit animals to deliver pollen and also to disperse their seeds.

- The **fungi** are distinguished by their mode of nutrition: they acquire their nutrients by absorption, digesting their food outside of their bodies. To accomplish this, their bodies consist of a **mycelium** composed of hairlike threads called **hyphae**.

- Fungal reproduction may be sexual or asexual. During reproduction, a **fruiting body** is formed that releases **spores** into the environment. Much as seeds do, these spores then develop into new individuals.

THE QUESTIONS

The Basics

1 Which of the following descriptions is *not* true of all eukaryotes?

(a) They are multicellular.

(b) They have cellular organelles.

(c) They have a larger cell size than prokaryotes have.

(d) They have a true nucleus.

(e) All of the above are characteristics of all eukaryotes.

2 Protists

(a) are the largest prokaryotes.

(b) are all single-celled.

(c) include plants and fungi.

(d) are an unnatural grouping, placed together for convenience.

(e) are all photosynthetic.

3 Fungi

(a) reproduce only sexually.

(b) reproduce only asexually.

(c) may have multiple mating types within a single species.

(d) are more closely related to plants than to animals.

(e) are more closely related to protists than to animals.

4 Which of these evolutionary innovations enabled larger cell size?

(a) autotrophic mode of nutrition

(b) multicellularity

(c) sexual reproduction

(d) subcellular compartmentalization

(e) all of the above

5 Select the correct terms:
Green algae belong in the kingdom (**Plantae** / **Protista**). They are (**aquatic** / **terrestrial**) and are (**autotrophs** / **heterotrophs**).

Challenge Yourself

6 Place the following adaptations in the correct order from earliest to most recent by numbering them from 1 to 5.

_____ a. vascular tissue

_____ b. flowers

_____ c. seeds

_____ d. multicellularity

_____ e. movement to land

7 Which of the following groups contains only multicellular species?

(a) algae

(b) protists

(c) eukaryotes

(d) fungi

(e) angiosperms

8 Of the kingdoms covered in this chapter, which, if any, are composed of only autotrophic species? Which, if any, have only heterotrophs?

Try Something New

9 Symbiosis is a long-term and intimate association between two different types of organisms. A symbiotic organism may live on or inside another species. For each of the following symbiotic relationships, (1) define the relationship in one to three sentences, (2) identify the domain—and for Eukarya, the kingdom—of each partner in the relationship, and (3) discuss whether the relationship evolved as a mutualism (both benefit), commensalism (one benefits, while the other is not affected), or parasitism (one benefits to the detriment of the other). You may need to read more than your textbook to answer this question.

a. mycorrhizae

b. lichens

c. hermit crabs/shells

d. malaria

10 Which plant groups produce pollen, and what is the adaptive value of pollen?

11 Plants are not the only organisms susceptible to fungal infections. Give an example of a fungal infection to which humans are susceptible.

Leveling Up

12 **Life choices** If fungi are more closely related to animals than to plants, should vegetarians and vegans refrain from eating fungi? Why or why not?

13 *Write Now* **biology: is a mass extinction under way?** The International Union for Conservation of Nature maintains what it calls its Red List, which identifies the world's threatened species. To be defined as such, a species must face a high to extremely high risk of extinction in the wild. The 2016 Red List contains almost 25,000 species threatened with extinction out of the approximately 79,000 species assessed.

Because this assessment accounts for only a small percentage of the world's 1.7 million described species, the total number of species threatened with extinction worldwide may actually be much larger. For example, only 6,051 of 1 million described insect species have been assessed in terms of their survival risk.

The Red List is based on an easy-to-understand system for categorizing extinction risk. It is also objective, yielding consistent results when used by different people. These two attributes have earned the Red List international recognition as an effective tool for assessing extinction risk.

a. If the threatened species listed by the IUCN do become extinct and the percentages of species under threat in other taxonomic groups turn out to be similar to those listed, then the percentages of species that will go extinct will approach the proportions lost in some of the previous mass extinctions. Does this mean that a mass extinction is under way? Why or why not? Explain your reasoning, using information about mass extinctions from Chapter 14 (in particular, Figure 14.15) to support your argument.

b. What are the causes of the current high extinction rates? Compare the causes of the current situation with those of previous mass extinction events. In what ways are they similar, and in what ways do they differ?

c. Why are some groups in more danger of extinction than others? Choose one group that has a large proportion of species threatened with extinction, and find out more about it (you can begin at http://www.iucnredlist.org):

● What kinds of habitat are they found in?

● Does a particular aspect of their ecology—for example, feeding or reproduction—make them more vulnerable to extinction?

● Is anything being done to protect them and/or their habitat?

● Has there been an increase or decrease in the number of species within the group that have been identified as vulnerable to extinction?

For more, visit digital.wwnorton.com/bionow2 for access to:

Neanderthal Sex

The relationship between modern humans and Neanderthals just got a whole lot spicier.

After reading this chapter you should be able to:

- Identify the key characteristics of animals, and give an example organism for each group within the chordates.

- Explain the significance of symmetry and segmentation for some animal groups.

- Compare and contrast the three types of mammals.

- Interpret an evolutionary tree of the primates.

- Illustrate the inheritance of mitochondrial DNA as compared to nuclear DNA.

- Describe the evidence suggesting that prehistoric hominin species reproduced with our direct ancestors.

In popular culture, Neanderthals are ugly, hairy cavemen with big brains but no wits. The first Neanderthal bones were discovered in Germany in 1856, and since then we've cultivated an image of our closest extinct human relatives as hulking brutes who communicated by grunting, walked like chimps, and hit each other over the head with clubs. Yet, over the years, paleontologists have dug up fossilized vocal bones, sophisticated tools, and other evidence suggesting that Neanderthals were a fairly advanced group—and not as different from our own species as we once believed.

The big revelation came in 2010, when scientists sequenced the Neanderthal genome and compared it to the modern human genome: Humans have some Neanderthal DNA. Our two species may have been a whole lot closer than we thought. "There could have been interbreeding between modern humans and Neanderthals," says Silvana Condemi, a researcher at the National Center for Scientific Research (CNRS) at Aix-Marseille University in France. "We can imagine they not only exchanged culture, but exchanged genes." That's right: mounting evidence suggests that Neanderthals and modern humans had sex.

This story begins where many stories end—with a pile of bones. In the early years of the twenty-first century, Laura Longo, a curator at the Civic Natural History Museum of Verona, Italy, decided to take a second look at a group of fossils excavated from a rock shelter called Riparo Mezzena. Riparo Mezzena is nestled in the Lessini Mountains in northern Italy, a wide-open landscape speckled with large rocks and evergreen trees. The region is snowy and silent in the winter but green and thriving in the summer, when paleontologists come to work.

In the 1950s, paleontologists had carefully collected fossils of Neanderthals that lived around Riparo Mezzena about 35,000 years ago, late in the history of Neanderthals. But the fossils had sat at the museum, untouched, for over 50 years (**Figure 17.1**). Now, Longo believed, the fossils could help answer a hotly debated question about human history: How closely did modern humans and Neanderthals interact?

Animal Kingdom

Modern humans and Neanderthals lived in some of the same areas of Europe at the same times. Because of this proximity, some paleontologists hypothesize that as modern humans expanded their territory, Neanderthals were

Figure 17.1

Fossil remains

A jawbone like this one was found at the Riparo Mezzena rock shelter in the Lessini Mountains in Italy. The individual it belonged to lived between 40,000 and 30,000 years ago.

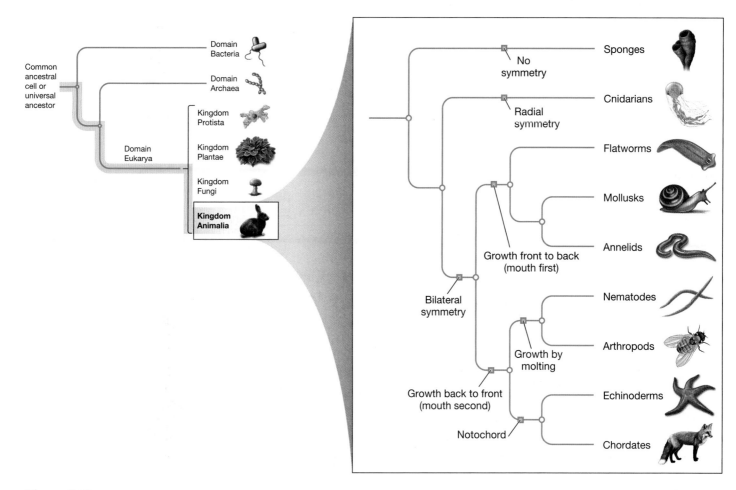

Figure 17.2

Animals are heterotrophic eukaryotes

Animals ingest their food, which can be organisms from any of the other domains and kingdoms, and then digest it internally.

Q1: Are mollusks more closely related to flatworms or to annelids? Explain your reasoning.

Q2: If you found an animal with no symmetry, to what group would it belong? Give an example of an animal that is radially symmetrical, and one that is bilaterally symmetrical.

Q3: How do we know that a sponge is an animal, and not a plant or an alga?

quickly driven to extinction, and therefore the two did not live side by side. Others claim the opposite: that Neanderthals, *Homo neanderthalensis*, were slowly incorporated into the population of newly incoming humans, *Homo sapiens*. But to know how humans evolved, it is important first to realize where we came from.

All species in the *Homo* genus, of which *H. sapiens* is the only one alive today, are in the Animalia kingdom. **Animals** are multicellular ingestive heterotrophs; that is, we obtain energy and carbon by ingesting food into our bodies and digesting it internally. Animals first evolved

some 700 million years ago, descended from flagellated protists, single-celled organisms with whiplike tails that thrived in wet habitats and began to use oxygen to break down food into energy. From those humble origins, countless animal species evolved—beginning with sponges and including mollusks such as snails and clams, the annelids (segmented worms), and arthropods like crustaceans, spiders, and insects (**Figure 17.2**).

Chordates make up a large phylum that includes all animals with backbones, such as fishes, birds, and mammals (see **Figure 17.3** and

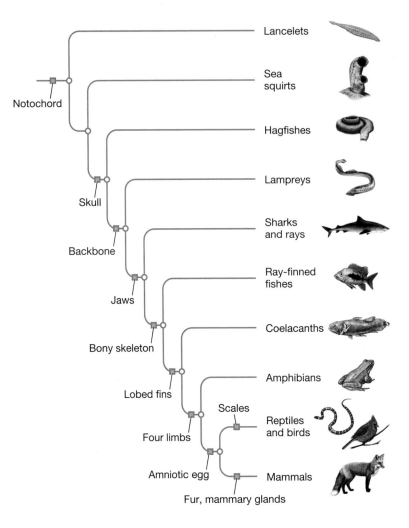

Figure 17.3

Not all chordates have a backbone

As with all eukaryotes except the protists, the evolutionary relatedness of the chordates is very well known. Not all chordates have a backbone, but all have a notochord, which is the precursor to part of the vertebrate backbone.

Q1: Do amphibians have amniotic eggs?

Q2: What group of animals has jaws but not a bony skeleton?

Q3: When people talk about animals, they are sometimes referring only to mammals. How would you explain to them their error?

"Get a Backbone!" on page 305). The phylum also includes several subgroups of less familiar animals, such as sea squirts and lancelets, that have a nerve cord along the back of the body but no backbone. Chordates that do possess a backbone are known as **vertebrates**. All other phyla of animals are informally lumped together as invertebrates. But keep in mind that "invertebrates" is not an evolutionarily meaningful label; it is an artificial grouping (like that of protists) of animals with varied evolutionary histories and different sets of evolutionary adaptations.

Body plan is another key factor that distinguishes one group of animals from another. All animals except sponges, the most ancient of the animal lineages, have a distinct body symmetry. The asymmetric sponges aside, animals can be divided into two main groups: those with radial

Get a Backbone!

Within the kingdom Animalia, humans belong to the phylum of chordates and the subphylum of vertebrates. All chordates have a dorsal **notochord**, a flexible rod along the center of the body that is critical for development. In vertebrates, the dorsal notochord has evolved to become the cushioning discs between **vertebrae** (singular "vertebra"), which are strong, hollow sections of the backbone, or vertebral column. Vertebrates include fishes, amphibians (frogs and salamanders), reptiles (snakes, lizards, turtles, and crocodiles), birds, and mammals.

The jawless fishes were the first vertebrates to evolve. Their skeletons—including the backbone—were made from a strong but flexible tissue called **cartilage**. Only a few groups of jawless fishes have survived to the present day, most notably the lampreys. The next great leap in vertebrate evolution was hinged jaws, which enabled predators to grab and swallow prey efficiently. The evolution of teeth made jaws even more effective because teeth enabled animals to seize and tear food.

Another major step in the evolution of vertebrates was the replacement of the cartilage-based skeleton with a denser tissue strengthened by calcium salts: bone. Although the descendants of cartilaginous fishes—sharks, skates, and rays—are still with us today, bony fishes are far more diversified and widespread in both marine and freshwater environments. With more than 30,000 species, bony fishes are the most diverse vertebrates today.

The advent of lungs was a crucial milestone in the transition of vertebrates onto land. Amphibians made this transition only partially; they can live on land but must return to the water to lay eggs and breed. The several thousand species of amphibians include frogs and salamanders.

Reptiles were the first vertebrates to head into drier environments, and they evolved a number of adaptive traits to deal with the risk of dehydration. These adaptations included skin covered in waterproof scales, a water-conserving excretory system, and the amniotic egg with its calcium-rich protective shell, which retards moisture loss while allowing the entry of life-giving oxygen and the release of waste carbon dioxide for the developing embryo. Reptiles dominated Earth during the age of the dinosaurs, and the dinosaurs' descendants (as we saw in Chapter 14) remain with us today in the form of birds. Like mammals, birds are warm-blooded, but they have feathers instead of fur for insulation. At least 10,000 different species of birds are living today.

symmetry and those with bilateral symmetry. The body of an animal with **radial symmetry** (**Figure 17.4**, left)—including cnidarians like jellyfish, sea anemones, and corals—can be sliced symmetrically along any number of vertical planes that pass through the center of the animal, like cutting a pie. Radial symmetry gives an animal sweeping, 360-degree access to its environment. The animal can snare food drifting in from any direction of the compass, and it can also sense and respond to danger from any side.

Animals with **bilateral symmetry** (**Figure 17.4**, right), on the other hand, can be divided by just one plane passing vertically from the top to the bottom of the animal into two halves that mirror each other. Bilateral animals, which include all chordates, have distinct right and left sides, with near-identical body parts on each side. The symmetrical arrangement of body parts on either side of a central body facilitates movement in bilateral animals. The paired arrangement of limbs or fins, for example, enables quick and efficient movement on land

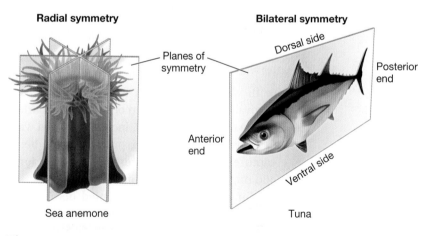

Figure 17.4

Body symmetry in animals

All animals other than sponges have symmetrical bodies that enable them to better sense and respond to the world around them.

Q1: Is a sea star radially or bilaterally symmetrical?

Q2: What advantage might a bilaterally symmetrical animal have over one that is radially symmetrical, and vice versa?

Q3: What kind of symmetry do you (a human) have?

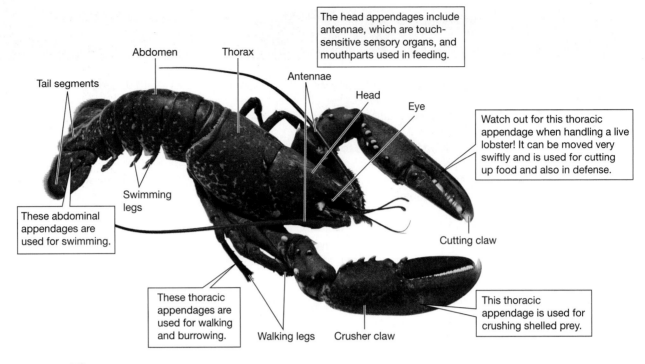

Tail segments

Abdomen

Thorax

Antennae

Head

Eye

The head appendages include antennae, which are touch-sensitive sensory organs, and mouthparts used in feeding.

Watch out for this thoracic appendage when handling a live lobster! It can be moved very swiftly and is used for cutting up food and also in defense.

Swimming legs

These abdominal appendages are used for swimming.

Cutting claw

These thoracic appendages are used for walking and burrowing.

Walking legs

Crusher claw

This thoracic appendage is used for crushing shelled prey.

Figure 17.5

Segmentation in animals

Segmentation, a body plan in which segments repeat, enabled the evolution of diverse uses of appendages, shown here in a lobster.

> **Q1:** List all of the lobster's thoracic appendages.
>
> **Q2:** Which of the lobster's appendages is most important for sensing the environment?
>
> **Q3:** What body segments do you (a human) have?

or in water. Locomotion is a key evolutionary innovation of animals and one that has sparked a wide range of behaviors, including varied ways of capturing prey, eating prey, avoiding being captured, attracting mates, caring for young, and migrating to new habitats.

Many animals have segmented bodies; that is, their body plan consists of repeated units known as segments (**Figure 17.5**). Specialized body parts, known as appendages, often originate in pairs from specific segments of the body, such as fins or limbs. Over evolutionary time, the segments and the appendages that spring from them have evolved diverse form and function, enabling the animal body to adapt to new habitats or acquire new modes of life.

The evolution of just the posterior segments of arthropods illustrates how evolution can take a basic body plan and modify it to produce many variations over time: The last segment in the

body of arthropods has evolved into the delicate abdomen of the butterfly, the piercing abdomen of the wasp, and the delicious tail of the lobster. The front appendage of vertebrates has evolved as an arm in humans, a wing in birds, a flipper in whales, an almost nonexistent nub in snakes, and a front leg in salamanders and lizards.

Mammals R Us

To get back to our egocentric focus—that is, humans—we are part of the kingdom Animalia and the class Mammalia. Humans share specific features with all other mammals, including body hair, sweat glands, and milk produced by mammary glands. And at the risk of tooting our own horn, **mammals** have been a highly successful class of animals, with over 5,000 species living in a variety of habitats. This is largely thanks to

Figure 17.6

Three kinds of mammals

All mammals have milk-producing mammary glands for suckling their young. Eutherians like the polar bear give birth to well-developed young. Kangaroos are marsupials, and give birth to immature young that finish developing in a pouch. Monotremes like the platypus lay eggs instead of giving birth to live young.

Q1: The Virginia opossum, or possum, is the only North American marsupial. Describe how its young are born and develop.

Q2: If you were a monotreme, would you still produce milk and nurse your young?

Q3: What kind of mammal is a cow? How about a human?

the extinction of dinosaurs; if dinosaurs still roamed Earth, it is likely that mammals would be nothing but dinner. Lucky for us, mammals have replaced dinosaurs as the top predators in most terrestrial habitats, and they thrive in both marine and freshwater environments. Only one type of mammal can fly—bats—although a few others can glide through the air.

Mammals can be divided into three broad categories, all of which feed their offspring with milk (**Figure 17.6**). More than 95 percent of mammals alive today are **eutherians**, including humans. A unifying characteristic of eutherians is that the offspring are nourished inside the mother's body through a special organ called the *placenta* and are therefore born in a relatively well-developed state. **Marsupials**, the second category of mammals, have a simple placenta, resulting in offspring born early, which then complete development in an external pocket or pouch. Marsupials are found mainly in Australia and New Zealand, with a few species in the Americas.

Making up the third category, **monotremes** are egg-laying mammals that lack a placenta altogether. The only living species of monotremes are just one platypus species and several echidna species, all confined to Australia and New Guinea.

Moving now from class to order, humans and Neanderthals are in the order of the **primates** (**Figure 17.7**). Like all other primates, we have flexible shoulder and elbow joints, five functional fingers and toes, thumbs that are **opposable** (that is, they can be placed opposite each of the other four fingers), flat nails (instead of claws), and brains that are large in relation to our body size.

Within the primates we are members of the ape family, the **hominids**. We are not just closely related to apes; we *are* apes. As such, we share many characteristics with other apes, especially chimpanzees, including the use of tools, a capacity for symbolic language, and the performance of deliberate acts of deception. But we are part of a distinct branch of apes called **hominins**—the "human" branch of the ape family that includes

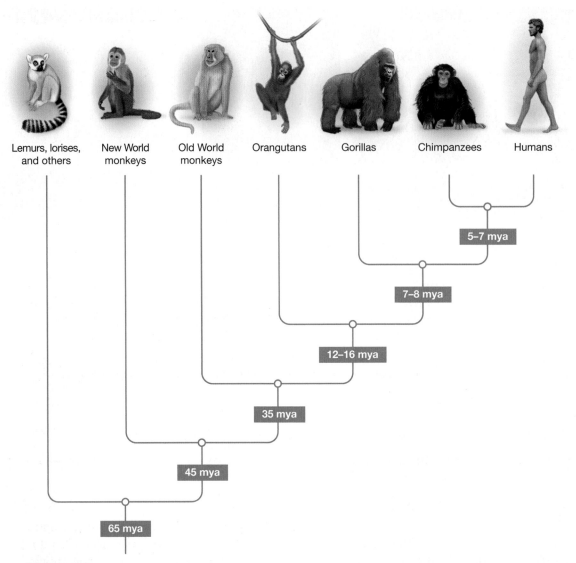

Figure 17.7

The primates include lemurs, monkeys, and apes

Genetic analyses and a series of spectacular fossil discoveries have led scientists to propose that the human lineage diverged from that of chimpanzees about 5–7 million years ago (mya). Similar evidence suggests that the evolutionary lineage leading to humans diverged from the lineage leading to gorillas about 7–8 mya, and from the lineage leading to orangutans about 12–16 mya.

Q1: According to this evolutionary tree, which primate group is most closely related to humans?

Q2: According to this evolutionary tree, which primate group is most distantly related to humans?

Q3: What characteristics are common to all the primates, including humans?

our extinct relatives, such as Neanderthals. The members of the hominin lineage have one or more humanlike features—for example, thick tooth enamel or upright posture—that set them apart from the other apes, like gorillas and chimpanzees.

Rise of the Apes

A major step in hominin evolution, and the main feature that distinguishes hominins from other hominids, was the shift from moving on four legs to being **bipedal**, walking upright on

Hereditary Heirlooms

Because of a shared common ancestor 1.6 billion years ago, humans share DNA with all animals, plants, and fungi. But how much? Take a look to see how much genetic material you have in common with other organisms.

Assessment available in smartwork5

Percentage of genes shared with humans

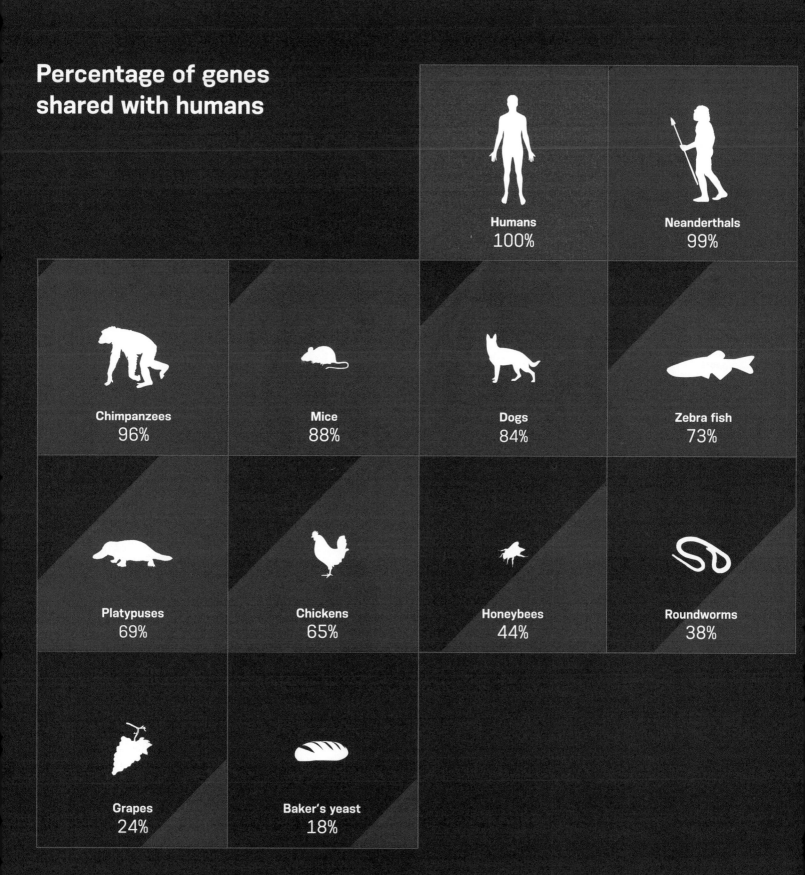

Humans
100%

Neanderthals
99%

Chimpanzees
96%

Mice
88%

Dogs
84%

Zebra fish
73%

Platypuses
69%

Chickens
65%

Honeybees
44%

Roundworms
38%

Grapes
24%

Baker's yeast
18%

Other apes

The skull and its relation to the spinal cord

The face is pulled forward.

The spinal cord enters from the rear of the skull.

Humans

The spinal cord enters from the center of the skull.

The face is relatively flat.

The pelvis—frontal view

Overall skeletal structure

Other apes have a posture supported by four limbs.

Humans have an upright posture.

The hip bones in humans allow an upright posture.

Pelvis

Shifting weight support

The angle of weight support in other apes...

...differs from that in humans.

The big toe is opposable.

The big toe is not opposable.

There is no arch.

The arch and heel are pronounced.

Structure of foot and function of big toe

Figure 17.8

Evolutionary differences between humans and other apes

The switch to walking upright required a drastic reorganization of primate anatomy, especially of the hip bones. ▶

two legs (**Figure 17.8**). Many skeletal changes accompanied the switch to walking upright, including the loss of opposable toes, as you will notice if you try touching your little toe with the big toe on the same foot.

The loss of opposable toes that accompanied walking upright would have been a handicap in trees, since opposable toes help grasp branches during climbing. It is therefore likely that bipedalism was an adaptation for living on the ground. Walking on two feet freed the hands to carry food, tools, and weapons, and it also elevated the head, enabling the walker to see farther and over more things.

The shift to life on the ground was probably not sudden or complete. The skeletal structure of some of the oldest fossil hominins (3–3.5 million years old) indicates that they walked upright. However, foot bones and fossilized footprints show that the hominins living at that time still had partially opposable big toes (**Figure 17.9**). Perhaps they still occasionally climbed trees.

The earliest known hominin is *Sahelanthropus tchadensis*, identified from a 6- to 7-million-year-old skull discovered in 2002. Other early hominins include *Ardipithecus ramidus*, who lived 4.4 million years ago, and several *Australopithecus* species that are 3–4.2 million years old, including the first full-time walker with the first modern foot: *Australopithecus afarensis*. All of these hominins are thought to have walked upright. Their brains were still relatively small (less than 400 cubic centimeters in volume), and their skulls and teeth were more similar to those of other apes than to those of humans (**Figure 17.10**). A typical modern human has a brain volume of about 1,400 cubic centimeters, about the same volume as a 1.5-liter soda bottle.

Within the hominin branch is the *Homo* genus. The fossils identified at Riparo Mezzena were believed to be *Homo neanderthalensis* bones, but they had never been closely studied. So, Longo, the curator from Verona, assembled a team of

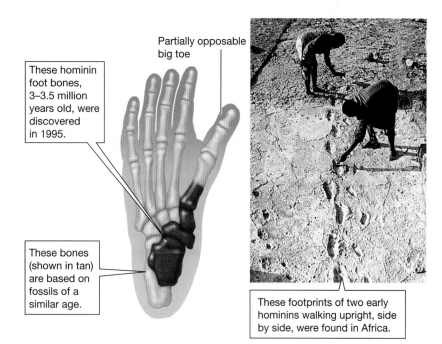

Partially opposable big toe

These hominin foot bones, 3–3.5 million years old, were discovered in 1995.

These bones (shown in tan) are based on fossils of a similar age.

These footprints of two early hominins walking upright, side by side, were found in Africa.

Figure 17.9

Early hominins had an upright stance and partially opposable big toes

Fossilized foot bones show that some hominins living between 3.5 and 3 million years ago walked upright but had partially opposable big toes.

Q1: What other reason besides continuing to use trees might explain why early hominins had partially opposable big toes?

Q2: In what way does the pattern of footprints in this figure suggest that the print makers were walking upright?

Q3: Why do you think we no longer have partially opposable big toes?

researchers to analyze them, including Condemi, the anthropologist. Condemi had long been interested in the movement of Neanderthals across Europe and how their populations overlapped with modern *Homo sapiens* populations, and she wanted to compare the Riparo Mezzena fossils to fossils from Neanderthal groups that had been dug up elsewhere around Europe, hoping to see what the fossils might say about how Neanderthals and modern humans interacted.

Q1: Through natural selection, deleterious traits will tend to disappear from a population over time. Which traits might have been deleterious for ground-dwelling early hominins?

Q2: Through natural selection, advantageous traits will tend to persist in a population over time. Which traits might have been advantageous for ground-dwelling early hominins?

Q3: Adaptations to upright walking also mean that human females have more difficulty giving birth than do females of other species. What adaptation would you predict has had the greatest impact on this result?

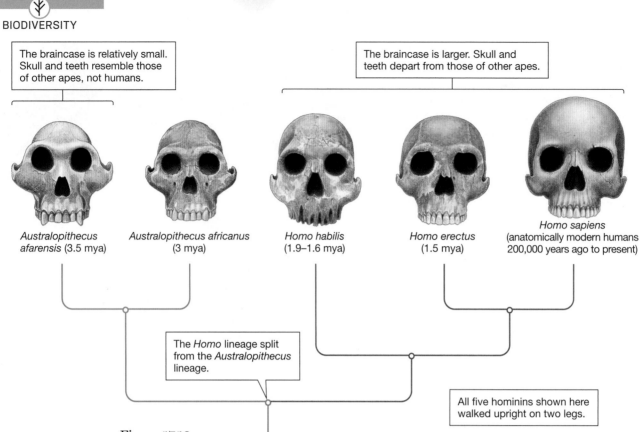

The braincase is relatively small. Skull and teeth resemble those of other apes, not humans.

The braincase is larger. Skull and teeth depart from those of other apes.

Australopithecus afarensis (3.5 mya)

Australopithecus africanus (3 mya)

Homo habilis (1.9–1.6 mya)

Homo erectus (1.5 mya)

Homo sapiens (anatomically modern humans 200,000 years ago to present)

The *Homo* lineage split from the *Australopithecus* lineage.

All five hominins shown here walked upright on two legs.

Figure 17.10

A gallery of hominin skulls
This tree shows the evolutionary relationships and the skulls of five hominin species. A complete evolutionary tree of hominins would be "bushier," with multiple side branches emerging at different times.

Q1: Where would the Neanderthal species branch be on this tree?

Q2: How would the Neanderthal skull differ from the *Homo erectus* skull?

Q3: How would the Neanderthal skull differ from the *Homo sapiens* skull?

Hominins United

"For years, it was a very simple story: Neanderthals either disappeared quickly when modern humans came or they integrated with humans," says Condemi. But she suspected that the dynamic was more complex. "In some regions, Neanderthals disappeared very quickly. In other regions, we have evidence the two [species] could have overlapped." Researchers, including Condemi, had often suggested that during the overlap, Neanderthals and humans interbred. But there was little physical evidence.

Even DNA evidence initially suggested that there was no interbreeding. DNA from the mitochondria of Neanderthals (isolated first from a single Neanderthal fossil in 1997 and then from

four Neanderthal fossils in 2004) was compared to modern human mitochondrial DNA, and the tests showed that there was no genetic overlap between the species. Mitochondrial DNA (mtDNA) is unique because it is passed down virtually unchanged from mother to child, so it can be tracked from one generation, or one species, to another (**Figure 17.11**). But modern *Homo sapiens* did not have Neanderthal mitochondrial DNA, so it appeared there was no interbreeding, at least not involving female Neanderthals.

That mitochondrial work was performed by Svante Pääbo and researchers at the Max Planck Institute for Evolutionary Anthropology in Leipzig, Germany. Pääbo is one of the founders of the effort to use genetics to study early

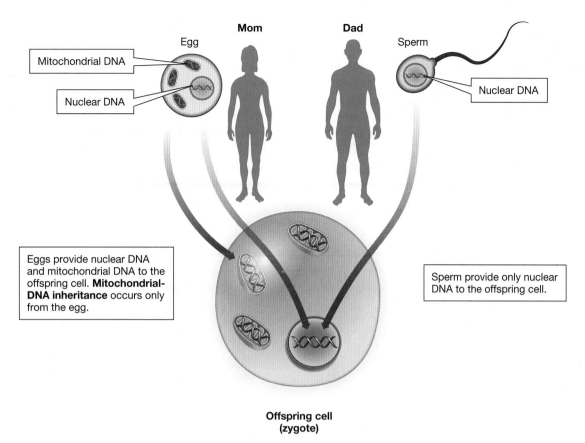

Mom **Dad**

Egg — Sperm

Mitochondrial DNA

Nuclear DNA

Nuclear DNA

Eggs provide nuclear DNA and mitochondrial DNA to the offspring cell. **Mitochondrial-DNA inheritance** occurs only from the egg.

Sperm provide only nuclear DNA to the offspring cell.

Offspring cell (zygote)

Figure 17.11

Mitochondrial DNA comes only from your mom (not from your dad)

In addition to the chromosomes in the nucleus (nuclear DNA), cells also have DNA in their mitochondria, called mitochondrial DNA. Mitochondria and their resident DNA all come from Mom.

Q1: Why does mitochondrial DNA come only from your mother?

Q2: If a Neanderthal-human hybrid was born to a human mother and a Neanderthal father, could you tell by mitochondrial-DNA sequencing?

Q3: If a Neanderthal-human hybrid was born to a human father and a Neanderthal mother, could you tell by mitochondrial-DNA sequencing?

humans and other ancient populations, and he has pioneered numerous techniques to extract delicate DNA from even the tiniest slivers of fossilized bones. At the time that he began his work, it was easier to find and extract mitochondrial DNA because cells contain hundreds of copies of this type of DNA, and only one copy of nuclear DNA. In addition, mitochondrial DNA can be isolated from cells and tissues that aren't so well preserved and from damaged DNA. Whole-genome sequencing, by contrast, requires well-preserved cells or tissues with fully intact nuclear DNA. But despite his initial findings that modern humans and Neanderthals did not share

mtDNA, Pääbo still thought there might be room for some small contribution. Being a diligent scientist, he decided to look even deeper—this time at the whole Neanderthal genome.

SVANTE PÄÄBO

Svante Pääbo is a Swedish geneticist who directs the Department of Genetics at the Max Planck Institute for Evolutionary Anthropology in Leipzig, Germany. He specializes in using genetics to study early humans and other ancient populations.

Pääbo spent 4 years sequencing the 1.5 billion base pairs in the Neanderthal genome, using DNA extracted from the femur bones of three 38,000-year-old females. Then he compared that long, composite genome sequence to the genomes of five living humans from China, France, Papua New Guinea, southern Africa, and western Africa.

According to the results, all modern ethnic groups, other than Africans, carry traces of Neanderthal DNA in their genomes—between 1 and 4 percent. Most of us have a little Neanderthal in us. But whether that nuclear DNA is the result of thousands of sexual encounters between humans and Neanderthals or a few one-night stands remains unknown (**Figure 17.12**). When Pääbo published his work in 2010, he and others admitted it was possible that the shared DNA wasn't necessarily a product of interbreeding. It could have been a remnant of DNA from a common shared ancestor.

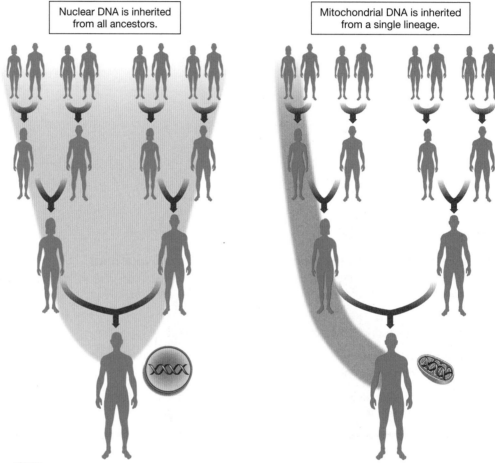

| Nuclear DNA is inherited from all ancestors. | Mitochondrial DNA is inherited from a single lineage. |

Figure 17.12

Nuclear-DNA inheritance

By sequencing the bases of nuclear DNA, scientists can determine how related an individual is to its ancestors, both male and female (see Chapter 11). Mitochondrial-DNA sequencing can determine only how related an individual is to the female ancestors on its mother's side.

Q1: If a human-Neanderthal hybrid was born to a human mother and a Neanderthal father, could you tell by whole-genome DNA sequencing that it was a hybrid?

Q2: If a human-Neanderthal hybrid was born to a Neanderthal mother and a human father, could you tell by whole-genome DNA sequencing that it was a hybrid?

Q3: Under what circumstances are scientists able to do whole-genome sequencing, and when are they restricted to mitochondrial-DNA sequencing?

Who would that ancestor have been? The oldest *Homo* fossil fragments were found in Ethiopia in 2015 and date between 2.80 and 2.75 million years ago (mya), suggesting that the earliest members of the genus *Homo* originated in Africa 2–3 mya. More complete early *Homo* fossils exist from the period 1.9–1.6 mya; these fossils have been given the species name *Homo habilis*. The oldest *H. habilis* fossils resemble those of *Australopithecus africanus*, yet more recent *H. habilis* fossils show a more rounded skull and a face that isn't pulled as far forward. *H. habilis* fossils therefore provide an excellent record of the evolutionary shift from ancestral hominins (in *Australopithecus*) to more recent species, such as *Homo erectus*, the most likely candidate for a shared common ancestor of Neanderthals and modern humans (**Figure 17.13**).

Figure 17.13

Meet the folks

Cartoon representations of seven hominin species depicting their average height in centimeters and their presumed features. *Homo sapiens* is represented as a 6-foot-tall male for reference.

Q1: Are you surprised by the interpretations of the hominins in this picture? Why or why not?

Q2: Describe the main differences that distinguish the hominin species.

Q3: From what you've learned about these species, do you think these representations are accurate? How can you find more information about each species to help you answer this question?

Taller and more robust than *H. habilis*, *H. erectus* also had a larger brain and a skull more like that of modern humans. It is likely that by 500,000 years ago, *H. erectus* could use, but not necessarily make, fire. In addition, *H. erectus* probably hunted large animals, as suggested by a remarkable 2010 discovery in Germany of three 400,000-year-old spears, each about 2 meters long and designed for throwing with a forward center of gravity (like a modern javelin). *H. erectus*, or one of the other *Homo* ancestors, migrated from Africa about 2 million years ago. From there, this ancestor species spread around the Middle East and into Asia. *Homo* fossils dating from the period 1.9–1.7 mya have been found in the central Asian republic of Georgia, in China, and in Indonesia.

Overall, current research on *H. habilis*, *H. erectus*, and other early *Homo* species indicates that there were more species of *Homo* than was once thought, and that several of these species existed in the same places and times. More research and evidence will be necessary before general agreement is reached regarding the exact number of early *Homo* species and their evolutionary relationships.

So, was the Neanderthal DNA found in modern human genomes simply a remnant of a common ancestor? In 2012, Pääbo's team and others were able to determine the age of the pieces of Neanderthal DNA in the human genome. They found that the DNA had been introduced into our genome between 90,000 and 40,000 years ago, around the same time that modern humans spread out of Africa and met the Neanderthals. A remnant of DNA from a common ancestor would have been 10 times older.

All in the Family

The fossil record indicates that the first *Homo sapiens*, called archaic *H. sapiens*, originated between 400,000 and 300,000 years ago.

Uniquely Human?

Yes, the frontal lobe of the brain is unique to human beings, and it enables us to reason like no other animal on our planet. But what about the rest of the attributes that we so commonly consider unique to us? Humans pride themselves on their intelligence and deep emotional connections to others, but are these really only human traits?

- **Language.** Researchers once believed language was an exclusively human trait. We now know that chimpanzees in the wild use sign language, with approximately 70 different signs for distinct words. Meanwhile, other primates, birds, whales, and bats have distinct, learned vocalizations that they use to communicate.

- **Memory.** Some have suggested humans alone possess the ability to store memories. But dogs easily learn and remember many commands, while crows can learn and remember shapes better than human adults can, and they can use causal reasoning, not trial and error, to unlock doors and find hidden objects.

- **Social culture.** Once thought to be strictly human, social culture is a learned trait that chimpanzees, Japanese macaques, and killer whales pass on throughout their populations. Tool use by dolphins, elephants, and octopi varies in its specifics from population to population—a sure sign of learned behavior.

- **Emotions.** Our emotions make us human, right? Others in the animal kingdom have been documented expressing empathy (elephants), grief (dolphins, elephants), jealousy (apes), curiosity (cats, lizards), altruism (apes), and gratitude (whales). Apes have been seen laughing at a clumsy fellow ape and using deception to outwit a family member.

- **Self-Awareness.** The ability to recognize oneself in the mirror, or show self-awareness, was once thought to be ours alone. As it turns out, all the apes, some gibbons, elephants, magpies, and some whales pass the mirror test of self-awareness.

- **Morality.** Finally, what about a sense of morality or an understanding of social norms? Monkeys and rats will not accept offered food if, in doing so, a fellow member of their species receives an electric shock.

To be sure, there *is* something unique about humans that lies at the intersection of all these abilities. However, our expanding knowledge of animal behavior can't help but make us feel more closely connected to the other species with whom we share this planet.

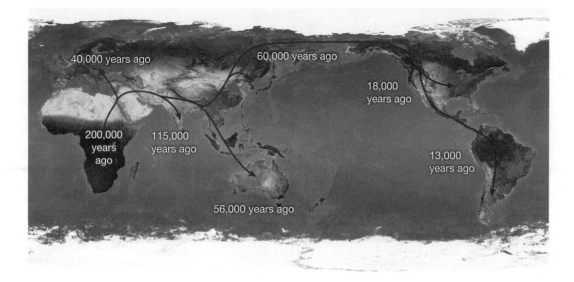

Figure 17.14

Anatomically modern humans evolved in Africa

The earliest known archaeological specimens of modern humans (*Homo sapiens*) come from Africa. The dates provided give the age of the earliest evidence that anatomically modern humans lived in different regions of the world. These dates are continually challenged by new fossil evidence that scientists must then work to confirm.

> **Q1:** What evidence suggests that Neanderthals never lived in Africa?
>
> **Q2:** How does the hypothesized origin of modern humans (*Homo sapiens*) differ from the hypothesized origin of Neanderthals (*Homo neanderthalensis*)?
>
> **Q3:** What species of hominins other than the Neanderthals may have commingled with modern humans?

Archaic *H. sapiens* bore features intermediate between those of *H. erectus* and those of "anatomically modern" *H. sapiens*—our species—which arose some 195,000–200,000 years ago. These ancestors of anatomically modern humans developed new tools and new ways of making tools, used new foods, and built complex shelters. (But humans are not the only organisms to do many of these things; see "Uniquely Human?" on page 316.)

Early populations of archaic humans eventually gave rise to both the Neanderthals (who lived from 300,000 to 28,000 years ago) and us—that is, anatomically modern humans. There is some debate as to whether Neanderthals are simply an odd form of archaic *H. sapiens*, or their own distinct species. That question has yet to be resolved.

According to the out-of-Africa hypothesis, anatomically modern humans first evolved in Africa about 195,000–200,000 years ago from a unique population of archaic *H. sapiens*, and then spread into other continents to live alongside other hominins (**Figure 17.14**). Evidence from the fossil record indicates that anatomically modern humans overlapped in time with *H. erectus* and Neanderthal populations, yet remained distinct from them. Neanderthals and modern humans coexisted in western Asia for about 80,000 years, and in Europe for some 10,000 years, until modern humans completely replaced all other *Homo* populations.

But what happened in that intervening time? Were modern humans and Neanderthals friendly neighbors, or were the latter quickly wiped out by the former?

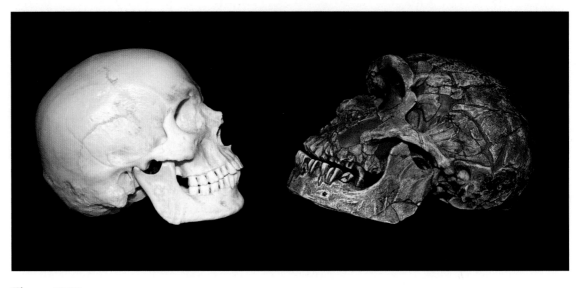

Figure 17.15

Neanderthal and human skulls

The skull on the right is from a Neanderthal; the one on the left, from a modern human. Notice the differences in the eyebrow ridge, forehead, and lower jaw.

Q1:   Describe the difference you observe between the modern human skull's chin and the Neanderthal skull's lower jaw.

Q2:   What other differences do you observe between the two skulls?

Q3:   Why would you expect a hybrid of Neanderthals and modern humans to have intermediate features?

With a previous team, Condemi had used fossil evidence from southern Italy to determine that modern humans arrived on the Italian peninsula between 45,000 and 43,000 years ago, *before* the disappearance of Neanderthals. So the two populations likely made contact in

Italy. During this period, it's possible that "there was a kind of interbreeding," says Condemi. But if the two species interbred and had children, what did those children look like? And why hadn't Condemi's team found their bones?

Then Condemi examined the Riparo Mezzena bones. One caught her attention. It was a jawbone, a mandible, from a late Neanderthal living in Italy at the same time that modern humans had already made their way into Europe (see **Figure 17.1**). But the jawbone didn't look like a Neanderthal's, which has no chin. Instead, the face of the Riparo Mezzena individual, when reconstructed with three-dimensional imaging, had an intermediate jaw, something between no chin and a strongly projected chin. Because chins are a feature unique to modern humans (**Figure 17.15**),

SILVANA CONDEMI

Silvana Condemi is an anthropologist and research director at the National Center for Scientific Research (CNRS) at Aix-Marseille University in France. She studies the movement of Neanderthals across Europe and how their populations overlapped with modern humans.

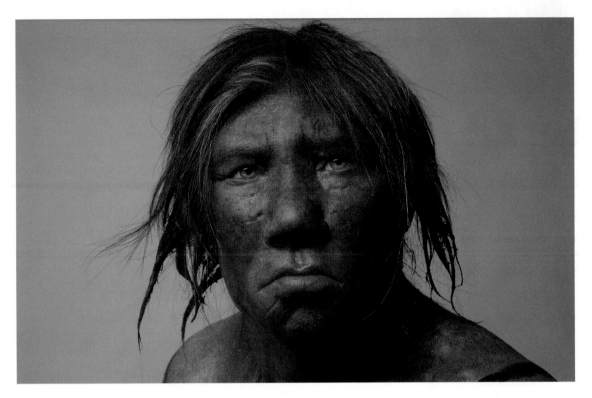

Figure 17.16

Neanderthal-human hybrid
Paleontological reconstruction of a Neanderthal-human hybrid.

the jaw appeared to be something of a hybrid between a Neanderthal and a modern human. "This, in my view, could only be a sign of interbreeding," says Condemi.

To back up her hypothesis, Condemi's team analyzed the fossil's DNA. The fossil had Neanderthal mitochondrial DNA, confirming that at least the individual's mother was a Neanderthal. From the DNA and imaging evidence, Condemi and her team concluded that it was the child of a "female Neanderthal who mated with a male *Homo sapiens*" (**Figure 17.16**). What's more, she added, this evidence supports the idea of a slow transition from Neanderthals to anatomically modern humans, in which the two species intermingled in both culture and sex, rather than the abrupt extinction of Neanderthals when modern humans arrived.

It is unlikely that Condemi's finding will be the last word on Neanderthal-human interbreeding. In 2016, the first in-depth genetic analysis of a Neanderthal Y chromosome revealed mutations in genes on the Neanderthal chromosome that would have been incompatible with comparable genes in the modern human Y chromosome. In other words, these Y chromosome mutations could have discouraged or prevented the formation of a hybrid Neanderthal–modern human species.

An even bigger question lingers for future scientists: While modern humans continued to develop their culture and populate the planet, Neanderthals went extinct. Why? Perhaps another pile of bones will someday reveal the answer.

REVIEWING THE SCIENCE

- **Animals** are multicellular ingestive heterotrophs. The **chordates** include all animals with a dorsal nerve cord. Chordates with a backbone are **vertebrates**, and all other phyla of animals are informally designated as invertebrates.

- Sponges lack distinct symmetry; some invertebrates have **radial symmetry**; all other animals, including all chordates, have **bilateral symmetry**.

- Many animal groups have a body plan consisting of repeated units known as **segments. Appendages,** body parts with specialized functions, develop in pairs from particular segments of the body. The evolution of segmentation and appendages has allowed animals to exploit new habitat and food sources in novel ways.

- **Mammals** are grouped into three categories: **Eutherians** are placental and born in a well-developed state. **Marsupials** have a simple placenta and finish development in a pouch. **Monotremes** lack a placenta and lay eggs.

- Humans are mammals (class), and more specifically **primates** (order), having flexible shoulder and elbow joints, five functional fingers and toes, **opposable** thumbs, flat nails (instead of claws), and large brains.

- **Hominins** are characterized by **bipedalism**, the ability to walk upright on two legs. They are a branch of the apes, the **hominids** (family).

- **Mitochondrial-DNA inheritance** occurs exclusively through the maternal egg; the sperm contributes essentially no mitochondria and none of the mitochondrial DNA. **Nuclear-DNA inheritance** occurs equally through both the eggs and sperm; all individuals have half maternal and half paternal nuclear DNA.

THE QUESTIONS

The Basics

1 Which of the following statements is *not* true?

(a) A single evolutionary line led from *Ardipithecus ramidus* to modern humans.

(b) Some hominid traits evolved more rapidly than others.

(c) Brain size increased greatly from early hominids to *Homo sapiens*.

(d) Toolmaking technology has improved greatly over the past 300,000 years.

2 The out-of-Africa hypothesis states that

(a) all new species of hominins arose in Africa and then migrated to the rest of the world.

(b) many new species of hominins arose outside of Africa and then migrated back to Africa later.

(c) all hominin speciation events occurred outside of Africa.

(d) all speciation events occurred in Africa, but only *Homo sapiens* distributed itself across the globe.

3 Which of the following features do sponges lack that some other animals have?

(a) segmented body plan

(b) symmetrical body plan

(c) backbone

(d) all of the above

4 Select the correct terms:

(**Mitochondrial DNA / Nuclear DNA**) is passed on only from the maternal line. (**Mitochondrial DNA / Nuclear DNA**) is inherited from both the mother and the father.

5 For each of the following cases, identify whether the animal group has bilateral symmetry (B), radial symmetry (R), or no distinct symmetry (N).

_____ a. sponges

_____ b. cnidarians

_____ c. arthropods

_____ d. chordates

_____ e. primates

Challenge Yourself

6 _____ specimens have features that are intermediate between those of *Australopithecus africanus* and *Homo erectus* and provide an amazing record of the evolutionary shift from ancestral hominin characteristics seen in *Australopithecus* fossils to more recent ones seen in *H. erectus* fossils.

(a) *Homo sapiens*

(b) *Homo neanderthalensis*

(c) *Homo habilis*

(d) *Ardipithecus ramidus*

7 Briefly describe the key differences that distinguish monotreme, marsupial, and eutherian mammals.

8 Place the following hominids in the correct order from earliest to most recent by numbering them from 1 to 5.

_____a. archaic *Homo sapiens*

_____b. *Australopithecus afarensis*

_____c. modern *Homo sapiens*

_____d. *Homo habilis*

_____e. *Homo erectus*

Try Something New

9 You visit your local museum of natural history and come upon an exhibit showcasing hominin fossils that date back 300,000–400,000 years ago. You notice that these fossils have features intermediate between those of *Homo erectus* and those of "anatomically modern" *Homo sapiens*. Who were these fossils?

(a) archaic *Homo erectus*

(b) archaic *Homo sapiens*

(c) *Homo neanderthalensis*

(d) *Homo habilis*

10 In 2004, scientists discovered the fossilized remains of an extinct species they named *Tiktaalik roseae*. The fossil appears to be a transitional species between (aquatic) fishes and (terrestrial) four-legged amphibians. In what environments do you predict this animal lived?

11 Give an example animal for some of the key adaptations of chordates: (a) backbone, (b) skull, (c) amniotic egg, (d) bony skeleton. Choose a different animal for each adaptation.

12 How does the fact that all ethnic groups *except* Africans contain some Neanderthal DNA (1–4 percent of their DNA) support the out-of-Africa hypothesis for the origin of modern humans (*Homo sapiens*)?

Leveling Up

13 *Write Now* **biology: if we were not alone** Fossil evidence indicates that in the relatively recent past (about 30,000 years ago), anatomically modern humans, or *Homo sapiens*, may have shared the planet with at least three other distinct hominins: *H. erectus*, *H. neanderthalensis*, and *H. floresiensis*. If one or more of these species were alive today, how would their existence affect the world as we know it?

14 **Is it science?** Watch the original 1968 movie *Planet of the Apes*. Document as many scientific problems with the movie as you can. Which of the apes' adaptations would be biologically possible, and which ones would be impossible, from your understanding of apes on Earth today? If this species did evolve to have the adaptive traits of *Homo*, would its members still be called apes?

For more, visit digital.wwnorton.com/bionow2 for access to:

Amazon on Fire

Can the world's largest rainforest survive a vicious cycle of wildfires and climate change?

After reading this chapter you should be able to:

- Define the biosphere and the role that humans play in it.
- Articulate the difference between a biotic factor and an abiotic factor in the study of ecology.
- Distinguish between climate and weather.
- Describe the greenhouse effect.
- Compare and contrast the water cycle and the carbon cycle.
- Explain how global warming contributes to climate change.
- List the consequences of climate change, and describe one consequence in detail.

CHAPTER

18

GENERAL
PRINCIPLES OF
ECOLOGY

...ment,
...includes
...organisms)
...gure 18.3).
...natural world
...e to change our
...ult, perhaps even

human exp... ...we...
and we use fire in our ca... ...
tion, on a daily basis," says B...
have an imperfect relationship...
vulnerable to fire; we don't com...
this tool."

Balch decided to try to control it...
to learn about fire's impact on th...
By performing a planned burn in...
area, her team could collect data tha...
obtained by studying an accidental...
fire: inventories of the plants in the...
and after a fire, a census of the ani...
area, and other measurements (F...
With a crew of other researchers, Ba...
trated an experiment using three...
kilometer plots of forest. One contro...
never be burned, one experiment...
...3 years, a...

...n the Amazon in
...t the Yale School of
...udies. She is an
...ow organ-
...d affect

323

Flames rise from the forest floor, licking at Jennifer Balch's heels. Balch walks carefully ahead of the heat, stoking it, encouraging it. She tips a large metal container, dripping flaming kerosene onto another pile of dead branches on the ground (**Figure 18.1**). The forest burns behind her.

Balch steps back to look at her work—the destruction of a small patch of the Amazon rainforest. Balch, an ecologist at the University of Colorado Boulder, is in the southern part of the Amazon basin in Brazil, in the state called Mato Grosso. There, on a small square kilometer and a half of land belonging to a soybean farmer, Balch is experimenting to see what happens when the Amazon burns.

The Amazon is the largest rainforest in the world and a critical part of our planet's biosphere. The **biosphere** consists of all of Earth's organisms, plus the physical space we all inhabit. It includes inorganic chemicals like water, our nitrogen-rich atmosphere, every living organism, and more (**Figure 18.2**). Put more simply, the biosphere is the integration of all the environments on Earth. It is crucial to our survival and well-being because humans depend on the biosphere for food and raw materials.

Within the biosphere, the Amazon is home to more than half of the world's millions of species of plants and animals. It also contains one-fifth of the world's freshwater. Yet the Amazon is under threat: since 1960, the human population of the Brazilian Amazon region has increased from 6 million to 25 million people, leading to a dramatic expansion of agriculture, including major deforestation as humans cut down trees to make room for roads, cattle, and fields. The forest cover in the Amazon has declined by 20 percent during this period, and the grasslands and pastures that have taken its place—including soybean fields like the one Balch works at—are threatening the rainforest that is left.

JENNIFER BALCH

Jennifer Balch is an ecologist at the University of Colorado Boulder who studies how fire disturbance affects ecosystems. In 2004, she initiated a one-of-a-kind experimental burn study in the Amazon to see how wildfires affect the tropical rainforest.

Figure 18.1

Jennifer Balch ignites a controlled burn in an experimental plot in the rainforest

It's traditional for farmers in Brazil to clear land for agriculture by setting fires to burn away trees and brush. In addition, fire is a useful tool for clearing out old grasses to encourage the growth of new grasses for cattle to feed on. Yet when these fires are not kept under control, they can spread to nearby rainforest and destroy it. Historically, this has not been a major concern, because the Amazon is wet and humid, so forest fires burned slowly, extinguishing themselves before spreading too far. Yet a dramatic shift in the planet's climate is making the Amazon drier and more vulnerable to wildfires.

Our planet is warming. As a result of global climate change, the temperature in Earth's atmosphere is increasing, causing less rainfall and more drought in the Amazon, leaving the forests primed for fire. And they've begun to burn. Even worse, that loss of trees adds to the warming, resulting in a vicious cycle in which the Amazon becomes hotter and drier, burns again, and contributes to further climate change.

Figure 18.2

The biosphere is Earth and all of its inhabitants
This is a view of Earth from space. The atmosphere, Earth's surface, and all of the organisms make up the biosphere.

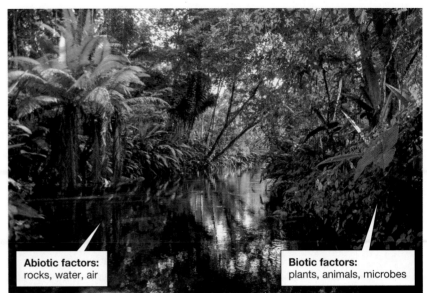

Abiotic factors: rocks, water, air

Biotic factors: plants, animals, microbes

Figure 18.3

Amazon rainforest ecology
Ecology is the study of how living organisms (biotic factors) interact with other organisms, and how they all interact with their nonliving (abiotic) environment.

Q1: List as many biotic and abiotic factors in this photograph as you can.

Q2: Is the forest part of the biotic or abiotic environment? Explain.

Q3: Is the river part of the biotic or abiotic environment? Explain.

Balch and other scientists have been investigating whether it might be possible to intervene in that cycle to slow or stop the loss of the Amazon rainforest. Their latest experiment, burning plots of Amazon rainforest to see what happens afterward, is an experiment in ecology. **Ecology** is the scientific study of interactions between organisms and their environment, where the environment of an organism includes both **biotic** factors (other living organisms) and **abiotic** (nonliving) factors (**Figure 18.3**). Ecology helps us understand the natural world we live in, but humans continue to change our biosphere in ways that are difficult, perhaps even impossible, to fix.

Hot and Dry

Balch began setting fires in the Amazon in 2004 as a graduate student at the Yale School of Forestry and Environmental Studies. She is an ecologist, a scientist fascinated by how organisms and environments interact with and affect each other.

Balch has always been interested in the link between humans and fire. "Fire is integral to the

human experience. We play with fire as kids, and we use fire in our cars, through combustion, on a daily basis," says Balch. "But humans have an imperfect relationship with fire. We are vulnerable to fire; we don't completely control this tool."

Balch decided to try to control it just enough to learn about fire's impact on the Amazon. By performing a planned burn in a restricted area, her team could collect data that cannot be obtained by studying an accidental or escaped fire: inventories of the plants in the area before and after a fire, a census of the animals in the area, and other measurements (**Figure 18.4**). With a crew of other researchers, Balch orchestrated an experiment using three 0.5-square-kilometer plots of forest. One control plot would never be burned, one experimental plot would be burned once every 3 years, and a second experimental plot would be burned once every year.

Figure 18.4

Scientists take measurements before a planned burn in the Amazon

To prevent fires in their two experimental plots from spreading, each plot was protected by a firebreak—a perimeter around the plot, several meters wide, cleared of all plants, sticks, grasses, and debris. Once the plots were prepared, it was time to set the fires. The team walked along 10 kilometers of trail cutting through the plots, dripping flaming kerosene from special tanks and watching the forest burn.

The experiment began not a moment too soon. In 2005, the Amazon experienced an extreme drought. Experts called it a "hundred-year drought," meaning that such a drought is expected to occur, on average, only once in a century. But just five years later, in 2010, there was another hundred-year drought, this one even more widespread and severe. And in both cases, extensive wildfires followed the drought, destroying over

MICHAEL COE

Michael Coe is a hydrologist and leader of the Amazon Program at the Woods Hole Research Center in Massachusetts. He investigates how humans affect the water and energy balance in tropical South America.

30,000 square miles of rainforest. In 2005 alone, fires increased 20-fold in the year following the drought. "It was a huge increase," says Michael Coe, a researcher at the Woods Hole Research Center in Massachusetts and one of Balch's collaborators. "And if humans are increasing drought frequency, that's going to be a big problem."

Humans, researchers suspect, are increasing the number and intensity of droughts in the Amazon through climate change. It's important to distinguish between "climate" and "weather." **Weather** refers to short-term atmospheric conditions in a limited geographic area, such as today's temperature, precipitation, wind, humidity, and cloud cover. **Climate** describes the prevailing weather of a region over relatively long periods of time (30 years or more). Organisms are more strongly influenced by climate than by any other feature of their environment. On land, for example, features of climate such as temperature and precipitation determine whether a particular region is desert, grassland, or tropical forest.

Climate change, then, is a large-scale and long-term *alteration* in Earth's climate, and it includes such phenomena as global warming, change in rainfall patterns, and increased frequency of violent storms. Although Earth has gone through many changes in its average climate over its 4.6-billion-year history, the speed of the change that has taken place in the past 100 years is without precedent in the climate record. Climate change in recent history has been caused to a large extent by human actions, and its consequences are likely to be negative for people and ecosystems around the world (**Figure 18.5**).

"There are a lot of indications that climate change causes increased drought events, causing increased fire events," says Balch. "It's hard to predict the future, but that's definitely the trend."

A Warmer World

The terms "global warming" and "climate change" are related but not synonymous. **Global warming** is a significant increase in the average surface temperature of Earth over decades or more. Temperature on Earth is generally determined by the angle at which sunlight strikes the planet. Sunlight strikes Earth most directly at the equator, but at a more slanted angle near the North and

Trift Glacier, Switzerland
1948   2002   2006

Figure 18.5

Consequences of climate change

Extreme weather and new rainfall patterns caused by climate change have led to increased flooding in some areas and drought in others, bringing about ecosystem and habitat destruction, and species extinction. The increase in global temperatures has led to melting of glaciers and polar ice, causing sea level rise, increase in ocean temperatures, and coral reef bleaching and death. Effects on our food sources include increased crop failure and depletion of important fisheries.

Q1: Name two ways in which climate change affects the frequency and severity of floods.

Q2: How has climate change caused a rise in sea level?

Q3: Give an example of an environmental effect of climate change in your state or region.

South Poles (**Figure 18.6**). Where sunlight hits the atmosphere more directly, its solar energy is more concentrated. For this reason, more solar energy reaches the equator, making it and neighboring tropical regions much warmer than the poles. Because Earth is tilted on its axis as it revolves around the sun on its annual orbit, sunlight hits different areas of the planet at different angles over the course of the year. This is the cause of our seasons: the Northern Hemisphere experiences summer when Earth is tilted toward the sun, while the Southern Hemisphere simultaneously experiences winter because it is tilted away.

Global warming, however, is not dependent on how sunlight strikes the Earth. Instead, it is caused by an increase in **greenhouse gases**. Some gases in Earth's atmosphere, such as carbon dioxide (CO_2), water vapor (H_2O), methane (CH_4), and nitrous oxide (N_2O), absorb heat that radiates away from Earth's surface, preventing it from being released into space. These gases are called greenhouse gases because they

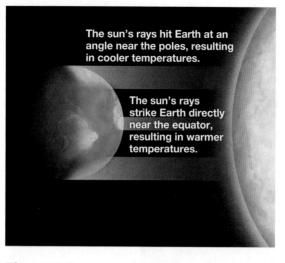

The sun's rays hit Earth at an angle near the poles, resulting in cooler temperatures.

The sun's rays strike Earth directly near the equator, resulting in warmer temperatures.

Figure 18.6

Sunlight strikes Earth most directly at the equator

The angle at which the rays of the sun strike Earth determines how much energy or heat reaches Earth's surface. The more direct the rays are when they strike Earth, the more heat they deliver.

Q1: Why is it colder at the poles than at the equator?

Q2: Why is it warmer at the equator than at the poles?

Q3: During part of the year the Northern Hemisphere is tilted at a more direct angle to the sun than the Southern Hemisphere is, and for the other part of the year the opposite is true. How does this tilt explain temperature differences in summer and winter?

function much as the walls of a greenhouse or the windows of a car do: they let in sunlight and trap heat in a process known as the **greenhouse effect** (**Figure 18.7**).

Greenhouse gases are not inherently bad; in fact, they have existed in Earth's atmosphere for more than 4 billion years, and they play an important role in maintaining temperatures that are warm enough for life to thrive on Earth. Yet human activities, primarily the burning of fossil fuels, have released an excess of greenhouse gases into the atmosphere, especially the infamous king of greenhouse gases: **carbon dioxide (CO_2)**.

Scientists have estimated atmospheric CO_2 levels for both the recent and the relatively distant past, up to hundreds of thousands of years ago, by measuring CO_2 concentrations in air bubbles trapped in ice. This evidence shows a near-perfect historical correlation between CO_2 levels and the surface temperature on Earth. Yet during the last 200 years, levels of atmospheric CO_2 have risen greatly—from roughly 280 to 400 parts per million (**Figure 18.8**). Measurements from ice bubbles show that this rate of increase is greater than even the most sudden increase that occurred naturally during the past 420,000 years. Carbon dioxide levels are now higher than those estimated for any time during that period.

The cause of the increasing levels of atmospheric CO_2 has been linked to the burning of fossil fuels like coal and oil, which releases CO_2 into the air. About 75 percent of the current yearly increase in atmospheric CO_2 is due to the burning of fossil fuels. Logging and burning of forests are responsible for most of the remaining 25 percent of the increase, but industrial processes also make a significant contribution. "Humans have a giant impact on the environment," says Coe. "With more than seven billion people on the planet, individual decisions add up to big global changes." (For more on the environmental impact of personal choices, see "How Big Is Your Ecological Footprint?" on page 332.)

All this is to say that human burning of fossil fuels has led to an increase in carbon dioxide in our atmosphere. That carbon dioxide acts as a greenhouse gas and traps additional heat as it leaves the atmosphere, causing temperatures on Earth to rise. Since the early twentieth century, Earth's mean surface temperature has increased by about 1.4°F (0.8°C), and it is estimated to rise another 2°F–11.5°F (1.1°C–6.4°C) in the future (**Figure 18.9**). As of the date this book went to publication, 2016 was the hottest year ever recorded; 2015 was previously the hottest year ever; 2014 was the winner before that. We keep beating our own record.

Our globe is warming. And the effects of that warming on the biosphere are now evident, especially in the Amazon, where trees dominate the landscape. Trees act as a layer connecting the atmosphere above them and the ground beneath them, absorbing water from the ground and releasing oxygen into the air. Any change in tree cover can change the local

Figure 18.7

How greenhouse gases warm the surface of Earth

Carbon dioxide (CO_2), water vapor (H_2O), methane (CH_4), and nitrous oxide (N_2O) are known as greenhouse gases because they absorb and trap heat that would otherwise radiate away from Earth. ▶

Q1: How much of the incoming solar energy is reflected back to outer space?

Q2: What kind of energy is reemitted to the atmosphere after being absorbed by Earth's surface?

Q3: How are greenhouse gases like a blanket on your bed at night?

climate, and increasing wildfires are devastating that tree cover.

"The big question is, What's the limit for these tropical forests? How much fire can they withstand before they convert to something else?" says Balch. That something else would be grasslands, also called savanna. If major parts of the Amazon rainforest were to convert to savanna, the climate would be significantly affected, and there would be no easy way—perhaps no way at all—to revert that land back to rainforest.

Figure 18.8

Atmospheric CO₂ levels are rising rapidly

Atmospheric CO_2 levels (measured in parts per million, or ppm) have increased greatly in the past 200 years.

Q1: What measurements do the green circles represent?

Q2: What measurements do the red circles represent?

Q3: For approximately how many years has the Mauna Loa Observatory been recording CO_2 levels?

Figure 18.9

Global temperatures are on the rise

Average global temperature has increased greatly over the 140 years since it has been recorded.

Q1: In what years were global temperatures the lowest?

Q2: In what years were global temperatures the highest?

Q3: What trend is apparent in this graph of actual global temperatures?

Fire and Water

During her first burn in 2004, Balch was surprised to see that the plants in the two experimental plots were fairly resistant to fire. The Amazon's tall, dense tree canopy creates humid air below it, protecting the forest from most fires, which tend to burn slowly under such conditions. For this reason, Balch's experimental plots initially suffered little damage.

That wasn't true in the following years. Remember that the Amazon saw a major drought in 2005, and droughts cause tree death, allowing more sunlight to reach the forest floor and dry the leaves and shrubs there, priming the forest for fire (**Figure 18.10**). During the planned burns after the drought in 2005, the two experimental plots experienced major damage. The majority of the trees on the plots died, and grasses took their place—particularly at the forest edge, where it was hottest and driest.

In both the 1-year and the 3-year burned plots, grasses invaded from adjacent pasturelands. This experimental result suggests that wildfires do indeed push the Amazon toward a drier,

Figure 18.10

Before and after the Amazon drought of 2005

The drought devastated the forest to this extent over much of the forest's range.

Forest Devastation

Agriculture is the main driver of tropical deforestation, and Brazil clears more land, by far, than any other country in the world. Since 1990, Brazil has cleared over 42 million hectares of forest, an area the size of California. Unless drastic measures are taken, deforestation will continue as farmers and companies clear land to meet rising global food demands.

Assessment available in smartwork5

Causes of deforestation
(Latin America)

- 1%
- 3%
- 2%
- 27%
- 67%

- ● Agriculture (commercial)
- ● Agriculture (local/subsistence)
- ● Infrastructure (e.g., roads)
- ● Mining
- ● Urban expansion

Drivers of forest degradation over time
(Latin America)

- 3%
- 8%
- 16%
- 73%

- ● Timber logging
- ● Uncontrolled fires
- ● Wood for fuel and charcoal
- ● Livestock grazing in forest

Top 5 countries that cleared the most forest between 2000 and 2005

Deforested area, in thousands of hectares

Brazil	Indonesia	Sudan	Myanmar	Zambia
15,515	9,357	2,945	2,332	2,224

How Big Is Your Ecological Footprint?

An action or process is **sustainable** if it can be continued indefinitely without causing serious damage to the environment. The current human impact on the biosphere is *not* sustainable.

Each of us can help build a more sustainable society. We can advocate legislation that fosters less destructive and more efficient use of natural resources, patronize businesses that take measures to lessen their negative impact on the planet, support sustainable agriculture, and modify our own lifestyles. For example, we can increase our use of renewable energy and energy-efficient appliances; reduce all unnecessary use of fossil fuels (for instance, by biking to work or using public transportation); buy seafood from sustainable fisheries; use "green" building materials; and reduce, reuse, and recycle waste. Experts estimate that more than 200 million women around the world wish to limit their family size but have no access to family planning. Those of us who live in developed countries can support aid efforts that provide education, health care, and family-planning services in developing countries.

One measure of sustainability is an **ecological footprint**, which is the area of biologically productive land and water that an individual or a population requires to produce the resources it consumes and to absorb the waste it produces. Scientists compute an ecological footprint using standardized mathematical procedures and express it in *global hectares* (*gha*). One gha is equivalent to one hectare (2.47 acres) of *biologically productive* space. Approximately one-fourth of Earth's surface is considered biologically productive; this definition excludes areas such as glaciers, deserts, and the open ocean.

According to recent estimates, the ecological footprint of the average person in the world is 2.7 gha, which is about 60 percent higher than the 1.7 gha that would be needed to support each of the world's 7.4 billion people in a sustainable manner. An ecological footprint can also be expressed in **Earth equivalents**, the number of planet Earths needed to provide the resources we use and absorb the wastes we produce. Currently, the global population uses 1.6 Earth equivalents each year (as shown at the bottom of the figure on the facing page).

Overall, such estimates suggest that, since the late 1970s, people have been using resources faster than they can be replenished—a pattern of resource use that, by definition, is not sustainable. As the world population grows, the amount of biologically productive land available per person continues to decline, increasing the speed at which Earth's resources are consumed.

The per capita consumption of Earth's resources by different countries is most directly related to energy demand, affluence, and a technology-driven lifestyle. As people in populous countries such as China and India become wealthier, their ecological footprints are growing rapidly.

What is *your* ecological footprint? If you are a typical American college student, your footprint is probably close to the U.S. average of 8.2 gha. It would take nearly five planet Earths to support the human population if everyone on Earth enjoyed the same lifestyle that you do (see the top row of the accompanying figure). Your ecological footprint depends on four main types of resource use:

1. *Carbon footprint*, or energy use

2. *Food footprint*, or the land and energy and water it takes to grow what you eat and drink

3. *Built-up land footprint*, which includes the building infrastructure (from schools to malls) that supports your lifestyle

4. *Goods-and-services footprint*, which includes your use of everything from home appliances to paper products

If you drive a gas guzzler, live in a large suburban house, routinely eat higher up on the food chain (more beef than grains or veggies/fruits), and do not recycle much, your footprint is likely to be higher than that of a person who uses public transportation, shares an apartment, eats mostly plant-based foods, and sends relatively little to the local landfill. Most of us can significantly reduce our ecological footprint with little reduction in our quality of life, while bestowing an outsized benefit on our planet.

---

grassier savanna-like ecosystem (**Figure 18.11**). "This could be the trigger for changing forest to savanna," says Balch. "The message is that repeated disturbances, such as multiple fires or drought and fire, cause forest to reach its tipping point, to shift into grassland."

While Balch and her team observed changes in plants after fire, ecologist Michael Coe investigated how fire affects its nemesis: water. "The forest is an incredibly important part of the **hydrologic cycle**," the circulation of water from the land to the sky and back again, says Coe (**Figure 18.12**).

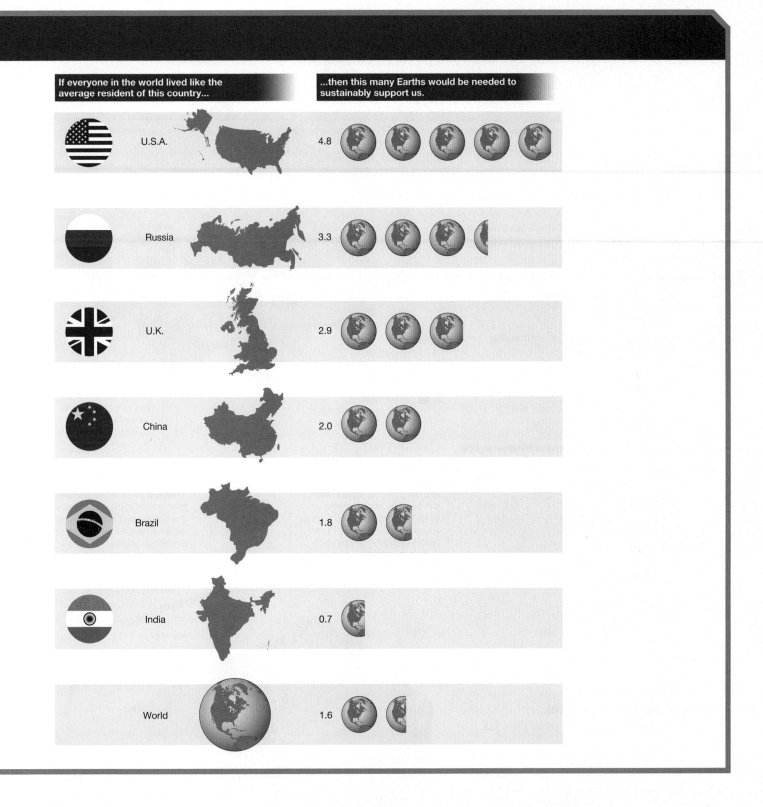

If everyone in the world lived like the average resident of this country...	...then this many Earths would be needed to sustainably support us.
U.S.A.	4.8
Russia	3.3
U.K.	2.9
China	2.0
Brazil	1.8
India	0.7
World	1.6

Near the equator, direct sunlight causes water to evaporate from Earth's surface. The warm, moist air rises because heat causes it to expand, making it less dense and lighter than air that has not been heated. But as it rises, the warm, moist air begins to cool. Because cool air cannot hold as much water as warm air can, much of the moisture from a cooling airmass is "wrung out" and falls as rain.

For this reason, most tropical regions, including the Amazon, receive ample rainfall. Earth has four giant **convection cells** in which warm,

Figure 18.11

A projection of the Amazon rainforest in 2030

This map of future Amazon rainforest coverage is based on the assumption that global warming will persist but not increase.

Q1: Where will fire most seriously affect the Amazon rainforest?

Q2: Where will fire be the least damaging to the Amazon rainforest?

Q3: This map does not include an increase in pasturelands for grazing animals. Do you think more or less pastureland will be needed in 2030? Explain.

The historical forest cover in this area is denoted by the dark black outline.

If global warming continues to increase, dry areas will likely increase as well.

■ Forest
□ Deforested
■ Dry and/or logged forest
■ Nonforested area (now grasslands and fields, due to climate change)

This depiction does not include the effect of fires in the dry areas. Fires will likely increase the deforested areas.

Arrows pointing toward the atmosphere show where evaporation occurs from water sources or is released from plants through transpiration to form humidity or clouds.

Transpiration is the process of plants absorbing water through their roots and releasing this water through their leaves into the atmosphere.

Figure 18.12

The hydrologic (water) cycle

Arrows indicate the direction of water flow.

Q1: What is transpiration?

Q2: Why is transpiration important to the water cycle?

Q3: If there are fewer plants and therefore less transpiration in a given area, what will happen to the humidity or cloud cover in this area?

Precipitation

Condensation

Transpiration from plants

Evaporation

Uptake by plants

Runoff

Ocean

Water percolation into soil

Groundwater flow

Arrows pointing toward Earth show where water is returned as precipitation.

moist air rises and cools, releasing moisture as rain or snow depending on temperature, and then sinks back to the ground as dry air (**Figure 18.13**). These convection cells, in combination with the angle of sunlight striking the Earth, play a large role in the creation of regional environments on Earth, such as rainforests and deserts.

In the Amazon, trees are very important players in the water cycle, adds Coe. "The trees pull water out of the soil and evaporate it into the atmosphere in the process of photosynthesizing (through transpiration), so they're the mediators between the rainfall and the streams," he says. "Burning the trees greatly reduces the amount of water getting back into the atmosphere." (See Chapter 5 for more on photosynthesis.)

Each year of Balch's study, Coe joined Balch at the test site. There, his team dug 10-meter-deep soil pits—long, dark caverns in which they inserted instruments to measure the moisture content of the soil. In a healthy forest ecosystem, trees absorb a lot of the water in the soil, leaving it nice and dry, with only minimal water runoff into streams. This is what Coe observed in the control plot where nothing was burned. "A healthy forest uses up almost all the water [in the soil]," says Coe.

But in the other two plots—burned every year or every 3 years—he found that the soil in the pits was wet to the touch. When the forest burned, trees died, so nothing absorbed the moisture from the soil. Consequently, nearby streams were overflowing with water—up to four times the volume of water seen in healthy forests. "That's not a good thing," says Coe. "We're circumventing the natural cycle. Instead of this water going back into the atmosphere, creating more rain and driving vegetation, it's flushing the water out of the system."

And on the burned plots, the invasive grasses that took the place of the trees have very shallow roots, absorb less moisture from the ground, and evaporate less water into the air. In this way, deforestation of large areas—whether through unintentional wildfires or the intentional cutting down of trees—results in less rainfall and hotter temperatures. "If you deforest enough of it, you're going to really decrease the rainfall over a broad swath of this region," says Coe.

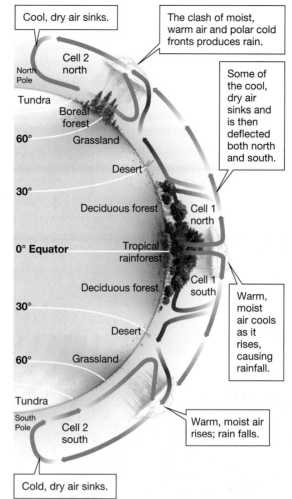

Figure 18.13

Earth has four giant convection cells
Two giant convection cells are located in the Northern Hemisphere and two in the Southern Hemisphere.

Q1: How do the patterns of rainfall in the Northern and Southern Hemispheres compare?

Q2: How do the patterns in the kinds of environments shown in the Northern and Southern Hemispheres compare?

Q3: What happens at the equator to make this region so wet?

The Carbon Games

In addition to measuring the hydrologic cycle and the expansion of grasses into the region, the teams led by Balch and Coe measured how much carbon was released into the atmosphere when the forests burned.

Carbon, in the form of CO_2 gas, makes up only about 0.04 percent of Earth's atmosphere, although that percentage has been creeping upward every year for the last 200 years and causing global warming, as we've seen. Carbon is also found in Earth's crust, where carbon-rich sediments and rocks formed from the remains of ancient marine and terrestrial organisms. Carbon is present in every living thing.

Living cells are built mostly from organic molecules—molecules that contain carbon atoms bonded to hydrogen atoms. After oxygen, carbon is the most abundant element in cells by weight; every one of the large biomolecules in an organism has a backbone of carbon atoms. Living organisms, in both aquatic and terrestrial ecosystems, acquire carbon mostly through photosynthesis. Aquatic producers, such as photosynthetic bacteria and algae, absorb dissolved CO_2 and convert it into organic molecules using sunlight as a source of energy. Plants, the most important producers in terrestrial ecosystems, absorb CO_2 from the atmosphere and transform it into food with the help of sunlight and water. This is how a plant builds itself—its leaves, stems and branches, flowers, and other structures.

The transfer of carbon within biotic communities and their physical surroundings, the abiotic world, is known as the global **carbon cycle** (**Figure 18.14**). (For more on nutrient

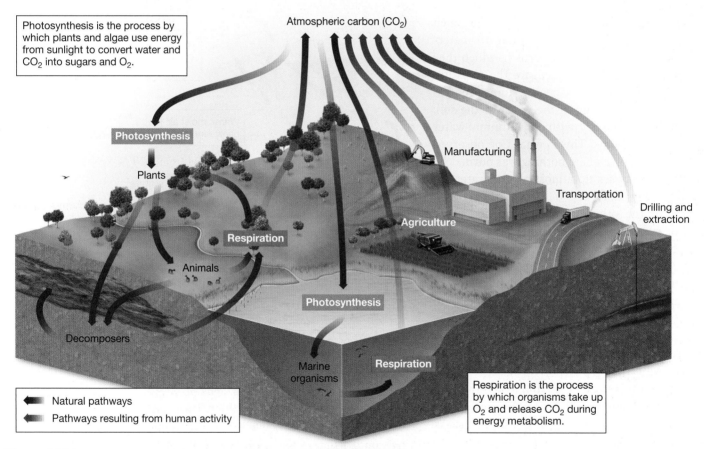

Photosynthesis is the process by which plants and algae use energy from sunlight to convert water and CO_2 into sugars and O_2.

Atmospheric carbon (CO_2)

Photosynthesis

Plants

Manufacturing

Transportation

Drilling and extraction

Agriculture

Respiration

Animals

Photosynthesis

Decomposers

Marine organisms

Respiration

Respiration is the process by which organisms take up O_2 and release CO_2 during energy metabolism.

Natural pathways

Pathways resulting from human activity

Figure 18.14

The carbon cycle

Arrows indicate the direction of carbon flow.

Q1: What are three ways that carbon is released into the atmosphere?

Q2: Are all of the pathways you listed for question 1 affected by human activity?

Q3: What are two biotic reservoirs of carbon?

cycling, see Chapter 21.) One way that carbon is transferred between the biotic and abiotic worlds is through combustion—the burning of carbon-rich materials, living or not.

Some of the organic matter from ancient organisms has been transformed by geologic processes into deposits of fossil fuels such as petroleum, coal, and natural gas. When we extract these fossil fuels and burn them to meet our energy needs, the carbon that was locked in these deposits for hundreds of millions of years is released into the atmosphere as carbon dioxide.

Plants also release carbon back into the atmosphere when they are burned. To assess the amount of carbon released by a burning rainforest, Balch and her team measured the amount of leaf litter and branches—the biomass—on the forest floor both before and after a burn, as well as the number of trees affected. Half of that biomass is carbon, so half of the difference between the biomass before and after the burn is the amount of carbon that was released into the atmosphere. It was a lot.

The very first burn, for example, released "about 20 tons of carbon per hectare," says Balch. That's approximately equal to the carbon emitted by a passenger car driven 864,000 miles (burning about 40,000 gallons of gasoline). And burns have another, indirect, detrimental effect that Balch wasn't able to directly measure: they destroy trees that normally absorb carbon dioxide from the air. In a normal year, trees in the Amazon absorb approximately 1.5 billion tons of carbon dioxide, helping to slow climate change. In this way, they act as a **carbon sink**, a natural or artificial reservoir that absorbs more carbon than it releases.

But during the 2005 drought, as trees died and rotted, the rainforest stopped absorbing more carbon dioxide than it gave off, and it actually released carbon *into* the atmosphere. In 2005, 5 billion tons of carbon were released. Instead of acting as a carbon sink, the Amazon became a **carbon source**, a reservoir that releases more carbon than it absorbs (**Figure 18.15**). And with increasing temperatures, decreasing rainfall, and increasing wildfires, ecologists fear that the Amazon could transition more and more into a carbon source instead of a carbon sink. If that happens, our planet will lose one of its greatest buffers against future climate change.

A carbon sink absorbs more CO_2 than it gives off.

A carbon source gives off more CO_2 than it absorbs.

Figure 18.15

Carbon sinks and sources

Q1: How does a carbon source contribute to global warming?

Q2: How does a carbon sink protect against global warming?

Q3: How can trees act as both a source and a sink?

Waiting and Watching

Balch and her crew have stopped burning the plots and have begun to record what happens to them as they recover. In addition to making direct observations of which plants grow back and how their growth affects the water and carbon cycles, the team is conducting new experiments on the burned plots to see whether it is possible to help prevent the invasion of grasses and encourage a return to rainforest.

In one area, the scientists are planning to plant different types of trees to see whether particular species can successfully reestablish themselves in a mat of grasses and retake that land. Another experiment, also still in the planning phase, will test different amounts of watering to see whether the lack of precipitation is preventing the growth of new trees. If any strategies work well on the small plots, perhaps they can be adapted to save larger areas of rainforest affected by fire.

"We're hoping we find ways to help the forest recover," says Balch. Still, the researchers worry about the future of the region. "Have we crossed a threshold where this part of the forest isn't going to grow back as rainforest?" asks Coe, worry in his voice. If that is true for the small experimental plots, it bodes danger for large regions of the Amazon that border agricultural land and are threatened by wildfires.

REVIEWING THE SCIENCE

- **Ecology** is the study of interactions between organisms and their environment. All ecological interactions occur in the **biosphere**, which consists of all living organisms on Earth (**biotic** factors), together with the physical environment they inhabit (**abiotic** factors).

- **Weather** consists of the short-term atmospheric conditions in a given area, such as temperature, precipitation, wind, humidity, and cloud cover. **Climate**, the prevailing weather of a specific region over relatively long periods of time, has a major effect on the biosphere. Climate is determined by incoming solar radiation, global movements of air and water, and major features of Earth's surface.

- **Climate change** is a large-scale and long-term alteration in Earth's climate, and it includes such phenomena as more extreme temperatures, change in rainfall patterns, and increased frequency of violent storms.

- Carbon dioxide acts as a **greenhouse gas** and contributes to the **greenhouse effect** by absorbing heat released at Earth's surface, rather than allowing it to leave the atmosphere. The greenhouse effect has led to a rapidly rising average global temperature, a phenomenon known as **global warming**.

- The concentration of **carbon dioxide** (CO_2) gas in the atmosphere is increasing at a dramatic rate because of an increase in **carbon sources**, such as the release of CO_2 through the burning of fossil fuels, and a loss of **carbon sinks**, such as the absorption of CO_2 through photosynthesis in large forests.

- Both the **hydrologic cycle** and the **carbon cycle** are affected by global warming, which disrupts the natural cycling of water and carbon molecules in the biosphere.

- The area of the biosphere required to produce the resources and to absorb the waste produced by an individual or population is known as the **ecological footprint**. The ecological footprint is often expressed in **Earth equivalents**, the number of planet Earths needed to provide the resources and absorb the waste of an individual or population.

THE QUESTIONS

The Basics

1 The biosphere is

(a) all organisms on Earth, together with their physical environments.

(b) crucial to human survival and well-being.

(c) a source of food and raw materials for human society.

(d) a web of interconnected ecosystems.

(e) all of the above

2 Greenhouse gases function by

(a) blocking sunlight but letting out heat from Earth to outer space.

(b) absorbing heat that radiates from Earth that would otherwise escape to outer space.

(c) absorbing heat that radiates from the sun toward Earth.

(d) releasing heat that radiates from Earth into outer space.

3 Which of the following is an abiotic component of the biosphere?

(a) algae

(b) insects

(c) lichen

(d) water

(e) none of the above

4 Place the following components of the hydrologic cycle in the correct order by numbering them from 1 to 5, beginning with precipitation.

_____ a. precipitation

_____ b. uptake by plants

_____ c. transpiration

_____ d. water percolation into soil

_____ e. condensation

5 Select the correct terms:

(Climate / Weather) is the short-term atmospheric conditions in an area, while (climate / weather) is the average atmospheric conditions in a given area over a long period of time. (Climate change / Global warming) consists of the long-term and large-scale changes in atmospheric conditions, which have been brought about by (climate change / global warming), an increase in the average global temperature.

Challenge Yourself

6 The carbon cycle and the water (hydrologic) cycle are similar in that

(a) both cycle molecules between abiotic and biotic components of the environment.

(b) photosynthesis is a critical process in both.

(c) both are involved in global warming and climate change.

(d) all of the above

(e) none of the above

7 An ecologist is studying an area of forested land that experiences a forest fire every 5 years. In the years between forest fires, the area of study absorbs 300 million tons of CO_2. If a forest fire releases 200 million tons of CO_2 in this area, does this forested area act as a carbon sink or a carbon source? Why?

8 Place the following events in the correct order of their onset by numbering them from 1 to 5, beginning with the earliest.

_____ a. Climate change, including more extreme weather

_____ b. Increase in human population and activities

_____ c. Increase in release of CO_2 and other greenhouse gases into the atmosphere

_____ d. Increase in calls to control the production and release of greenhouse gases

_____ e. Global warming

Try Something New

9 Which of the following behaviors could *decrease* the production of greenhouse gases, and thus perhaps curb global warming?

(a) using ethanol or biodiesel rather than fossil fuels to power autos

(b) using coal rather than gas to warm homes

(c) using solar or wind power rather than fossil fuels to produce electricity

(d) all of the above

(e) none of the above

10 Reviewing Figure 18.13, identify the type of environment at your latitude, and explain how it is influenced by Earth's convection cells.

11 Discuss briefly (in one to two sentences) how the more extreme weather caused by global warming may affect where you live.

Leveling Up

12 **Looking at data** Review Figures 18.8 and 18.9. Describe the change in atmospheric CO_2 levels from 1880 to the present, and the change in average global temperature for the same time period. What were the CO_2 level and the average temperature in 1880? in 1960? in 2016? Is there a relationship between the two variables? How would you explain the pattern you see here?

13 **Life choices** You can estimate your impact on the planet by using one of the many "footprint calculators" on the Internet. Take three online quizzes available from organizations such as the Global Footprint Network, the Nature Conservancy, and the U.S. Environmental Protection Agency. Each site calculates your ecological footprint a little differently. Compare and contrast the different sites you used. Which one do you think is the most accurate? Why? Which one do you think is the most superficial? Why? Write a list of ways you can decrease your ecological footprint, and try to adhere to your new lifestyle!

14 *Write Now* **biology: climate denial** Watch the documentary *Frontline's Climate of Doubt* (2012), produced by PBS. Write an essay reflecting on what you learned from the film and from this chapter about global warming. Use your knowledge of your ecological footprint to reflect on how important it is for all citizens to be aware of this global crisis.

For more, visit digital.wwnorton.com/bionow2 for access to:

Zika-Busting Mosquitoes

Genetically engineered insects help stop the spread of dengue, malaria, and Zika.

After reading this chapter you should be able to:

- Articulate the difference between population size and population density.
- Interpret graphical population data to determine whether the population exhibits logistic or exponential growth.
- Explain the concept of carrying capacity, and give examples of factors that can increase or decrease carrying capacity for a population.
- Differentiate between density-dependent and density-independent changes in population size.
- Illustrate population cycles of predators and prey.

CHAPTER

19

GROWTH OF
POPULATIONS

It is impossible to know when an infected mosquito bit the first person in Miami-Dade County, Florida. But on Friday, July 29, 2016, health officials confirmed that a woman and three men in a small area north of downtown Miami had been infected with the Zika virus through the bite of a local mosquito. It was the moment everyone had been dreading: the first known cases of the virus being transmitted by mosquitoes in the United States. "As we have anticipated, Zika is now here," Tom Frieden, the director of the U.S. Centers for Disease Control and Prevention (CDC), told reporters that afternoon (**Figure 19.1**).

By July 2016, Zika had become a household name. Zika fever, the illness that results from infection with Zika virus, is a mild sickness that typically lasts less than a week and involves fever, red eyes, headache, and a possible rash. But Zika has another, much more dire effect on health. A year before the Miami outbreak, doctors in Brazil had begun noticing an increasing number of infants born with a serious birth defect known as microcephaly, in which the head and brain are smaller than expected; they have not developed properly. Microcephaly can cause many problems in growing infants, including seizures, speech and intellectual delays, feeding and movement problems, and hearing and vision loss. Brazil's new cases of microcephaly were soon linked to a rapidly spreading virus passed through the bite of a mosquito: Zika.

The scientific evidence linking Zika virus and microcephaly is strong: infected mothers pass the virus to their fetuses, where it may stunt brain growth (**Figure 19.2**). Studies suggest that infants exposed to the Zika virus in utero are more than 50 times more likely to be born with microcephaly than those who are not exposed.

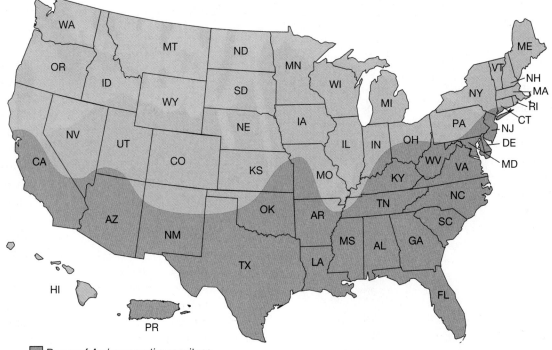

Range of *Aedes aegypti* mosquitoes

Figure 19.1

Range of the mosquito species most likely to carry the Zika virus
Aedes aegypti mosquitoes spread viruses like Zika, dengue, and chikungunya more often than other mosquito species do.

Q1: What parts of the United States are within the range of the mosquito that carries Zika?

Q2: What areas are *not* within the mosquito's range? Why do you think that is?

Q3: Find your own state, and where you are in the state. Are you at risk of contracting Zika?

In February 2016, the World Health Organization declared the spread of Zika virus to be a "public health emergency of international concern."

Since the outbreak in Brazil, governments around the world have begun looking for ways to stop Zika, and the local transmission in Florida galvanized U.S. officials and scientists. "It's scary," says Matthew DeGennaro, a biologist who studies mosquito genetics and behavior at Florida International University in Miami. "But people in other countries face these things all the time, and not much is being done about it. It is possible this [outbreak] will focus Americans' attention on mosquito-borne illness."

The human species has a long history of battling mosquito-borne viruses, including dengue virus and West Nile virus. Along the way, we've learned that when a vaccine is lacking, the best way to stop the virus is to stop the insect. Yet despite efforts to eradicate mosquito populations around the world, the pests continue to roar back no matter what we throw at them. Mosquitoes quickly develop resistance to common insecticides, for example, and in at least one case they showed an adapted behavior, attacking farmers outdoors in the early morning rather than indoors during the night when the farmers were protected by mosquito nets.

But now there's a new weapon in the arsenal against mosquito-borne diseases. Taking a technological approach to the problem, scientists have begun to target mosquito populations by using the insects themselves as weapons. It's a new kind of war—mosquito versus mosquito.

Population Control

One month after the first four reported cases of Zika transmission in Florida, officials captured Zika-infected mosquitoes at seven different locations on Miami Beach—the smoking gun to show that local disease transmission was indeed occurring. By October 2016, mosquitoes bearing Zika virus had spread another 3 miles north.

Zika is spread by females of both the *Aedes aegypti* species and the less common *Aedes albopictus* species, although the latter is less likely to bite humans because it breeds in rural areas and feeds on other animals in addition to humans. These two types of mosquitoes also

Figure 19.2

A Brazilian mother holds her microcephalic baby
The incidence of microcephaly in newborns has skyrocketed, and the increase has been linked to Zika infection of the mother during pregnancy.

serve as carriers, or *vectors*, for dengue, chikungunya, and other viruses. (Malaria, which is caused by parasitic protists, is transmitted among humans by female mosquitoes of the genus *Anopheles*.) Active primarily during the day, a female mosquito feeds on human blood via a tubelike mouthpart (which male mosquitoes lack) that pierces the skin of a host. During that moment, virus particles in the mosquito's saliva can be transferred to the person's skin. Once on the skin, the virus is able to replicate in skin cells and then can spread to the lymph nodes and bloodstream.

In addition to being transmitted by mosquitoes, Zika can spread from person to person via body fluids: blood, tears, semen, and saliva. In some cases, the virus can remain in those fluids for months after

MATTHEW DEGENNARO

Matthew DeGennaro is a neurogeneticist at Florida International University in Miami, studying mosquito genetics and behavior. As a Florida resident, he witnessed the outbreak of Zika in Miami and advocates the use of genetically modified mosquitoes to control mosquito populations and stop the spread of disease.

infection (**Figure 19.3**). Slowing the spread of Zika, therefore, involves the screening of donated blood and the proper use of condoms during sex. But true prevention ultimately requires decreasing the incidence of mosquito bites, which can be accomplished by getting rid of the mosquitoes.

Scientists have taken many approaches to trying to reduce the mosquitoes' **population size**, the total number of individuals in the population. A **population** is a group of organisms of the same species in a defined area. Population size tends to change over time—sometimes increasing, sometimes decreasing. Whether the size of a population increases or decreases depends on the number of births and deaths in the population, as well as on the number of individuals that enter (immigrate) or

Figure 19.3

How Zika is transmitted

The CDC produces posters such as this for public health issues including Zika, sexually transmitted diseases, and the flu.

Q1: What is the main way by which someone is infected with the Zika virus?

Q2: Judging by the poster and your knowledge of mosquito behavior, what can you do to decrease your risk of being infected with the Zika virus?

Q3: Besides the transmission methods shown on the poster, what are some other ways you could become infected with the Zika virus?

leave (emigrate) the population. Birth and immigration increase population size; death and emigration reduce it. Environmental factors also have a strong impact on population size. Mosquito populations, for example, increase in warmer, wetter weather when the conditions are ripe for reproduction.

To target mosquito populations, scientists rely on **population ecology**, the study of the size and structure of populations and how they change over time and space. Population ecology can help us determine where mosquitoes live, eat, and breed. Mosquitoes congregate and lay their eggs at water sources, and many prefer urban environments with trash receptacles and concentrated groups of people. At these locations we find the highest **population density**, or number of individuals per unit of area.

To calculate population density, total population size is divided by the corresponding area of interest. For example, in an urban neighborhood of Rio de Janeiro, Brazil, where people were being regularly infected with dengue virus, scientists captured mosquitoes in mosquito traps and calculated the neighborhood's population of female mosquitoes to be 3,505 pregnant females. The neighborhood has an area of 911 acres, so the female mosquito population density was 3.85 mosquitoes per acre (3,505 mosquitoes divided by 911 acres). Keep in mind that population density is often difficult to measure, because it depends on an accurate count of the population size. Individuals may be hard to detect, may move between populations, or may inhabit a complex, hard-to-define area.

Scientists can track population density to see how well eradication efforts work against mosquitoes. As mentioned earlier, one of the most popular ways of reducing mosquito populations is poisoning them with insecticides. But that's not ideal, scientists agree. "Insecticides affect the entire insect population usually, and that leads to overall imbalances in the environment," says DeGennaro. "So putting a chemical into the environment is something you need to do very carefully."

An alternative approach, thanks to advanced genetic technologies, is to tweak the genes of the mosquitoes to sabotage their reproduction. If the mosquitoes can't reproduce, population size will plummet, and fewer mosquitoes means fewer people being bitten and infected. This gene tweaking is called **genetic modification**, or **GM**—altering the genes of an organism for a specific purpose. It is also referred to as genetic engineering, or GE. In contrast to insecticides, GM mosquitoes can directly target just one insect species, says DeGennaro. "You're only affecting one species of mosquito, the species that is the primary vector for Zika, dengue, and more."

Early field tests of GM mosquitoes in Brazil appeared to stop the spread of dengue. But could they do the same for Zika?

Rapid Spread

Before it hit Florida, Zika virus struck Brazil like a tsunami. In a span of 3 months in early 2015, Brazil documented nearly 7,000 cases of mild illness similar to dengue fever, though no one yet suspected Zika. That's the problem with a newly emerging infectious disease: no one is looking for it.

Then, in May, a national laboratory reported that Zika virus was circulating around the country. Two months later, Brazilian health workers began reporting neurological disorders associated with Zika infection, including brain inflammation. In October came the worst news of all: reports of an unusual spike in newborns with microcephaly.

That dramatic spread and link to microcephaly put Zika on the map, but it isn't actually a new virus. Zika virus—a round particle with a dense core packed with RNA (**Figure 19.4**)—was first

Zika virus particle

Figure 19.4

Zika is a spherical RNA virus

isolated in 1947 in the Zika Forest of Uganda, for which it is named. The first human cases of Zika were detected in Uganda in 1952, and the little-studied virus subsequently spread across central Asia, the South Pacific, and into South and Central America and the Caribbean.

Humans are no strangers to mosquito-borne diseases (see "Mosquito-Borne Diseases"). While Zika and other diseases were once isolated to specific areas of the globe, global travel and human population growth, along with climate change, have increased mosquitoes' ability to proliferate and spread disease on a grand scale. In the second half of the twentieth century, our planet saw the fastest rate of human population growth in its history. **Population doubling time**—the time it takes a population to double in size—is a good measure of how fast a population is growing. For example, the doubling time for the U.S. human population at our current reproductive rate is about 100 years. In stark contrast, the population doubling time for mosquitoes can be as little as 30 days in an optimal environment.

Reaching Capacity

Now back to mosquitoes, and how to get rid of them. Scientists argue strongly that understanding mosquito population ecology is a prerequisite for eliminating mosquito-borne diseases. For example, many mosquitoes lay their eggs directly on the surface of water. That knowledge led to one of the most important public health efforts to reduce mosquito-borne diseases: the removal of pools of standing water. Eliminating these breeding sites makes it possible to reduce the **carrying capacity** of the environment, the maximum population size that can be sustained in a given area.

In most species, the growth rate of a population decreases as the population size nears the carrying capacity, because resources such as food and water begin to run out. If there are only a few locations of standing water in a given area, for example, a limited number of eggs can be laid in that area, thereby limiting the size of the population. Any limiting resource needed for survival, such as habitat, food, or water, will determine an environment's carrying capacity for a specific population. Because different species have unique needs, the same environment can have different carrying capacities for different resident species. For the same reason, two different environments in which the same species lives will have different carrying capacities. At the carrying capacity, the population growth rate is zero.

If a population has no constraints on its resources, it will experience **exponential growth**, which occurs when a population increases by a constant proportion over a constant time interval, such as one year. Exponential growth is represented by a **J-shaped growth curve** (**Figure 19.5**). A population that approaches its carrying capacity because of constrained resources will experience **logistic growth**, in which the population grows nearly exponentially at first but then stabilizes at the maximum population size that can be supported indefinitely by the environment. Logistic growth is represented by an **S-shaped growth curve**.

Mosquito-Borne Diseases

Mosquitoes cause more human suffering than any other organism (with the exception of our own species). Here are just a few of the deadly or debilitating diseases they carry:

- *Dengue virus* is a huge problem for the population of Brazil. Spread by the same mosquitoes as Zika, dengue also has no vaccine or cure, and it causes a disease chillingly known as "breakbone fever." Symptoms include continuous joint pain, vomiting, high fever, and severe rashes. In 2015, Brazil chalked up a record 1.6 million cases and 863 deaths to dengue—an 82.5 percent increase over 2014.

- *Malaria*, an ancient disease that has been traced as far back as 2700 BCE, affects an estimated 40 percent of the world's population. Hundreds of thousands of deaths are still reported annually from malaria.

- *Yellow fever*, against which there is a vaccine, still leads to an estimated 30,000 deaths per year.

- *West Nile virus*, which emerged in the 1930s, can invade the nervous system and cause death.

- *Chikungunya* is found mainly in Africa and Asia, but cases have been reported in Florida. The main symptoms are fever, rash, and joint pain that can last for years. Only one in 1,000 cases results in death.

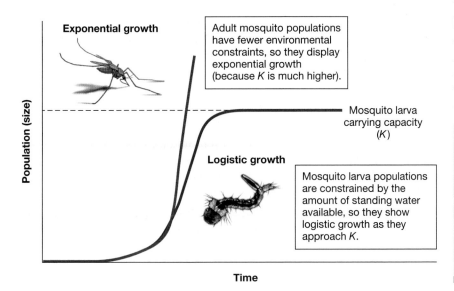

Exponential growth

Adult mosquito populations have fewer environmental constraints, so they display exponential growth (because *K* is much higher).

Mosquito larva carrying capacity (*K*)

Logistic growth

Mosquito larva populations are constrained by the amount of standing water available, so they show logistic growth as they approach *K*.

Population (size)

Time

Figure 19.5

Populations can experience exponential growth or logistic growth

A population that is not constrained by resources or by the environment can grow exponentially, whereas one that is constrained by a set carrying capacity will show logistic growth.

Q1: Which form of population growth displays a J-shaped curve?

Q2: Which form of population growth displays an S-shaped curve?

Q3: Describe a situation in which a population initially shows exponential growth and later shows logistic growth.

Little research has been done on growth curves for mosquito populations, primarily because the insects are small and airborne, making them hard to track. But we do know how the human population has changed: Over the last 500 years, Earth's human population exhibited both logistic and exponential growth. At the end of the last ice age, in approximately 10,000 BCE, there were only 5 million people on Earth. With the advent of agriculture in about 8000 BCE, the world population began to rise logistically, until about 200 years ago. Then, alongside the use of fossil fuels and the industrial revolution, human population growth exploded exponentially. Modern populations have continued to show exponential growth to the detriment of the environment. Current estimates of the carrying capacity of Earth range from 2 billion to over 1,000 billion people, with the majority of studies insisting that 9–10 billion people is the maximum number that Earth can support. At current population growth rates, we will reach this number by the year 2050 (**Figure 19.6**).

Seeking Change

With human populations increasing in number and concentrating in urban areas, we can expect the continued spread of mosquito-borne illnesses. Thankfully, recent results from field

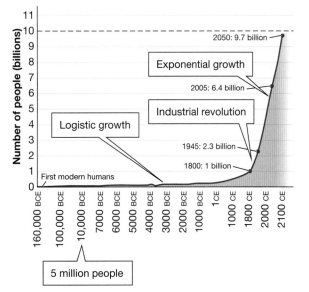

2050: 9.7 billion

Exponential growth

2005: 6.4 billion

Industrial revolution

Logistic growth

1945: 2.3 billion

1800: 1 billion

First modern humans

5 million people

Figure 19.6

Curves of logistic and exponential growth in the world human population

The dashed line indicates the United Nations' estimated carrying capacity of Earth.

Q1: According to this graph, approximately when did exponential growth begin?

Q2: What milestone corresponds to the transition from logistic to exponential population growth?

Q3: What is the UN's projected carrying capacity of Earth, and when will we reach it?

tests of GM mosquitoes in Brazil suggest that we may have a new way to stop that spread.

Starting in 2002, British company Oxitec began producing a GM line of the *Aedes aegypti* mosquito. The company inserts a single gene into a line of lab-bred mosquitoes. This gene, designed to work only in insect cells, produces a protein that prevents the mosquito from transitioning between two of its life stages: larva and adult. In the lab, the mosquitoes are exposed to an antibiotic (tetracycline) that neutralizes the protein so that the mosquitoes can grow to be adults. Then, only males are released into the wild, because the males don't bite and so will not contribute to the spread of disease. Once free, the male GM mosquitoes mate with wild females and produce offspring with the lethal gene—and

no access to antibiotics—so the offspring die off before becoming adults and therefore do not reproduce. In this way the size of the population is reduced (**Figure 19.7**).

Populations—mosquito, human, or otherwise—can change in a density-dependent or density-independent manner. **Density-dependent population change** occurs when birth and death rates change as the population density changes. The number of offspring produced and the death rate are often density-dependent. Food shortages, lack of space, and habitat deterioration—all these factors influence a population more strongly as it increases in density (**Figure 19.8**).

In addition, when a population has many individuals, disease spreads more rapidly (because individuals tend to encounter one another more

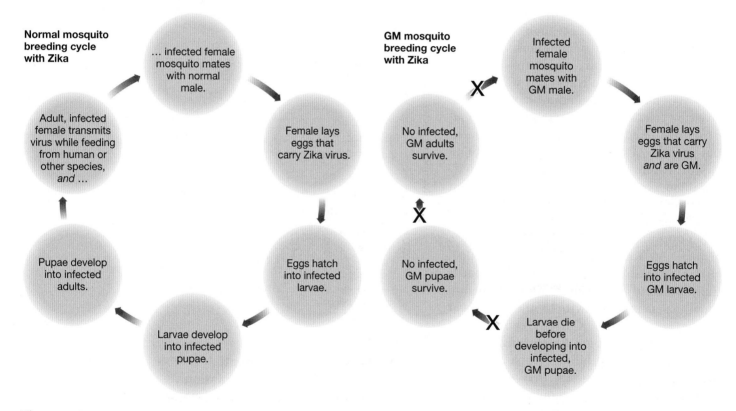

Normal mosquito breeding cycle with Zika

... infected female mosquito mates with normal male.

Female lays eggs that carry Zika virus.

Eggs hatch into infected larvae.

Larvae develop into infected pupae.

Pupae develop into infected adults.

Adult, infected female transmits virus while feeding from human or other species, *and* ...

GM mosquito breeding cycle with Zika

Infected female mosquito mates with GM male.

Female lays eggs that carry Zika virus *and* are GM.

Eggs hatch into infected GM larvae.

Larvae die before developing into infected, GM pupae.

No infected, GM pupae survive.

No infected, GM adults survive.

Figure 19.7

How mosquitoes are genetically modified to stop the Zika virus

Genetically modified (GM) male mosquitoes have been created and released in Brazil and Florida, where they mate with wild females and produce offspring that cannot grow to adulthood.

Q1: List all the stages of the mosquito life cycle.

Q2: Which life cycle stage is vulnerable to the GM treatment?

Q3: Why are only male GM mosquitoes released into the wild, rather than both males and females?

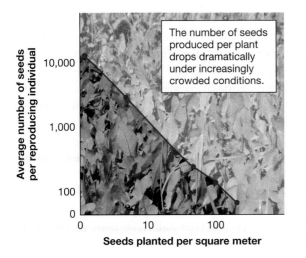

Figure 19.8

Overcrowded conditions result in density-dependent population change

Overcrowding affects many species. The plantain, shown here, is a small, herbaceous plant that has decreased reproduction in crowded conditions.

Q1: What factors may be limiting growth and reproduction in the plantain's crowded conditions?

Q2: Why are overcrowded conditions considered density-dependent population changes?

Q3: Relate this example of overcrowded conditions to the human population growth shown in Figure 19.6. How do you think the situations are similar? How are they different?

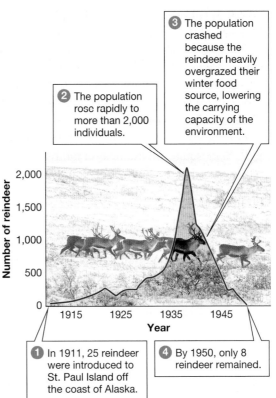

3 The population crashed because the reindeer heavily overgrazed their winter food source, lowering the carrying capacity of the environment.

2 The population rose rapidly to more than 2,000 individuals.

1 In 1911, 25 reindeer were introduced to St. Paul Island off the coast of Alaska.

4 By 1950, only 8 reindeer remained.

Figure 19.9

Habitat destruction results in density-dependent population change

Q1: In what year did the reindeer's numbers begin to rise exponentially?

Q2: In what years was the reindeer's population growth logistic?

Q3: How do you predict the graph (population size) would change if someone had begun bringing in supplemental food for the reindeer in 1940? Draw a sketch of your prediction.

often), and predators may pose a greater risk (because most predators prefer to hunt abundant sources of food). Disease and predators obviously increase the death rate. These changes are also density-dependent. A 2014 study found that the mosquito *Culex pipiens*, which spreads the West Nile virus, undergoes strong density-dependent population changes arising from competition for resources within the watery habitats where it lays its eggs. In other words, larvae compete with each other for food. But once they become adults, the population is no longer density-dependent, because the mosquitoes become airborne and have wider access to food.

For some organisms, if a population exceeds the carrying capacity of its environment by depleting its resources, it may damage that environment so badly that the carrying capacity is lowered for a long time. A drop in the carrying capacity means that the habitat cannot support as many individuals as it once could. Such habitat deterioration may result in widespread starvation and death, causing the population to decrease rapidly (**Figure 19.9**).

Not all population changes are due to density. **Density-independent population change** occurs when populations are held in check by factors unrelated to population density. Density-independent factors can prevent populations from reaching high densities in the first place. Year-to-year variation in weather, for example, may cause conditions to be suitable for rapid population growth. On the other hand, poor weather conditions may reduce the growth of a population directly (by freezing the eggs of a mosquito, for example) or indirectly (such as by decreasing the number of food plants available to an animal).

Other natural disturbances, such as fires and floods, can also limit the growth of populations in a density-independent way. Finally, the effects of environmental pollutants, such as the insecticide DDT, are density-independent; such pollutants can threaten natural populations with extinction (**Figure 19.10**).

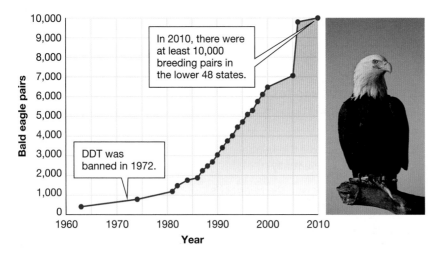

Figure 19.10

Banning the use of the pesticide DDT removed a density-independent population limit

DDT poisoning was directly responsible for declining eagle populations by the middle of the twentieth century. By the early 1960s, population counts indicated that only 417 breeding pairs of bald eagles remained in the lower 48 states—a huge drop from the estimated 100,000 breeding pairs present in 1800. Bald eagle populations increased dramatically after DDT was banned.

Q1: In what year did the bald eagle population rise to more than 2,000 breeding pairs?

Q2: Give some examples of possible density-dependent limits on bald eagle populations.

Q3: Is the population growth of bald eagles more like logistic or exponential growth? Explain why you think so.

Populations of many species rise and fall unpredictably over time. These **irregular fluctuations** in population size are far more common in nature than a smooth rise to a stable population size. In the 1950s, for example, Brazil mounted a massive antimosquito campaign to combat yellow fever—a disease also transmitted by *Aedes aegypti*—that included spraying insecticides and encouraging citizens to get rid of standing water. The success of the program led officials to declare in 1958 that *Aedes aegypti* had been eradicated. If only that had been true. By the 1970s the mosquitoes had come roaring back, breeding and spreading like crazy—an unwelcome irregular fluctuation. And now the pests are resistant to many chemical attacks.

Populations can also exhibit **cyclical fluctuations**, predictable patterns that occur with seasonal changes in temperature and precipitation or when at least one of two species is strongly influenced by the other. The Canadian lynx, for example, depends on the snowshoe hare for food, so lynx populations increase when hare populations rise, and they decrease when hare populations drop. In this example, the population cycles are also density-dependent population changes because each population is affected by the other's numbers. They cycle together in response to each other's density (**Figure 19.11**).

Friendly Fight

The Oxitec GM mosquitoes—trademarked "Friendly" *Aedes aegypti*—were first released in Brazil in February 2011, under the direction of biochemist Margareth Capurro of the University of São Paulo. In a densely populated suburb in northeastern Brazil, the released mosquitoes reduced the wild population of *Aedes aegypti* by 85 percent, says Capurro. In 2013, a similar test in the village of Mandacaru resulted in a 96 percent reduction of the wild mosquito population in the area after only 6 months.

"For the first time, we demonstrated that transgenic mosquitoes can work," says Capurro, who is now developing alternate GM mosquito lines in her own laboratory, attempting to make the GM insects even more potent against their wild counterparts. For example, Capurro's lab is trying out different genetic mutations to create

Figure 19.11

Populations of two species may increase and decrease together
The Canadian lynx depends on the snowshoe hare for food, so the number of lynx is strongly influenced by the number of hare.

Q1: During which years did the hare likely have the greatest food supply?

Q2: Besides the number of hare, what other factors might contribute to the number of lynx?

Q3: Can you draw an average carrying-capacity line on these graphs? Why or why not?

The number of lynx rises and falls with the number of hare.

a line of mosquitoes that does not require the use of an antibiotic to be kept alive in the laboratory, which would be less expensive. "We are finishing the lines now. They are very promising," says Capurro. Other genetic modification techniques are being attempted in labs elsewhere to make *Anopheles* incapable of carrying the malaria parasite, for example.

More recently, Oxitec released Friendly *Aedes aegypti* in a neighborhood of 5,000 residents in the city of Piracicaba, Brazil, to combat the spread of dengue virus. The company released 3–4 million mosquitoes per month over a period of 10 months. The experiment reduced the number of wild *A. aegypti* larvae in the area by 82 percent, and the neighborhood experienced a

MARGARETH CAPURRO

Margareth Capurro is a biochemist at the University of São Paulo in Brazil. Starting in 2011, she led two field releases of Oxitec's genetically modified mosquitoes in Brazil, and today she continues to explore the best way to use GM mosquitoes to control mosquito-borne viruses.

91 percent reduction in cases of dengue, according to local officials.

In August 2016, shortly after the first cases of Zika infection from local mosquitoes were reported, the U.S. Food and Drug Administration (FDA) approved a field test of the Oxitec mosquitoes in Key Haven, Florida. Critics argue that the approach is too expensive, because mosquitoes have to be bred and released multiple times. There are also concerns that the engineered gene won't correctly insert in the genome or will become somehow deactivated in the wild. Release of such mosquitoes would simply exacerbate the problem, adding reproducing mosquitoes into the environment. After examining such claims, the FDA concluded that the proposed field trial in Florida "would not result in significant impacts on the environment." In a 3–2 vote in November 2016, the Florida Keys Mosquito Control District approved a trial of genetically modified mosquitoes, yet local opposition continues, and the district continues to seek a location for the trial.

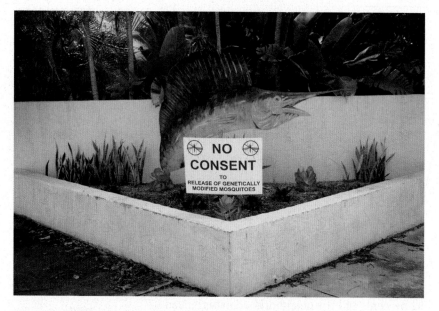

Figure 19.12

Some Florida residents oppose the release of GM mosquitoes in their community

Misinformation about genetically modified organisms, including mosquitoes, has caused concerns about their use.

Some reject the plan solely on the grounds that it involves genetic modification (**Figure 19.12**). "The public fears genetic engineering. Nearly all politicians don't understand it," said bioethicist Arthur Caplan at the New York University School of Medicine, as quoted in a 2016 article in the *Atlantic*. "I don't think the issue is economic. It is ignorance, distrust, fear of the unknown." Recent polls suggest that the tide may be changing: three independent polls found that between 53 and 78 percent of Americans support the release of GM mosquitoes in the United States. "I think it will work well," says DeGennaro. "It's time we use modern methods here in Florida like they are using in Brazil to reduce mosquito populations."

Just the Beginning?

As of October 2016, Zika infections had been reported in every U.S. state, with the exception of Alaska. These infections are due primarily to travel outside of the country. In Florida, however, local infections continue; as of this writing, 214 people have locally acquired the virus since 2015.

The CDC continues to warn pregnant women away from traveling to any area with Zika, which now includes dozens of countries; three U.S. territories; Brownsville, Texas; and, of course, parts of Florida. As of the date this book went to publication, there were no approved vaccines or therapies for Zika virus infection—but not for lack of trying. In June 2016, California-based Inovio Pharmaceuticals dosed the first humans with a prospective Zika vaccine. Even if the trials go well, however, it could take years for a vaccine to reach the market.

It remains too soon to tell whether GM mosquitoes will stop the spread of Zika. Success will depend on population ecology: on mosquito populations, the human population, and how much virus is circulating among both. But history suggests that in the fight against mosquitoes, it's time to try something new.

World's Deadliest Animals

You may have heard that humans are the deadliest animals on the planet. It's true that we, as a species, kill hundreds of thousands of humans. But there's one family of animals that has us beat: mosquitoes. Many species of these small, pesky insects transmit harmful infections, including Zika fever, malaria, West Nile disease, dengue fever, and many more.

Assessment available in smartwork**5**

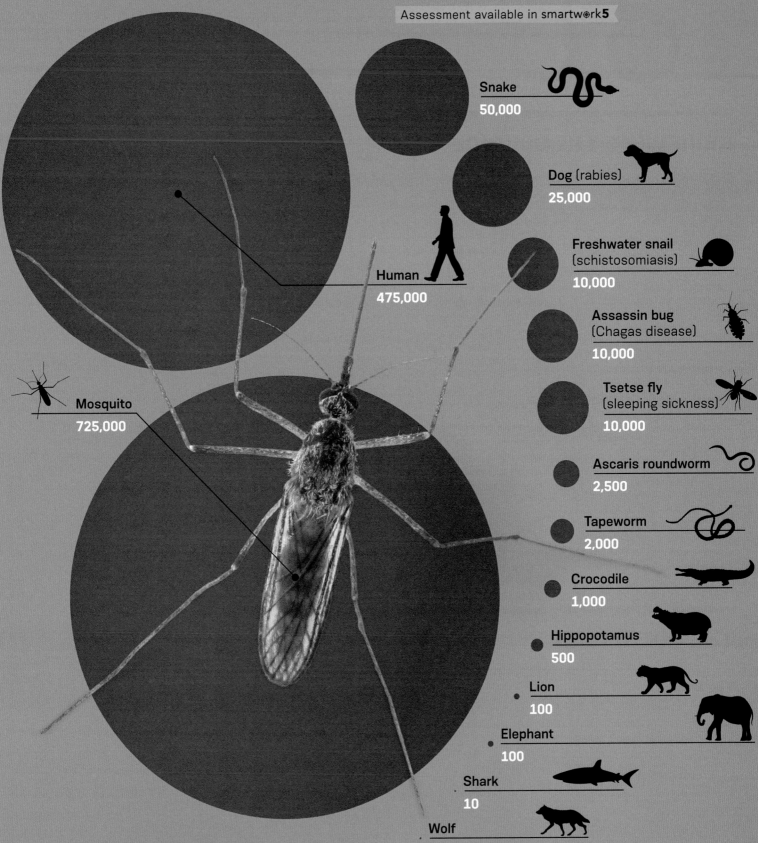

Snake
50,000

Dog (rabies)
25,000

Human
475,000

Freshwater snail
(schistosomiasis)
10,000

Assassin bug
(Chagas disease)
10,000

Tsetse fly
(sleeping sickness)
10,000

Ascaris roundworm
2,500

Tapeworm
2,000

Crocodile
1,000

Hippopotamus
500

Lion
100

Elephant
100

Shark
10

Wolf
10

Mosquito
725,000

[These 15 deadliest animals are ranked in order of the average number of deaths they are responsible for in a year, both directly and through the diseases they transmit.]

REVIEWING THE SCIENCE

- A **population** is a group of individuals of a single species located within a particular area.

- **Population size** is the total number of individuals in a population. **Population density** is the population size divided by the area covered by that population.

- Environmental factors such as lack of space, food shortages, predators, disease, and habitat deterioration limit populations. These factors affect the **carrying capacity**, the number of individuals that can live in an environment indefinitely.

- In **logistic growth**, a population grows exponentially at first but then stabilizes after reaching the carrying capacity. It is associated with an **S-shaped growth curve** when graphed.

- In **exponential growth**, a population increases by a constant proportion from one generation to the next generation. It is associated with a **J-shaped growth curve** when graphed.

- **Density-dependent population change** occurs when birth and death rates are affected by population density, which is the case when many individuals occupy the same space and therefore compete for resources.

- **Density-independent population change** occurs when population size is affected by factors that have nothing to do with the density of the population.

- Populations may rise and fall over time, exhibiting **irregular fluctuations** or **cyclical fluctuations**.

THE QUESTIONS

The Basics

1 A group of interacting individuals of a single species located within a particular area is called

(a) a biosphere.

(b) an ecosystem.

(c) a community.

(d) a population.

2 A population that is growing exponentially increases

(a) by the same number of individuals in each generation.

(b) by a constant proportion in each generation.

(c) in some years and decreases in other years.

(d) none of the above

3 In a population with an S-shaped (logistic) growth curve, after an initial period of rapid increase the number of individuals

(a) continues to increase exponentially.

(b) drops rapidly.

(c) remains near the carrying capacity.

(d) cycles regularly.

4 The growth of populations can be limited by

(a) natural disturbances.

(b) weather.

(c) food shortages.

(d) all of the above

5 The maximum number of individuals in a population that can be supported indefinitely by the population's environment is called the

(a) carrying capacity.

(b) J-shaped curve.

(c) sustainable size.

(d) exponential size.

6 Select the correct terms:

(**Density-independent / Density-dependent**) factors limit the growth of populations more strongly at high densities than at low densities. One example of a (**density-independent / density-dependent**) factor is a natural disaster.

Challenge Yourself

7 A population of plants has a density of 12 plants per square meter and covers an area of 100 square meters. What is the population size?

(a) 120

(b) 1,200

(c) 12

(d) 0.12

(e) none of the above

8 Which of the following would *not* cause the carrying capacity of a mosquito population to change?

(a) lower availability of standing water in which to lay eggs

(b) fewer animals for female mosquitoes to feed on

(c) warmer weather that increases overwintering survivorship

(d) All of the above would change carrying capacity.

(e) None of the above would change carrying capacity.

9 Place the following elements in the strategy to control mosquito populations in the correct order from earliest to latest by numbering them from 1 to 5.

_____ a. Scientists genetically modify male mosquitoes in the lab.

_____ b. Mosquito offspring produced by the matings do not survive to adulthood.

_____ c. GM male mosquitoes mate with normal wild females.

_____ d. The mosquito population decreases dramatically.

_____ e. Scientists release GM male mosquitoes into the wild.

Try Something New

10 Draw a diagram that illustrates the possible fluctuating population cycles of a predator, such as the mountain lion, and its prey species, such as deer. The prey regularly experiences population crashes when it overgrazes its food source. Label your axes.

11 An artificial pond at a college campus was populated with several mating pairs of pond slider turtles. After many years of slow population growth, campus administrators installed a coin-operated food dispenser at the pond's edge. Students and community members bought and fed the turtles an enormous amount of food each day, causing the population of turtles in the pond to increase exponentially. How did the installation of the food dispenser affect the carrying capacity of the turtle population at this pond?

12 Suppose that population ecologists at the college in question 11 determined that the number of turtles in the pond had increased from 6 individuals to 24 individuals in just the first year after installation of the coin-operated food dispenser. What is the population doubling time for the turtles in this population?

(a) 6 years

(b) 6 months

(c) 1 year

(d) 1 month

Leveling Up

13 **What do _you_ think?** Some Florida residents do not want to allow genetically modified (GM) mosquitoes to be released, even though they have been demonstrated to significantly reduce mosquito-borne illnesses in Brazil.

a. Should Floridians be allowed to decide not to have GM mosquitoes released in their neighborhoods? Should this question be decided by vote? Explain your reasoning.

b. Currently, the Florida Department of Health may require people to remove standing water from their yards in order to decrease breeding habitat for mosquitoes. Should people be allowed not to comply with this order, given that the yard is their private property? Explain your reasoning.

14 **Life choices** Do you feel that people should choose to limit the number of children they have, given the increasing human population? Alternatively, should people focus on decreasing their own ecological footprint—given that, for example, one American uses the same amount of resources as do four Chinese citizens? Explain your reasoning.

15 **Is it science?** Your best friend found a Reddit thread stating that GM mosquitoes do not suppress the population size of Zika-bearing mosquitoes, and instead claimed that Zika was _caused_ by the release of GM mosquitoes. Putting aside the idea that Zika was caused by GM mosquitoes, which is refuted by multiple studies as discussed on many legitimate websites, you decide to simply show your friend some data comparing mosquito population sizes in areas with and without GM mosquitoes. In your own words, write a paragraph that describes the results shown in the figure here and explains to your friend how they relate to the scientific claims he read.

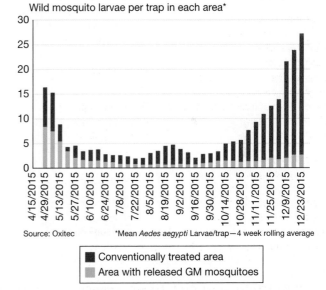

Mosquito larvae population numbers in areas with and without released GM mosquitoes

Wild mosquito larvae per trap in each area*

Source: Oxitec   *Mean _Aedes aegypti_ Larvae/trap—4 week rolling average

◼ Conventionally treated area
◼ Area with released GM mosquitoes

For more, visit digital.wwnorton.com/bionow2 for access to:

Of Wolves and Trees

The extermination of wolves in Yellowstone National Park had unforeseen effects on the park's ecosystem. Can the return of this top predator restore order?

After reading this chapter you should be able to:

- Explain the concept of an ecological community and how it relates to ecological diversity.
- Articulate the role of relative species abundance and species richness in defining ecological diversity.
- Distinguish between a food chain and a food web.
- Illustrate how the removal of a keystone species disrupts an ecological community.
- Describe the four major kinds of species interactions, and give an example of each.
- Compare and contrast primary and secondary succession.

CHAPTER

20

COMMUNITIES
OF ORGANISMS

Robert Beschta will always remember the day he visited Yellowstone National Park's Lamar Valley in 1996. Beschta, a hydrologist, was there to observe the Lamar River, which winds through the valley's lush lowland of grass and sage (**Figure 20.1**). But on that day, as he walked toward the waterway, he noticed something odd: all around the valley, there were not many trees. The few tall, white aspens he saw looked haggard, their bark eaten away. And there were no young saplings to be seen.

Beschta had studied forestry and rivers for decades and knew what a healthy valley was supposed to look like. This was not it. Beschta approached the river and observed that its banks were also devoid of trees. The leafy green cottonwoods and wide willows that had once arched gracefully from the riverbank over the water were absent. And with no tree roots to hold the soil in place, the riverbanks themselves were jagged and eroding. "I was dumbstruck," recalls Beschta. Something unprecedented was happening in Yellowstone.

Beschta returned to Oregon State University (OSU), where he worked, and described his observations in a seminar. He showed pictures of aspens with their bark stripped away and empty riverbanks where saplings should have been growing. In the audience, William Ripple sat up a little straighter. Also a scientist at OSU, Ripple studied forest ecology. He was particularly interested in aspen trees, which grow as tall as 70 feet and live up to 150 years. From what Beschta was saying, aspens were no longer growing in Yellowstone. Ripple wanted to know why.

"It was a scientific mystery as to what was the cause of the decline," he recalls. In that moment, listening to Beschta, Ripple knew exactly what his next research project would be: to document the extent of the aspen decline and determine why it was occurring.

Within a year, Ripple and one of his graduate students, Eric Larsen, traveled to the Lamar Valley. There, they drilled small holes into the centers of aspens and removed from each a plug of wood about the diameter of a pencil. Then they counted the growth rings in each plug—one for every year the tree has been in existence—to determine the age of the tree. They found that most of the aspens had begun to grow *prior* to 1920. After 1920, almost no new trees had begun growing.

"We started scratching our heads at that point," says Ripple. He, Beschta, and Larsen began to brainstorm reasons why the trees might have stopped regenerating. They looked for environmental changes in the 1920s that could have done it: a fire that killed off saplings, or a change in climate that reduced the trees' ability to reproduce. But nothing lined up—until an ordinary moment gave Ripple an extraordinary idea.

Standing in a gift shop in Grand Teton National Park, just south of Yellowstone, Ripple looked up at a poster on the wall. It featured a grove of tall, white aspen trees in the winter. In the middle of the trees, its paws covered in snow, stood a large gray wolf. "That was an 'aha' moment," says Ripple. "I thought, 'Maybe the wolf protects the aspen.'"

ROBERT BESCHTA

Robert Beschta is a professor emeritus at Oregon State University. He studies water processes in forest and rangeland ecosystems.

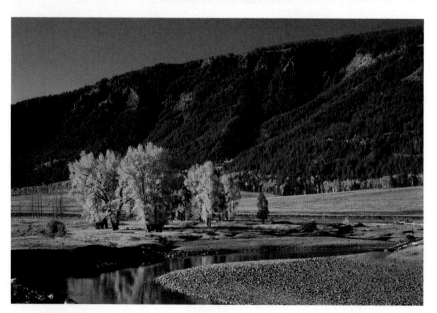

Figure 20.1

The Lamar River flows through Yellowstone National Park

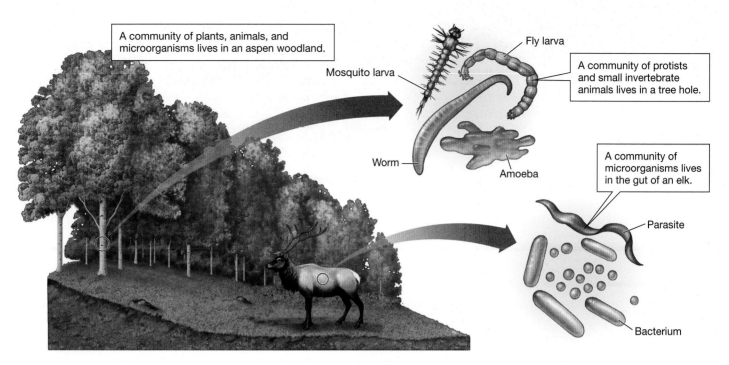

Figure 20.2

Ecological communities come in all sizes

Smaller communities can be nested within a larger community. This aspen woodland community contains the smaller communities of a temporary pool of water in a tree hole and an elk's gut, among others.

Q1: List another species that is part of this community.

Q2: Of which community could this aspen woodland be a smaller part?

Q3: Which other small communities could be found within this larger community?

The idea of wolves protecting aspens initially seems nonsensical. How would meat-eating predators protect trees? They wouldn't—at least not directly. Ripple surmised that wolves in Yellowstone might have had an indirect effect on the community. As an ecologist, Ripple had long studied **ecological communities**, associations of species that live in the same area. Communities vary in size and complexity, from a small group of microorganisms in a temporary pool of water to the whole of Yellowstone Park, home to an estimated 322 species of birds, 67 species of mammals, 1,349 species of plants, and an uncounted number of insects (**Figure 20.2**).

Ripple knew that an ecological community is characterized by the diversity of species that live there, and that diversity is governed by two things: the **relative species abundance** (how common one species is when compared to another) and the **species richness** (the total number of different species that live in the community; **Figure 20.3**). Ripple also knew that communities are subject to constant change, and that something must have changed in Yellowstone.

Ecological communities change naturally as a result of interactions between and among species, and as a result of interactions between species and their physical environment. Ripple knew that both the relative abundance and the

WILLIAM RIPPLE

William Ripple is director of the Trophic Cascades Program at Oregon State University. There, he leads a research project investigating how gray wolves affect other species in the Yellowstone ecosystem.

This community has higher relative species abundance for white-barked trees than the community shown below.

This community has higher species richness than the community shown above.

Figure 20.3

Two measures of species diversity

The diversity of an ecological community is determined by relative species abundance and also species richness. High relative species abundance means that one species dominates the community (so the community is less diverse), and high species richness means that many species are present in the community (so it is more diverse).

Q1: If relative species abundance increased in the first community, how would the figure look different from now?

Q2: If species richness decreased in the second community, how would the figure look different from now?

Q3: How do relative species abundance and species richness define the species diversity of a forest community?

richness of species had changed in Yellowstone during the previous century. The relative species abundance had changed because aspen trees and willows were in decline, elk and coyote numbers had increased, and bison and beaver populations had decreased. Many of the common inhabitants of Yellowstone had changed in abundance; therefore, the relative species abundance of this community had changed.

The species richness had also changed because early in the twentieth century, a *keystone species*, the gray wolf, had gone missing. The total number of species in the community had gone down by only one, but that seemingly small change had a big impact. A **keystone species** is a species that has a disproportionately large effect on a community, relative to the species' abundance. There are few wolves compared to, say, rabbits, yet the wolves have a stronger effect on the community. Keystone species in other communities include prairie dogs in the American western plains, hummingbird pollinators in the Sonoran Desert, and sea stars in intertidal waters (**Figure 20.4**).

Keystone species are often recognized only when they go missing and their disappearance results in dramatic changes to the rest of the community. And that is exactly what happened in Yellowstone in the 1920s: the mighty gray wolf disappeared. Or, to be accurate, the mighty gray wolf was exterminated.

A *Pisaster* sea star feeding on a mussel. Without this keystone species, the mussels crowd out other species in their community.

Loss of a keystone species reduces diversity

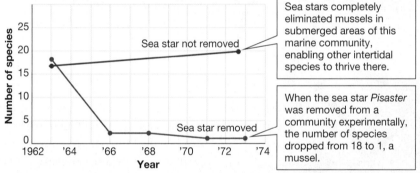

Sea stars completely eliminated mussels in submerged areas of this marine community, enabling other intertidal species to thrive there.

Sea star not removed

Sea star removed

When the sea star *Pisaster* was removed from a community experimentally, the number of species dropped from 18 to 1, a mussel.

Figure 20.4

The star of the community

The sea star *Pisaster ochraceus* is an example of a keystone species. In a classic experiment conducted along the Pacific coast of Washington State in 1963, sea stars were removed from one site while an adjacent site was left undisturbed.

Q1: How many species were left in 1966 in the community where sea stars were *not* removed?

Q2: How many species were left in 1966 in the community where sea stars *were* removed?

Q3: How do your answers to questions 1 and 2 demonstrate the importance of a keystone species for the maintenance of diversity in a community?

A Key Loss

In the early 1900s, ranchers and homesteaders killed wolves all across the United States and eliminated them in many eastern states. Then, in 1915, the U.S. government began subsidizing wolf extermination programs all over the country, and a systematic slaughter began. Under the national program, states paid bounties of up to $150 for individual wolf pelts. Wolves, feared and hated by private landowners for killing livestock, were trapped, shot, and skinned. The extermination happened quickly: the last known wolf den in Yellowstone was destroyed in 1926. At least 136 wolves, maybe more, were killed during the eradication campaign in Yellowstone. Once it was done, the park was wolf-free.

Seven decades later, staring at the poster of a wolf standing among the aspens, Ripple wondered whether the loss of that keystone species had contributed to the decline of aspen trees. He immediately looked up the historical records to see whether the timing matched. Lo and behold, the last wolf had been killed at

about the same time the aspens stopped regenerating, in the mid-1920s. Suddenly, it seemed obvious why the aspens had declined: wolves kill elk and elk eat aspens—three species in the same food chain.

A **food chain** is a simple list of who eats whom. In scientific terms, it is the direct path by which nutrients are transferred through the community. A **food web**, on the other hand, is a more complex diagram of all the food chains in a single ecosystem and how they interact and overlap (**Figure 20.5**).

In the wolf-elk-aspen food chain, aspen are the **producers**, the organisms at the bottom of the food chain that use energy from the sun to produce their own food through photosynthesis. In Yellowstone and on land all over Earth, photosynthetic plants like trees, grasses, and shrubs are the major producers (**Figure 20.6**). In aquatic biomes, photosynthetic plankton are the major producers, as we'll see in Chapter 21.

Further up the food chain are **consumers**, organisms that obtain energy by eating all or parts of other organisms or their remains. Elk and wolves are both consumers: elk eat aspens, and wolves eat elk. In the Yellowstone food chain, elk are **primary consumers**: they eat producers. Wolves are **secondary consumers** because they eat primary consumers. This sequence of organisms eating organisms can continue: a bird that eats a spider that ate a beetle that ate a plant is a **tertiary consumer**; a killer whale that eats a leopard seal that ate a sea bass that ate a krill that ate a phytoplankton is a **quaternary consumer**. We will explore the flow of energy up the food chain in Chapter 21.

The more they discussed it, the more Ripple, Larsen, and Beschta believed that the loss of wolves in Yellowstone had allowed the elk population to flourish and eat so many young trees that the aspen population could not regenerate. "We developed a hypothesis that maybe the killing of wolves actually affected the reproduction of aspen trees," says Ripple. Any change in species diversity will have a ripple effect (no pun intended) throughout the community, and the wolves of Yellowstone were no exception. Ripple and Larsen published their hypothesis in 2000, suggesting that the loss of wolves had led to increased elk

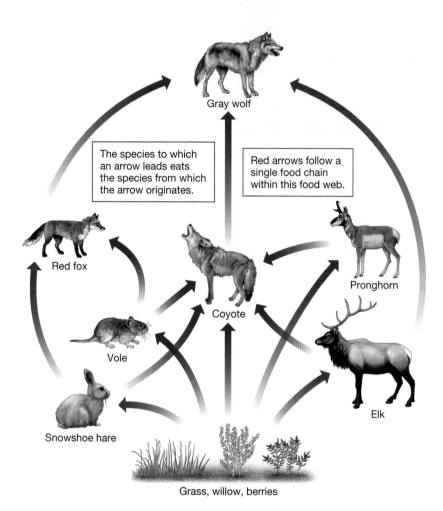

Figure 20.5

Food webs show how energy moves through a community

Food webs are composed of many food chains that show one species eating another. ▶️

Q1: What species eats the coyote?

Q2: What species does the coyote eat?

Q3: What do you think would happen to a community that lost its coyotes?

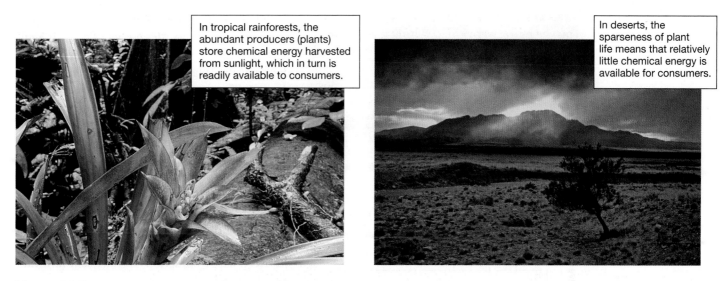

In tropical rainforests, the abundant producers (plants) store chemical energy harvested from sunlight, which in turn is readily available to consumers.

In deserts, the sparseness of plant life means that relatively little chemical energy is available for consumers.

Figure 20.6

Producers are the energy base of a food chain

All communities and ecosystems have producers, but they vary in abundance, depending on the circumstances.

Q1: Where do producers acquire the energy they need to perform their function in the food chain?

Q2: Why are producers necessary for life on Earth?

Q3: Given the lower abundance of producers in desert regions, compared to tropical rainforests, what would you predict about the abundance of consumers in the two environments?

populations and altered elk movements and feeding patterns. In other words, with wolves gone, elk were free to find and eat young aspen shoots whenever they wanted, with no fear of wolves.

In 2001, Beschta returned to the Lamar Valley and collected data on another species of tree, the cottonwood, which can live more than 200 years. He documented the same trend as the aspens showed: in the 1920s, cottonwoods suddenly stopped generating new, young trees. In fact, since the 1970s, not a single new cottonwood had been established. "This was dramatic," recalls Beschta. "It's a big deal when you can't have a single cottonwood in this large valley make it to a mature tree."

It is possible for a consumer to eat a species to extinction, so if elk populations had continued to grow unchecked, they could have reduced the aspen and cottonwood populations to zero, disrupting the ecological community permanently. But before that could happen, something dramatic occurred in Yellowstone: humans brought the wolves back.

A Second Ripple Effect

In 1973, wolves became the first animals to be protected under the Endangered Species Act. It was the dawn of the modern conservation movement, and the idea of returning wolves to Yellowstone grew in popularity. It took some time, but eventually, lawmakers agreed to the plan, and between 1995 and 1997, 41 wild wolves were captured in Canada and released in Yellowstone. It didn't take long for the wolf populations to recover: by 2007, an estimated 170 wolves lived in and around Yellowstone; today that estimate is 528. As of January 2016, at least 98 wolves in 10 packs were living within the park limits (**Figure 20.7**).

The loss, and then return, of the wolves had significant impacts on **species interactions** in the park. There are four central ways in which species in a community interact: *mutualism, commensalism, predation,* and *competition.*

Figure 20.7

Wolves today in Yellowstone

Q1: When were wolves first reintroduced to Yellowstone?

Q2: What was the highest number of wolves observed in Yellowstone? In what year did that occur?

Q3: Why did it take a few years after the reintroduction of wolves for the aspen population to increase, as well as the beavers and bears?

The classification is based on whether the interaction is beneficial, harmful, or neutral to each species involved. These interactions affect where organisms live and how large their populations grow. Species interactions also drive natural selection and evolution, thereby changing the composition of communities over short and long periods of time.

With wolves back in the park, Ripple suddenly had a way to test his hypothesis about aspen decline. If the loss of wolves was responsible for aspen loss, then the return of the wolves should incite a revival of aspens (and other woody plants,

including cottonwoods). But he needed a way to quantify that change. "It's a scientifically difficult task to connect a wolf to a plant. Obviously wolves don't consume plants, so instead we had to connect the dots. There are data that wolves affect elk, and other data that elk affect plants," says Ripple.

Ripple and Beschta went into the field in 2006 and again in 2010 to take measurements. In addition to recording the ages of the trees, they looked for and documented signs of elk feeding on aspens—such as scars where branches or buds had been bitten off—and measured the heights of young aspens. Beschta did the same with the cottonwoods. They were eager to see which species interactions would occur now that wolves were back.

The first type of species interaction, **mutualism**, occurs when two species interact and both benefit. For example, Yellowstone is home to 4,600 bison, the largest land mammals in North America. Bison have a mutualistic relationship with the black-billed magpie. Pests such as ticks burrow into a bison's short, dense hair to suck the beast's blood, but hungry little magpies perch on top of the bison and eat those ticks. Thus, both the bison and the magpie benefit from close interaction with one another. Mutualism is common and important in ecosystems all over Earth: many species receive benefits from, and provide benefits to, other species, as the clownfish-anemone partnership in **Figure 20.8** illustrates. These benefits increase the survival and reproduction of both interacting species.

When they aren't perched atop bison, black-billed magpies can be found in large nests atop deciduous or evergreen trees, where they reproduce once a year. These trees, another member of the community, share a commensal relationship with the magpies. **Commensalism** happens when one partner benefits while the other is neither helped nor harmed—in this case, the magpie benefits from having a safe place to lay eggs, and the interaction has no effect on the tree. Another example of a commensal relationship is barnacles living on a whale (**Figure 20.9**).

As you might have guessed, not all species interactions are as pleasant as those among bison, birds, and trees. In two other types of interactions, at least one of the two species is harmed: *competition* (which we will return to in a moment) and *predation*.

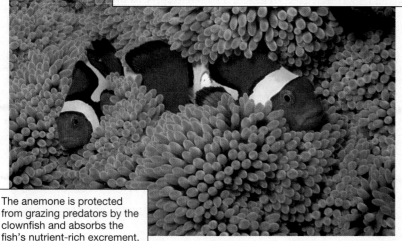

The clownfish gains protection from predators by hiding within the sea anemone's stinging tentacles. The clownfish is not harmed, because a thick mucus covers its body.

The anemone is protected from grazing predators by the clownfish and absorbs the fish's nutrient-rich excrement.

Figure 20.8

Mutualism: friends in need
Both the clownfish and the sea anemone benefit from their relationship.

> **Q1:** What might happen to an anemone without a resident clownfish?
>
> **Q2:** What would happen to a clownfish that did not produce a mucous coating?
>
> **Q3:** In the example of mutualism given in the text, what would happen if ticks were no longer able to feed on bison?

In **predation**, one species benefits and the other is harmed, and **predators** are defined as consumers that eat part or all of other organisms. A **parasite** is a kind of predator that lives in or on the organism it harms, its **host**. An important group of parasites is pathogens, which cause disease in their hosts. The bacteria that cause strep throat, tuberculosis, and pneumonia are pathogens, for example. Many organisms have evolved mechanisms to avoid being hosts, such as immune systems to help fight off parasitic diseases and infections.

Other types of predators are distinguished by what they eat. Elk are **herbivores**, animals that eat plants. Yellowstone elk feed on the shoots, saplings, and new branches of woody plants like aspen, cottonwood, and willow, especially in winter, when other plants are scarce. Wolves are **carnivores**, animals (and, in rare cases, plants) that kill other animals for food. Yellowstone wolves predominantly eat elk, especially in winter, but they also eat deer and any small mammals they can catch, notably beaver. Other animals, such as raccoons and coyotes, eat both animals and plants, so we call them **omnivores**. Both raccoons and coyotes are also **scavengers**, animals that eat dead or dying plants and animals. Of the species that are scavengers, **decomposers**, such as fungi, dissolve their food; and **detritivores**, including worms and millipedes, mechanically break apart and consume their food.

The barnacles on a whale's snout enjoy a continuous stream of nutrient-rich water flowing across them from which they can feed. The whale is neither harmed nor helped by these stowaways.

Figure 20.9

Commensalism: a whale of a ride!
This gray whale's snout is covered in barnacles.

> **Q1:** How do barnacles benefit from living on a whale?
>
> **Q2:** Do you think a whale could avoid being colonized by barnacles? Why or why not?
>
> **Q3:** Explain why detritivores are considered commensal to the organisms they consume.

Cause and Effect

Wolves were reintroduced to Yellowstone National Park in 1995, and their return had a major impact on other species in the park, especially elk and aspens. Elk began avoiding areas where wolves could easily prey upon them, such as near riverbanks, especially riverbanks with downed logs, where escape would be difficult. With decreasing elk presence, aspens flourished in these areas. The effects depicted here are a powerful example of the influence of a keystone species on an ecological community.

Assessment available in smartwork5

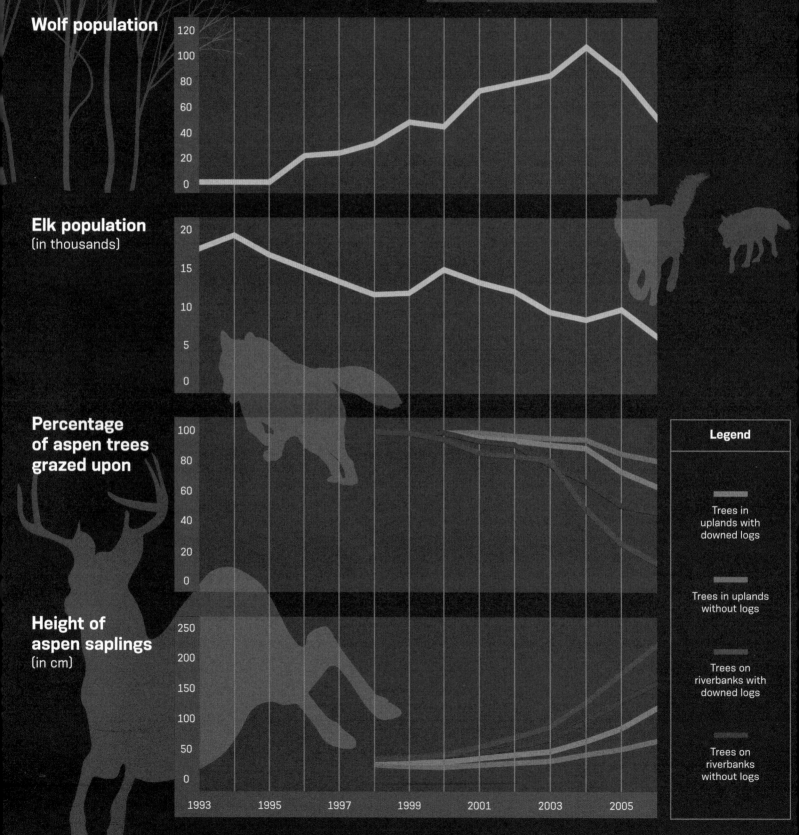

Wolf population

120
100
80
60
40
20
0

Elk population
(in thousands)

20
15
10
5
0

Percentage of aspen trees grazed upon

100
80
60
40
20
0

Height of aspen saplings
(in cm)

250
200
150
100
50
0

1993   1995   1997   1999   2001   2003   2005

Legend

Trees in uplands with downed logs

Trees in uplands without logs

Trees on riverbanks with downed logs

Trees on riverbanks without logs

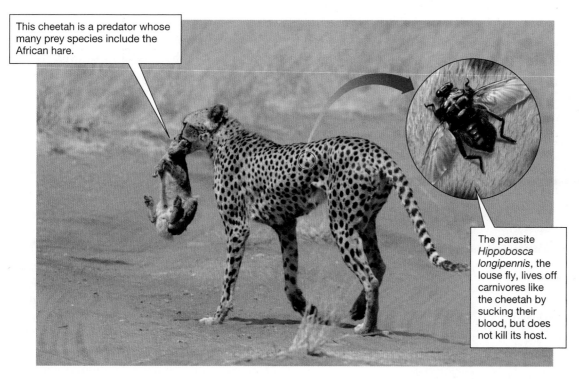

This cheetah is a predator whose many prey species include the African hare.

The parasite *Hippobosca longipennis*, the louse fly, lives off carnivores like the cheetah by sucking their blood, but does not kill its host.

Figure 20.10

Predators can also be prey

While the cheetah is what we traditionally think of as a predator, it can also be prey to parasites, and eventually to decomposers and detritivores.

Q1: What kind of predator is the cheetah in the figure?

Q2: What kind of predator is the louse in the figure?

Q3: What kind of predator are the elk that graze on aspen tree saplings?

The animals eaten by predators are called **prey**. All of the animal residents of Yellowstone (and most plants) are eaten by other species, except for grizzly bears, mountain lions, eagles, and gray wolves. These four animals are all at the top of the food chain. They are top predators, although they are themselves prey to parasites (**Figure 20.10**).

Back in the Park

Now that wolves, a top predator in the community, were back in Yellowstone, how would their prey, the elk, react? And how would that reaction affect the elk's food, the trees?

Using the plant measurements taken in the park and comparing those measurements to historical data, Beschta and Ripple found that between 1998 and 2010, as the wolf population in the park grew, the elk population decreased and therefore fewer aspens were eaten. In 1998, essentially 100 percent of the young aspen plants were being preyed upon, but by 2010, only 18–24 percent were being eaten. In addition, average aspen heights increased for all areas that the scientists observed. Cottonwoods experienced the same revival. In the 1970s, cottonwoods had entirely stopped adding new young saplings, but by 2012, some 4,660 young cottonwoods had grown to over 2 meters high.

Together, the aspen and cottonwood data sets convinced Ripple and Beschta that the reintroduction of wolves was responsible for a cascade of species interactions leading to the restoration of aspen and cottonwood populations. "With

Warning coloration

The bright colors of the poison dart frog warn potential predators of the deadly chemicals contained in its tissues.

Mimicry

The viceroy butterfly (left) mimics the color and pattern of the monarch butterfly (right), which contains toxic compounds.

Camouflage

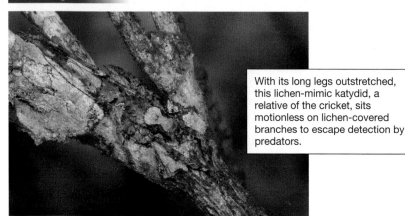

With its long legs outstretched, this lichen-mimic katydid, a relative of the cricket, sits motionless on lichen-covered branches to escape detection by predators.

Figure 20.11

Adaptive coloration responses to predation

Prey species have adapted many elaborate strategies to avoid being eaten by predators, including warning coloration, mimicry, and camouflage.

Q1: How do predators know that brightly colored prey are usually toxic?

Q2: Do you think mimicry works if the toxic species is in low abundance? Why or why not?

Q3: Why is camouflage considered an adaptive response to predation?

wolves back, young, woody plants are doing better and growing taller," says Ripple. "Plant communities are beginning to recover." Different species are growing and spreading at different rates of recovery, he adds, "but there's enough new growth that we suggest it is in support of our basic hypothesis, that the presence or absence of the top predator—the wolf in this case—makes a difference in these plant populations."

Since Ripple and Beschta's discovery, the scientific community has been debating two potential reasons why the plants are flourishing with the return of the wolves. The most straightforward possibility is that the elk population has decreased: wolves kill elk, so there are fewer elk to consume the plants. Yet some of the tree populations seemed to recover faster than the drop in the elk populations would suggest. So a second possibility is that the presence of wolves led to a change in elk behavior called a *fear effect*. Often, the presence of a predator in an area can affect the behavior of its prey. In this case, it is possible that elk stopped grazing in areas where they could easily be seen by wolves, such as along the banks of the Lamar River.

"These two mechanisms, the population density and fear behavior, are difficult to tease apart, and we're working on that," says Ripple. "Many today believe it [the change in plant populations in Yellowstone] is due to a combination of the two." Beschta agrees: "In my opinion, they've both been going on."

Safety in Numbers and Colors

Elk may avoid lingering at streams as a way to evade their predators, but other prey have far more elaborate strategies to avoid being consumed. The poison dart frog, for example, is among the most toxic animals on Earth, and it evolved bright colors as **warning coloration** to alert potential predators to the dangerous chemicals in its tissues (**Figure 20.11**, top). Such warning coloration can be highly effective. Young blue jays, for example, quickly learn not to eat brightly colored monarch butterflies, which contain chemicals that cause nausea and, at high doses, death.

The success of goshawk attacks on wood pigeons decreases greatly when there are many pigeons in a flock.

Although a single musk ox may be vulnerable to predators such as wolves, a group forming a circle makes a difficult target.

Goshawk attack success (%)

Number of pigeons in flock

Figure 20.12

Safety in numbers

Animals that live in groups are better able to warn each other and sometimes fend off attacking predators.

Q1: What percentage of pigeons are caught when they are alone and not in a flock?

Q2: For wood pigeons, what is the minimum number of individuals that provides protection from goshawks?

Q3: Why do you think a group of musk oxen versus a lone musk ox would be safer from a pack of wolves?

Then there are species that, though not poisonous, evolved coloration to make them look as if they were. Through **mimicry**, the viceroy butterfly, which is not poisonous, imitates the color and pattern of the monarch butterfly (**Figure 20.11**, middle). That "borrowed" coloration scares away blue jays and other birds that may have felt sick the last time they ate a monarch. Another mechanism to avoid being eaten is **camouflage**, any type of coloration or appearance that makes an organism hard to find or hard to catch (**Figure 20.11**, bottom). Finally, many prey, from musk oxen to wood pigeons, have evolved a different strategy to avoid becoming dinner: living together. By group living, these animals are able to act together to warn each other when a predator is about to attack and even to repel attacks as a united front (**Figure 20.12**).

In predation, the predator benefits. And each of the other two types of species interactions we've discussed—mutualism and commensalism—benefits at least one of the species involved. But in one final type of species interaction, **competition**, no one benefits. Instead, both interacting species are negatively affected.

Competition most often occurs when two species share an important but limited resource, such as food or space. In Yellowstone, both beavers and elk eat woody plants. For both species, woody plants are part of the **ecological niche**, the set of conditions and resources that a population needs in order to survive and reproduce in its habitat. Because the niches of beavers and elk overlap, these species compete. When two (or three or more) species compete, each has a negative effect on the other because one is using resources that the other then cannot access. (If resources are abundant, however, there may be no competition between species, even if their niches overlap.)

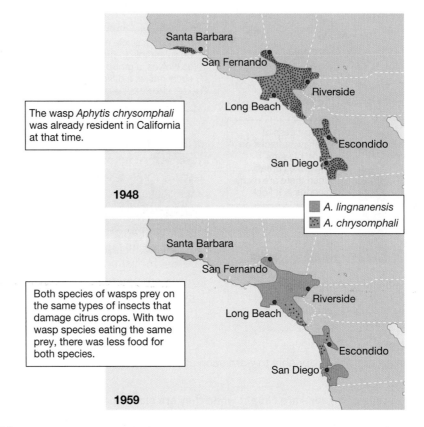

The wasp *Aphytis chrysomphali* was already resident in California at that time.

1948

A. lingnanensis
A. chrysomphali

Both species of wasps prey on the same types of insects that damage citrus crops. With two wasp species eating the same prey, there was less food for both species.

1959

Figure 20.13

Exploitative competition: a new species moves in

Two different species of wasps—*Aphytis lingnanensis* and *Aphytis chrysomphali*—feed on the same resource but do not directly compete for access to it.

Q1: Which species appears to be the superior competitor?

Q2: Why is this example considered exploitative competition?

Q3: What would you predict if these species had undergone competitive exclusion? (You will need to read the next paragraph to answer this question.)

There are two main categories of competition: *exploitative* and *interference*. In **exploitative competition**, species compete indirectly for shared resources, such as food. In this case, each species reduces the amount of the resource that is available for the other species, but they do not directly interact or come in contact with each other (**Figure 20.13**). When wolves returned and elk populations declined, beavers had less competition for food, especially willow trees, and the number of beaver colonies in Yellowstone rose from 1 in 1996 to 12 in 2009. By 2015, Yellowstone had an estimated 100 beaver colonies.

Elk and bears also interact through exploitative competition. Elk eat the leaves and branches on shrubs, resulting in a decrease in the number of berries the shrubs produce. That's not good for grizzly bears, which love to eat berries. Knowing of this relationship, Ripple hypothesized that a decrease in elk would result in an increase in berry-producing shrubs, and that bears would be eating more berries. To test this hypothesis, Ripple, Beschta, and others spent 2 years analyzing grizzly scat that had been collected in the park. They compared the percentage of fruit in current scat to that of scat data that had been collected and saved before 1995, prior to the wolf return. Over a 19-year period, they found that the percentage of berries in the grizzly diet went up as elk populations went down. This is an example of the **competitive exclusion principle**, which predicts that different species that use the same resource can coexist only if one of the species adapts to using other resources. In this case, grizzlies adapted to the presence of elk by finding food sources other than berries, and in the absence of elk they increased their consumption of berries.

Organisms also compete through **interference competition**, in which one organism directly excludes another from the use of a resource. In Yellowstone, for example, bears and wolves often fight over the carcass of an animal (**Figure 20.14**). In interference competition, one individual physically interferes with another individual that is trying to obtain a resource.

A Community Restored

Today, the return of the wolf is having a clear and significant impact on the ecological community of Yellowstone, from the rebirth of aspens and cottonwoods to growing beaver populations. These are signs of **succession**, the process by which the species in a community change over time. "We're on a very important upward trend," says Beschta.

All ecological communities change over time, sometimes because of human intervention, as in Yellowstone, but also because of natural changes in species composition—the number of individuals in a population often changes as the seasons change, for example—and natural disturbances such as fires, floods, and windstorms. In addition, communities

Figure 20.14

Interference competition can be dramatic

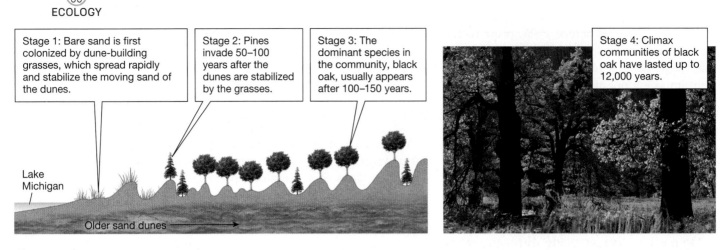

Stage 1: Bare sand is first colonized by dune-building grasses, which spread rapidly and stabilize the moving sand of the dunes.

Stage 2: Pines invade 50–100 years after the dunes are stabilized by the grasses.

Stage 3: The dominant species in the community, black oak, usually appears after 100–150 years.

Stage 4: Climax communities of black oak have lasted up to 12,000 years.

Lake Michigan

Older sand dunes

Figure 20.15

Primary succession: from nothing to climax community
Sand becomes woodland near Lake Michigan.

Q1: What species represents the first colonizers of the sand dunes?

Q2: What species is the intermediate species, and how does it become the dominant species?

Q3: What species is the mature, climax community species, and how does it become the dominant species?

can broadly change by the slow loss or gain of populations of species over long periods of time through natural selection.

In nature, ecologists have observed two major types of succession: primary and secondary. **Primary succession** occurs in newly created habitats, such as an island that just emerged from the sea, the soil left behind a retreating glacier, or new sand dunes (**Figure 20.15**). A new habitat begins with no species. Often the first species to colonize an area alters the habitat in ways that enable later-arriving species to thrive. A specific type of flowering plant may grow on a new island, for example, and then a species of bee that feeds on that flower subsequently joins the habitat.

Secondary succession occurs after a disturbance within a community—such as the loss of a keystone species like the wolves in Yellowstone or the plant loss caused by a forest fire—reduces the number of species in that community. During secondary succession, communities usually regain the successional state that existed before the disturbance. This type of succession does not take as long as primary succession, because some species still exist in the community.

Luckily, communities can and do bounce back from disturbances, but the time required to regain a previous state varies from years to decades to centuries. Yellowstone is currently experiencing a secondary succession as wolf, aspen, and beaver populations slowly return. It will likely still be a while before the park returns to its previous state, says Ripple. "We were 70 years without wolves, and now we're less than 20 years since wolves have returned, so this is going to take time." But he and others are hopeful that Yellowstone will return to its status as a **climax community**, a mature community whose species composition remains stable over long periods of time (**Figure 20.16**).

Yet even as Yellowstone's recovery is under way, ecological communities around the planet are being threatened by the loss of other keystone species, especially large carnivores. In 2013, Ripple and colleagues analyzed 31 carnivore species around the globe, including leopards, lions, cougars, and sea otters. They found that more than 75 percent of those 31 species are declining, and that 17 of them now occupy less than half of their former ranges. In 2016, they found that bushmeat hunting, in particular, is

In 1988, a large fire destroyed a portion of the mature lodgepole-pine forest in Yellowstone National Park.

By 1992, the lodgepole-pine forest regrowth was gaining momentum.

An example of a mature, climax community lodgepole-pine forest.

Figure 20.16

Secondary succession: from disturbance to climax community
Lodgepole-pine forest has been slowly but steadily regrowing in Yellowstone National Park.

Q1: What other types of disturbances could you imagine destroying a forest?

Q2: How is secondary succession different from primary succession?

Q3: What is a climax community?

driving many mammal species to extinction, including 113 species in Southeast Asia. Because of all the species interactions discussed here, it is clear that those changes will have major effects on ecological communities. "Humans are affecting predators around the globe in a major way," says Ripple. "It's a worldwide issue."

REVIEWING THE SCIENCE

- An **ecological community** can be characterized by its species composition, or diversity. This diversity has two components: **relative species abundance** (the number of individuals of each species that exist in the community) and **species richness** (the total number of different species that live in the community).

- **Keystone species** have a disproportionately large effect, relative to their own abundance, on the richness and abundances of the other species in a community. The removal or disappearance of these keystone species results in dramatic changes to the rest of the community.

- A **food chain** is a single direct line of who eats whom among species in a community. A **food web** depicts how overlapping food chains of a community are connected.

- **Producers**, organisms found at the bottom of a food chain that use light energy to produce their own food, are eaten by **consumers**.

Consumers are classified as **carnivores**, **herbivores**, or **omnivores**, depending on whether they eat animals or plants or both, respectively.

- Dead and dying organisms are prey for many kinds of consumers. **Scavengers** are omnivores or carnivores that hunt for dead and dying prey. **Decomposers** dissolve their food, and **detritivores** mechanically break apart and consume their food.

- **Species interactions** in a community can be beneficial, harmful, or without benefit or harm to each of the interacting species.

  - In **mutualism**, both species benefit.

  - In **commensalism**, one species benefits at no cost to the other.

  - In **predation**, one species benefits and the other is harmed. **Parasites** are predators that live in or on their **hosts**.

- In **competition**, both species are harmed. Competition occurs when **ecological niches** overlap and includes **exploitative competition** and **interference competition**. The **competitive exclusion principle** predicts that different species that use the same resource can coexist only if one of the species adapts to using other resources.

- **Succession** establishes new communities (**primary succession**) and replaces disturbed communities (**secondary succession**). In stable environments without disturbances, called mature communities or **climax communities**, species composition remains stable over long periods of time.

THE QUESTIONS

The Basics

1 A single sequence of feeding relationships describing who eats whom in a community is a

(a) life history.

(b) keystone relationship.

(c) food web.

(d) food chain.

2 The process of species replacement over time in a community is called

(a) global climate change.

(b) succession.

(c) competition.

(d) community change.

3 Organisms that can produce their own food from an external source of energy without having to eat other organisms are called

(a) suppliers.

(b) consumers.

(c) producers.

(d) keystone species.

4 A low-abundance species that has a large effect on the composition of an ecological community, especially when removed from that community, is called a

(a) predator.

(b) herbivore.

(c) keystone species.

(d) dominant species.

5 Select the correct terms:

A cheetah eats an antelope that ate some grass. The cheetah is a (**secondary consumer / primary consumer**), while the antelope is a (**secondary consumer / primary consumer**). The grass is a (**tertiary consumer / producer**).

Challenge Yourself

6 Wolves are considered a keystone species in Yellowstone because

(a) their removal in the early twentieth century caused many changes to the Yellowstone ecological community.

(b) their reintroduction in the late twentieth century caused many changes to the Yellowstone ecological community.

(c) when they were removed, elk populations increased, leading to increased competition with beavers and bears, which then declined.

(d) when they were reintroduced, aspen populations began to increase because of decreased predation from elk.

(e) all of the above

7 Link each species interaction with an example of the interaction.

MUTUALISM

COMMENSALISM

PREDATION

COMPETITION

1. Elk graze on aspen.

2. Bison allow magpies to perch on them; the birds eat ticks they find on the bison's bodies.

3. Birds nest in aspen trees.

4. Beavers and elk eat the same trees.

8 Place the following elements of the scientists' study of the relationship between aspen populations and wolf populations in the correct order from earliest to latest by numbering them from 1 to 5.

_____ a. Beschta hypothesized that the decimation of wolf populations in the early twentieth century allowed elk populations to grow, thus increasing grazing of the aspen by elk.

_____ b. Beschta observed there were few trees in the river valley.

_____ c. Ripple heard Beschta describe his observations in a talk.

_____ d. Ripple and his graduate student collected data showing that no new trees had grown in the valley since the 1920s.

_____ e. With the reintroduction of wolves into the park, elk populations declined and aspen populations rebounded.

9 Which ecological community would be more diverse: one with high relative species abundance or one with low relative species abundance? Which community would be more diverse: one with high species diversity or one with low species diversity?

Try Something New

10 When a female cat comes into heat and is ready to mate, she urinates more frequently and in a large number of places. Male cats from the neighborhood congregate near urine deposits and fight with each other for the female's attention and breeding rights. In what type of interaction are the male cats engaging?

(a) commensalism

(b) predation

(c) interference competition

(d) exploitative competition

(e) mutualism

11 Rabbits can eat many plants, but they prefer some plants over others. Assume that the rabbits in a grassland community containing many plant species prefer to eat a species of grass that happens to be a superior competitor. If the rabbits are removed from the region, predict whether relative species abundance will increase or decrease, and whether species richess will increase or decrease.

12 Identify whether each of the following is an example of primary succession (P) or secondary succession (S) in Rocky Mountain National Park.

_____a. A mountain slope was cleared of evergreen trees and is now sprouting aspen trees.

_____b. A mountainside with mature evergreen trees has dominated the landscape for generations.

_____c. An area of soil, sand, and rocks remains after a dam burst and flooded the area.

_____d. Lichens and mosses grow on bare rock at upper elevations because of an increase in average yearly temperatures.

13 Analyze the food web shown in the accompanying figure and answer the following questions:

a. Which species do *not* have a predator shown on the food web?

b. Which species shown on the food web performs photosynthesis (captures light energy to make its own food)?

c. Which species have only one predator and only one prey shown on the web?

d. Which species has the most predators shown on the web?

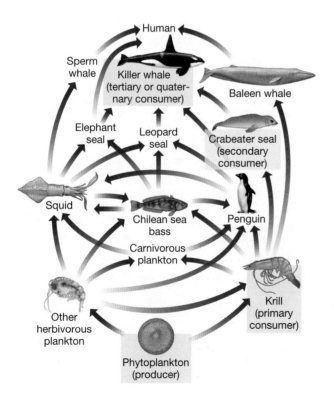

Leveling Up

14 **Doing science** Citizen science is an amazing way for anyone to get involved in scientific research. Search the Internet for citizen science projects relevant to this chapter's topics, using keywords like "parasite," "predator," "prey," and "group living," to name a few. Examples include Project Monarch Health (http://www.monarchparasites.org), in which volunteers sample wild monarch butterflies for a protozoan parasite to track its prevalence across North America. Participate in a project and, in writing, reflect on what you learned.

15 *Write Now* **biology: saving a species** Write a letter to your congressperson in which you use the concepts learned in this chapter to explain how wolves (or a different species near you) are beneficial to our wilderness ecosystems. Explain what role the species plays in the environment, what has caused it to be endangered or at risk now, what the likely effect will be if it goes extinct (locally or globally), and what could be done to ensure its survival.

For more, visit digital.wwnorton.com/bionow2 for access to:

Here and Gone

Researchers discover an alarming decline of plankton in the ocean.

- -

After reading this chapter you should be able to:

◆ Define an ecosystem and explain how it differs from an ecological community.

◆ Compare and contrast nutrient and energy cycles.

◆ Place the members of a specified ecosystem into an energy pyramid at the appropriate trophic level.

◆ Identify where on Earth a specified biome is found, and compare its relative productivity to other biomes.

◆ Define net primary productivity (NPP) and explain its significance to studies of ecosystems.

CHAPTER

21

ECOSYSTEMS

As a teenager, Daniel Boyce made extra cash working as a deckhand on fishing boats, so he was no stranger to the ocean. Eventually, Boyce turned that interest in marine environments into a career, and in 2007 he joined the lab of marine biologist Boris Worm at Dalhousie University in Nova Scotia, Canada.

Several years earlier, Worm had discovered that the industrialized fishing boom that began in the 1950s had decimated predatory fish communities. Boyce decided to follow up on his mentor's work by studying how that loss of predatory fish reverberated down the food chain. Had the disappearance of ocean predators affected plankton? Plankton, a diverse group of free-floating organisms, are the base of the ocean food web, supporting virtually all marine animals. Although they are tiny, they play a mighty role in marine environments.

When Boyce proposed this study, little did he know that it would uncover a profound shift in oceans around the world—a finding so shocking that just suggesting it would plunge him and his collaborators into a heated public controversy.

Going Green

Boyce's initial goal, similar to that of William Ripple's work studying wolves in Yellowstone National Park (see Chapter 20), was to show how the loss of a top predator affects the environment in which it lives. In Ripple's story, a variety of species living in Yellowstone interacted to form what we call an *ecological community*. A group of communities interacting with one another and with the physical environment they share is an **ecosystem** (**Figure 21.1**). To say it another way, an ecosystem is characterized by interactions of organisms in the biotic (living) world with each other, as well as with the abiotic (nonliving) world.

An ecosystem may be small or large; a puddle teeming with protists is an ecosystem, as is the Atlantic Ocean. And smaller ecosystems can be nested inside larger, more complex ecosystems. This variety means that ecosystems do not always have sharply defined physical boundaries. Instead, ecologists often define an ecosystem by the

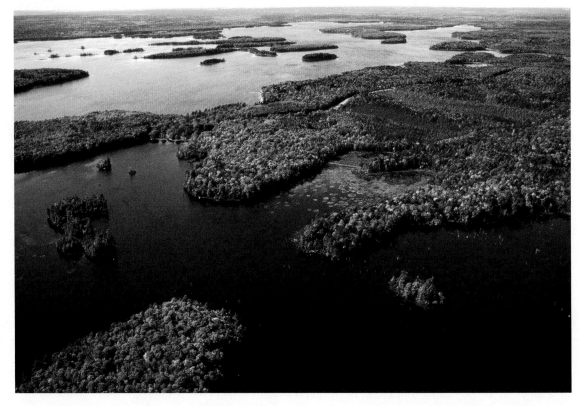

Figure 21.1

Ecosystems in Nova Scotia, Canada
Overfishing has decimated fish populations in Nova Scotia, affecting the entire ecosystem in which these fish reside. Both biotic and abiotic elements of the ecosystem have been affected.

Figure 21.2

Phytoplankton bloom

The turquoise area in this aerial photo is a phytoplankton bloom occurring off the coast of Norway. When a population of phytoplankton (inset) increases rapidly, it discolors the water in which it resides.

distinctive ways in which it functions, especially the means by which *energy* and *nutrients* are acquired and distributed by the biotic community.

The activity of primary producers, in particular, profoundly influences the characteristics of an ecosystem. Ecologists often describe an ecosystem according to the types of producers it contains and the consumers that the producers support. A duckweed-covered pond, a tallgrass prairie, and a beech-maple woodland are all examples of ecosystems that can be defined by the specific types of producers that capture and supply energy to consumers.

To see how overfishing was affecting food chains in the ocean, Boyce first looked to the primary producers of the ocean ecosystem: phytoplankton. These small, floating microalgae come in a fantastic array of shapes and sizes, from smooth orbs to segmented spirals to pointy crescents (**Figure 21.2**). Phytoplankton are

primarily microscopic, but in large groups they form the green color often seen in water. The more phytoplankton in the water, the greener the water is; the less phytoplankton, the bluer it is.

Phytoplankton are green because they are photosynthetic: they convert light energy from the sun into chemical energy using chlorophyll, the green pigment critical to the process

DANIEL BOYCE AND BORIS WORM

Daniel Boyce (left) is a postdoctoral researcher at Queen's University in Ontario, Canada. As a graduate student with Boris Worm (right), he studied the amount of phytoplankton biomass in the ocean. Boris Worm is a marine biologist at Dalhousie University in Nova Scotia, Canada, studying global marine biodiversity.

of photosynthesis. Because they are photosynthetic, these water-living organisms inhabit the top layer of water in the ocean (and almost every body of freshwater as well), a location that gives them easy access to sunlight. Thanks to their ability to photosynthesize using sunlight, phytoplankton are primary producers and the central means through which energy enters the ocean ecosystem. If overfishing was affecting phytoplankton levels, Boyce worried, the whole ocean ecosystem could be in danger.

Bottom of the Pyramid

Energy and nutrients flow through ecosystems in distinctive patterns. First, let's consider the path of energy. Producers, like phytoplankton, capture

Only 10% of the energy captured and stored at each trophic level is transferred to the next trophic level. The remainder is lost through metabolic heat and inefficient consumption.

Of each 10,000 Kilocalories of energy captured from the sun by producers, primary consumers capture and store only about 10%.

Trophic level

Tertiary consumer = fourth trophic level

Secondary consumer = third trophic level

Primary consumer = second trophic level

Producer = first trophic level

Energy (Kilocalories)

10

100

1,000

10,000 Kcal

Figure 21.3

Energy pyramid

The levels of the energy pyramid correspond to steps in a food chain. ▶

Q1: What percentage of the original 10,000 Kilocalories is available to a shark that might eat the tuna in this figure?

Q2: What trophic level and term would describe a predator of tuna?

Q3: Give an example of a primary consumer in a terrestrial environment.

---

energy from the sun and transform it into fuel energy. That fuel is passed up the food chain as one organism eats another. An **energy pyramid** represents the amount of energy available to organisms in an ecosystem. Each level of the pyramid corresponds to a step in a food chain and is called a **trophic level**. In the ocean, for instance, phytoplankton are on the first trophic level. Zooplankton, larger, multicellular plankton that feed on phytoplankton, are the second trophic level. Small fishes such as herring are the third level, and large fishes such as tuna are the fourth (**Figure 21.3**).

At each trophic level, a portion of the energy captured by producers is lost as **metabolic heat**, the heat released as a by-product of chemical reactions within a cell, especially during cellular respiration. Organisms lose a lot of energy as metabolic heat, as revealed by the fact that a small room crowded with people rapidly becomes hot; that warmth is the result of metabolic heat leaving our bodies. On average, roughly 10 percent of the energy at one trophic level is transferred to the next trophic level. The remaining 90 percent of the energy that is not transferred is either not consumed (for example, when we eat an apple, we eat only a small part of the apple tree), is not taken up by the consumer's body (for example, we cannot digest the cellulose contained in the apple), or is lost as metabolic heat.

Because of this steady loss of heat, energy flows in *only one direction* through ecosystems. It enters Earth's ecosystems from the sun (in most cases) and leaves them as metabolic heat. Every unit of energy captured by producers is eventually lost from the biotic world as heat. Therefore, energy cannot be recycled within an ecosystem. It travels up an energy pyramid, never down.

In contrast, **nutrients**—chemical elements required by living organisms—*are* recycled and reused within and across ecosystems. Although Earth receives a constant stream of light energy from the sun, our planet does not acquire more nutrients on a daily basis; rather, a constant and finite pool of nutrients cycles through the land, water, and air. If nutrients were not cycled between organisms and the physical environment, life on Earth would not exist.

Nutrients pass through the abiotic world, from rocks and mineral deposits into soil, water, and air, and then on to the biotic world via absorption by producers. Once in the biotic world, they are cycled among consumers for varying lengths

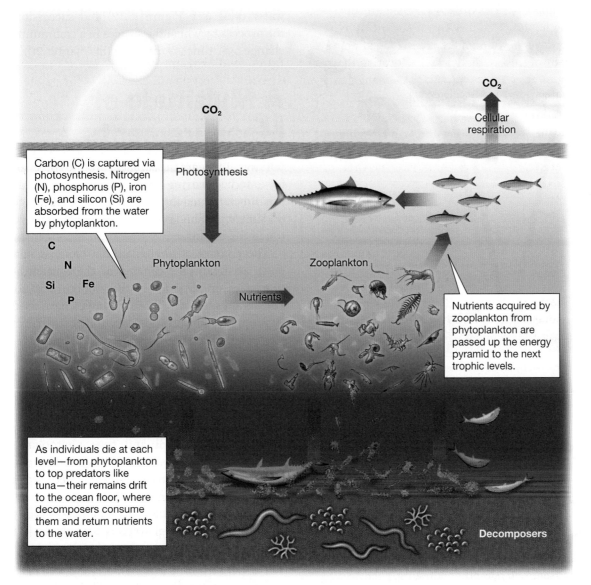

CO₂

Cellular respiration

Carbon (C) is captured via photosynthesis. Nitrogen (N), phosphorus (P), iron (Fe), and silicon (Si) are absorbed from the water by phytoplankton.

Photosynthesis

CO₂

C

N

Si   Fe

P

Phytoplankton

Zooplankton

Nutrients

Nutrients acquired by zooplankton from phytoplankton are passed up the energy pyramid to the next trophic levels.

As individuals die at each level—from phytoplankton to top predators like tuna—their remains drift to the ocean floor, where decomposers consume them and return nutrients to the water.

Decomposers

Figure 21.4

Nutrients cycle within and beyond the ocean ecosystem

In the ocean, phytoplankton absorb nutrients from the abiotic world. Within the biotic world, these nutrients are cycled among consumers, beginning with zooplankton and moving up through trophic levels. The nutrients then return to the abiotic world when dead or dying organisms are broken down by decomposers into their constituent elements.

Q1: Which organisms are the producers in this ecosystem?

Q2: How do nutrients flow from the abiotic to the biotic components of the ecosystem?

Q3: How do nutrients flow from the biotic to the abiotic components of the ecosystem?

of time. Phytoplankton, for example, require the nutrients nitrogen, phosphorus, iron, and silicon for growth. When zooplankton eat phytoplankton, they take up those nutrients, and so on up the food chain (**Figure 21.4**).

Nutrients are eventually returned to the abiotic world when **decomposers** break down the dead bodies of other organisms. In some ecosystems, decomposers break down 80 percent of the *biomass*, or biological material,

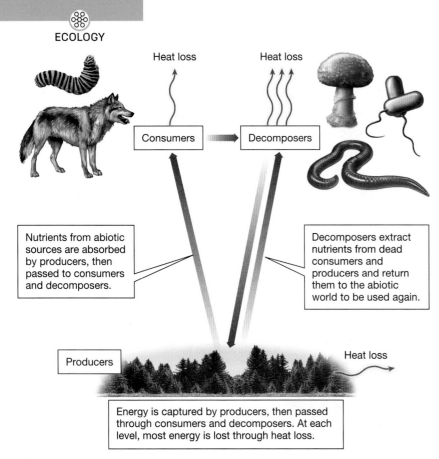

Heat loss

Heat loss

Consumers → Decomposers

Nutrients from abiotic sources are absorbed by producers, then passed to consumers and decomposers.

Decomposers extract nutrients from dead consumers and producers and return them to the abiotic world to be used again.

Producers

Heat loss

Energy is captured by producers, then passed through consumers and decomposers. At each level, most energy is lost through heat loss.

Figure 21.5

Energy flow and nutrient cycling

Unlike energy, which moves up and out of ecosystems, nutrients are constantly cycled between the abiotic and biotic worlds. Important nutrients for the biotic world include carbon (C), potassium (K), phosphorus (P), and nitrogen (N).

Q1: How is a decomposer different from a more typical consumer?

Q2: What is the difference between how carbon is brought into the biotic portion of the ecosystem, and how other nutrients, such as phosphorus, are brought in?

Q3: Describe all the points at which heat is lost in this figure.

made by producers. Without decomposers, nutrients could not be repeatedly reused, and life would cease because all essential nutrients would remain locked up in the bodies of dead organisms. In this way, decomposers are the "cleaners" of an ecosystem. Bacteria and fungi are important decomposers in the ocean, as are hagfishes, worms, and others.

Ecologists and earth scientists use the term "nutrient cycle" to describe the passage of a chemical element through an ecosystem (**Figure 21.5**). The nutrient cycle and the flow of energy are two of four processes that link the biotic and abiotic worlds in an ecosystem. These **ecosystem processes** also include the

water cycle (see Figure 18.12) and succession, the process by which the species in a community change over time (as discussed in Chapter 20).

A Multitude of Measurements

For over 100 years, researchers around the globe have studied ecosystems containing phytoplankton. Boyce tapped into that wealth of research to document past and present levels of phytoplankton in the oceans.

The amount of phytoplankton biomass in a given area can be estimated by the concentration of chlorophyll found there (**Figure 21.6**). For decades, nearly all ocean studies have used chlorophyll concentration as a reliable metric of phytoplankton biomass. Chlorophyll concentration is measured by detecting the color of water. Water takes on deeper shades of green as the amount of chlorophyll increases. When there is no chlorophyll, water appears clear.

Ideally, Boyce would have used satellite data to detect chlorophyll and thus phytoplankton concentrations, since satellites today take high-resolution color measurements of the ocean surface. Yet Boyce planned to review phytoplankton levels over the past 100 years, and high-quality satellite data have been available for only the last decade. He needed another source of data.

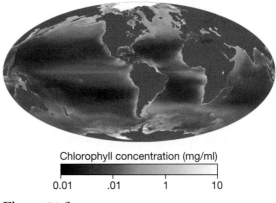

Chlorophyll concentration (mg/ml)

0.01   .01   1   10

Figure 21.6

Average chlorophyll concentration in the oceans

Phytoplankton are most abundant in high latitudes, along coastlines and continental shelves, and along the equator in the Pacific and Atlantic Oceans (yellow)—but are scarce in remote oceans (dark blue).

In a first-of-its-kind analysis, Boyce, together with his adviser Boris Worm and the oceanographer Marlon Lewis, combined two types of chlorophyll measurements. The first type, dating all the way back to 1899, were recorded with nothing more than a rope and a disk.

In 1865, the pope asked priest and astronomer Pietro Angelo Secchi to measure the clarity of water in the Mediterranean Sea for the purposes of the papal navy. Secchi designed one of the simplest measurement devices ever used: a dinner plate–sized disk painted with black and white stripes attached to a rope. The disk is lowered into water until the white stripes disappear (as they become obscured by chlorophyll from phytoplankton), and the depth at that point is recorded (**Figure 21.7**; see also the chapter-opening photo). Chlorophyll concentrations derived from Secchi disk measurements have been corroborated by satellite data, so scientists know they are reliable.

In addition to gathering Secchi disk data, scientists at sea use lab tools to directly measure the quantity of chlorophyll in the water (as opposed to observing its color and relating that to chlorophyll concentration). Boyce found hundreds of thousands of these direct chlorophyll measurements online in open-source databases. "There's been a huge increase in the amount of publicly available oceanographic data out there," says Boyce.

But to use the data, Boyce first had to separate the wheat from the chaff. "With any big database, there are bound to be measurements that are entered incorrectly, for whatever reason," says Boyce. He, Worm, and Lewis ruled out measurements that were inappropriate for their study, such as those taken where the ocean floor was less than 25 meters deep, because changes of water transparency in those cases could be caused by sediment or runoff from landmasses nearby rather than by phytoplankton.

The team analyzed each data set separately—Secchi disk measurements and direct chlorophyll measurements—and then together. To combine the two, they converted all the Secchi measurements into the same units as those used for direct chlorophyll concentrations. In total, the blended data set included 445,237 chlorophyll measurements collected between 1899 and 2008.

What they found made them pause. With those two different methods of analysis, they identified a significant decline in phytoplankton levels—a whopping 60–80 percent—in Earth's oceans where data were available during the

Figure 21.7

Secchi disks indirectly measure chlorophyll concentrations
The Secchi disk is lowered into water until its white stripes become obscured by the chlorophyll in phytoplankton.

last century. Overall, phytoplankton appeared to have declined by about 1 percent of the global average each year.

One percent sounds like a small number, but that amount every year just since 1950 translates into a staggering total phytoplankton decrease of 50 percent in the world's oceans.

The Precious 1 Percent

A 50 percent loss in the main producer in any ecosystem is a worrisome number, but especially with respect to phytoplankton. Phytoplankton support fisheries, produce half the oxygen we breathe, and take in carbon dioxide from the atmosphere, which helps offset the greenhouse effect and global warming.

An ocean with less phytoplankton will function differently because ecosystems depend on **energy capture**, the trapping and storing of solar energy by the producers at the base of the ecosystem's energy pyramid. Herbivores, carnivores, and detritivores all depend indirectly on energy capture. If an ecosystem has an abundance of producers, it can often support more consumers at higher trophic levels. In a tropical forest, for example, an abundance of plants capture energy from the sun, and the forest teems with life. On the flip side, relatively little energy is captured in an environment with few producers. In tundra or desert regions, for example, less food is available, and fewer animals can live there. These significant differences have prompted ecologists to categorize large areas of Earth's surface into distinct regions, called **biomes**, that are defined by their unique climatic and ecological features (**Figure 21.8**).

Assessing the overall amount of energy captured by producers is important in determining how an ecosystem works, because energy

The **boreal forest** is dominated by coniferous trees that grow in northern or high-altitude regions with cold, dry winters and mild summers. The soil is thin and nutrient-poor, so while plants generally receive adequate moisture during the growing season, plant diversity is relatively low. Large herbivores include elk and moose. Small carnivores, such as weasels, wolverines, and martens, are common. Larger carnivores include lynx and wolves.

The **tundra** biome is found at the poles and on mountaintops. Trees are absent or scarce because of the short growing season. The vegetation is dominated by low-growing flowering plants, and the boggy landscape is covered in mosses and lichens, important food sources of herbivores. Rodents provide food for carnivores like foxes and wolves. Bears and musk oxen are among the few large mammals.

The **chaparral** is a shrubland biome dominated by dense growths of drought-resistant plants in regions with cool, rainy winters and hot, dry summers. These conditions make the chaparral exceptionally susceptible to wildfires. Common vegetation in the California chaparral includes scrub oak, pines, mountain mahogany, manzanita, and the chemise bush. Small mammals such as jackrabbits and gophers are common, and there are many species of lizards and snakes.

Terrestrial biomes:
- Tundra
- Boreal forest
- Temperate forest
- Grassland
- Chaparral
- Desert
- Tropical forest

Ecosystems within the **freshwater biome** are heavily influenced by the terrestrial biomes that they border or through which their water flows. Lakes are landlocked bodies of standing freshwater. Rivers are bodies of freshwater that move continuously in a single direction. Wetlands are characterized by standing water that is shallow enough for rooted plants to emerge above the water surface. A bog is a freshwater wetland with stagnant, oxygen-poor water, and low productivity and species diversity. In contrast, grassy marshes and tree-filled swamps are highly productive wetlands with a high diversity of organisms.

The **tropical forest** biome is characterized by warm temperatures, about 12 hours of daylight each day, and either seasonally heavy or year-round rains, which tend to leach nutrients from the soil. Soils in this biome further tend to be nutrient-poor because a large percentage of nutrients are locked up in the living tissues (biomass) of organisms. Tropical rainforests, which may receive in excess of 80 inches (200 cm) of rain annually, are some of the most productive ecosystems on Earth, with a rich diversity of organisms.

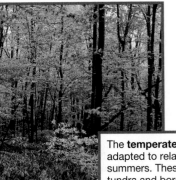

The **grassland** biome cannot sustain vigorous tree growth, but its moisture levels are not as low as in deserts. Grasslands are found in both temperate and tropical latitudes and are dominated by grasses and herbaceous plants such as coneflower and shooting star. Scattered trees are found in some, such as the tropical grasslands known as the savanna. Burrowing rodents like voles and prairie dogs may aerate the soil, thereby improving growing conditions. Many grasslands have been converted to agriculture.

The **temperate forest** biome is dominated by trees and shrubs adapted to relatively rich soil, snowy winters, and moist, warm summers. These forests display greater species diversity than do the tundra and boreal forest biomes: oak, maple, hickory, beech, and elm are common. Herbivores include squirrels, rabbits, deer, raccoons, and beavers, while bobcats, mountain lions, and bears make up the carnivores. Amphibians and reptiles are common.

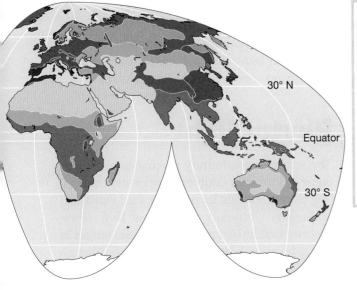

30° N

Equator

30° S

Because air in the **desert** biome lacks moisture, it does not retain heat well. As a result, temperatures can be above 113°F (45°C) in the daytime and then plunge to near freezing at night. Desert plants have small leaves to minimize water loss. Succulents, such as cacti, store water in their fleshy stems or leaves. Most animals in the desert are nocturnal, hiding in burrows during the heat of the day and emerging at night to feed.

Estuaries are the shallowest but most productive of the aquatic biomes. They are tidal ecosystems where rivers flow into the ocean. They have a constant ebb and flow of fresh and salt water, and organisms must be able to tolerate daily changes in saltwater concentrations. The plentiful light, the abundant supply of nutrients delivered by the river system, and the regular stirring of nutrient-rich sediments by water flow create a rich and diverse community of photosynthesizers. Grasses and sedges are the dominant vegetation in most estuaries.

The **marine biome**, characterized by salt water, is the largest biome on our planet. The *coastal region* stretches from the shoreline to the edge of the continental shelf and is highly productive because of the availability of nutrients and oxygen. A majority of Earth's marine species live in the coastal region. The intertidal zone, closest to the shore, is a challenging environment where organisms such as seaweeds, worms, crabs, sea stars, sea anemones, and mussels are submerged and exposed to dry air on a twice-daily basis. The relatively nutrient-poor *open ocean* begins about 40 miles offshore and is much less productive than coastal waters.

Figure 21.8

Amazing biomes

Biomes do not begin and end abruptly, but rather transition into one another. Terrestrial biomes are categorized by temperature, precipitation, and altitude; aquatic biomes are determined by proximity to shorelines.

capture influences the amount of food available to other organisms. The **net primary productivity** (**NPP**) of an ecosystem is the energy, acquired through photosynthesis over a particular time period, that is available for the growth and reproduction of producers. NPP is the amount of energy captured by photosynthetic organisms minus the amount they expend on cellular respiration and other maintenance processes. NPP is typically determined by estimating the amount of carbon captured during photosynthesis. This can be done by measuring the amount of new biomass produced by the photosynthetic organisms in a given area during a specified period of time.

According to scientists' estimates, the NPP of all producers on Earth exceeds 100 billion tons of carbon biomass per year. Roughly half of this productivity comes from phytoplankton in the ocean. Therefore, phytoplankton capture approximately 50 billion tons of carbon per year. So if Boyce's calculations are right, a loss of 1 percent of that biomass is 500 million tons of organic matter lost from the oceans *each year*. That's a lot of biomass to lose.

Net primary productivity relies on four things: sunlight, water, temperature, and the availability of nutrients. The most productive ecosystems on land are tropical forests; the least productive are deserts and tundra (including some mountaintop communities). The most productive ecosystems in water are estuaries—regions where rivers empty into the sea—because nutrients drained off the land stimulate the growth and reproduction of phytoplankton and other producers, which in turn nourish large populations of consumers. The least productive aquatic biome is the deep ocean, where sunlight does not penetrate.

Despite similarities between the NPP requirements on land and in the ocean, the global pattern of NPP differs between the two. On land, the NPP is highest at the equator and decreases toward the poles. But in the ocean, the general pattern relates not to latitude but to distance from shore: the productivity of marine ecosystems is typically high in ocean regions close to land and relatively low in the open ocean (**Figure 21.9**). This is because nutrients needed by aquatic photosynthetic organisms are in better supply near land, thanks to delivery from streams and rivers. Wetlands such as swamps and marshes, which trap soil sediments rich in nutrients and organic matter, can be so productive that they match the productivity levels of tropical forests.

A loss of 500 million tons of phytoplankton each year could potentially affect the ocean's NPP and ocean life. "Almost all biological life in the ocean depends on phytoplankton. A reduction in the biomass of phytoplankton will result in less secondary production in the oceans," says Boyce. That means fewer sharks, whales, fishes, eels—you name it.

The team's discovery was shocking, to say the least. No one else had documented a global decline in phytoplankton before. So, to be confident in their results, Boyce and Worm did several more rounds of data analysis, checking again and again to make sure they were using the right numbers in ways that correctly represented what was happening in the natural world. And over and over, they came back with the same results: global phytoplankton declined over the last century. In 2010, they published that finding in the peer-reviewed scientific journal *Nature*.

Phyto-Fight

The scientific community reacted immediately. Some researchers doubted Boyce's conclusions; others were outright incredulous. Paul Falkowski at Rutgers University told a *New York Times* reporter that he had not found the same trend in a long-term analysis of the North Pacific (though Boyce contends that their trends were very similar), and another team had actually seen an increase in phytoplankton starting around 1978 in the central North Pacific. Falkowski called Boyce's paper "provocative" but said he would "wait another several years" to see whether satellite data would back up the finding.

Then, in 2011, three separate research teams went so far as to publish formal critiques of the work. One suggested that the declining trend was an error resulting from the use of two different types of measurements: Secchi disk readings and direct chlorophyll measurements. A second team echoed that idea, reanalyzed the data in a way that showed an increase in phytoplankton, and then bluntly concluded, "Our results indicate that much, if not all, of the century-long decline reported by [Boyce] is attributable to this [sampling bias] and not to a global decrease in phytoplankton biomass."

The third team noted that Boyce's finding conflicted with eight decades of data on phytoplankton biomass collected by a large project called the Continuous Plankton Recorder (CPR)

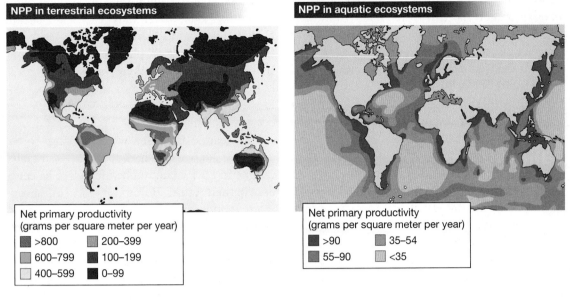

NPP in terrestrial ecosystems

Net primary productivity
(grams per square meter per year)

- >800
- 600–799
- 400–599
- 200–399
- 100–199
- 0–99

NPP in aquatic ecosystems

Net primary productivity
(grams per square meter per year)

- >90
- 55–90
- 35–54
- <35

Figure 21.9

Global variation in net primary productivity

NPP can be measured as the number of grams of new biomass made by producers each year in a square meter of each biome's area. NPP varies greatly across both terrestrial and aquatic ecosystems.

Q1: Which terrestrial biome has the lowest NPP? Which aquatic biome?

Q2: Where are the most productive terrestrial biomes located?

Q3: Give a possible reason for your answer to question 2.

survey, which monitors the Northeast Atlantic Ocean. The CPR survey, started in 1931, employs a unique instrument pulled through the ocean by commercial fishing vessels to collect millions of samples of plankton. The CPR survey found that over the last 20–50 years, phytoplankton biomass increased in the Northeast Atlantic, says Abigail McQuatters-Gollop, a former researcher at the Sir Alister Hardy Foundation for Ocean Science, which operates the survey.

In response, Boyce, Worm, and Lewis didn't get angry; they got focused. After reading the critiques, the three researchers went back to the data. First they applied a correction factor suggested by the critics, in the hopes of removing any bias between the two types of data. "We did that, and the trends remained similar," says Boyce. Next, they again estimated changes over time individually for the two data sources. "That didn't change the trends either," says Boyce. After that, they incorporated additional suggestions by their peers and created a new, expanded database of chlorophyll measurements to work from. Finally, they reestimated changes in chlorophyll

using this new database and their revised analysis methods, but the phytoplankton still seemed to be declining, independently of the type of data or how the data were analyzed. The three researchers published their reanalysis in a series of three papers in 2011, 2012, and 2014, demonstrating the same decline again and again.

But the additional work has not silenced the critics. "It's still pretty hotly debated," admits Boyce. "The story is not over by any means." In 2011, in fact, Worm traveled to an international plankton conference where he, McQuatters-Gollop, and others debated the topic in front of a live audience. "It was an amicable meeting that generated loads of discussion," says

ABIGAIL MCQUATTERS-GOLLOP

Abigail McQuatters-Gollop was a science and policy researcher at the Sir Alister Hardy Foundation for Ocean Science, home of the Continuous Plankton Recorder survey, from 2010 to 2015. She is now a lecturer in marine conservation at Plymouth University in the United Kingdom.

Productive Plants

Data on net primary productivity (NPP)—the total energy available in an ecosystem for the growth and reproduction of primary producers—is sparse for whole biomes, but researchers have been able to estimate the NPP of those listed here using information about different vegetation types and carbon sources in each area.

Assessment available in smartwork**5**

Total energy available from primary producers

(in grams per square meter per year)

Desert
150

Tundra
240

Lakes and streams
440

Grassland (temperate)
790

Grassland (tropical)
1,770

Wetlands
2,620

McQuatters-Gollop. Still, they could not agree on whether phytoplankton populations have increased or decreased in the ocean. At the meeting, the researchers agreed that the best thing to do was to combine as many data sets as possible.

Heating Up

In 2015, the long-awaited satellite data came in—and they weren't good. Ocean color measurements from two NASA satellites led scientists to conclude that diatoms, the largest type of phytoplankton, had declined more than 1 percent per year from 1998 to 2012 globally, with major losses in the North Pacific, North Indian, and Equatorial Indian Oceans.

It looks like Boyce, Worm, and Lewis may be right, but the important question of why a decline is occurring remains. As part of his research, Boyce investigated possible causes of the downward trend. In one study, he compared changes in sea surface temperature to changes in chlorophyll levels and noted a strong correlation: over the last 100 years, chlorophyll concentrations declined *and* ocean temperatures increased, in line with increases in global warning. This correlation has been closely followed in recent years (**Figure 21.10**).

But correlation does not prove causation, and much more work needs to be done to support the idea that global warming caused the decline. It is a logical hypothesis, however, because as the planet warms, water in the oceans mixes less, limiting the nutrients delivered to the surface from decomposers in the deep sea. As a result, phytoplankton do not receive the nutrients they need for growth and reproduction. In 2014, Boyce published experimental data supporting the hypothesis. Working with marine scientists in Germany, he found that warming the water in a controlled, experimental ocean water enclosure led to reduced phytoplankton biomass.

A 2014 study by a large, international group of researchers predicted that an increase in ocean temperature due to global warming would cause phytoplankton biomass to decrease by 6 percent by the end of this century. Consequences of continued phytoplankton decline could include altering the carbon cycle between the ocean and the atmosphere, changing heat distribution in the ocean, and causing a decrease in the supply of food in the ocean. Whatever the case, if phytoplankton populations are decreasing, says Boyce—and he is sure they are—there will be profound effects, which have only just begun.

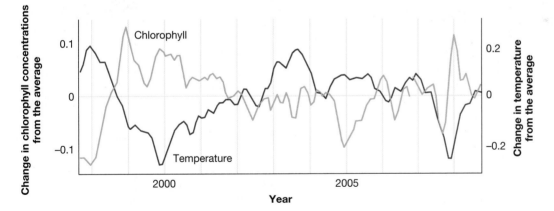

Figure 21.10

As ocean temperature increases, chlorophyll decreases
Between late 1997 and mid-2008, satellites observed that warmer-than-average temperatures (red line) were correlated with below-average chlorophyll concentrations (green line).

Q1: In what years were chlorophyll levels the highest?

Q2: In what years were the temperature changes from the average the greatest?

Q3: How do you predict this graph will look 10 years from now?

REVIEWING THE SCIENCE

- An **ecosystem** consists of communities of organisms and the physical environment in which those communities live. It is the sum of all biotic elements interacting with all abiotic elements. Energy, materials, and organisms can move from one ecosystem to another.

- An **energy pyramid** represents the amount of energy available to each **trophic level** of a food chain in an ecosystem.

- Energy enters an ecosystem when producers capture it from an external source, such as the sun. A portion of the energy captured by producers is lost as **metabolic heat** at each trophic level. As a result, energy flows in only one direction through ecosystems.

- **Nutrients** are the chemical elements required by living organisms. Unlike energy, nutrients are recycled and reused within and across ecosystems. Earth has a fixed amount of nutrients.

- **Decomposers** break down the dead bodies of other organisms, both consumers and producers, releasing the nutrients in the bodies of dead organisms back to the physical environment.

- Four **ecosystem processes** link the biotic and abiotic worlds in an ecosystem: nutrient cycling, energy flow, water cycling, and succession.

- Ecosystems depend on **energy capture**, the trapping of solar energy by producers via photosynthesis, and the storage of that energy as chemical compounds in their bodies.

- Earth is categorized into 10 major **biomes**, regions defined by their climatic and ecological features.

- **Net primary productivity (NPP)** is the energy acquired through photosynthesis that is available for growth and reproduction to producers in an ecosystem. NPP is estimated by the amount of biomass produced in a given area during a specified period of time.

THE QUESTIONS

The Basics

1 The movement of nutrients between organisms and the physical environment is called

(a) nutrient cycling.

(b) ecosystem services.

(c) net primary productivity.

(d) decomposition.

2 How much energy is transferred up the energy pyramid from one trophic level to the next?

(a) 90%

(b) 50%

(c) 10%

(d) 10%–50%

3 Which organisms are considered the "recyclers" of our planet?

(a) consumers

(b) producers

(c) phytoplankton

(d) decomposers

4 The terrestrial biome that receives the most consistent year-round rainfall is

(a) wetland.

(b) boreal forest.

(c) tropical forest.

(d) chaparral.

5 Link each term with the correct definition.

BIOME

ECOSYSTEM

NET PRIMARY PRODUCTIVITY

ECOLOGICAL COMMUNITY

1. The energy acquired through photosynthesis by producers of an ecosystem.

2. A group of communities interacting with one another and with the physical environment they share.

3. A large, distinct region defined by its unique climatic and ecological features.

4. The populations of different species that live and interact with one another in a particular place.

Challenge Yourself

6 Select the correct terms:
The biome characterized by shrubs and nonwoody plants that grow in regions with cool, rainy winters and hot, dry summers is (tundra / chaparral). Another biome with few trees, but in this case dominated by grasses and nonwoody plants, is (grassland / tundra). The most productive aquatic biome is (freshwater / estuaries).

7 Which of the following is a component of an ecosystem but *not* of an ecological community?

(a) a producer

(b) water

(c) a secondary consumer

(d) a primary consumer

8 Place the following elements of the scientists' study of oceanic phytoplankton levels in the correct order from earliest to latest by numbering them from 1 to 5.

_____ a. Boyce analyzed data from multiple sources showing that ocean phytoplankton levels have declined dramatically over the last 100 years.

_____ b. Worm demonstrated that industrialized fishing had decimated predatory fish communities.

_____ c. Secchi designed a device to measure phytoplankton levels in water.

_____ d. Boyce's finding of declining phytoplankton levels was challenged by fellow scientists, because of his methodology.

_____ e. Boyce, a graduate student in Worm's lab, decided to study how the loss of predatory fish reverberated through the oceanic ecosystem—in particular, its phytoplankton populations.

9 How is an ecosystem different from an ecological community?

Try Something New

10 In the energy pyramid shown here, an owl, a cardinal, and a grasshopper are the fourth, third, and second trophic levels, respectively.

a. If each grasshopper passes 100 Kilocalories to the cardinal when eaten, how many grasshoppers would the cardinal have to eat to obtain 10,000 Kilocalories?

b. How many Kilocalories of grass are required to produce 10,000 Calories' worth of grasshopper?

c. Where does the lost energy go?

11 Is the water cycle (see Figure 18.12) more similar to the movement of energy or to the movement of nutrients through an ecosystem? Justify your answer.

12 While sloshing around in the swampy wetlands of a nearby forest preserve, a small child notices a rotting tree branch covered in fungi. What are these organisms doing on this tree branch?

(a) These are producers acquiring energy through photosynthesis.

(b) These are consumers acquiring energy from the wood and releasing nutrients back to the earth.

(c) These are decomposers acquiring energy from the wood and releasing nutrients back to the earth.

(d) These are decomposers releasing nutrients back to the earth, but acquiring no energy for themselves in the process.

Leveling Up

13 *Write Now* biology: human-caused biome shifts The location of Earth's different biomes depends on climate and altitude, for the most part. However, human activities play a role in the conversion of one biome to another, as has been seen many times in history. Research one major change in a biome category based on human activity, and describe how and why this change happened. Speculate on how this change could have been specifically avoided. (*Note:* Do not analyze a change via deforestation to agricultural land, since agricultural land is not a natural biome. *Hint:* Take a look at Easter Island as one example.)

14 Doing science Join forces with millions of others by classifying phytoplankton on your computer. Do an Internet search for "citizen science phytoplankton" and sign in as a citizen scientist, complete the tutorial, and start helping researchers quantify the phytoplankton in our oceans.

15 What do *you* think? Some people think the current U.S. Endangered Species Act should be replaced with a law designed to protect ecosystems, not species. The intent of such a law would be to focus conservation efforts on what its advocates think really matters in nature: whole ecosystems. Given how ecosystems are defined, do you think it would be easy or hard to determine the boundaries of what should and should not be protected if such a law was enacted? Give reasons for your answer.

For more, visit digital.wwnorton.com/bionow2 for access to:

Answers

CHAPTER 1

END-OF-CHAPTER ANSWERS

1. a, c, e. Viruses meet no criteria except for evolving as groups. The diamond meets no criteria.

2. b

3. observation, hypothesis, predictions

4. (a) 1, (b) 4, (c) 5, (d) 3, (e) 6, (f) 2, (g) 7

5. (a) organ, (b) organism, (c) population, (d) ecosystem, (e) organ system, (f) community

6. Observation: Some bats have white noses.
   Hypothesis: Bats with white noses are infected with a fungus.
   Experiment: Inject bats with a fungicide and observe whether they are less likely to develop white-nose syndrome than are bats who are given sham injections.

7. b

8. c

9. (a) reproducing autonomously, (b) responding to the environment, (c) obtaining energy from the environment, (d) evolving as a group, (e) consisting of one or more cells

10. c

11. (a) hypothesis, (b) result, (c) experiment, (d) observation, (e) result

12. (a) No; a question is not a hypothesis. (b) Yes; this is a plausible and falsifiable explanation. (c) Yes; this is a plausible and falsifiable explanation. (d) No; a question is not a hypothesis. (e) No; proposing the existence of a "mysterious cloud" does not generate testable predictions.

ANSWERS TO FIGURE QUESTIONS

Figure 1.1

Q1: What were the original observation and question of the scientists studying the sick bats?

A1: They observed many dead bats and many bats with white noses. They questioned whether the high death rates were in some way related to the white noses.

Q2: At what point in the scientific method would a scientist decide on the methods she should use to test her hypothesis?

A2: After predictions are generated from a hypothesis, and before experiments are run to test those predictions.

Q3: How might you explain the scientific method to someone who complains that "scientists are always changing their minds; how can we trust what they say?"

A3: Science is a process, and nothing is ever "proved" in science, so we have to expect that our best understanding of nature will change as science proceeds.

Figure 1.2

Q1: Which step(s) in the scientific method does this photograph illustrate?

A1: Observation, and testing the predictions of the hypothesis through observational study (either descriptive or analytical).

Q2: What types of environmental data might the researchers have collected?

A2: Examples of environmental data would include temperature and humidity readings within the cave, as well as soil and air samples to see whether fungal spores are present.

Q3: Why do you think the researchers are wearing protective gear?

A3: To avoid coming into contact with pathogens.

Figure 1.4

Q1: State the hypothesis that this advertisement is claiming was scientifically tested.

A1: Lucky Strike cigarettes taste milder than other leading brands of cigarettes.

Q2: State a prediction that comes from this hypothesis. Is it testable? Why or why not?

A2: Prediction: Subjects who smoke a Lucky Strike cigarette will say it is milder than cigarettes from other brands that they are asked to smoke. Yes, this prediction is testable, because it can be measured and repeated.

Q3: Explain in your own words why the hypothesis cannot be "proved."

A3: Nothing in science can be proved. In this case, only a small number of subjects can be tested, and there might always be people (who weren't tested) who think another brand is milder.

Figure 1.5

Q1: Give a possible hypothesis that could be tested by weighing the bats.

A1: Hypothesis: Bats with WNS weigh less than uninfected bats.

Q2: State the hypothesis being tested in the photo on the bottom right.

A2: Hypothesis: Healthy bats injected with a fungicide will have lower rates of infection by WNS than will bats that are sham injected (no fungicide).

Q3: Explain in your own words why an experimental study is the only way to show a cause-effect relationship.

A3: One possible explanation: If you find a relationship between two variables in an observational study, you can't know which one causes the other, or even whether a third variable is causing both to occur. In an experimental study, it is possible to manipulate one variable and see whether that manipulation causes a second variable to change.

Figure 1.6

Q1: Which is the control group in this experiment, and what are the three treatment groups?

A1: Control group: Housed alone, with no exposure to WNS. Treatment groups: Housed in air contact but not physical contact with infected bats; housed in physical contact with infected bats; with fungus applied to wings.

Q2: What is the hypothesis being tested in this experiment?

A2: Hypothesis: WNS is caused by contact with the fungus *Geomyces destructans*.

Q3: In one or two sentences, state the conclusions you can draw from the experiment. Was the hypothesis supported? Why or why not?

A3: Bats that come into physical contact with *Geomyces destructans* are highly likely to develop WNS, whereas those exposed only through the air do not develop WNS. The hypothesis was supported in part (physical versus air contact).

Figure 1.7

Q1: What is the control group in this experiment, and what are the two treatment groups?

A1: Control group: Sham injection. Treatment groups: Injected with *Geomyces destructans* from North America; injected with *G. destructans* from Europe.

Q2: At day 40, approximately how many individuals were alive in each treatment group? At day 80? At day 100?

A2: Day 40: 100% of all groups survived. Day 80: 100% of control and North American *G. destructans* (Gd) groups survived; 14 of 18 in the European Gd group survived. Day 100: 100% of control group survived; 13 of 18 in the North American Gd group survived; only 2 of 18 in the European Gd group survived (although the European Gd experiment was stopped at about day 90).

Q3: In one or two sentences, state the conclusions you can draw from the experiment. Was the hypothesis supported? Why or why not?

A3: Conclusion: *Geomyces destructans* causes WNS and leads to higher mortality. The hypothesis was supported. The study found that WNS was, in fact, caused by the fungus.

Figure 1.8

Q1: Give one *fact* about bats that you learned from this chapter.

A1: One possible fact: Bats can develop fungal infections.

Q2: What is another example of evidence for the *germ theory of disease*? (*Hint:* Think about human diseases.)

A2: One example would be strep throat, caused by *Streptococcus* bacteria and cured by antibiotics.

Q3: Explain in your own words the difference between a fact and a hypothesis, and between a hypothesis and a theory.

A3: Hypotheses are not as certain and are more complex than facts; they are simpler and less well documented than theories.

Figure 1.9

Q1: Give examples of other kinds of organs that mammals such as bats have. (*Hint:* Think of the organs in your own body.)

A1: Examples include kidney, liver, heart, lungs.

Q2: Are bats in California part of the community of bats in upstate New York, if they are of the same species? Why or why not?

A2: No, they are not, because they do not interact with each other.

Q3: Is the soil in a cave where bats live a part of the bats' population, community, or ecosystem? Explain your reasoning.

A3: Soil is part of the ecosystem; populations and communities are composed only of living things, and soil is part of the physical environment.

CHAPTER 2

END-OF-CHAPTER ANSWERS

1. b

2. c

3. d

4. scientific literacy: 2, basic research: 5, applied research: 1, secondary literature: 4, primary literature: 3

5. credentials, bias, secondary literature

6. (a) 1, (b) 6, (c) 3, (d) 2, (e) 4, (f) 5

7. a

8. (a) primary, (b) neither, (c) neither, (d) secondary

9. c

10. (a) Pseudoscience. Dr. Oz has relevant credentials but, from the success of his TV show, appears to have an agenda. In addition, the idea of a "fat-burning" dietary aid does not align with current scientific understanding.

    (b) Pseudoscience. The idea of a scientific conspiracy is a red flag, as is the rejection of scientific consensus. And though the

Wall Street Journal is a respected newspaper, it tends to have a conservative bias, especially since being bought by Rupert Murdoch.

(c) Real science. Practicing scientists have reported experimental findings to their peers, but they do not yet appear to have published in a peer-reviewed journal, which will need to be the next step.

(d) Pseudoscience. Astrologers are not scientists. There is no scientific evidence that date of birth has any effect on personality.

(e) Real science. The journal *Diabetes* is a peer-reviewed and well-established scientific journal. We do not know, however, the design of the study or the sample size.

11. (a) The first graph shows the incidence of pertussis (whooping cough) cases in the United States from 1922 to 2012, with the inset showing a zoomed-in view of 1990–2012. The x-axis shows the year from which the pertussis cases were counted; the y-axis shows the counts. Any point on the line shows the total number of cases reported in the corresponding year on the x-axis. The general trend of the graph is a decrease in cases of pertussis over time.

(b) The reason for the increase in cases in the first decade of the twenty-first century is probably that parents were choosing not to have their children vaccinated.

(c) The second graph shows the incidence of pertussis (whooping cough) cases in the United States by age group from 1990 to 2012. The x-axis shows again the year from which the pertussis cases were counted, and the y-axis shows the counts. This graph's y-axis shows not the total number of cases, but rather the number of cases per 100,000 individuals in the population of that particular age group. One point on each line shows the number of cases per 100,000 individuals in the population of that particular age group reported in the corresponding year on the x-axis. The general trend is an increase in cases of pertussis by age group over time. The lines differ by age group. The younger the age group, the greater the increase in pertussis cases per year.

(d) The incidence of pertussis cases per 100,000 individuals in the population of children under 1 year old is very high, and in people over the age of 20 it is very low. Increase over time is much greater in children under 1 year old and is very minimal in people over the age of 20.

(e) It appears to be very risky not to vaccinate. I would vaccinate!

12. (1) Evaluate the credentials of the group members recommending the vaccine (in this case it would be members of the ACIP, FDA, and CDC).

(2) Assess any possible biases of the people in step 1.

(3) Read the secondary literature (for example, recent articles in *Popular Science* or the *New York Times*) for an overview of the issue.

(4) Read recent primary literature that was cited in the secondary literature (for example, articles in *JAMA* or the *New England Journal of Medicine* or *Vaccine*).

(5) Review papers from the primary literature for credentials and biases of the authors, good research design, reasonable sample size, and conclusions.

ANSWERS TO FIGURE QUESTIONS

Figure 2.1

Q1: Describe in one sentence how a vaccine creates immunity to a virus.

A1: A vaccine causes the immune system to produce antibodies in response to an inactive or harmless virus. Those antibodies then immediately identify and attack any live virus (of that type) that enters the body.

Q2: Why is it impossible to become infected with a virus from a vaccine composed of viral proteins?

A2: Proteins don't replicate; DNA or RNA is needed for a virus to replicate.

Q3: Natural immunity occurs without a vaccine, just by exposure to a particular stimulus, like the chicken pox virus. Explain why people don't get chicken pox twice.

A3: Your body creates antibodies when you are exposed to, for example, chicken pox virus. If you are exposed a second time, those antibodies immediately identify and attack the chicken pox virus (just as in question 1 above).

Figure 2.2

Q1: Before vaccinations, which diseases had the highest and lowest mortality rates? What are these mortality rates?

A1: Highest: Tetanus, 90%. Lowest: Pertussis, 5%.

Q2: After vaccinations, which diseases have the highest and lowest mortality rates? What are these mortality rates?

A2: Highest: Tetanus, 14%. Lowest: Polio, diphtheria, and smallpox, 0%.

Q3: If there were a sudden outbreak of whooping cough at a university where pertussis vaccinations were not required and no one was protected, how many students would die? What is the probability or chance that you would die if infected?

A3: At California State University, Northridge, for example, which has 40,000 students total, 40 students would die—a 0.1% chance of death for any individual student.

Figure 2.5

Q1: Why are we less confident of scientific claims made over social media?

A1: Claims made over social media are not subject to peer review before being "published." Therefore, people can—and do—make ridiculous scientific claims in social media.

Q2: Where would you place a blog in this figure? Would it matter whether or not it was written by a practicing scientist? Explain your reasoning.

A2: A blog could be listed either under social media or under secondary literature, depending on the quality of the content. In the case of a science blog written by a practicing scientist, the choice would depend on the subject matter, but it would likely be listed under secondary literature. (Counterargument: The physicist who thinks global climate change is a scientific conspiracy.)

Q3: Give an example of when you would rely on secondary literature to evaluate a scientific claim and an example of when you would go to the primary literature. What is the basis of that decision?

A3: You might go to the secondary literature only for something that is not life-threatening—for example, what kind of exercise to do, or whether to turn down the thermostat in your bedroom. Primary literature is challenging reading, so it is usually left for life-critical choices—for example, whether to vaccinate your children, or possibly whether to eat a vegetarian or vegan diet.

Figure 2.6

Q1: How much did organic food sales grow during the period covered in the graph? How much did the incidence of autism grow?

A1: Organic food sales grew from $5 billion to $25 billion (a fivefold increase). Autism diagnoses grew from about 50,000 individuals to over 250,000 (also a fivefold increase).

Q2: Why might both organic food sales and autism prevalence have increased during this time period? A Reddit user in the original discussion thread suggested that both might be affected by increasing wealth in the United States. How might increased wealth affect these variables?

A2: People with more disposable income are able to spend more on food (hence the rise in organic food sales), and they are also better able to take their children in for advanced medical care (possibly the reason for increased identification of autism disorder).

Q3: In what way has the vaccine-autism debate been confused by people misinterpreting correlation as causation?

A3: Because vaccination was increasing at the same time that autism rates were rising (correlation), people suggested that the former caused the latter (causation). In addition, the time at which children are typically vaccinated is about the same age at which autism symptoms typically appear.

Figure 2.7

Q1: State the hypothesis of the people who believe that vaccines cause autism. What is an alternative hypothesis?

A1: Hypothesis: Vaccination harms a child's immune system and stimulates the development of autism. One alternative hypothesis: Autism is caused by a genetic predisposition, and symptoms begin to show up in the second year of a child's life.

Q2: What part(s) of the figure show where Wakefield's study failed to meet the standards of the scientific method?

A2: Wakefield's study really falls apart in steps 4–6. It was not carefully designed and definitely not reproducible. The sample studied was not of sufficient size and did not include a random control group. The conclusions did not follow from the analysis of the experimental results, and the study was published without adequate review from scientists in the field.

Q3: Why does only one arrow point to "real science," whereas multiple arrows point to pseudoscience?

A3: All of the criteria listed in the figure must be met for a study to meet the expectations of the scientific method.

Figure 2.8

Q1: Why is it important to know the education and expertise of a person making a scientific claim?

A1: The opinion of a person who doesn't really understand the science behind a claim is not valid.

Q2: List at least five possible biases that people making scientific claims might have.

A2: (1) They could make money if you buy a product related to the claim. (2) They might win a lawsuit if the judge believes the claim. (3) They could become famous if people accept the claim. (4) Their religious beliefs might be supported if the claim is true. (5) Their political beliefs might be supported if the claim is true.

Q3: Describe a situation in which you might not dismiss the scientific claim of a person who did not have appropriate credentials, or who had a bias toward the claim.

A3: Paul Offit's case provides an example: He made a great deal of money by selling a rotavirus vaccine he created. However, his expertise is extremely strong, and he does not have an ongoing financial interest in supporting vaccination.

Figure 2.10

Q1: Why do vaccine manufacturers begin with tests on animals or cell lines before moving on to adult human subjects?

A1: They need to be sure of the safety of the vaccine before exposing people to it.

Q2: What ongoing testing and reporting are vaccines subjected to?

A2: Manufacturers test all vaccine lots, the FDA regularly inspects manufacturing facilities, the Advisory Committee on Immunization Practices (ACIP) and the CDC director review all test results before approving vaccines, and vaccine safety is continually monitored through the Vaccine Adverse Event Recording System (VAERS) and the Vaccine Safety Datalink (VSD).

Q3: What do ACIP, FDA, and CDC stand for, and what is the role of each in evaluating vaccines?

A3: ACIP = Advisory Committee on Immunization Practices; FDA = Food and Drug Administration; CDC = Centers for Disease Control and Prevention. All are involved in the initial approval and ongoing monitoring of vaccines.

Figure 2.11

Q1: What happens to an immunized person when a disease spreads through a population? (*Hint:* In the graphic, follow an immunized individual before and after a disease spreads.)

A1: The immunized person does not contract the disease.

Q2: Explain why a disease is less likely to spread to vulnerable members of a population if most people are immunized.

A2: "Herd immunity" means that fewer people contract the disease, and therefore vulnerable people are less likely to be exposed to a contagious person.

Q3: How does vaccination help an individual person? How does it help that person's community?

A3: An individual who is vaccinated is much less likely to become ill and is therefore less likely to pass on a disease to others in the community.

CHAPTER 3

END-OF-CHAPTER ANSWERS

1. a

2. d

3. ion: 3, matter: 6, solution: 9, element: 5, chemical compound: 8, molecule: 4, isotope: 7, polymer: 2, atom: 1

4. a

5. polymers, sugar, nucleotides, are not, carbon

6. a, c, e

7. Carbon can be the basis of more complex molecules than can hydrogen or oxygen because carbon can form four bonds (versus one for hydrogen and two for oxygen).

8. Six of the amino acids produced contained sulfur; these could not have formed in the absence of a sulfur-containing reactant.

9. b

10. b

11. (a) oven, (b) coffee maker, (c) neither, (d) oven, (e) coffee maker

12. The nonpolar end of the detergent will bond to the oil in the salad dressing while the polar end bonds to the water molecules, lifting the oil into the wash water. Vinegar is a polar molecule, so it will dissolve in the wash water; you don't need detergent to remove vinegar.

ANSWERS TO FIGURE QUESTIONS

Figure 3.1

Q1: How many protons, electrons, and neutrons does the hydrogen atom shown here have? What are the atomic number and the atomic mass number of the hydrogen atom?

A1: 1 proton, 1 electron, 0 neutrons. The atomic number (number of protons) and the atomic mass number (number of protons plus neutrons) are both 1.

Q2: What are the atomic number and the atomic mass number of the carbon isotope shown?

A2: The atomic number is 6 (6 protons); the atomic mass number, 12 (6 protons plus 6 neutrons).

Q3: Nitrogen-11 is an isotope of nitrogen that has 7 protons and 4 neutrons. What are the atomic number and atomic mass number of nitrogen-11?

A3: The atomic number is 7 (7 protons); the atomic mass number, 11 (7 protons plus 4 neutrons).

Figure 3.2

Q1: Before the experiment was run, the apparatus was sterilized and then carefully sealed. Why was this an important thing to do?

A1: Having a spotlessly clean apparatus was important to ensure that any amino acids were produced by the experimental conditions, not from contamination.

Q2: Why is inclusion of methane in the gas flasks an essential part of the hypothesis that complex organic molecules were formed in the early atmosphere of Earth? (*Hint:* What makes a molecule organic?)

A2: Amino acids contain carbon, and methane (a simple organic molecule) is the only chemical in the mixture that contains carbon.

Q3: Answer this question after reading about Miller's "steam injection" experiments: Where was the steam injected in the experimental apparatus?

A3: The steam was injected straight into the gas chamber, allowing the gases to interact with the water (and the other gases) more directly.

Figure 3.4

Q1: Where are the covalent bonds in this figure?

A1: The covalent bonds are located at the electrons shared between the oxygen atom and the two hydrogen atoms.

Q2: This figure shows a water molecule (H_2O). A hydrogen molecule (H_2) consists of two hydrogen nuclei that share two electrons. Draw a simple diagram of a hydrogen molecule indicating the positions of the two electrons.

A2: The electrons are equidistant from the two hydrogen nuclei because there is no difference in the strength of their attraction to electrons.

Q3: When table salt (sodium chloride, NaCl) dissolves in water, it separates into a sodium ion (Na^+) and a chloride ion (Cl^-). Which portion of a water molecule would attract the sodium ion, and which portion would attract the chloride ion?

A3: The Na^+ ions are attracted to the partial negative charges on the oxygen atoms, and the Cl^- ions are attracted to the partial positive charges on the hydrogen atoms.

Figure 3.5

Q1: Describe what will happen to the molecules of olive oil if you shake the bottle and then leave it alone for an hour. What about the molecules of vinegar?

A1: When you shake the bottle, the olive oil will form small beads of oil. As the bottle rests, these beads of oil will float back to the top of the bottle and coalesce into a layer of oil; that is, the bottle will return to its starting condition. Shaking the bottle will distribute the vinegar molecules through the water, and they will remain in solution.

Q2: What would happen if you added another fat to the bottle, such as bacon grease, and shook it?

A2: Bacon grease is a fat and, like the olive oil, would form small beads of fat that would float back to the top of the bottle and coalesce into a layer of grease.

Q3: Given how sugar behaves when it is mixed into coffee or tea, would you predict that it is hydrophobic or hydrophilic?

A3: Sugar dissolves in coffee and tea; therefore, it is hydrophilic.

Figure 3.6

Q1: Identify where in the picture water can be seen in its liquid, solid, and gas states.

A1: The water in the hot springs is in a liquid state. The snow and ice are solid water. Water vapor (steam) in the air is a gas.

Q2: In the gas state, water molecules move too rapidly and are too far apart to form hydrogen bonds. Compare the volumes occupied by an equal number of water molecules in the liquid, solid, and gas states.

A2: The same number of molecules will occupy more volume in both ice (solid) and water vapor (gas) than in liquid water. The volume of ice is defined (9% greater than the volume in water) because the water molecules form an ordered array. The volume of water vapor depends on its temperature.

Q3: Explain in your own words how ice floats on water.

A3: Water molecules are farther apart in ice than they are in liquid water. Because the same number of water molecules occupies a greater volume in ice than in water, the density of ice is lower than that of water, and ice floats.

Figure 3.7

Q1: Suppose you were going to repeat Miller's experiments. How would you decide how much of each gas to include in the chamber?

A1: Miller wanted to replicate the atmosphere of early Earth, so he based his mixtures on estimates of the proportions of the gases in the atmosphere at that time.

Q2: Why did the addition of steam to the gases in Miller's second set of experiments increase the yield of amino acids?

A2: The steam (hot water vapor) added heat energy to the mixture, and the water enabled more kinds of chemical reactions.

Q3: Miller used electrical energy in his experiment. What other forms of energy were present in the early atmosphere of Earth that could have led to the formation of complex molecules?

A3: Heat and ultraviolet light were present in early Earth's atmosphere.

Figure 3.8

Q1: How did the NASA scientists find the fragments of the meteorite that exploded over eastern Africa?

A1: The fragments from the asteroid were dark and stood out against the light-colored sand in Sudan.

Q2: What piece of evidence suggests that amino acids found in the meteorite fragments originated in outer space?

A2: Some of the amino acids from the meteorite fragments have a "right-handed" orientation, whereas all naturally occurring amino acids on Earth are in a "left-handed" orientation.

Q3: Speculate on the significance of finding extraterrestrial amino acids.

A3: The existence of extraterrestrial amino acids suggests that at least some of the building blocks of life are found in outer space.

Figure 3.9

Q1: Which has a higher concentration of free hydrogen ions: vinegar, pH 2.8; or milk, pH 6.5?

A1: A *high* concentration of free hydrogen ions corresponds to a *low* pH. Thus, vinegar has a higher concentration of free hydrogen ions than milk has.

Q2: What happens to the concentration of free hydrogen ions in your stomach when you drink a glass of milk?

A2: The concentration of free hydrogen ions decreases (that is, the pH increases).

Q3: Black coffee has a pH of 5. Does adding coffee to water (pH 7) increase or decrease the concentration of free hydrogen ions in the liquid?

A3: The concentration of free hydrogen ions increases.

Figure 3.10

Q1: In methane gas, how many electrons is each hydrogen atom sharing? How many is the carbon atom sharing?

A1: Each hydrogen is sharing one electron with carbon. Carbon is sharing four electrons.

Q2: In carbon dioxide, how many electrons is each oxygen atom sharing? How many is the carbon atom sharing?

A2: Each oxygen is sharing two electrons with carbon. Carbon is sharing four electrons.

Q3: Draw a molecule of formaldehyde (CH_2O). How many electrons is the oxygen atom sharing with the carbon atom? How many is the carbon atom sharing with the oxygen atom and with each hydrogen atom?

A3: Oxygen is sharing two electrons with carbon. Carbon is sharing two electrons with oxygen and one electron with each of the hydrogens.

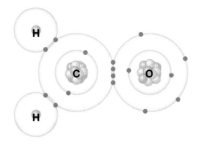

CHAPTER 4

END-OF-CHAPTER ANSWERS

1. c

2. d

3. receptor-mediated endocytosis: 2, phagocytosis: 1, pinocytosis: 4, exocytosis: 3

4. chloroplast: 7, Golgi apparatus: 4, lysosome: 5, mitochondrion: 6, nucleus: 1, rough endoplasmic reticulum: 2, smooth endoplasmic reticulum: 3

5.

		Eukaryotes	
Component	Prokaryotes	Animals	Plants
Plasma membrane	X	X	X
Cellulose cell wall			X
Nucleus		X	X
Endoplasmic reticulum		X	X
Golgi apparatus		X	X
Ribosomes		X	X
Cytoskeleton		X	X
Mitochondria		X	X
Chloroplasts			X

6. a

7. The dashed curve represents simple diffusion because as the concentration of solute increases, the rate of transport increases in a directly proportional way. The solid curve represents facilitated diffusion because it reaches the transport protein saturation line but cannot exceed it, since facilitated diffusion relies on these proteins to send the solutes across the membrane. If the transport proteins are fully in use, adding a higher concentration of solute will not increase the rate of transport.

8. b

9. right side, more, fewer

10. c

11. (a) isotonic, (b) neither gain nor lose, (c) equal to, (d) hypertonic, (e) lose, (f) higher than, (g) hypotonic, (h) gain, (i) lower than

ANSWERS TO FIGURE QUESTIONS

Figure 4.2

Q1: What was the purpose of inserting the gene that codes for blue pigment into the synthetic DNA?

A1: The blue color identified the cells that contained the DNA from *M. mycoides.*

Q2: What part of the transformed bacterium is synthetic?

A2: Only the DNA is synthetic; all of the structural components of the cells are from the *M. capricolum* cells into which the synthetic DNA was inserted.

Q3: Did this experiment create life?

A3: Although some articles in the popular press refer to the synthetic bacterium as a new life-form, it is better described as "repackaged life." The DNA is synthetic, but all the intracellular components that enable the DNA to function were already present in the cell.

Figure 4.3

Q1: Why is it important that the phosphate head of a phospholipid is hydrophilic?

A1: The fact that the phosphate head is attracted to water (hydrophilic) and also to other phosphate heads means that a bilayer will form.

Q2: What essential component of a cell do liposomes lack, and why is that omission important?

A2: Liposomes lack genetic material (DNA), so the characteristics of a liposome are not transmitted to its descendants.

Q3: Could the tendency of phospholipid bilayers to spontaneously form spheres have played a role in the origin of life? (*Hint:* Refer to "The Characteristics of Living Organisms" on page 6 of Chapter 1.)

A3: Yes. Once phospholipids formed (how that happened is still an open question), they would have formed liposomes spontaneously, trapping substances in their interiors.

Figure 4.4

Q1: In what ways is the plasma membrane a barrier, and in what ways is it a gatekeeper?

A1: It is a barrier in that it keeps out many molecules. It is a gatekeeper in that it selectively allows in other molecules.

Q2: Why can't ions (such as Na⁺) cross the plasma membrane without the help of a transport protein?

A2: The electrical charges on ions make them hydrophilic, so they cannot cross the lipid portion of the bilayer, which is hydrophobic, without the help of a transport protein.

Q3: If no energy were available to the cell, what forms of transport would not be able to occur? What forms of transport could occur? (*Hint:* Look ahead at Figures 4.5 and 4.6.)

A3: No form of active transport is possible without the input of energy. Any form of passive transport, including diffusion and osmosis, could occur even in the absence of an energy source.

Figure 4.5

Q1: Is the dye at equilibrium in any of these glasses? Describe how the first glass will look when the dye is at equilibrium with the water.

A1: Yes, the dye is at equilibrium in the third glass. If the first glass reaches equilibrium, it will look like the third glass.

Q2: Will diffusion mix the molecules of dye evenly through the water, or is it necessary to shake the container to get a uniform mixture?

A2: Diffusion is sufficient to mix the dye thoroughly, but it is a very slow process and you would probably get tired of waiting.

Q3: Will diffusion mix the dye faster in hot water than in cold water? Why or why not? (*Hint:* Review the discussion of the behavior of water molecules at different temperatures in Chapter 3.)

A3: Diffusion is faster at higher temperatures because the water molecules have more energy, form and break hydrogen bonds at a higher rate, and hence move around more rapidly.

Figure 4.6

Q1: What would the second diagram look like if the pores in the semipermeable membrane were too small to allow water molecules to pass through?

A1: The second diagram would look the same as the first diagram, since neither the sugar molecules nor the water molecules would be able to pass through the membrane.

Q2: What would the second diagram look like if the pores were large enough to let both water molecules and sugar molecules through?

A2: If the pores were large enough for sugar molecules to pass through, some sugar molecules would diffuse down the concentration gradient of sugar from the left side to the right while water molecules were diffusing from right to left down the concentration gradient of water. At equilibrium, the concentration of sugar and water would be the same on both sides of the membrane, and the depth of the solution would also be the same on both sides.

Q3: The fluid in an IV bag is isotonic to blood. What change would you see in the red blood cells of a patient if a bag of hypertonic solution was used in error?

A3: A hypertonic solution would be more dilute than blood, so an IV of hypertonic solution would dilute the blood—that is, increase the concentration of water in the blood. Now the red blood cells would have a lower concentration of water than the blood surrounding them, so water would move by osmosis into the red blood cells, causing them to swell.

Figure 4.7

Q1: If endocytosis itself is nonspecific, how does receptor-mediated endocytosis bring only certain molecules into a cell?

A1: The receptor protein embedded in the plasma membrane attracts and holds only specific molecules. When they attach, the plasma membrane bulges inward to engulf the molecule.

Q2: What sorts of molecules could be moved by endocytosis or exocytosis, but not by diffusion?

A2: Endocytosis and exocytosis can move molecules that are too large to pass through the plasma membrane through diffusion.

Q3: How does the fluid that enters a cell via pinocytosis differ from the fluid that enters by osmosis?

A3: The fluid that fills the inward bulge of a cell during pinocytosis may contain molecules that are in solution in the fluid outside the cell, whereas osmosis allows only water molecules to enter a cell.

Figure 4.8

Q1: What structures do prokaryotic and eukaryotic cells have in common?

A1: A plasma membrane, ribosomes, and DNA.

Q2: What cellular processes occur in both prokaryotic and eukaryotic cells?

A2: Both prokaryotic and eukaryotic cells regulate their internal concentration and the movement of substances in and out. Both require a supply of energy and carry out metabolic processes. Both respond to changes in their internal state and to conditions in the environment around them.

Q3: Both plants and animals are eukaryotes, but there are differences in their cellular structure. What are those differences?

A3: Plants have a cell wall and chloroplasts, while animals do not. Most animal cells also do not contain vacuoles.

CHAPTER 5

END-OF-CHAPTER ANSWERS

1. c
2. a
3. c
4. b
5. d
6. opposite, catabolism, produces

7. (See figure below)

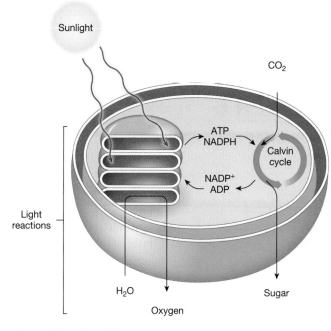

8. (a) 3, (b) 2, (c) 1, (d) 4

9. The Calvin cycle cannot be maintained indefinitely, because it requires a source of energy to produce ATP. This energy source (glucose) is created from sunlight energy during the light reactions.

10. e

11. (a) The reaction without an enzyme requires more energy to proceed, because its line on the graph shows a higher peak on the y-axis representing how much energy is needed.

    (b) This reaction is catabolic because there is less energy in the products than in the reactants, as shown by the line's lower position on the y-axis at the end of the reaction. Catabolic reactions release energy; therefore, their products have less energy than their reactants.

12. c

13. If metabolic processes occur at a higher rate than can be sustained by cellular respiration, then it is possible to die from lack of usable energy.

14. a

ANSWERS TO FIGURE QUESTIONS

Figure 5.2

Q1: Why is photosynthesis called "primary production"?

A1: Because photosynthesis converts two inorganic molecules (carbon dioxide and water) into an organic molecule (glucose).

Q2: How does animal life depend on photosynthesis?

A2: Only photosynthetic organisms can create organic molecules from inorganic molecules, and animals cannot carry out photosynthesis. (There are a very few exceptions to that generalization.)

Q3: Explain how photosynthesis and cellular respiration are "complementary" processes.

A3: They use each other's products and reactants.

Figure 5.3

Q1: What source of energy would plants use for anabolic reactions? Would an animal use the same kind of energy?

A1: Plants would use light energy from the sun for anabolism. Animals, in contrast, would use energy from food or glucose.

Q2: What source of energy would plants release in catabolic reactions? Would an animal release the same kind of energy?

A2: All living things release energy carriers, usually ATP, during catabolic reactions.

Q3: Create a mnemonic or jingle that helps you remember the difference between anabolism and catabolism.

A3: This question is just for fun and to aid in learning. Here's one possible answer: "Anabolism synthesizes; catabolism cannibalizes" (breaks down molecules).

Figure 5.4

Q1: Define ATP in your own words.

A1: Something like this: "ATP is a molecule that cells use to store and deliver energy."

Q2: How is ATP involved in anabolism and catabolism? (*Hint:* Review Figure 5.3.)

A2: Anabolism requires energy, which is provided most often by ATP. Catabolism releases energy, and would be involved in creating ATP from ADP.

Q3: Arsenic disrupts ATP production. Why would this characteristic cause it to be a potent poison?

A3: If ATP stopped being produced, our cells could no longer function and we would quickly die.

Figure 5.6

Q1: Is chlorophyll found only within chloroplasts?

A1: No. In bacteria it is embedded in membranes within the cell.

Q2: What could be an advantage of concentrating chlorophyll molecules in the membranes of chloroplasts?

A2: The location of the chlorophyll molecules maximizes the efficiency of the metabolic pathway of photosynthesis.

Q3: What is the advantage of having multiple chloroplasts per cell?

A3: Each chloroplast provides membranes needed for more and more embedded chlorophyll molecules.

Figure 5.7

Q1: What is the source of the carbon dioxide used for photosynthesis?

A1: Atmospheric CO_2.

Q2: Which products of the light reactions of photosynthesis does the Calvin cycle use?

A2: ATP and NADPH.

Q3: What are the two major products of photosynthesis?

A3: Glucose and oxygen.

Figure 5.8

Q1: Why is it important that enzymes are not permanently altered when they bind with substrate molecules?

A1: The fact that they are not permanently altered means that they can continue to perform their function without having to be present in great numbers or needing to be continually produced.

Q2: How would a higher temperature or higher salt concentration make it more difficult for an enzyme to function effectively?

A2: Both would change the shape of the enzyme so that it would not be able to bind to its specific molecules.

Q3: If a cell was unable to produce a particular enzyme necessary for a metabolic pathway, describe how the absence of that enzyme would affect the cell.

A3: The metabolic pathway would either proceed very slowly or, possibly, come to a complete halt, so the cell would not be able to function.

Figure 5.11

Q1: What are the products of cellular respiration?

A1: Carbon dioxide and water.

Q2: Considering the inputs and products of each process, why is cellular respiration considered the reciprocal process to photosynthesis?

A2: Photosynthesis uses carbon dioxide and produces oxygen, whereas cellular respiration does the opposite, using oxygen and producing carbon dioxide.

Q3: Which of the three stages of cellular respiration—glycolysis, the Krebs cycle, or oxidative phosphorylation—could organisms have used 4 billion years ago, before photosynthesis by cyanobacteria released oxygen into the atmosphere?

A3: Glycolysis, because it does not depend on oxygen. As the text says, "Glycolysis was probably the earliest means of producing ATP from food molecules, and it is still the primary means of energy production in many prokaryotes."

Figure 5.12

Q1: Which product released by fermentation accounts for the bubbles in beer?

A1: Carbon dioxide (CO_2).

Q2: Bakers of yeast breads also rely on fermentation, allowing bread to "rise" before baking. Describe what is occurring with the yeast as the bread rises.

A2: During bread rising, yeasts are performing metabolic functions—and growing and reproducing—using energy created through fermentation. The fermentation produces CO_2 as a waste product, which is trapped within the bread dough and causes it to rise.

Q3: Explain in your own words why lactic acid builds up in your muscles during strenuous physical activity.

A3: Something like this: "During strenuous exercise, my muscles can't get enough oxygen to produce all the needed ATP. So, glycolysis produces ATP anaerobically, which generates lactic acid as a by-product."

CHAPTER 6

END-OF-CHAPTER ANSWERS

1. b

2. a

3. (a) 4, (b) 1, (c) 2, (d) 3

4. Meiosis, binary fission, homologous chromosomes, sister chromatids

5. (a) 1, (b) 2, (c) 5, (d) 3, (e) 4

6. a

7. b

8. b

9. b

10. S phase. Chromosomes replicate during S phase; at some point during S phase, there will be the original DNA strands and half of the newly synthesized DNA strands. In all other phases, there is either the same amount of DNA as in the G_1 phase, or twice as much DNA as in G_1.

11. (a) 54, (b) 27, (c) 27

12. c

13. The G_1 checkpoint ensures that the cell is ready to divide—for example, that it is large enough and has enough energy to produce two normal daughter cells. The G_2 checkpoint ensures that the cell's DNA has been replicated and packed into pairs of sister chromatids. Bypassing the G_1 checkpoint could allow cells to divide before they're ready; bypassing the G_2 checkpoint could lead to the production of daughter cells with defective chromosomes.

ANSWERS TO FIGURE QUESTIONS

Figure 6.2

Q1: When is DNA replicated during the cell cycle?

A1: In interphase.

Q2: When in the cell cycle does DNA separate into the two genetically identical daughter cells?

A2: Late in the mitotic phase of cell division.

Q3: If a cell is not destined to separate into daughter cells, what phase does it enter? Is this part of the cell cycle?

A3: It enters the G_0 phase, which is not part of the cell cycle.

Figure 6.3

Q1: Name one similarity between cell division in prokaryotes and cell division in eukaryotes.

A1: Both prokaryotes and eukaryotes replicate DNA, separate the DNA to opposite poles of the cell, and then physically split the cell into two parts.

Q2: Why is binary fission referred to as "asexual reproduction"?

A2: Because it does not involve sharing DNA between individuals, as in sexual reproduction.

Q3: Name one difference between cell division in prokaryotes and cell division in eukaryotes.

A3: Prokaryotic cells are smaller and simpler than eukaryotic cells, so cell division happens more rapidly and is less complex; for example, prokaryotes do not have organelles, so they do not have to be divided up among daughter cells.

Figure 6.4

Q1: Do all cells in an organism enter each stage of mitosis at the same time? (*Hint:* See image of onion root tip at far left in the figure.)

A1: No.

Q2: What happens between the end of interphase and early prophase that changes the appearance of the chromosomes?

A2: The DNA within the chromosomes is condensed to prepare for division.

Q3: Explain in your own words the role of the mitotic spindle in mitosis.

A3: The mitotic spindle is responsible for accurately separating chromatids into daughter cells.

Figure 6.5

Q1: Why is it important for a chromosome to be copied before mitosis?

A1: There need to be two copies of each chromosome, so that each daughter cell has identical genetic material.

Q2: Are sister chromatids attached at the centromere considered to be one or two chromosomes?

A2: Sister chromatids are, when attached, one chromosome. When split apart, they are two separate but identical chromosomes.

Q3: Why is the chromosome's DNA tightly packed for mitosis and cytokinesis? (*Hint:* Think about what would happen if it were unpackaged, as during interphase.)

A3: Tightly packed chromosomes are more easily divided into daughter cells.

Figure 6.6

Q1: What could happen if the cell's checkpoints are disabled?

A1: The cell cycle could occur more quickly, because it will not stop to ensure that the checkpoint conditions are met.

Q2: What is the advantage of stopping the cell cycle if the cell's DNA is damaged?

A2: If the nutrient supply is inadequate, the daughter cells cannot grow. By stopping the cell cycle, the parent cell can wait until the nutrient supply increases.

Q3: Which part of the cell cycle may have been influenced in Soto and Sonnenschein's breast tumor cell experiments?

A3: The G_0 or G_1 checkpoint.

Figure 6.7

Q1: Is a zygote haploid or diploid?

A1: Diploid.

Q2: What cellular process creates a baby from a zygote?

A2: Mitosis.

Q3: If a mother or father was exposed to BPA prior to conceiving a child, how might that explain potential birth defects in the fetus?

A3: If the parent's BPA exposure caused mutation in sex cells (eggs or sperm), that could lead to a birth defect in the fetus.

Figure 6.8

Q1: Is a daughter cell haploid or diploid after the first meiotic division? How about after the second meiotic division?

A1: After meiosis I, a daughter cell is haploid (has one of each homologous chromosome). After meiosis II, each daughter cell is still haploid (has one of each homologous chromosome), and sister chromatids have split into separate daughter cells.

Q2: What is the difference between homologous chromosomes and sister chromatids?

A2: Sister chromatids are identical DNA molecules, replicated from a single DNA molecule, that remain bound to each other. They exist in a cell only from the S phase until anaphase of mitosis, or anaphase II of meiosis. A homologous chromosome pair consists of the two copies—one maternal, the other paternal—of the same type of chromosome. The pair is present at all times in diploid cells, but a haploid cell has just the paternal or the maternal copy.

Q3: If the skin cells of house cats contain 38 homologous pairs of chromosomes, how many chromosomes are present in the egg cells they produce?

A3: 38.

Figure 6.9

Q1: Why is the term "crossing-over" appropriate for the exchange of DNA segments between homologous chromosomes?

A1: Segments of DNA physically "cross over" between homologous chromosomes.

Q2: At what stage of meiosis (I or II) does crossing-over occur?

A2: Meiosis I.

Q3: What would be the effect of crossing-over between two sister chromatids?

A3: There would be no effect, because sister chromatids are genetically identical.

Figure 6.10

Q1: During meiosis, does random assortment occur before or after crossing-over?

A1: After.

Q2: What would be the effect on genetic diversity if homologous chromosomes did not randomly separate into the daughter cells during meiosis?

A2: Genetic diversity would decrease.

Q3: With two pairs of homologous chromosomes, four kinds of gametes can be produced. How many kinds of gametes can be produced with three pairs of homologous chromosomes? What does this suggest for the 23 homologous pairs of chromosomes in human cells?

A3: Three pairs of homologous chromosomes can produce eight kinds of gametes. Therefore, 23 homologous pairs could produce huge variation in gametes. To be more precise, it could produce 2^{23} (8,388,608) different combinations of chromosomes in gametes.

CHAPTER 7

END-OF-CHAPTER ANSWERS

1. genotype: 4, phenotype: 5, heterozygote: 1, homozygote: 2, dominant: 6, recessive: 3

2. gene, alleles

3. meiosis, segregation, independent assortment

4. (a) M, (b) C, (c) C, (d) C, (e) M

5. e

6. e

7. incomplete dominance

8. d

9. pleiotropic

10. b, c, e

11. The first-generation result suggests that the round-shape allele is dominant to the oval-shape allele. The next cross should be to breed offspring with themselves to create an F_2 generation, and the proportion of ovals would be 1 in 4, or 25 percent (3:1 round to oval).

12. Orange color must be dominant to black color. You could breed their offspring to test the hypothesis.

13. (a) *llff* (homozygous recessive for both long hair and no furnishings)
    (b) *LlFf, LLFF, LlFF, LLFf*
    (c) 9:3:3:1 short furnished to short unfurnished to long furnished to long unfurnished.

ANSWERS TO FIGURE QUESTIONS

Figure 7.2

Q1: What is the physical structure of a gene?

A1: A gene is a strand of DNA within a chromosome.

Q2: How many copies of each gene are found in the diploid cells in a woman's body?

A2: Two.

Q3: With 46 chromosomes in a human diploid cell, how many chromosomes come from the person's mother and how many from the father?

A3: Each parent contributes half of the chromosomes, so 23 chromosomes from the mother and 23 from the father.

Figure 7.3

Q1: Which might you observe directly: the genotype or the phenotype?

A1: The phenotype may be directly observable.

Q2: Which poodle could be heterozygous: the one with the black coat or the one with the brown coat?

A2: Only the black-coated poodle may be heterozygous.

Q3: Can you identify with certainty the genotype of a black poodle? A brown poodle?

A3: No for a black poodle; yes for a brown poodle.

Figure 7.4

Q1: What would you predict about the color of the F_1 plants' flowers?

A1: They should all be purple.

Q2: Why was it important that Mendel begin with pea plants that he knew bred true for flower color? Why couldn't he simply cross a purple-flowered plant and a white-flowered plant?

A2: The purple-flowered plant might be heterozygous *or* homozygous.

Q3: Over the years, Mendel experimented with more than 30,000 pea plants. Why did Mendel collect data on so many plants? Why didn't he study just one cross? *Hint:* Read "What Are the Odds?" on page 124 before answering.

A3: With data from more plants, there is a better chance that the results will accurately reflect reality.

Figure 7.5

Q1: Why did Mendel's entire F_1 generation look the same?

A1: All of the F_1 plants were heterozygous, so they all had a purple phenotype.

Q2: The phenotype ratio in the F_2 generation is 3:1 purple-to-white flowers. What is the genotype ratio?

A2: 1:2:1.

Q3: Draw a Punnett square for a genetic cross of two heterozygous, black-coated dogs. What are the phenotype and genotype ratios of their offspring?

A3: Phenotype ratio is 3:1 black to brown. Genotype ratio is 1:2:1 *BB* to *Bb* to *bb*.

Figure 7.6

Q1: List all the possible offspring genotypes and phenotypes.

A1: Round, yellow (dominant dominant): *RRYY*, *RrYY*, *RRYy*, *RrYy*.
Round, green (dominant recessive): *RRyy*, *Rryy*.
Wrinkled, yellow (recessive dominant): *rrYY*, *rrYy*.
Wrinkled, green (recessive recessive): *rryy*.

Q2: What is the offspring phenotype ratio?

A2: 9:3:3:1.

Q3: Complete a Punnett square for a genetic cross of two true-breeding Portuguese water dogs—one with a black, wavy coat (homozygous dominant, *BBWW*) and one with a brown, curly coat (homozygous recessive, *bbww*). What is the phenotype ratio of their offspring (F_1)? Now fill out another Punnett square, crossing two of the offspring. What is the phenotype ratio of the F_2 generation?

A3: F_1-generation phenotype ratio is 3:1 black wavy to brown curly. F_2-generation phenotype ratio is 9:3:3:1 black wavy to black curly to brown wavy to brown curly.

Figure 7.7

Q1: Boxers are far more inbred than poodles. Why does that inbreeding make the former a better target for genetic studies of disease than the latter?

A1: Since they are more inbred, boxers are more likely than poodles to be homozygous for traits of interest.

Q2: Explain why a geneticist interested in finding a gene linked to cancer would want to look at the DNA of senior golden retrievers with *and* without cancer?

A2: To know whether there was a genetic difference between goldens with and without cancer, the DNA of cancer-free dogs would have to be known.

Q3: Obsessive-compulsive disorder (OCD) in humans is characterized by obsessive thoughts and compulsive behavior, such as pacing. Canine compulsive disorder (CCD) is characterized by compulsive behavior such as "flank sucking," sometimes seen in Doberman pinschers. Would you predict that the medications given to humans with OCD would decrease compulsive behaviors in CCD dogs? Why or why not?

A3: One would predict that they might, since dogs and humans share many genes and therefore OCD and CCD may share a common genetic basis and be treatable by the same means.

Figure 7.8

Q1: What are the genotypes of a large and a small dog?

A1: *LL* (large) and *ll* (small).

Q2: Is it possible to have a heterozygous large dog? Explain why or why not.

A2: No, because *Ll* is a medium-sized dog.

Q3: Crossing a Great Dane and a Chihuahua is likely to be unsuccessful, even though they are members of the same species (and thus have compatible sperm and egg). Why is that? What are some potential risks of such a cross?

A3: It would be difficult for a Great Dane and a Chihuahua to mate, because of the size difference. One risk would be that if a female Chihuahua were to become pregnant this way, the pups would be too large for her to carry safely to term.

Figure 7.9

Q1: What are the possible genotypes (at both genes) of the black dog? The yellow dog? The brown dog?

A1: Black dog: *BBEE*, *BbEE*, *BBEe*, *BbEe*.
Yellow dog: *BBee*, *Bbee*, *bbee*.
Brown dog: *bbEe*, *bbEE*.

Q2: Draw a Punnett square showing possible matings between the black dog (assuming it is heterozygous at both genes) and the yellow dog (assuming it is heterozygous at the *B* gene). List all the possible phenotypes of their offspring. (See Figure 7.6 for an example of a Punnett square made with two traits.)

A2: There are three possible offspring phenotypes: black, yellow, and brown.

Q3: If you wanted the most variable litter possible, what colors of Labrador retrievers would you cross?

A3: If they were true-breeding (homozygous), you would cross yellow and brown.

Figure 7.10

Q1: The gene that brings about the pale Siamese body fur is also responsible in part for the typical blue eyes of the species. What is the term for this type of inheritance?

A1: Pleiotropy.

Q2: Siamese kittens that weigh more tend to have darker fur on their bodies. Why might this be?

A2: It could be that larger kittens have lower core temperatures, thus allowing more melanin to be produced on their bodies.

Q3: The Siamese cat pictured is called a "seal point" because it has seal-colored (dark brown) extremities. Some Siamese cats show the same color pattern, but the dark areas are of a lighter color or even a different shade—for example, lilac point, red point, blue point. What results would you predict if the experiments described in the text (shaving the cat and then increasing or decreasing temperature) were conducted on cats with these color patterns?

A3: The results should be the same, since even though the color of the melanin has changed, presumably the mechanism for laying down the melanin (based on temperature) has not.

CHAPTER 8

END-OF-CHAPTER ANSWERS

1. chromosomes, genes, loci, alleles

2. d

3. gene therapy: 3, in vitro fertilization: 5, preimplantation genetic diagnosis: 4, chorionic villus sampling: 1, amniocentesis: 2

4. The sex chromosomes are the two in the final pair. Since there is one large sex chromosome and one small one, the individual is a male.

5.
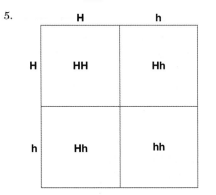

	H	h
H	HH	Hh
h	Hh	hh

6. d

7. e

8. c

9. b, d, e

10. a (where allele *T* represents stumpy tails and allele *t* represents long tails)

11. (a) from their mother, (b) no, (c) yes, (d) two types, (e) no, no, yes, female only

12. (a) $X^g X^g$, (b) $X^G X^g$, (c) $X^g Y$, (d) $X^g Y$, (e) $X^G X^g$

13. (a) *gg*, (b) *Gg*, (c) *Gg*, (d) *gg*

ANSWERS TO FIGURE QUESTIONS

Figure 8.2

Q1: Which two children in this pedigree have cystic fibrosis? How do you know?

A1: Individuals III-2 and III-3 have cystic fibrosis. They are depicted by filled symbols.

Q2: Does either parent of these two children have cystic fibrosis? If so, which one(s)? How do you know?

A2: Neither parent II-1 nor II-2 has cystic fibrosis. They are both depicted by open symbols.

Q3: Do any of the grandparents of these two children have cystic fibrosis? If so, which one(s)? How do you know?

A3: Yes. I-2, the paternal grandmother, has cystic fibrosis. She is depicted by a filled symbol.

Figure 8.3

Q1: How many male and how many female descendants (individuals that did not join the family by marriage) does generation IV of Aldrich's pedigree contain?

A1: 8 males and 4 females.

Q2: What proportions of the male and female descendants in generation IV were affected by the disorder?

A2: 50% (4/8, or 0.50) of males, and 0% (0/4, or 0.00) of females.

Q3: Why did Aldrich hypothesize that the disease was X-linked? (You will need to read ahead to answer this question.)

A3: The disease was observed only in males, but their mothers' male relatives were also affected.

Figure 8.4

Q1: Is this the karyotype of a male or a female?

A1: Male.

Q2: How would the karyotype of a person with Down syndrome differ from this karyotype?

A2: There would be three copies of chromosome 21.

Q3: The size of a chromosome correlates roughly with the number of genes residing on it. Why are an extra copy of chromosome 21 and a missing Y chromosome two of the least damaging chromosomal abnormalities?

A3: Chromosome 21 and the Y chromosome are two of the smallest chromosomes, so fewer genes would be affected by there being too many or too few of them (compared to a larger chromosome).

Figure 8.5

Q1: What are the odds that a given egg cell will contain an X chromosome? A Y chromosome? What are those odds for a sperm cell?

A1: An egg cell has a 100% chance of containing an X chromosome and a 0% chance of containing a Y chromosome. For the sperm cell, the odds are 50/50 for both cases.

Q2: If a couple has two daughters, does that mean their next two children are more likely to be sons? Explain your reasoning. (*Hint:* Refer back to "What Are the Odds?" on page 124.)

A2: No, the probability for each event is independent of prior and future events.

Q3: Sisters share the same X chromosome inherited from their father, but they may inherit different X chromosomes from their mother. What is the probability that brothers share the same Y chromosome? What is the probability that brothers share the same X chromosome?

A3: Brothers have a 100% probability of sharing the same Y chromosome, and a 50% probability of sharing the same X chromosome.

Figure 8.6

Q1: Why are changes in chromosome *number* almost always more severe than changes in chromosome structure?

A1: There are so many genes on an individual chromosome that deleting or adding an entire chromosome has massive effects on an individual.

Q2: In which part of meiosis would you predict that chromosomal abnormalities are produced? (Refer back to Chapter 6 if necessary.)

A2: During metaphase, when paired chromosomes are separated.

Q3: Create a mnemonic to help remember the four kinds of structural changes (for example, Doctors Improve Treatment Daily).

A3: Answers will vary.

Figure 8.7

Q1: How do we know whether two chromosomes are homologous?

A1: If the chromosomes carry alleles of the same genes and align during cell division, they are homologous.

Q2: In one sentence, explain how the terms "gene," "locus," and "chromosome" are related.

A2: A gene is found at a particular location—a locus—on a chromosome.

Q3: If hair color were determined by a single gene, what would be an example of the gene's alleles?

A3: *B* for brown hair, *b* for blonde hair.

Figure 8.9

Q1: Which of the children specified in this Punnett square represents Felix? What is his genotype?

A1: Felix is the "affected son," with genotype X^{a}Y.

Q2: Explain why Felix is neither homozygous nor heterozygous for the *WAS* gene.

A2: The *WAS* gene is found on the X chromosome, and Felix carries only one X chromosome because he is a male. Only with two chromosomes are homozygous and heterozygous genotypes possible.

Q3: Create a Punnett square to illustrate the offspring that could result if Felix had children with a noncarrier woman. What is the probability that a son would have WAS? What is the probability that a daughter would be a carrier of WAS?

A3: A son would have a 0% probability of having WAS. A daughter would have a 100% probability of being a carrier.

Figure 8.10

Q1: Which chromosome contains the gene for cystic fibrosis? For Tay-Sachs disease? For sickle-cell disease?

A1: Cystic fibrosis: chromosome 7. Tay-Sachs disease: chromosome 15. Sickle-cell disease: chromosome 11.

Q2: No known genetic disorders are encoded on the Y chromosome. Why do you think this is?

A2: The Y chromosome has very few genes; further, any disorder would always be expressed and therefore selected against.

Q3: In your own words, explain why most single-gene disorders are recessive rather than dominant.

A3: Dominant, single-gene disorders experience heavier selection than recessive disorders because they are always expressed (there are no carriers).

Figure 8.12

Q1: Which of the children in this Punnett square represents Zoe? What is her genotype?

A1: Zoe is the "affected child," with genotype *aa*.

Q2: If Zoe's parents had another child, what is the probability that the child would have cystic fibrosis? That the child would be a CF carrier?

A2: The child would have a 25% (¼) probability of having CF, and a 50% (½) probability of being a carrier.

Q3: If Zoe is able to have a child of her own someday, and the other parent is not a carrier of cystic fibrosis (he would likely be tested before they chose to have children), what is the probability that the child would have cystic fibrosis? That the child would be a carrier?

A3: The child would have a 0% probability of having CF, and a 100% probability of being a carrier.

Figure 8.13

Q1: What is the probability that a child with one parent who has an autosomal dominant disorder will inherit the disease?

A1: 50%.

Q2: Why are there no carriers with a dominant genetic disorder?

A2: Anyone with the gene would express the disorder.

Q3: Because dominant genetic disorders are rare, it is extremely rare for both parents to have the condition (genotype *Aa*). Draw a Punnett square with two *Aa* parents. What proportion of the offspring would have the disorder? What proportion would be normal?

A3: 75% of the offspring would have the disorder; 25% would be normal.

Figure 8.14

Q1: Which gene was missing or damaged in Felix's case? From what chromosome would a healthy copy be taken?

A1: The gene missing or damaged in Felix was the *WAS* gene. A healthy copy of the gene would be taken from an X chromosome.

Q2: Why did Dr. Klein's group first conduct gene therapy on mice rather than on humans? What are the advantages and limitations of this approach?

A2: In case of unforeseen problems or dangers, a human would not be hurt during the early trials. The limitation is that mice are not humans, so the results are not directly transferable.

Q3: If Felix has children of his own someday, will they run the risk of inheriting his disorder, or has gene therapy removed that possibility? Explain your reasoning.

A3: Gene therapy does not replace the damaged gene; it simply enables a missing gene product to be made. Felix remains the same genetically, so there will be a risk of his children inheriting the disorder.

CHAPTER 9

END-OF-CHAPTER ANSWERS

1. c

2. d

3. a

4. d

5. nucleotide: 4, base pair: 1, DNA molecule: 3, base: 2

6. (See figure below)

7. PCR, CRISPR

8. c

9. (a) 1, (b) 5, (c) 3, (d) 2, (e) 4

10. (a) D, (b) S, (c) I

11. ATGCAAATCCTGG
    TACGTTTAGGACC

12. (a) 20% T, (b) 30% G, (c) 30% C

13. It is not possible to delete the PERV sequences in each cell of the adult pig. Deleting the PERVs in the DNA of pig embryos ensures that none of the resulting cells will contain PERV sequences.

14. Yes, it matters. Noncoding DNA sequences still perform important functions and can easily be disrupted by mutations. Regulatory regions rely on specific nucleotide sequences, and they could become nonfunctional if those sequences are changed.

ANSWERS TO FIGURE QUESTIONS

Figure 9.2

Q1: Name one reason why, for a potential transplant, matching the size of organs shown would be important.

A1: The pig organs must fit into the space where the human organ was removed. Organs that are too big will not fit. Organs that are too small will not function at the level necessary to sustain life.

Q2: Name a tissue transplant for which size matching would not be as important.

A2: Tissue from the cornea of the eye need not be exactly the same size.

Q3: Name a tissue transplant for which size matching would not be important at all.

A3: Bone marrow will replenish in any organisms regardless of size matching.

Figure 9.3

Q1: Name two base pairs.

A1: Adenine-thymine (A-T) and cytosine-guanine (C-G).

Q2: Why is the DNA structure referred to as a "ladder"? What part of the DNA represents the rungs of the ladder? What part represents the sides?

A2: The structure looks like a ladder. The rungs are hydrogen bonds, and the sides are the bases.

Q3: Is the hydrogen bond that holds the base pairs together a strong or weak chemical bond? (*Hint:* Refer to Chapter 3 to review chemical bonds, if necessary.)

A3: Weak.

Figure 9.5

Q1: If all genes are composed of just four nucleotides, how can different genes carry different types of information?

A1: Each gene is composed of different numbers of nucleotides in different arrangements, thus allowing for many different types of information to be conveyed.

Q2: Would you expect to see more variation in the sequence of DNA bases between two members of the same species (such as humans) or between two individuals of different species (for example, humans and chickens)? Explain your reasoning.

A2: More variation would be expected between two individuals of different species. Within a species, individuals have the same genes but different alleles; across species, there are different genes *and* different alleles.

Q3: Do different alleles of a gene have the same DNA sequence or different DNA sequences?

A3: Different.

Figure 9.6

Q1: What common mechanism is employed by the guide RNA to find its target DNA sequence?

A1: Complementary base-pairing between the guide RNA and one strand of the target DNA.

Q2: How many strands of DNA must Cas9 cut to be effective?

A2: Both strands must be cut.

Q3: Does Cas9 also cause the deletion of DNA from the genome?

A3: No. Normal DNA repair proteins cause the deletion of DNA from the genome.

Figure 9.7

Q1: What are the structures that result from the first level of coiling around proteins called?

A1: Nucleosomes.

Q2: What makes up a "bead" and what makes up the "string" in the beads-on-a-string structure of DNA?

A2: The beads are the nucleosomes; the string is the double-stranded DNA linking them together.

Q3: What is the name for the structure that is more compact than the beads-on-a-string structure but less compact than an actual chromosome?

A3: Chromatin fiber.

Figure 9.8

Q1: Which of the DNA strands here are the template strands? Why are they called "template" strands?

A1: The strands from the original double helix are the template strands. They are called this because new DNA is built using the information from them as a template.

Q2: What must be broken before replication can begin?

A2: Hydrogen bonds between base pairs.

Q3: In your own words, explain why replication is described as "semiconservative."

A3: It is semiconservative because each new double helix of DNA is composed of one "conserved" strand from the original DNA molecule and one new strand.

Figure 9.9

Q1: PCR replicates DNA many times to increase the amount available for analysis. Why is this process called "amplification"?

A1: Because PCR substantially increases, or "amplifies," the quantity of DNA.

Q2: During the PCR cycle, what causes the DNA strands to separate?

A2: Heat.

Q3: Identify a difference between how PCR and DNA replication are accomplished.

A3: PCR is the targeted amplification of a specific region of DNA, while DNA replication is the duplication of an entire chromosome.

Figure 9.10

Q1: Summarize how DNA repair works and why the repair mechanisms are essential for the normal function of cells and whole organisms.

A1: An error is detected and tagged, then the damaged section of DNA is removed and replaced. Without DNA repair, mutations would persist and potentially result in cell death.

Q2: Is DNA repair 100 percent effective?

A2: No.

Q3: What would happen to an organism if its DNA repair became less effective?

A3: Its cells would have more trouble operating properly as the DNA's genetic instructions became less accurate.

Figure 9.11

Q1: What are the three types of point mutations?

A1: Substitution, insertion, and deletion.

Q2: Sickle-cell disease is an autosomal recessive genetic disorder. How many mutated hemoglobin alleles do people with sickle-cell disease have?

A2: Two.

Q3: Because of improved treatments, individuals with sickle-cell disease are now living into their forties, fifties, or longer. How might this extension of life span affect the prevalence of sickle-cell disease in the population?

A3: Sickle-cell disease will increase because people will be able to survive long enough to reproduce and pass on the sickle-cell allele.

Figure 9.12

Q1: Name a step in this process that is similar to the original CRISPR method that removed the PERVs from the pig genome. (*Hint:* Review Figure 9.6.)

A1: CRISPR is employed to remove a gene from the genome.

Q2: Name a step in this process that is not included in the original CRISPR method that removed the PERVs from the pig genome.

A2: Human stem cells are added to the pig embryos.

Q3: What parts of this process would scientists need to change in order to develop several different human organs in a single pig?

A3: Scientists would need to use CRISPR to target only the genes that are required for the development of each organ in the pig. Those organs would then be replaced by human organs via use of the same human stem cells added to create a single human kidney in the pig.

CHAPTER 10

END-OF-CHAPTER ANSWERS

1. gene expression: 2, gene regulation: 3, transcription: 1, translation: 4

2. (a) tRNA, (b) rRNA, (c) mRNA, (d) tRNA, (e) rRNA

3. redundancy, ambiguity

4. (See figure below)

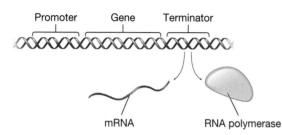

5. (a) 2, (b) 4, (c) 6, (d) 8, (e) 1, (f) 9, (g) 5, (h) 3, (i) 7

6. a, b, c, d

7. (a) asparagine, (b) stop codon, (c) isoleucine, (d) glycine, (e) proline

8. (a) CGU, CGC, CGA, CGG, AGA, or AGG; (b) GCU, GCC, GCA, or GCG; (c) AUG; (d) GGU, GGC, GGA, or GGG

9. b

10. e

11. At control point 2 (transcription), there might be an error in up-regulation. At control point 3 (breakdown of mRNA) there might be an error in down-regulation.

12. a

13. They are similar in that both involve building one molecule from another. They differ in that gene expression is DNA to RNA, while DNA replication is RNA to protein.

ANSWERS TO FIGURE QUESTIONS

Figure 10.3

Q1: Which is the faster way to produce vaccines: biopharming with plants or creating vaccines in eggs? Why is this important?

A1: Biopharming in plants is faster, making it possible to bring greater quantities of vaccine to the public to stem an outbreak before it becomes an epidemic.

Q2: How much cheaper is biopharming with plants than creating vaccines in eggs? Why is this important?

A2: Biopharming with plants is $364 million cheaper. The lower cost means that the vaccines will be marketed to the public at a much lower price. In addition, more insurance companies will cover the lower costs, and more individuals will receive the vaccine.

Q3: Why must tobacco-derived vaccines, or any new medications for that matter, be approved by the FDA?

A3: The FDA is responsible for ensuring the safety of drugs (and food). FDA approval is necessary because vaccines are drugs.

Figure 10.5

Q1: In which of the step(s) illustrated here does DNA replication occur? In which step(s) does gene expression occur?

A1: Replication: steps 4 and 5. Expression: step 6.

Q2: Why do vaccine producers not simply replicate the entire viral genome, instead isolating the gene for one protein and replicating only that gene?

A2: Two reasons: It would take more time to replicate the entire genome, and more important, the added time would allow the vaccine to replicate and act like the infectious virus, causing disease rather than preventing it.

Q3: What role do the bacteria play in this process? Why are they needed?

A3: Because they can produce proteins much more quickly than eukaryotes can, the bacteria rapidly generate the proteins needed for the vaccine.

Figure 10.6

Q1: Why is only one strand of DNA used as a template?

A1: Because the other strand would code for the *opposite* mRNA.

Q2: If a mutation occurred within the promoter or terminator region, do you think it would affect the mRNA transcribed? Why or why not?

A2: Yes, because it wouldn't be clear where the gene started or stopped, so the mRNA might not be transcribed at all, or it would grow too long.

Q3: The template strand of part of a gene has the base sequence TGAGAAGACCAGGGTTGT. What is the sequence of RNA transcribed from this DNA, assuming that RNA polymerase travels from left to right on this strand?

A3: ACUCUUCUGGUCCCAACA.

Figure 10.7

Q1: In your own words, define RNA splicing. When during gene expression does it occur?

A1: RNA splicing removes the introns from the RNA transcript. It occurs after transcription, but before the mRNA transcript is translated into a protein.

Q2: What do you predict would happen if the introns were not removed from RNA before translation? Why would it be a problem if the introns were not removed?

A2: If the introns were not removed, they would be translated, and the protein would be much larger and presumably nonfunctional.

Q3: Where is the mRNA destined to go once it has been transported out of the nucleus?

A3: Into the cytoplasm.

Figure 10.8

Q1: Which amino acid always begins an amino acid chain? Which codon and anticodon are associated with that amino acid?

A1: Methionine always begins the chain. The associated codon and anticodon are AUG and UAC, respectively.

Q2: Each of the codons for stopping translation binds to a tRNA molecule that does not carry an amino acid. How would the binding of a stop codon cause the completed amino acid chain to be released?

A2: Because there is no amino acid, the growing protein becomes detached from the mRNA.

Q3: Given the partial mRNA sequence that you specified in Figure 10.6's question 3 as being transcribed from the DNA template strand, what is the amino acid sequence that would be translated?

A3: The partial mRNA sequence ACUCUUCUGGUCCCAACA yields the amino acid sequence threonine-leucine-leucine-valine-proline-threonine.

Figure 10.9

Q1: How many codons code for isoleucine? For tryptophan? For leucine?

A1: Isoleucine: 3. Tryptophan: 1. Leucine: 6.

Q2: What codons are associated with asparagine? With serine?

A2: Asparagine: AAU, AAC. Serine: AGU, AGC.

Q3: From the partial mRNA sequence that you specified in Figure 10.6's question 3 as being transcribed from the DNA template strand, remove only the first A. What amino acid sequence would be translated as a result of this change? How does that sequence compare to the amino acid sequence you translated from the original mRNA sequence? *Bonus:* What kind of mutation is this? (*Hint:* See Chapter 9.)

A3: The partial mRNA sequence CUCUUCUGGUCCCAACA yields the amino acid sequence leucine-phenylalanine-tryptophan-serine-glutamine. There is one less amino acid, and each of the amino acids is different. *Bonus:* Deletion.

Figure 10.10

Q1: Why is an insertion or a deletion in a gene more likely to alter the protein product than a substitution, such as A for C, would?

A1: Because an insertion or deletion causes a "frameshift," so every single amino acid from that point on is likely to be different, as opposed to a substitution, where, at most, one amino acid changes.

Q2: Which would you expect to have more impact on an organism: a point mutation as shown here, or the insertion or deletion of a whole chromosome (discussed in Chapter 8)?

A2: The loss or addition of an entire chromosome, with all the genes on it, is likely to have far more impact on an organism than a mutation within a single gene.

Q3: Which mechanisms in a cell prevent mutations? (*Hint:* Refer back to Chapter 6 if needed.)

A3: Checkpoints in the cell cycle prevent (or at least repair) mutations.

Figure 10.11

Q1: As illustrated here, at what control point is transcription regulated?

A1: Control point 2.

Q2: What is a possible advantage of regulating gene expression before transcription, versus after?

A2: The cell does not waste time and energy producing mRNA transcripts that it will not use.

Q3: If you wanted to up-regulate production of the hemagglutinin protein in a tobacco plant carrying the hemagglutinin gene, at which control point(s) would that be possible? Justify your reasoning.

A3: Control points 1–4 all could have an impact on the amount of hemagglutinin being produced by a cell. The levels of DNA compaction, transcription, mRNA degradation, and translation all work together to increase or decrease the production of hemagglutinin. Modifying any one of these or any in combination can have a large impact on production.

CHAPTER 11

END-OF-CHAPTER ANSWERS

1. c

2. b

3. c

4. biogeography: 4, fossil record: 1, DNA sequence similarity: 2, embryonic similarity: 5, homologous traits: 3

5. d

6. b

7. adaptation, natural selection

8. In artificial selection, humans choose which organisms survive and reproduce. In natural selection, the environment determines which organisms survive and reproduce, by selecting for individuals with more beneficial inherited traits and against individuals with less useful ones.

9. d

10. species Y

11. d

12. All (or most) of the current continents.

ANSWERS TO FIGURE QUESTIONS

Figure 11.3

Q1: What is selective breeding, and how does it work?

A1: Selective breeding is the process by which humans allow only individuals with certain inherited characteristics to breed. Generation after generation, the selective breeder chooses which individuals mate and pass their traits to offspring.

Q2: Explain how selective breeding leads to artificial selection.

A2: Over time, the population resulting from generations of selective breeding can change significantly, and we can then say that artificial selection has occurred.

Q3: Name as many organisms as you can whose current characteristics are due to artificial selection.

A3: Examples include any domesticated animal, including common pets and agriculturally important animals. All agricultural plants are also products of artificial selection.

Figure 11.4

Q1: What is natural selection?

A1: Natural selection is the process by which individuals with genetic characteristics that are advantageous for a particular environment survive and reproduce at a higher rate than do individuals that have other, less useful characteristics.

Q2: If heavy rains caused an abundance of small, tender seeds and fewer large seeds, what do you predict would happen to the average beak size of the finches?

A2: Average beak size would become smaller, more like the beak of the bird on the left.

Q3: Compare and contrast artificial selection and natural selection. Name two ways in which they are similar. How are they different?

A3: They are similar in that both occur in populations, not individuals; that there must be a change or changes in the population; and that they occur over time—usually many, many generations. They are different in that artificial selection results from selective breeding performed by human beings, while natural selection occurs by the breeding of individuals in a population that survive in a particular environment.

Figure 11.6

Q1: What is the general definition of a fossil?

A1: Fossils are the mineralized remains of formerly living organisms or the impressions of formerly living organisms.

Q2: How are the ancestors of modern whales different from their present form?

A2: They became larger, they lost their back limbs, and their front limbs became proportionally smaller.

Q3: What is meant by the term "intermediate fossil" when referring to the fossil record?

A3: Intermediate forms or fossils are fossils of species with some similarities to the known extinct ancestral group and some similarities to the descendant or currently living species. They can be thought of as "missing links" in evolution.

Figure 11.9

Q1: Why do water-dwelling animals have thicker bones than land-dwelling animals?

A1: Thick bone is an adaptation to living in water. Thick bones are heavier and help water-dwelling animals control their buoyancy (ability to float).

Q2: Why does this thick-bone adaptation suggest a water-dwelling lifestyle?

A2: Fossil mammals with thick bones are presumed to have been water dwellers because almost all currently living water-dwelling mammals have thick bones.

Q3: How did this adaptation likely increase survival or reproduction in *Indohyus*?

A3: Thick bones probably enabled *Indohyus* to forage on the bottoms of lakes or ponds more efficiently than could species with lighter bones that had to work harder to stay submerged. A feeding advantage could have enabled *Indohyus* to eat more, live longer, and have a higher reproductive rate (producing more babies that survived) than those with lighter bones.

Figure 11.11

Q1: What is meant by the term "common ancestor"? Give an example.

A1: A common ancestor is the species from which at least two currently living species both descended, ancestral species that two or more new species arose from through a change in the traits of a population over time.

Q2: Why are homologous structures among organisms evidence for evolution?

A2: Homologous structures are parts of an organism that have changed in size or specific form over time but are easily determined to be the same structure in the ancestral species from which the organism evolved. Species with homologous traits are all related by originally coming from an ancestor with those specific structures.

Q3: Aside from skeletal structural similarities, what other commonalities among organisms might be considered homologous?

A3: Any traits that are shared by related organisms and also shared in an ancestor could be homologous traits. Some examples include mammary glands, egg laying, structures to extract oxygen from air or water, and the use of DNA as the genetic material.

Figure 11.12

Q1: Why are vestigial structures among organisms evidence for evolution? Give an example of another vestigial structure.

A1: Vestigial structures are evidence for evolution because they are shared among related species that all have a common ancestor. Goose bumps in humans are one example. In our furry ancestors, goose bumps fluffed the fur, thereby increasing its insulating effects and helping the animals to keep warm.

Q2: Are vestigial structures also homologous structures? Explain.

A2: Yes. Vestigial traits are shared in organisms that have a common ancestor. For example, all organisms that descended from a furry ancestor have goose bumps when they are chilled or cold.

Q3: Why do vestigial structures still exist if they are no longer useful?

A3: Only traits that harm an organism's ability to survive and reproduce disappear from the fossil record, because organisms having these traits die and do not reproduce. Traits that are merely useless and not harmful will persist in the organisms that have them. They may diminish because these structures are no longer needed and do not give organisms a selective advantage, but the organisms survive and reproduce just as well with or without them.

Figure 11.13

Q1: If a sequence from another species showed a 96 percent sequence similarity to humans, would that species be more or less closely related to humans than chimpanzees are?

A1: The hypothetical species would be less similar to humans than chimpanzees are (96% similar versus 98.4% in chimps).

Q2: Should similarities in the DNA sequences of genes be considered evolutionary homology? Explain.

A2: All living organisms use DNA as their genetic material, suggesting that the first true ancestral cell used DNA as its genetic material. DNA sequence similarity is a homologous trait because all descendant cells use DNA; exactly the same four nucleotide molecules, in different orders, make up the DNA in all cells on Earth.

Q3: How is the increased similarity in the DNA sequences of genes between more closely related organisms—and the decreased similarity between less closely related organisms—evidence for evolution? Use the example in this figure to support your answer.

A3: DNA sequence similarity is the gold standard for determining species relatedness because it is the genetic material in all cells on Earth—the best homologous trait that exists. The changes in DNA sequences in populations over time create the changes in traits that drive evolution. We can map the changes that occur in populations or species by looking at sequence similarity and re-creating a family tree. All other evidence for evolution supports the theory that all organisms derived from an ancestral cell that used DNA as its genetic material. The more related a species is to another species, the more similar the DNA sequences are. Humans and chimps are primates, mammals, and vertebrates, and they are more similar in DNA sequence than are humans and mice, which are only both vertebrates and mammals, but not both primates. Of the examples in this figure, chickens are the least similar to humans because although they are birds, which are vertebrates, they are not mammals or primates. A nonvertebrate animal like a jellyfish or worm would be even less similar to humans than the species named here, but it would likely still show some similarity.

Figure 11.14

Q1: Why should we expect to find *N. fosteri* fossils all over the world, given that it first evolved in Pangaea?

A1: We expect to find *N. fosteri* fossils all over the world because these organisms existed before the breakup of the mass continent Pangaea. During the breakup, these organisms traveled with their continent to the current locations.

Q2: Can we use biogeographic evidence to support evolution without using fossil evidence? Give examples.

A2: Yes. We can use the current locations of living organisms that are related to support evolution by biogeography. For example, members of the primate family are found on almost all the continents on Earth, suggesting that their common ancestor lived at the time of Pangaea.

Q3: How might we use DNA sequence similarities together with biogeography as evidence for evolution?

A3: We can couple DNA sequencing with the locations of fossil or living organisms to support evolution by common descent. We can perform DNA sequence analysis on all the primates on Earth, coupled with their current locations, to reenact the history of primate evolution.

Figure 11.15

Q1: Why are the similarities among organisms during early development evidence for evolution? Give an example.

A1: Similarities between organisms during early development suggest that they have a common ancestor whose early development occurred in the same or similar manner. All vertebrates go through similar stages of development in the early embryo. Many invertebrate organisms also share the same steps in embryonic development.

Q2: Are the similar structures among vertebrate species during embryogenesis homologous structures? Explain.

A2: Yes. The similar structures are homologous traits because they are shared with a common ancestor.

Q3: Why do embryonic structures still exist at points during embryogenesis if they are not used after birth?

A3: These structures can be considered vestigial traits, since they are now useless to the organism in which they still exist embryonically. Remember, vestigial traits still exist because they do not harm the organisms' ability to survive and reproduce.

CHAPTER 12

END-OF-CHAPTER ANSWERS

1. genetic drift, establish a new distant population

2. b

3. b

4. c

5. b

6. b

7. An individual that survives well but is unattractive to potential mates or is unable to compete for access to mates will not reproduce and will not pass on genes.

8. Gene flow is the most likely mechanism because mutations are random, natural selection would have caused populations in different environments to diverge, and genetic drift is most relevant for small populations.

9. Population bottlenecks cause individuals in the resulting population to be more genetically similar to each other. In this case, the two individuals are so similar that each individual does not distinguish another devil's cells as different from its own cells.

10. b

11. b

ANSWERS TO FIGURE QUESTIONS

Figure 12.2

Q1: What is natural selection selecting for here?

A1: Methicillin-resistant *S. aureus* (MRSA).

Q2: Why do bacteria that are not genetically resistant to antibiotics die out when exposed to antibiotics?

A2: After entering a bacterium, an antibiotic generally blocks or poisons one or more processes of the bacterium's life cycle so that it cannot survive or reproduce. Bacteria that have a mechanism to pump out the poison generally survive the poison and live to reproduce; they are termed "resistant."

Q3: Why is the antibiotic represented by a kitchen strainer in this figure?

A3: The antibiotic is depicted as a kitchen strainer because antibiotics act like strainers: they can "catch" or kill most bacteria in a population, but there will always be at least one bacterium that can survive the antibiotic assault and slip through the strainer.

Figure 12.4

Q1: What is the difference between MRSA and VRSA?

A1: Both of these bacteria are members of the species *Staphylococcus aureus*. The species name is the "SA" part of both names. The "MR" in MRSA stands for "methicillin-resistant," and the "VR" in VRSA stands for "vancomycin-resistant." The only difference between these two populations is that MRSA survives in the presence of the antibiotic methicillin but can be killed by vancomycin, while VRSA survives in the presence of both methicillin and vancomycin.

Q2: Why is there a clear zone (the "zone of inhibition") around the paper disk in the top dish but not in the bottom dish?

A2: The clear zone represents the area where the antibiotic has seeped into the medium from the antibiotic-soaked paper disk and killed off the bacteria. The bacteria cannot grow here, so all you see is the growth medium in the dish, with no bacteria growing on it. The rest of the dish is covered by bacteria and appears opaque.

Q3: Why is the lack of a clear zone around the paper disk in the bottom dish so alarming?

A3: The lack of a clear zone in the bottom dish suggests that the antibiotic of last resort, vancomycin, cannot kill the bacteria and they grow just as well in the antibiotic area as in the areas away from the vancomycin-soaked paper disk. If the last-resort antibiotic doesn't kill these bacteria, there is no current antibiotic that will. VRSA is a deadly bacterial infection against which we have no good defense.

Figure 12.5

Q1: What would the white-fur-pigment allele frequency be if three of the homozygous black allele mice (having two black alleles) were heterozygous (having one white and one black allele) instead?

A1: 16/30 = 53%.

Q2: What would the white-fur-pigment allele frequency be if all of the white mice died and were therefore removed from the population? Would the black-fur-pigment allele frequency be affected? If so, how?

A2: The white-fur-pigment allele frequency would be 3/20 = 15%. Yes, the black-fur-pigment allele frequency would be affected; there would then be 17 black alleles out of a total of only 20 alleles: 17/20 = 85%.

Q3: What would the white-fur-pigment allele frequency be if all of the gray mice died and were therefore removed from the population?

A3: 10/10 = 100%.

Figure 12.6

Q1: Why does the population of *S. aureus* bacteria *not* pose a life-or-death health threat outright?

A1: These are the bacteria that normally live on our skin and do not harm us unless there is a major skin disturbance like a burn or a large scrape that is not cleaned and kept protected.

Q2: Why do vancomycin-resistant bacteria have a higher frequency in the population after treatment with vancomycin?

A2: All of the bacteria that do not contain the resistance allele are killed by the vancomycin and therefore no longer exist. The only bacteria left are the vancomycin-resistant bacteria (VRSA).

Q3: If this figure used the mouse example of allele frequency from Figure 12.5, and the white mice increased in numbers like the vancomycin-resistant bacteria here did, what would happen to the frequencies of the white-fur-pigment and black-fur-pigment alleles?

A3: The white-fur-pigment allele would increase in frequency, and the black-fur-pigment allele would decrease in frequency.

Figure 12.7

Q1: If one extreme phenotype makes up most of a population after directional selection, what happened to the individuals with the other phenotypes?

A1: They were killed and eaten by predators.

Q2: What do you think would happen to the phenotypes of the peppered moth if the tree bark was significantly darkened again by disease or pollution?

A2: Since the moths that survive are more similar to the color of the bark and thus are protected from birds, which cannot as easily see them to kill and eat them, the phenotype of the population of peppered moths would become darker like the trees.

Q3: What do you think would happen to the phenotypes of the peppered moth if the tree bark became a medium color, neither light nor dark? (You will need to read the next paragraph to answer this question.)

A3: Stabilizing selection would likely occur, and only medium-colored moths would not be killed and eaten by birds.

Figure 12.8

Q1: Think of another example of stabilizing selection in human biology. Has modern technology or medicine changed its impact on the resulting phenotypes?

A1: Stabilizing selection probably affected many human traits before modern technology and medicine played a major role in survival and quality of life. Examples include adult height and weight, which would be affected by many hormone levels and overall metabolism.

Q2: How do you think a graph of birth weight versus survival for a developing country with little health care would compare to the graph shown here?

A2: This graph would be even sharper, with less survival at either end. Evolution would be more stabilizing than in the example shown here.

Q3: How do you think a graph of birth weight versus survival for an affluent city in the United States today would compare to the graph shown here?

A3: This graph would be much wider, with more survival at both ends. Evolution would be much less stabilizing than in the example shown here.

Figure 12.9

Q1: Almost all birds starved during the dry season depicted here. What type of selection would have been present if only the intermediate-beaked birds had survived (instead of the small- and large-beaked birds)?

A1: Stabilizing selection.

Q2: Describe a scenario in which African seed crackers would experience directional selection for either smaller- or larger-beaked birds. What kind of environmental conditions might bring about such a situation?

A2: If only small seeds were produced in a particular year, then only small-beaked birds would survive and the population would evolve toward smaller beaks. Similarly, if only large seeds were produced, then only large-beaked birds would survive and the population would evolve toward larger beaks. An environmental condition that might bring about smaller seeds would be a situation in which the faster germination time of smaller seeds was an advantage—for example, a very brief growing period. A condition that might favor larger seeds would be a situation in which the greater resources contained in the seed would allow it to survive longer—for example, an extended drought.

Q3: Of the three patterns of natural selection presented in this discussion, which one always results in two different phenotypes in the following generations?

A3: Disruptive selection.

Figure 12.10

Q1: How is convergent evolution different from evolution by common descent?

A1: Convergent evolution is essentially the opposite of evolution by common descent. Convergent evolution begins with two distantly related organisms that, over many generations, end up with similar phenotypes because they have adapted to similar environments. Evolution by common descent begins with an original common ancestor and, over many generations, may split into many different populations that are phenotypically different.

Q2: What is the main difference between a homologous trait (see Figure 11.11) and an analogous trait?

A2: A homologous trait is shared between organisms because a common ancestor had that trait; an analogous trait performs a similar function in different organisms but is not shared by a common ancestor. Analogous traits form through convergent evolution.

Q3: Why are convergent traits considered evidence for evolution (see Chapter 11)?

A3: Convergent traits occur through changes in allele frequencies over time—essentially the definition of evolution. Convergent evolution results in organisms that are better adapted to their environment—again, evolution.

Figure 12.13

Q1: If a goose with genotype *AA* had migrated instead of the goose with genotype *aa*, would the scenario described here still be considered gene flow? Why or why not?

A1: No, this is technically not gene flow. Although alleles are being exchanged, they are the same as the existing alleles in the population and will not change allele frequencies over time and many generations.

Q2: If a goose with genotype *Aa* had migrated instead of the goose with genotype *aa*, would the scenario still be considered gene flow? Why or why not?

A2: Yes, this is gene flow. Although the effect is not as extreme as with the *aa* genotype, the *Aa* genotype introduces a new allele into an existing population, creating offspring that can be *Aa*, and thereby changing allele frequencies over time and many generations.

Q3: If the goose with genotype *aa* had migrated to population 2 as shown but had failed to mate with any of the *AA* individuals, would the scenario still be considered gene flow? Why or why not?

A3: No, this is not gene flow. Just adding an individual with different alleles to a population does not count as gene flow. There must be an exchange of alleles between the newcomer and an existing individual.

Figure 12.14

Q1: Why do you think a genetic bottleneck is more likely to occur in a small population than in a large population?

A1: In a large population, it is less likely that a chance event can kill off almost all of the individuals, leaving only a few behind that randomly represent only one of multiple phenotypes. In a small population, a tsunami, hurricane, volcanic eruption, or other natural disaster could easily kill off all but a few individuals. All subsequent offspring would arise from these few individuals, whatever phenotype they might have, regardless of which phenotypes in the original population were best adapted.

Q2: Genetic drift is often described as a "chance event." Give other examples of chance events that could cause a genetic bottleneck.

A2: Examples that could cause a genetic bottleneck include deadly viruses, famine, drought, immigration of many predators, habitat loss, tsunami, or other natural disaster.

Q3: Which resulting population has more genetic diversity?

A3: The population on the left.

CHAPTER 13

END-OF-CHAPTER ANSWERS

1. b
2. a
3. c
4. c
5. genetic divergence, allopatric
6. Postzygotic barriers, an infertile hybrid
7. d
8. b, c, d
9. (a) 5, (b) 4, (c) 1, (d) 2, (e) 3
10. b
11. b
12. Any large geographic barrier such as a river, canyon, or mountain range might bring about allopatric speciation. In addition to the barrier, genetic divergence between the separated populations is necessary for speciation.
13. The ten species must have differed significantly in traits that were unobservable (or at least were not observed by the scientists) and that meant the fish were unable to mate and produce viable offspring.

ANSWERS TO FIGURE QUESTIONS

Figure 13.4

Q1: How do we know that these rattlesnakes are members of the same species?

A1: The caption specifies that they successfully mated and the resulting offspring survived and reproduced.

Q2: How would you design an experiment to determine whether two populations are distinct species according to the biological species concept?

A2: Your experiment would require mixing individuals from the two populations under conditions conducive to sexual reproduction. If the individuals did mate, the offspring would then need to be raised to maturity and be set up to also mate and produce live offspring.

Q3: For which types of populations does the biological species concept *not* work as a way of determining how they are related?

A3: Because it requires sexual reproduction, the biological species concept cannot be applied to populations that reproduce asexually (such as bacteria).

Figure 13.5

Q1: List three kinds of information that scientists use to distinguish between species.

A1: Biogeographic information, DNA sequence similarity, and physical characteristics.

Q2: What differences can you observe between the individuals in the photos? Why are these differences not enough to confirm that they are from two different species?

A2: There are several striking differences in coloration (for example, along the side of the body) between the frogs in the two photos. These differences could simply be within-species variation, much like hair and eye color differences between humans of the same species.

Q3: How is genetic divergence among populations determined?

A3: Genetic divergence between two populations is determined by examination of the DNA sequences of many individuals in the two populations. A lot of similarity between DNA sequences suggests little genetic divergence, while the existence of many changes between the gene sequences suggests much greater genetic divergence.

Figure 13.6

Q1: What is the definition of gene flow? How was gene flow blocked between these species?

A1: Gene flow is defined as the passing of alleles between different populations of the same species. Squirrels from the populations on either side of the river could not mate because of the geographic barrier, and thus gene flow was blocked.

Q2: Name as many types of geographic barriers as you can. Which do you think would be the best at blocking gene flow?

A2: Examples of geographic barriers include but are not limited to rivers, lakes, oceans, glaciers, mountains, canyons, brick walls, freeways, fences. The larger the barrier, the better it blocks two individuals from finding each other and mating.

Q3: Are geographic barriers universal for all species? If not, name a geographic barrier that might block gene flow for one species but not another.

A3: No, geographic barriers are not universal. A river, for example, might block gene flow between two lizard populations but not two bird populations.

Figure 13.7

Q1: What factors must be present for allopatric speciation to occur?

A1: A geographic barrier.

Q2: If a geographic barrier is removed and the two reunited populations intermingle and breed, what attributes must the offspring have in order for the two populations, according to the biological species concept, to be considered still the same species?

A2: The offspring must be viable (alive) and fertile (be able to reproduce).

Q3: If the two populations in question 2 are determined to still be the same species, did allopatric speciation occur?

A3: If they are still the same species, no speciation occurred.

Figure 13.11

Q1: Describe how coevolution, as with the hummingbird bill and hummingbird-pollinated flowers, is different from the kind of evolution described in Chapters 11 and 12.

A1: In coevolution, a species evolves directly to interact better with another species. Coevolution can be both species evolving to interact better with each other, or it can be just one of the species adapting to the other species.

Q2: Is coevolution the same thing as convergent evolution, described in Chapter 12? Why or why not?

A2: No. In convergent evolution, two genetically different species look more alike over time because they are adapting to similar environments. In coevolution, two different species adapt to each other's adaptations over time.

Q3: Do you think one species' adapting over time to feed specifically and extremely successfully on another species is an example of coevolution? Why or why not?

A3: Yes. Coevolution does not have to be reciprocal.

Figure 13.12

Q1: What is the main difference between allopatric and sympatric speciation?

A1: Allopatric speciation requires a geographic barrier, and sympatric speciation cannot include a geographic barrier.

Q2: Name two events that must happen for both allopatric speciation and sympatric speciation to occur.

A2: The two populations must be reproductively isolated, and genetic change must occur.

Q3: Do you think all of the 500 species in Lake Victoria arose through sympatric speciation? Why or why not?

A3: Yes, because there are no geographic barriers to separate the populations in the lake. There would have to be a human-made wall or fence for allopatric speciation to occur.

Figure 13.13

Q1: What does "prezygotic" mean?

A1: "Prezygotic" means "before a zygote" or "before fertilization of an egg by a sperm"—in other words, no fusion of egg and sperm.

Q2: How is the booby's ritual dance a prezygotic reproductive barrier?

A2: This dance happens before mating. If the dance is not correct, no mating happens. No mating means no fusion of egg and sperm.

Q3: What are some other prezygotic reproductive barriers besides a mating dance?

A3: Examples include all geographic barriers, inability of egg and sperm to fuse for genetic reasons (gamete incompatibility and isolation), inability to mate because the genitalia are physically incompatible (mechanical incompatibility and isolation), and ecological isolation in which two species breed in different portions of their habitat, at different seasons, or at different times of day.

Figure 13.14

Q1: Which species is/are sympatric with *Echinometra lucunter*?

A1: *E. viridis*, because it is found on the same side of the Panama landmass as *E. lucunter*.

Q2: Which species is/are allopatric with *E. lucunter*?

A2: *E. vanbrunti*, because it is found on the opposite side of the Panama landmass as *E. lucunter*.

Q3: If two individuals have incompatible gametes, what will be the result of a mating event between them?

A3: No zygote will form. Egg and sperm will not fuse.

CHAPTER 14

END-OF-CHAPTER ANSWERS

1. c

2. b

3. a

4. d

5. clade: 2, node: 3, lineage: 5, evolutionary tree: 4, shared derived trait: 1

6. prokaryotes, Eukarya, Plantae, Animalia

7. d

8. e

9. (a) Eukarya, Animalia; (b) Eukarya, Fungi; (c) Eukarya, Plantae; (d) Eukarya, Plantae; (e) Eukarya, Animalia

10. (a) kingdom, phylum, class, order, family, genus, species; (b) domain; (c) answers will vary

11. (a) 3, (b) 1, (c) 4, (d) 2, (e) 5

12. (See figure below)

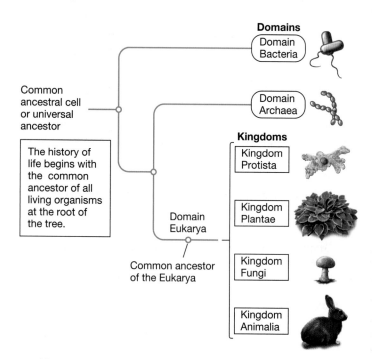

13. (a) Bacteria, Archaea, Protista; (b) Bacteria, Archaea; (c) Bacteria, Archaea; (d) Animalia; (e) Plantae, Bacteria

14. **Xu:** We hypothesize that *Archaeopteryx* and *Xiaotingia* are dinosaurs, closely related to the deinonychosaurs.
Godefroit: We hypothesize that *Archaeopteryx* and *Xiaotingia* are early birds, and *Aurornis* is an earlier bird.

ANSWERS TO FIGURE QUESTIONS

Figure 14.2

Q1: Why is there a shared line from the universal ancestor for Archaea and Eukarya?

A1: Archaea and Eukarya share a common ancestor with each other more recently than either one does with Bacteria.

Q2: Where would birds be found within this figure? What about humans?

A2: Birds and humans would both be found as branches of Eukarya.

Q3: To which domain would you expect a disease-causing organism to belong? What if the organism was multicellular?

A3: Disease-causing organisms could be found within Bacteria, but also within Eukarya (kingdom Fungi). Multicellular organisms would be within Eukarya only.

Figure 14.3

Q1: During what geologic period did life on Earth begin?

A1: The Precambrian.

Q2: How long ago did species begin to move from water to land? What period was this?

A2: About 480 mya, in the Ordovician.

Q3: In what period would *Archaeopteryx* have been alive?

A3: The Jurassic.

Figure 14.5

Q1: In what ways were theropods the same as modern birds? Give at least two similarities.

A1: They ran on two legs and had hollow, thin-walled bones.

Q2: In what ways did theropods differ from modern birds? Give at least two differences.

A2: They were more variable in size and in skin covering.

Q3: Birds are often referred to as "living dinosaurs." Is this accurate? Why or why not?

A3: Birds are direct descendants of dinosaurs, so they could be argued to *be* dinosaurs.

Figure 14.6

Q1: In the traditional tree, identify the node showing the common ancestor for early birds and dinosaurs.

A1: The common ancestor came after the split from the theropods.

Q2: What do both the traditional tree and Xu's tree suggest about troodontids and dromaeosaurids?

A2: Both trees suggest that these two groups were closely related.

Q3: In both trees, identify the node for the common ancestor of *Archaeopteryx* and other birds. In what way are the nodes different in the two trees?

A3: The traditional tree shows *Archaeopteryx* on the bird side of the split between birds and dinosaurs; the common-ancestor node is the point where the birds split into two groups. Xu's tree shows *Archaeopteryx* on the dinosaur side of the split; the common-ancestor node is the theropods.

Figure 14.9

Q1: What group of organisms shares the most recent common ancestor with plants?

A1: Green algae.

Q2: Are fungi more closely related to plants or to animals? Does the answer surprise you? Why or why not?

A2: Fungi are more closely related to animals.

Q3: If you were to create an evolutionary tree in which amoebas were included within the kingdom of organisms to which they were the most closely related (rather than with protists, where they are currently placed), where would you put them?

A3: Either with Animalia or with Fungi.

Figure 14.10

Q1: Within which category are individuals most closely related to one another?

A1: Species.

Q2: Within which category are individual species most distantly related?

A2: Kingdom.

Q3: Are individual species more closely related within the same order or within the same family?

A3: Family.

Figure 14.14

Q1: Is *Xiaotingia* an earlier or later bird than *Archaeopteryx* in this tree?

A1: *Xiaotingia* is later than *Archaeopteryx*.

Q2: If a future study, based on more fossils or new measurements, placed *Archaeopteryx* back with dinosaurs, would this suggest that birds are not related to dinosaurs? Why or why not?

A2: No. What is in question is not *whether* birds are related to dinosaurs, but when exactly they split off from related dinosaur species and which species is the first example of that split.

Q3: If you were to create an evolutionary tree of modern birds, where would you expect to place the roadrunner (judging by its appearance in this figure) as compared to a house sparrow or pigeon?

A3: The roadrunner looks more like a dinosaur, so it might be argued that it belongs closer to the base of the tree.

Figure 14.15

Q1: What extinction event occurred about 200 mya? What animal groups were most affected by this event?

A1: Triassic; reptiles.

Q2: Which of the mass extinctions appears to have removed the most animal groups? How long ago did this extinction occur?

A2: Permian; about 250 mya.

Q3: The best studied of the mass extinctions is the Cretaceous extinction. Why do you think it has been better studied than the other extinctions?

A3: Because it is more recent, more fossils may be available from this extinction than from earlier ones.

CHAPTER 15

END-OF-CHAPTER ANSWERS

1. d
2. e
3. c
4. b
5. Archaea, Bacteria, Prokaryotes, eukaryotes
6. b
7. e
8. (a) 1, (b) 5, (e) 3, (d) 2, (c) 4
9. a
10. Each individual eukaryote weighs much more than an individual prokaryote.

11. Prokaryotes are smaller and have simpler cellular structures, so they take less time to divide into two. Because they replicate so much more quickly, it is hard for our immune systems to respond to them in a timely way.

ANSWERS TO FIGURE QUESTIONS

Figure 15.3

Q1: If an individual prokaryote divides every 20 minutes, how many individuals will there be after an hour?

A1: 8.

Q2: If the generation time is 20 minutes, how much time will have gone by when the final generation shown has doubled?

A2: 20 minutes.

Q3: Many bacteria are able to reproduce more quickly in warmer conditions. What does this suggest to you about the importance of refrigerating foods?

A3: Refrigerating food slows generation time, so dangerous bacteria like *E. coli* have less of a chance to increase to threatening levels.

Figure 15.4

Q1: In which of the three ways did the navel microbiome participants contribute?

A1: They were experimental subjects.

Q2: Which of the advantages listed above do you think the navel microbiome citizen scientists received?

A2: They may have felt a sense of contribution and purpose.

Q3: Would you be willing to contribute to the navel microbiome project? Why or why not?

A3: Various answers possible; some might express privacy concerns.

Figure 15.5

Q1: The figure (and thus the study) demonstrates that Archaea and Eukarya are more closely related to each other than to Bacteria. How is that illustrated?

A1: The first split seen is Bacteria, apart from the other groups.

Q2: Where in the figure would you place the first life found on Earth?

A2: At the base of the tree.

Q3: Where in the figure is the earliest split between Archaea and Eukarya? When did that occur?

A3: The earliest split occurred where the cluster identified as Eukarya begins. This would have been about 2.7 million years ago.

Figure 15.7

Q1: Which of these shapes do you think *Streptococcus* would take?

A1: Sphere, or coccus.

Q2: From the micrographs here, does it appear that all prokaryotes have a flagellum?

A2: No.

Q3: Which one of these shapes is most clearly capable of self-motility? Why?

A3: The comma, or vibrio, because of the flagellum.

Figure 15.8

Q1: Which shape in Figure 15.7 corresponds to the archaeans from deep-sea thermal vents?

A1: Rod, or bacillus.

Q2: Why are many archaeans referred to as "extremophiles"?

A2: They are found in extreme environments (at extremes of heat, acidity, and salt level).

Q3: Is there anywhere you think archaeans could *not* survive? Justify your answer.

A3: Various answers are possible.

Figure 15.9

Q1: How do individual bacteria know that they have a "quorum"?

A1: They detect increasing concentrations of signaling molecules.

Q2: There is a well-known biofilm found in your mouth. What is it?

A2: Dental plaque.

Q3: Under what conditions might bacteria want to coordinate (via quorum sensing) to increase their reproductive rate?

A3: If environmental conditions are very good, then it makes sense for the population of bacteria to grow as quickly as possible to take advantage of those conditions.

Figure 15.10

Q1: What source of energy would you expect a cave-dwelling prokaryote to use?

A1: It could not rely on light, so it would most likely use chemical energy.

Q2: In which of these categories would you place the bacteria responsible for nitrogen fixing? Why?

A2: Chemoautotroph, because they use nitrogen from the atmosphere and carbon from carbon dioxide.

Q3: In which of these categories do decomposers belong? Explain your reasoning.

A3: Chemoheterotroph, because they receive both carbon and energy from dead or dying organisms.

Figure 15.11

Q1: From the prokaryotic structures shown in Figure 15.7, what shape would you assign to drawing number 8 in van Leeuwenhoek's illustration?

A1: The comma, or vibrio, because of the flagellum.

Q2: Which of the large prokaryote drawings has the coccus shape?

A2: 24.

Q3: Do you think all of these "animalcules" drawn by van Leeuwenhoek are prokaryotes? Why or why not?

A3: No. Some have such an elaborate structure that they are most likely eukaryotes.

CHAPTER 16

END-OF-CHAPTER ANSWERS

1. a
2. d
3. c
4. d
5. Protista, aquatic, autotrophs
6. (a) 3, (b) 5, (c) 4, (d) 1, (e) 2
7. e
8. The kingdom Plantae contains only autotrophs. The kingdom Fungi contains only heterotrophs.
9. For each case, answers to the first part of the question will vary; here's an example for part (b): Lichens are a symbiotic relationship between a fungus and intracellular bacteria. The photosynthetic bacteria produce food for the fungus, and the fungus provides protection and anchoring for the bacteria. Answers to the second and third parts of the question are as follows:

   (a) The partners of mycorrhizae are plants (domain Eukarya, kingdom Plantae) and fungi (domain Eukarya, kingdom Fungi). The relationship is a mutualism.

   (b) The partners of a lichen are fungi (domain Eukarya, kingdom Fungi) and either algae (domain Eukarya, kingdom Protista) or cyanobacteria (domain Bacteria). The relationship is a mutualism.

   (c) The partners in this relationship are hermit crabs and the (nonliving) shells of various species of snails (both domain Eukarya, kingdom Animalia). The relationship is a commensalism.

   (d) The partners in this relationship include the malaria protozoan, its insect host, and its vertebrate host (all domain Eukarya; kingdom Protista for the malaria protozoan, kingdom Animalia for the two host types). It is a parasitic relationship.

10. Gymnosperms and angiosperms produce pollen, which is able to travel via wind or a pollinator to other individuals of the same species, thereby enabling sexual reproduction in a nonaquatic environment.

11. Possible answers include athlete's foot, fungal pneumonia, and yeast infections.

ANSWERS TO FIGURE QUESTIONS

Figure 16.3

Q1: Are ciliates more closely related to euglenoids or to diatoms? To euglenoids or to forams?

A1: Ciliates are more closely related to both diatoms and forams than they are to euglenoids.

Q2: Are all the algae groups (red, green, and brown) equally related?

A2: No. Red and green algae are more closely related to each other than either is to brown algae.

Q3: Which protist group do you think is most closely related to plants? Justify your answer.

A3: Green algae, because they share a common ancestor with plants.

Figure 16.5

Q1: What evolutionary innovation separates all land plants from their aquatic ancestors?

A1: They are terrestrial.

Q2: How do ferns differ from bryophytes? Do they share this difference with other plant groups? (You will need to read ahead to answer this question.)

A2: Ferns have lignin and vascular systems, while bryophytes do not. Gymnosperms and angiosperms also have these adaptations.

Q3: What group(s) might a plant with seeds belong to? What about a plant with flowers?

A3: Seeds are found in gymnosperms or angiosperms; flowers are found only in angiosperms.

Figure 16.6

Q1: In what ways are terrestrial plants and their aquatic ancestors the same? Give at least two similarities.

A1: Both photosynthesize and must absorb nutrients.

Q2: In what ways do terrestrial plants and their aquatic ancestors differ? Give at least two differences.

A2: Unlike their aquatic ancestors, terrestrial plants must protect themselves from dehydration and absorb nutrients from the soil.

Q3: Would you predict that aquatic plants (which have secondarily evolved to live in water) would be more like plants in a rainforest or more like desert plants? Explain your reasoning.

A3: They should be more like rainforest plants, because water is not as limiting a resource in the rainforest as in the desert.

Figure 16.8

Q1: What feature(s) of the ginseng plant tell you that it is not a bryophyte?

A1: It has roots, and also it is able to grow higher than a bryophyte.

Q2: What feature(s) of the ginseng plant tell you that it is not a fern or gymnosperm?

A2: It has flowers, which only angiosperms have.

Q3: Because of the CITES classification of ginseng, you are not allowed to sell plants younger than 5 years even if they grew on your own land. Do you agree with that law? Why or why not?

A3: Some students will argue they should be able to do what they want on their own land; others will agree it is best to maintain the species.

Figure 16.9

Q1: What group of fungi most resembles the mushrooms you buy in a grocery store?

A1: Basidiomycetes.

Q2: Are sac fungi more closely related to molds or to club fungi?

A2: Sac fungi are more closely related to club fungi, with which they share a common ancestor, than to molds.

Q3: How do we know that fungi are eukaryotes rather than prokaryotes?

A3: The simplest reason is that fungi are multicellular, but they also are composed of eukaryotic cells (with organelles, larger size, etc.).

Figure 16.11

Q1: Why is it important that the fruiting body is aboveground?

A1: Because belowground, spores would not be able to travel by wind.

Q2: What part of a fungus is the mushroom that you can buy in the grocery store?

A2: A fruiting body.

Q3: Write a sentence in your own words that uses the terms "mycelium," "fruiting body," and "spore" correctly.

A3: One possible answer: The mycelium is the main body of a mushroom, while the fruiting body develops to produce and release spores into the environment, for reproduction.

CHAPTER 17

END-OF-CHAPTER ANSWERS

1. a

2. a

3. d

4. Mitochondrial DNA, Nuclear DNA

5. (a) N, (b) R, (c) B, (d) B, (e) B

6. c

7. Monotremes lay eggs rather than developing young internally via a placenta. Marsupials have a simple placenta and so give birth to relatively undeveloped young, which then develop externally in a pouch. Eutherians have a well-developed placenta and so can support their young internally until they are more fully developed

8. (a) 4, (b) 1, (c) 5, (d) 2, (e) 3

9. b

10. Shallow water, oxygen-poor water, areas with seasonal flooding, and nutrient-poor water but nutrient-rich land.

11. Possible answers include (a) shark; (b) hagfish; (c) bird; (d) frog.

12. It suggests that Neanderthals evolved and left Africa before modern humans evolved in Africa, and that modern humans intermingled with Neanderthal populations only after leaving Africa.

ANSWERS TO FIGURE QUESTIONS

Figure 17.2

Q1: Are mollusks more closely related to flatworms or to annelids? Explain your reasoning.

A1: Mollusks share a more recent common ancestor with annelids than with flatworms, so they are more closely related to annelids.

Q2: If you found an animal with no symmetry, to what group would it belong? Give an example of an animal that is radially symmetrical, and one that is bilaterally symmetrical.

A2: Animals with no symmetry are sponges. Animals with radial symmetry include cnidarians and echinoderms (secondarily). Animals with bilateral symmetry include flatworms, mollusks, annelids, nematodes, arthropods, and chordates.

Q3: How do we know that a sponge is an animal, and not a plant or an alga?

A3: Sponges are heterotrophic, not autotrophic like plants and algae.

Figure 17.3

Q1: Do amphibians have amniotic eggs?

A1: No.

Q2: What group of animals has jaws but not a bony skeleton?

A2: Sharks and rays.

Q3: When people talk about animals, they are sometimes referring only to mammals. How would you explain to them their error?

A3: Mammals are a subset of a much larger group; animals include invertebrates, birds, and lizards, among other things.

Figure 17.4

Q1: Is a sea star radially or bilaterally symmetrical?

A1: Radially symmetrical.

Q2: What advantage might a bilaterally symmetrical animal have over one that is radially symmetrical, and vice versa?

A2: Animals with bilateral symmetry move more efficiently. Those with radial symmetry have 360-degree access to their environment.

Q3: What kind of symmetry do you (a human) have?

A3: Bilateral.

Figure 17.5

Q1: List all of the lobster's thoracic appendages.

A1: Walking legs, crushing claw, cutting claw.

Q2: Which of the lobster's appendages is most important for sensing the environment?

A2: The head appendage, which includes eyes and antennae.

Q3: What body segments do you (a human) have?

A3: Head, thorax, and abdomen.

Figure 17.6

Q1: The Virginia opossum, or possum, is the only North American marsupial. Describe how its young are born and develop.

A1: They are born relatively undeveloped and crawl into the mother's pouch, where they nurse and grow.

Q2: If you were a monotreme, would you still produce milk and nurse your young?

A2: Yes; all mammals do.

Q3: What kind of mammal is a cow? How about a human?

A3: Cows and humans are eutherians.

Figure 17.7

Q1: According to this evolutionary tree, which primate group is most closely related to humans?

A1: Chimpanzees.

Q2: According to this evolutionary tree, which primate group is most distantly related to humans?

A2: Lemurs, along with others in this branch.

Q3: What characteristics are common to all the primates, including humans? (You will need to read the next paragraph to answer this question.)

A3: The use of tools, a capacity for symbolic language, and the performance of deliberate acts of deception.

Figure 17.8

Q1: Through natural selection, deleterious traits will tend to disappear from a population over time. Which traits might have been deleterious for ground-dwelling early hominins?

A1: Opposable toes; walking on four limbs.

Q2: Through natural selection, advantageous traits will tend to persist in a population over time. Which traits might have been advantageous for ground-dwelling early hominins?

A2: Upright posture; arched feet.

Q3: Adaptations to upright walking also mean that human females have more difficulty giving birth than do females of other species. What adaptation would you predict has had the greatest impact on this result?

A3: Reorganization of the pelvic structure.

Figure 17.9

Q1: What other reason besides continuing to use trees might explain why early hominins had partially opposable big toes?

A1: This trait may have taken a long, long time to lose. A major change in structure cannot occur within a few generations. It was likely a gradual change over thousands of generations.

Q2: In what way does the pattern of footprints in this figure suggest that the print makers were walking upright?

A2: No hand or knuckle prints accompany the footprints.

Q3: Why do you think we no longer have partially opposable big toes?

A3: The opposable big toes would have made walking upright or running more difficult. If we no longer returned to the trees, there was no selective advantage to having them. In fact, individuals with fully opposable toes were at a disadvantage.

Figure 17.10

Q1: Where would the Neanderthal species branch be on this tree?

A1: The *Homo neanderthalensis* lineage would branch either from the common ancestor of *H. erectus* and *H. sapiens*, or from *H. erectus* along with *H. sapiens*. The exact hereditary line for these two closely related species is still unclear.

Q2: How would the Neanderthal skull differ from the *Homo erectus* skull?

A2: The skull would be larger, more like the *H. sapiens* skull size or larger.

Q3: How would the Neanderthal skull differ from the *Homo sapiens* skull?

A3: The skull would be longer, the forehead would slope more, and there would be no chin.

Figure 17.11

Q1: Why does mitochondrial DNA come only from your mother?

A1: Mitochondria are found in the egg but not the sperm. Eggs are from your mom, and sperm are from your dad.

Q2: If a Neanderthal-human hybrid was born to a human mother and a Neanderthal father, could you tell by mitochondrial-DNA sequencing?

A2: No. All the mitochondrial DNA would be from *Homo sapiens*.

Q3: If a Neanderthal-human hybrid was born to a human father and a Neanderthal mother, could you tell by mitochondrial-DNA sequencing?

A3: Yes. In this case, all the mitochondrial DNA would be from *Homo neanderthalensis*.

Figure 17.12

Q1: If a human-Neanderthal hybrid was born to a human mother and a Neanderthal father, could you tell by whole-genome DNA sequencing that it was a hybrid?

A1: Yes. The offspring would have nuclear DNA from both the egg and the sperm.

Q2: If a human-Neanderthal hybrid was born to a Neanderthal mother and a human father, could you tell by whole-genome DNA sequencing that it was a hybrid?

A2: Yes. The offspring would have nuclear DNA from both the egg and the sperm.

Q3: Under what circumstances are scientists able to do whole-genome sequencing, and when are they restricted to mitochondrial-DNA sequencing?

A3: Whole-genome sequencing requires well-preserved cells or tissues with fully intact DNA. Mitochondrial DNA can be isolated from cells and tissues that aren't so well preserved and from damaged DNA.

Figure 17.13

Q1: Are you surprised by the interpretations of the hominins in this picture?' Why or why not?

A1: This question asks for an opinion. One possible answer: *Homo sapiens* looks more primitive than expected, and the other species look strikingly like us.

Q2: Describe the main differences that distinguish the hominin species.

A2: Height, musculature, size of skull, slope of forehead, amount of hair and its location.

Q3: From what you've learned about these species, do you think these representations are accurate? How can you find more information about each species to help you answer this question?

A3: The first part of the question asks for an opinion. One possible answer for the second part: I can do much more extensive research about all of our family members online and at museums of natural history.

Figure 17.14

Q1: What evidence suggests that Neanderthals never lived in Africa?

A1: Modern humans of African descent have neither mitochondrial nor nuclear Neanderthal DNA sequences in their genomes.

Q2: How does the hypothesized origin of modern humans (*Homo sapiens*) differ from the hypothesized origin of Neanderthals (*Homo neanderthalensis*)?

A2: Modern humans evolved from archaic humans in Africa and spread to the rest of the world. Neanderthals are thought to have evolved from archaic humans living in the Middle East.

Q3: What species of hominins other than the Neanderthals may have commingled with modern humans?

A3: *Homo erectus.*

Figure 17.15

Q1: Describe the difference you observe between the modern human skull's chin and the Neanderthal skull's lower jaw.

A1: The human chin is higher and juts out further.

Q2: What other differences do you observe between the two skulls?

A2: The Neanderthal has a more pronounced eyebrow ridge and more sloping forehead.

Q3: Why would you expect a hybrid of Neanderthals and modern humans to have intermediate features?

A3: In hybrids, half of the DNA would be from *Homo sapiens* and half from *Homo neanderthalensis*. Hybrids have features of both parents. Some might have intermediate features, while others could have mixed features (such as a sloping forehead but a prominent chin).

CHAPTER 18

END-OF-CHAPTER ANSWERS

1. e

2. b

3. d

4. (a) 1, (b) 3, (c) 4, (d) 2, (e) 5

5. Weather, climate, Climate change, global warming

6. d

7. Carbon sink, because more carbon is absorbed in the years between the fires than is produced by the fires.

8. (a) 4, (b) 1, (c) 2, (d) 5, (e) 3

9. c

10. Answers will depend on student location.

11. Answers will depend on student location.

ANSWERS TO FIGURE QUESTIONS

Figure 18.3

Q1: List as many biotic and abiotic factors in this photograph as you can.

A1: Biotic: all living things, such as plants, as well as all animals, including microscopic bacteria and algae. Abiotic: all nonliving things, such as rocks and water.

Q2: Is the forest part of the biotic or abiotic environment? Explain.

A2: It can be considered both. The plants and microorganisms are biotic, while the dirt and minerals are abiotic.

Q3: Is the river part of the biotic or abiotic environment? Explain.

A3: It can be considered both. The algae, plants, and aquatic organisms like fishes living in the river are biotic, while the water and rocks of the river are abiotic.

Figure 18.5

Q1: Name two ways in which climate change affects the frequency and severity of floods.

A1: Changes in rainfall patterns and melting ice can cause rivers and lakes to overflow.

Q2: How has climate change caused a rise in sea level?

A2: Melting of glaciers and polar ice has caused sea levels to rise.

Q3: Give an example of an environmental effect of climate change in your state or region.

A3: Answers may include drought or flooding.

Figure 18.6

Q1: Why is it colder at the poles than at the equator?

A1: The sun's rays are spread wider at the poles and strike Earth less directly than at the equator. Less direct sunlight = cooler.

Q2: Why is it warmer at the equator than at the poles?

A2: All of the sun's energy is directed at the equator, and the rays strike Earth at a direct angle there. The result is more intense heat at the equator.

Q3: During part of the year the Northern Hemisphere is tilted at a more direct angle to the sun than the Southern Hemisphere is, and for the other part of the year the opposite is true. How does this tilt explain temperature differences in summer and winter?

A3: When a hemisphere is tilted toward the sun, the sun's rays strike it more directly, causing the warmer temperatures of summer. The opposite is true in the winter.

Figure 18.7

Q1: How much of the incoming solar energy is reflected back to outer space?

A1: About a third.

Q2: What kind of energy is reemitted to the atmosphere after being absorbed by Earth's surface?

A2: Infrared radiation.

Q3: How are greenhouse gases like a blanket on your bed at night?

A3: Greenhouse gases absorb the heat around Earth and hold it near the surface just as a blanket absorbs body heat, prevents it from escaping, and holds it near your body.

Figure 18.8

Q1: What measurements do the green circles represent?

A1: The green circles indicate CO_2 levels measured from bubbles of air trapped in ice that formed many hundreds of years ago.

Q2: What measurements do the red circles represent?

A2: The red circles are direct measurements of CO_2 levels at the Mauna Loa Observatory in Hawaii.

Q3: For approximately how many years has the Mauna Loa Observatory been recording CO_2 levels?

A3: Directly for about 60 years.

Figure 18.9

Q1: In what years were global temperatures the lowest?

A1: The years around 1910.

Q2: In what years were global temperatures the highest?

A2: The years around the late 1990s through the present.

Q3: What trend is apparent in this graph of actual global temperatures?

A3: Average global temperatures are rising.

Figure 18.11

Q1: Where will fire most seriously affect the Amazon rainforest?

A1: Fire will most likely affect areas labeled "Dry and/or logged forest" (shown in orange).

Q2: Where will fire be the least damaging to the Amazon rainforest?

A2: Fire will least affect the old-growth forest (shown in green).

Q3: This map does not include an increase in pasturelands for grazing animals. Do you think more or less pastureland will be needed in 2030? Explain.

A3: As populations increase, the need for more pastureland to raise cattle for human consumption will increase as well.

Figure 18.12

Q1: What is transpiration?

A1: Transpiration is the process of plants absorbing water through their roots and releasing this water through their leaves to the atmosphere.

Q2: Why is transpiration important to the water cycle?

A2: Transpiration returns water from the soil back to the atmosphere to form clouds and eventually precipitation.

Q3: If there are fewer plants and therefore less transpiration in a given area, what will happen to the humidity or cloud cover in this area?

A3: The humidity and cloud cover will decrease where there is a substantial decrease in transpiration.

Figure 18.13

Q1: How do the patterns of rainfall in the Northern and Southern Hemispheres compare?

A1: The same patterns emerge as you move away from the equator either northward or southward.

Q2: How do the patterns in the kinds of environments shown in the Northern and Southern Hemispheres compare?

A2: As you move away from the equator, the major biomes are equivalent distances from the equator either north or south.

Q3: What happens at the equator to make this region so wet?

A3: The density of plants is very high, and therefore the amount of transpiration is very high, resulting in cloud cover and high precipitation. High precipitation results in high plant growth and high transpiration rates.

Figure 18.14

Q1: What are three ways that carbon is released into the atmosphere?

A1: Respiration from animals; the burning of organic matter, including fossil fuels and wood; and the decomposition of dead organic material.

Q2: Are all of the pathways you listed for question 1 affected by human activity?

A2: Almost everything on our planet is affected by human activity in some way. Of the three pathways of carbon release, the one most affected by humans is the burning of organic materials for energy.

Q3: What are two biotic reservoirs of carbon?

A3: Plants and animals.

Figure 18.15

Q1: How does a carbon source contribute to global warming?

A1: A carbon source is a reservoir of carbon that releases more than it absorbs, thereby dumping CO_2 into the environment, increasing greenhouse gases, and causing global warming.

Q2: How does a carbon sink protect against global warming?

A2: A carbon sink absorbs more carbon than it releases, thereby removing CO_2 from the environment, decreasing greenhouse gases, and blocking global warming.

Q3: How can trees act as both a source and a sink?

A3: Trees act as a carbon sink when photosynthesizing and absorbing CO_2 and as a carbon source when they are burned for fuel or in a wildfire.

CHAPTER 19

END-OF-CHAPTER ANSWERS

1. d

2. b

3. c

4. d

5. a

6. Density-dependent, density-independent

7. b

8. c

9. (a) 1, (b) 4, (c) 3, (d) 5, (e) 2

10. (See figure below)

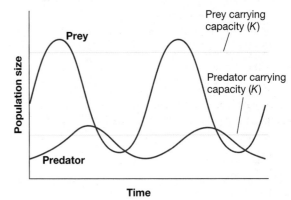

11. The food dispenser increased the carrying capacity significantly.

12. b

ANSWERS TO FIGURE QUESTIONS

Figure 19.1

Q1: What parts of the United States are within the range of the mosquito that carries Zika?

A1: Roughly, the southern United States and the eastern coastline.

Q2: What areas are *not* within the mosquito's range? Why do you think that is?

A2: The northern and northwestern states, probably because it is too cold for mosquitoes to overwinter there.

Q3: Find your own state, and where you are in the state. Are you at risk of contracting Zika?

A3: Answers will depend on student location. If you travel to areas with Zika-infected mosquitoes, your risk increases. If local Zika infections have not yet been reported in your state, you are at less risk than, for example, citizens of Florida.

Figure 19.3

Q1: What is the main way by which someone is infected with the Zika virus?

A1: By being bitten by a mosquito that is infected.

Q2: Judging by the poster and your knowledge of mosquito behavior, what can you do to decrease your risk of being infected with the Zika virus?

A2: Cover up when you're outside, don't go out at dusk, wear insect repellent, and use protection when having sex with someone who has been exposed.

Q3: Besides the transmission methods shown on the poster, what are some other ways you could become infected with the Zika virus?

A3: From kissing an infected individual, or possibly from sharing a razor or toothbrush with someone. (Other answers are possible.)

Figure 19.5

Q1: Which form of population growth displays a J-shaped curve?

A1: Exponential growth.

Q2: Which form of population growth displays an S-shaped curve?

A2: Logistic growth.

Q3: Describe a situation in which a population initially shows exponential growth and later shows logistic growth.

A3: If a population enters a new area—an island, for example— it will initially grow exponentially as it expands into the environment. Later, as it reaches the carrying capacity of the new area, it will begin to display logistic growth.

Figure 19.6

Q1: According to this graph, approximately when did exponential growth begin?

A1: Around the year 1800 CE.

Q2: What milestone corresponds to the transition from logistic to exponential population growth?

A2: The onset of the industrial revolution.

Q3: What is the UN's projected carrying capacity of Earth, and when will we reach it?

A3: The UN projects Earth's carrying capacity to be 9–10 billion people, which we will reach in about 2050 CE.

Figure 19.7

Q1: List all the stages of the mosquito life cycle.

A1: Egg, larva, pupa, adult.

Q2: Which life cycle stage is vulnerable to the GM treatment?

A2: The larvae, which do not survive to become pupae.

Q3: Why are only male GM mosquitoes released into the wild, rather than both males and females?

A3: Males do not feed on blood, so they are not able to transmit Zika virus, whereas females do feed and thus can transmit Zika.

Figure 19.8

Q1: What factors may be limiting growth and reproduction in the plantain's crowded conditions?

A1: Possible factors include availability of nutrients, water, sunlight, and room for root and shoot growth.

Q2: Why are overcrowded conditions considered density-dependent population changes?

A2: Because the more organisms there are in the environment or the more densely packed they are, the more competition there is for resources and the less each individual will likely receive.

Q3: Relate this example of overcrowded conditions to the human population growth shown in Figure 19.6. How do you think the situations are similar? How are they different?

A3: Increasing human populations may lead to an increase in competition for jobs, housing, and food. So far, humans have been able to increase their carrying capacity through technology, but that may no longer be possible.

Figure 19.9

Q1: In what year did the reindeer's numbers begin to rise exponentially?

A1: About 1933.

Q2: In what years was the reindeer's population growth logistic?

A2: Between about 1911 and 1932.

Q3: How do you predict the graph (population size) would change if someone had begun bringing in supplemental food for the reindeer in 1940? Draw a sketch of your prediction.

A3: Supplemental food would have caused the population decline to end, and the population would have either remained constant or again increased, depending on the amount of supplemental food provided.

Figure 19.10

Q1: In what year did the bald eagle population rise to more than 2,000 breeding pairs?

A1: About 1986 or 1987.

Q2: Give some examples of possible density-dependent limits on bald eagle populations.

A2: Examples include numbers of prey available, adequate habitat for nesting and hunting, and availability of water.

Q3: Is the population growth of bald eagles more like logistic or exponential growth? Explain why you think so.

A3: The population growth of bald eagles is more like logistic growth in that it is more of an S-shaped curve—at least between 1960 and 2005, where it seemed to level off for a few years. On the other hand, between 2005 and 2010 the growth appears to be exponential.

Figure 19.11

Q1: During which years did the hare likely have the greatest food supply?

A1: 1865 and 1888.

Q2: Besides the number of hare, what other factors might contribute to the number of lynx?

A2: Other possible factors include the quality of habitat for building dens and raising young, the availability of freshwater, and the competition with other lynx and other predators of the hare.

Q3: Can you draw an average carrying-capacity line on these graphs? Why or why not?

A3: It would be difficult to draw a line representing a carrying capacity for both of these animals because of the way they go through cycles of "boom and bust" in population size. They do not show logistic growth, where the leveling off of the S-shaped curve provides an obvious carrying capacity.

CHAPTER 20

END-OF-CHAPTER ANSWERS

1. d
2. b
3. c
4. c
5. secondary consumer, primary consumer, producer
6. e
7. mutualism: 2, commensalism: 3, predation: 1, competition: 4
8. (a) 4, (b) 1, (c) 2, (d) 3, (e) 5
9. More diverse communities have low relative species abundance and high species diversity.
10. c
11. Relative species abundance will decrease, and species richness will increase.
12. (a) S, (b) S, (c) P, (d) P
13. (a) humans and killer whales, (b) phytoplankton, (c) baleen whale and sperm whale, (d) other herbivorous plankton

ANSWERS TO FIGURE QUESTIONS

Figure 20.2

Q1: List another species that is part of this community.

A1: One possible answer: the wolf.

Q2: Of which community could this aspen woodland be a smaller part?

A2: A larger deciduous forest community.

Q3: Which other small communities could be found within this larger community?

A3: Many other communities could exist, including soil communities (soil-dwelling insects, other invertebrates, and microbes), communities of plants and animals residing in the undergrowth of the forest, and canopy communities of animals that live in the treetops, among others.

Figure 20.3

Q1: If relative species abundance increased in the first community, how would the figure look different from now?

A1: There would be even more trees of the white-trunked species, and fewer of the other species.

Q2: If species richness decreased in the second community, how would the figure look different from now?

A2: There would be fewer different species of trees.

Q3: How do relative species abundance and species richness define the species diversity of a forest community?

A3: High species diversity relies on having each species in reasonable abundance, and not on one species taking up the majority of space in an area (relative abundance), and on there being many different species in a given area (richness).

Figure 20.4

Q1: How many species were left in 1966 in the community where sea stars were *not* removed?

A1: About 17 or 18.

Q2: How many species were left in 1966 in the community where sea stars *were* removed?

A2: Only 2 or 3.

Q3: How do your answers to questions 1 and 2 demonstrate the importance of a keystone species for the maintenance of diversity in a community?

A3: The species diversity plummeted as a result of the loss of this sea star. Without this species, the entire mussel community changed from many species to only a few.

Figure 20.5

Q1: What species eats the coyote?

A1: The gray wolf.

Q2: What species does the coyote eat?

A2: The vole, the snowshoe hare, berries, the pronghorn, and the elk.

Q3: What do you think would happen to a community that lost its coyotes?

A3: The snowshoe hare and the vole would overeat the grasses and other plants, and once those food sources were destroyed, their unavailability would cause the snowshoe hare and vole populations to crash.

Figure 20.6

Q1: Where do producers acquire the energy they need to perform their function in the food chain?

A1: From the sun. In combination with CO_2 and water, they produce glucose through photosynthesis.

Q2: Why are producers necessary for life on Earth?

A2: Without producers, there would be no influx of energy into Earth's biosphere—no energy source for consumers to acquire.

Q3: Given the lower abundance of producers in desert regions, compared to tropical rainforests, what would you predict about the abundance of consumers in the two environments?

A3: Consumer species would be less abundant in deserts than in rainforests.

Figure 20.7

Q1: When did wolves begin to be seen again in Yellowstone?

A1: In the mid- to late 1980s.

Q2: What was the highest number of wolves observed in Yellowstone? In what year did that occur?

A2: About 174 individuals, in 2003.

Q3: Why did it take a few years after the reintroduction of wolves for the aspen population to increase, as well as the beavers and bears?

A3: Each individual parent plant can produce only a certain number of offspring at a time, and then those offspring have to reach maturity before they themselves can reproduce.

Figure 20.8

Q1: What might happen to an anemone without a resident clownfish?

A1: An anemone unaccompanied by clownfish, and therefore unprotected from anemone-eating fish, could be grazed extensively and be unable to support its growth needs without the nutrients in the clownfish excrement.

Q2: What would happen to a clownfish that did not produce a mucous coating?

A2: It would be stung by the anemone's tentacles, so the mutualism would fall apart.

Q3: In the example of mutualism given in the text, what would happen if ticks were no longer able to feed on bison?

A3: There would be nothing for the birds to eat off of the bison, and no reason for the bison to allow the birds on them. The mutualism would end.

Figure 20.9

Q1: How do barnacles benefit from living on a whale?

A1: The whale provides a home and a constant stream of water passing over the barnacles to bring them the tiny particles of food that they filter from the sea.

Q2: Do you think a whale could avoid being colonized by barnacles? Why or why not?

A2: It is probably unlikely that a whale could avoid being covered by barnacles, although since the barnacles do not help or hurt the whale, there would be no reason for a whale to try to avoid them.

Q3: Explain why detritivores are considered commensal to the organisms they consume. (You will need to read ahead to answer this question.)

A3: Since the organism is already dead, the detritivore is not harming it by eating it. The detritivore, though, benefits from the association.

Figure 20.10

Q1: What kind of predator is the cheetah in the figure?

A1: Carnivore.

Q2: What kind of predator is the louse in the figure?

A2: Parasite.

Q3: What kind of predator are the elk that graze on aspen tree saplings?

A3: Herbivore.

Figure 20.11

Q1: How do predators know that brightly colored prey are usually toxic?

A1: Most predators have experienced through the trial and error of tasting prey and becoming sick that brightly colored prey are toxic. Predators must learn by trying one to avoid others later. If the organism were so toxic that the predator died from eating one, then bright displays would not help prey avoid predation.

Q2: Do you think mimicry works if the toxic species is in low abundance? Why or why not?

A2: Mimicry is similar to warning coloration in that its benefit is usually accomplished through trial and error. If there are many more nontoxic mimics than real toxic individuals, predators that successfully eat the mimics will not learn to avoid them. Only if it is more likely that a predator will encounter an actual toxic and bad-tasting prey will it learn to avoid anything that looks similar to the prey it encountered.

Q3: Why is camouflage considered an adaptive response to predation?

A3: Camouflage enables prey species to blend in with their surroundings so that they're difficult for predators to detect. Random variation in coloration of a population enables those that arc bcttcr hidden from predators to survive and reproduce, while those that do not match the surroundings well are easily seen and eaten. Over time, only well-camouflaged prey survive to reproduce, passing the camouflage coloration on to their offspring.

Figure 20.12

Q1: What percentage of pigeons are caught when they are alone and not in a flock?

A1: Very close to 80%.

Q2: For wood pigeons, what is the minimum number of individuals that provides protection from goshawks?

A2: 11 individuals.

Q3: Why do you think a group of musk oxen versus a lone musk ox would be safer from a pack of wolves?

A3: The group of musk oxen could form a circle in which they faced out and use their large horns to impale the wolves, keeping their sides and rear ends protected from attack. A lone ox would be completely unprotected from a pack of wolves working together.

Figure 20.13

Q1: Which species appears to be the superior competitor?

A1: *Aphytis lingnanensis.*

Q2: Why is this example considered exploitative competition?

A2: These wasp species feed on the same foods in the same place but never physically come in contact with each other. They are in direct competition but do not physically interact.

Q3: What would you predict if these species had undergone competitive exclusion? (You will need to read the next paragraph to answer this question.)

A3: The first species would now be exploiting another resource—for example, apples rather than citrus.

Figure 20.15

Q1: What species represents the first colonizers of the sand dunes?

A1: Dune-building grasses.

Q2: What species is the intermediate species, and how does it become the dominant species?

A2: The pine is the intermediate species. It likely becomes dominant by outcompeting the grasses for water, nutrients, and sunlight. Once established, the pines' shade further inhibits the growth of grasses.

Q3: What species is the mature, climax community species, and how does it become the dominant species?

A3: The black oak is the mature, climax community species. It likely becomes the dominant species because it outcompetes the pines for water, nutrients, and sunlight.

Figure 20.16

Q1: What other types of disturbances could you imagine destroying a forest?

A1: Other types of disturbances could include clear-cutting or heavy logging, abandonment of agricultural fields, flooding, clearing by tornado or hurricanes, and the creation of empty lots in a newly built subdivision, among others.

Q2: How is secondary succession different from primary succession?

A2: Secondary succession starts with some existing producers that survived a disturbance, whereas primary succession starts from nothing—pure rock or sand. A disturbance that would result in primary succession would be a lava flow that covered an entire area; succession would have to start from pure lava rock and no existing producers.

Q3: What is a climax community?

A3: A climax community is the final step in succession, in which the species in the community remain and are not replaced by any other species as time goes by.

CHAPTER 21

END-OF-CHAPTER ANSWERS

1. a

2. c

3. d

4. c

5. biome: 3, ecosystem: 2, net primary productivity: 1, ecological community: 4

6. chaparral, grassland, estuaries

7. b

8. (a) 4, (b) 2, (c) 1, (d) 5, (e) 3

9. An ecosystem is composed of multiple ecological communities, and it also includes physical environmental factors (communities do not).

10. (a) 100. (b) 100,000. (c) The energy is expended in metabolism, and since predation is not 100% efficient (that is, predators don't eat every last bit of the calories in their prey, leaving, for example, the bones), some energy is lost there too.

11. The water cycle is more similar to the movement of nutrients, because water cycles through the biotic and abiotic components of an ecosystem, whereas energy flows in only one direction through ecosystems because of the steady loss of heat.

12. c

ANSWERS TO FIGURE QUESTIONS

Figure 21.3

Q1: What percentage of the original 10,000 Kilocalories is available to a shark that might eat the tuna in this figure?

A1: 1%.

Q2: What trophic level and term would describe a predator of tuna?

A2: A predator of tuna would be at the fifth trophic level and a quaternary consumer.

Q3: Give an example of a primary consumer in a terrestrial environment.

A3: Possible answers include deer, a grasshopper, a seed-eating bird, a mouse.

Figure 21.4

Q1: Which organisms are the producers in this ecosystem?

A1: Phytoplankton.

Q2: How do nutrients flow from the abiotic to the biotic components of the ecosystem?

A2: Via photosynthesis and uptake from the surrounding water.

Q3: How do nutrients flow from the biotic to the abiotic components of the ecosystem?

A3: Via decomposers.

Figure 21.5

Q1: How is a decomposer different from a more typical consumer?

A1: Decomposers feed off of only dead organic matter (dead plants or animals). All other types of consumers feed off of live plants and animals.

Q2: What is the difference between how carbon is brought into the biotic portion of the ecosystem, and how other nutrients, such as phosphorus, are brought in?

A2: Carbon is pulled from the air during photosynthesis, whereas other nutrients are pulled in from the soil.

Q3: Describe all the points at which heat is lost in this figure.

A3: During the survival and reproduction of producers, consumers, and decomposers.

Figure 21.9

Q1: Which terrestrial biome has the lowest NPP? Which aquatic biome?

A1: The lowest terrestrial NPP appears to occur in the desert and tundra biomes in the terrestrial map, and in the marine biome of the open ocean in the aquatic map.

Q2: Where are the most productive terrestrial biomes located?

A2: They appear to be mainly near the equator.

Q3: Give a possible reason for your answer to question 2.

A3: More sunlight reaches Earth at the equator, so larger populations of producers can be supported.

Figure 21.10

Q1: In what years were chlorophyll levels the highest?

A1: 1999 and 2008.

Q2: In what years were the temperature changes from the average the greatest?

A2: 2000 (low), 2003–4 (high), and 2008 (low).

Q3: How do you predict this graph will look 10 years from now?

A3: If the world's oceans are warming, then the chlorophyll levels may continue to decrease.

Glossary

A

abiotic Nonliving. Compare *biotic*.

acid A chemical compound that loses hydrogen ions (H⁺) in aqueous surroundings. Compare *base* (definition 1).

active site The location within an enzyme where substrates are bound.

active transport The movement of a substance in response to an input of energy. Compare *passive transport*.

adaptation 1. An evolutionary process by which a population becomes better matched to its environment over time. 2. See **adaptive trait**.

adaptive radiation The expansion of a group of organisms to take on new ecological roles and to form new species and higher taxonomic groups.

adaptive trait Also called an *adaptation*. A feature that gives an individual improved function in a competitive environment.

adenine (A) One of the four nucleotides that make up DNA. The other three are thymine (T), guanine (G), and cytosine (C).

ADP Adenosine diphosphate.

aerobic Requiring oxygen. Compare *anaerobic*.

algae One of two main groups of protists, whose members are photosynthetic and may or may not be motile. Compare *protozoans*.

alleles Different versions of a given gene.

allopatric speciation The formation of new species from geographically isolated populations. Compare *sympatric speciation*.

amino acid Any of a class of small molecules that are the building blocks of proteins.

amniocentesis A prenatal genetic screening technique in which amniotic fluid is extracted from the pregnancy sac that surrounds a fetus by means of a needle that is inserted through the abdomen into the uterus. Compare *chorionic villus sampling*.

anabolism Metabolic pathways that create complex biomolecules from smaller organic compounds. Compare *catabolism*.

anaerobic Not requiring oxygen. Compare *aerobic*.

analogous trait A feature that is shared across species because of convergent evolution, not because of modification by descent from a recent common ancestor. Compare *homologous trait*.

analytical study An observational study that looks for patterns in the information collected and addresses how or why those patterns came to exist. Compare *descriptive study*.

anchorage dependence The phenomenon in which cells stop dividing when detached from their surroundings.

angiogenesis The formation of new blood vessels.

angiosperms Flowering plants. One of two main groups of seed-bearing plants, characterized by seeds enclosed in an ovary and by flowers. Compare *gymnosperms*.

Animalia The animals. One of the six kingdoms of life, in the domain Eukarya, encompassing all animals, including humans, birds, and dinosaurs.

anticodon A unique sequence of three nitrogenous bases at one end of a tRNA molecule that binds to the corresponding *codon* on an mRNA molecule.

applied research Research in which scientific knowledge is applied to human issues and often commercial applications. Compare *basic research*.

Archaea One of the three domains of life (compare *Bacteria* and *Eukarya*) and also one of the six kingdoms of life. The domain and kingdom Archaea consists of single-celled organisms best known for living in extremely harsh environments.

artificial selection The process by which individuals display specific traits through selective breeding. Compare *natural selection*.

ascomycetes Sac fungi, one of three main groups of fungi. Compare *basidiomycetes* and *zygomycetes*.

asexual reproduction The process by which clones, offspring that are genetically identical to the parent, are generated. Compare *sexual reproduction*.

atom The smallest unit of an element that retains the element's distinctive properties. Atoms are the building blocks of all matter.

atomic mass number The sum of the number of protons and the number of neutrons in an atom's nucleus. Compare *atomic number*.

atomic number The number of protons in an atom's nucleus. Compare *atomic mass number*.

ATP Adenosine triphosphate, a small, energy-rich organic molecule that is used to store energy and to move it from one part of a cell to another. Every living cell uses ATP.

autosome Any chromosome that is not one of the *sex chromosomes*.

autotroph A metabolic producer, an organism that makes food on its own. *Chemoautotrophs* acquire their energy from inorganic chemicals in their environment. *Photoautotrophs* absorb the energy of sunlight and take in carbon dioxide to conduct photosynthesis. Compare *heterotroph*.

AV node See **atrioventricular node**.

B

Bacteria One of the three domains of life (compare *Archaea* and *Eukarya*) and also one of the six kingdoms of life. The domain and kingdom Bacteria includes familiar disease-causing bacteria.

base 1. A chemical compound that accepts hydrogen ions (H^+) in aqueous surroundings. Compare *acid*.

2. The nitrogen-containing component of nucleotides. The four nitrogenous bases of DNA are adenine, cytosine, guanine, and thymine. In RNA, thymine is replaced by uracil.

base pair Also called *nucleotide pair*. Two nucleotides that form one rung of the DNA ladder.

base-pairing rules The rules that govern the pairing of nucleotides in DNA. Adenine (A) on one strand can pair only with thymine (T) on the other strand, and cytosine (C) can pair only with guanine (G).

basic research Research that is intended to expand the fundamental knowledge base of science. Compare *applied research*.

basidiomycetes Club fungi, one of three main groups of fungi. Compare *ascomycetes* and *zygomycetes*.

behavioral trait A characteristic of an individual's behavior, such as shyness or extroversion. Compare *biochemical trait* and *physical trait*.

benign Referring to a tumor, confined to one site.

bias A prejudice or opinion for or against something.

bilateral symmetry An animal body plan in which the body can be divided by just one plane passing vertically from the top to the bottom of the animal into two halves that mirror each other. Compare *radial symmetry*.

binary fission A type of cell division in which a cell simply divides into two equal halves, resulting in daughter cells that are genetically identical to each other and to the parent cell.

biochemical trait A characteristic due to specific chemical processes of an individual, such as the level of a particular enzyme. Compare *behavioral trait* and *physical trait*.

biodiversity The variety of all the world's life-forms, as well as their interactions with each other and the ecosystems they inhabit.

biogeography The geographic locations where organisms or the fossils of a particular species are found.

biological hierarchy A way to visualize the breadth and scope of life, from the smallest structures to the broadest interactions between living and nonliving systems that we can comprehend.

biological species concept The idea that a species is defined as a group of natural populations that can interbreed to produce fertile offspring and cannot breed with other such groups.

biome A large region of the world defined by shared physical characteristics (especially climate) and a distinctive community of organisms.

biomolecule Also called *macromolecule*. A large organic molecule that is critical for living cells. Four major classes of biomolecules are proteins, carbohydrates, nucleic acids, and lipids.

biosphere All the world's living organisms and the physical spaces where they live.

biotic Living. Compare *abiotic*.

bipedal Walking upright on two legs.

boreal forest A terrestrial biome that is found in northern or high-latitude regions, has cold, dry winters and mild summers, and is dominated by coniferous trees.

bryophytes A group of nonflowering plants that includes liverworts and mosses.

C

Calvin cycle See **light-independent reactions**.

Cambrian explosion The burst of evolutionary activity, occurring about 540 million years ago, that resulted in a dramatic increase in the diversity of life. Most of the major living animal groups first appear in the fossil record during this time.

camouflage Any type of coloration or appearance that makes an organism hard to find or hard to catch. Compare *mimicry* and *warning coloration*.

cancer cell Also called *malignant cell*. A tumor cell gains anchorage independence and starts invading other tissues.

capsule An additional protective layer that surrounds the cell wall in prokaryotes.

carbohydrate Any of a major class of biomolecules, including sugars and starches, built of repeating units of carbon, hydrogen, and oxygen.

carbon cycle The movement of carbon within biotic communities, between living organisms and their physical surroundings, and within the abiotic world.

carbon dioxide (CO_2) The most abundant and consequential of the greenhouse gases.

carbon fixation See **light-independent reactions**.

carbon sink A **natural or artificial** reservoir that absorbs more carbon than it releases. Compare *carbon source*.

carbon source A **natural or artificial** reservoir that releases more carbon than it absorbs. Compare *carbon sink*.

carnivore An animal (or, rarely, plant) that eats other animals. Compare *herbivore* and *omnivore*.

carrying capacity The maximum population size that can be sustained in a given environment.

cartilage A dense tissue of the skeleton that combines strength with flexibility. It is found almost everywhere that two bones meet and prevents them from grinding together.

catabolism Metabolic pathways that release chemical energy in the process of breaking down complex biomolecules. Compare *anabolism*.

causation A statistical relation indicating that a change in one aspect of the natural world causes a change in another aspect. Compare *correlation*.

cell The smallest and most basic unit of life—a microscopic, self-contained unit enclosed by a water-repelling membrane.

cell cycle The sequence of events that make up the life of a typical eukaryotic cell, from the moment of its origin to the time it divides to produce two daughter cells.

cell division The final stage of the cell cycle. Cell division includes the transfer of DNA from the parent cell to the daughter cells.

cell theory One of the unifying principles of biology; a theory stating that every living organism is composed of one or more cells, and that all cells living today came from a preexisting cell.

cellular respiration The reciprocal process to *photosynthesis*, in which sugars are broken down into energy usable by the cell.

cellulose A chemical substance that provides structural strength in plant cells. Compare *lignin*.

centromere The central region of a chromosome that attaches sister chromatids together.

chaparral A terrestrial shrubland biome characterized by cool, rainy winters and hot, dry summers, and dominated by drought-resistant plants.

chemical bond A force that holds two atoms together.

chemical compound A molecule that contains atoms from two or more different elements.

chemical reaction The process of breaking existing chemical bonds and creating new ones.

chemoautotroph See **autotroph**.

chemoheterotroph See **heterotroph**.

chlorophyll A green pigment that is specialized for absorbing light energy.

chloroplast An organelle of plant cells and some protist cells that captures energy from sunlight and uses it to manufacture food molecules via photosynthesis.

chordates A large phylum that encompasses all animals with backbones, such as fishes, birds, and mammals.

chorionic villus sampling (CVS) A prenatal genetic screening technique in which cells are extracted by gentle suction from the villi (a cluster of cells that attaches the pregnancy sac to the wall of the uterus). Ultrasound is used to guide the narrow, flexible suction tube through a woman's vagina and into her uterus. Compare *amniocentesis*.

chromatin fiber A chromosome in relaxed form, made up of DNA and nucleosomes.

chromosomal abnormality Any change in the chromosome number or structure, compared to what is typical for a species.

chromosome A DNA double helix wrapped around spools of proteins.

chromosome theory of inheritance The theory that genes are located on chromosomes, and that these chromosomes are the basis for all inheritance.

citizen science Research that is assisted by members of the public, who participate by collecting and sometimes analyzing data in cooperation with professional scientists.

citric acid cycle See **Krebs cycle**.

clade A branch of an evolutionary tree, consisting of an ancestor and all its descendants.

class The unit of classification in the Linnaean hierarchy above order and below phylum.

climate The prevailing *weather* of a specific place over relatively long periods of time (30 years or more).

climate change A large-scale and long-term alteration in Earth's climate, including such phenomena as *global warming*, change in rainfall patterns, and increased frequency of violent storms.

climax community A mature community whose species composition remains stable over long periods of time.

codominance An interaction between two alleles of a gene that causes a heterozygote to display a phenotype that clearly displays the effects of both alleles. Compare *incomplete dominance*.

codon A unique sequence of three mRNA bases that either specifies a particular amino acid during translation or signals the ribosomes where to start or stop translation. Compare *anticodon*.

coevolution The tandem evolution of two species that results because interaction between the two so strongly influences their survival.

commensalism A species interaction in one species benefits at no cost to the other species. Compare *competition*, *mutualism*, and *predation*.

common ancestor An organism from which many species have evolved.

common descent The sharing of a common ancestor by two or more different species.

community The populations of different species that live and interact with one another in a particular place.

competition A species interaction in which both species may be harmed. Compare *commensalism*, *mutualism*, and *predation*.

competitive exclusion principle The idea that different species that use the same resource can coexist only if one of the species adapts to using other resources.

complementary base-pairing The relationship between two nucleic acid strands in which each purine on one strand hydrogen-bonds with a specific pyrimidine on the opposite strand. A pairs with T or U, and C pairs with G.

complex trait A genetic trait whose pattern of inheritance cannot be predicted by Mendel's laws of inheritance.

condensation The transition from gas to liquid. Compare *evaporation*.

conjugation The physical process of transferring genetic material through direct contact.

consumer An organism that obtains energy by eating all or parts of other organisms or their remains. Compare *producer*.

control group The group of subjects in an experiment that is maintained under a standard set of conditions with no change in the independent variable. Compare *treatment group*.

controlled experiment An experiment that measures the value of a dependent variable for two groups of subjects that are comparable in all respects except that one group (the treatment group) is exposed to a change in the independent variable and the other group (the control group) is not.

convection cell A large and consistent atmospheric circulation pattern in which warm, moist air rises and cool, dry air sinks. Earth has four giant convection cells.

convergent evolution Evolution that results in organisms that have different genetics but appear very much alike.

correlation A statistical relation indicating that two or more aspects of the natural world behave in an interrelated manner: if one shows a particular value, we can predict a particular value for the other aspect. Compare *causation*.

covalent bond The sharing of electrons between two atoms. Compare *hydrogen bond* and *ionic bond*.

credentials Evidence of qualifications and competence to be recognized as an authority on a subject. Such evidence would include education and accomplishments.

CRISPR Clustered regularly interspaced short palindromic repeats. An RNA sequence that guides molecular machinery to a DNA sequence through complementary base-pairing.

cross See **genetic cross**.

crossing-over The physical exchange of chromosomal segments between homologous chromosomes. Compare *genetic recombination*.

CVS See **chorionic villus sampling**.

cyclical fluctuation A relatively predictable pattern of change in population size that occurs when at least one of two species is strongly influenced by the other. Compare *irregular fluctuation*.

cytokinesis Division of the cytoplasm, the second step of mitotic division, resulting in two self-contained daughter cells. Compare *mitosis*.

cytosine (C) One of the four nucleotides that make up DNA. The other three are adenine (A), thymine (T), and guanine (G).

cytoskeleton The network of protein cylinders and filaments that forms the framework of a cell.

D

data (sing. datum) Information collected in a scientific study.

decomposer A scavenger that dissolves the dead bodies of other organisms to consume them. Compare *detritivore*.

deletion A mutation in which a base is deleted from the DNA sequence of a gene. Compare *insertion* and *substitution*.

density-dependent population change A change in population size that occurs when birth and death rates change as the population density changes. Compare *density-independent population change*.

density-independent population change A change in population size that occurs when populations are held in check by factors that are not related to the density of the population. Compare *density-dependent population change*.

deoxyribonucleic acid See **DNA**.

dependent variable Any variable that responds, or could potentially respond, to changes in an *independent variable*.

descriptive study An observational study that reports information about what is found in nature. Compare *analytical study*.

desert A terrestrial biome characterized by a lack of moisture and temperature fluctuation from very hot during the day to very cold at night, and dominated by succulent plants such as cacti.

detritivore A scavenger that mechanically breaks apart the dead bodies of other organisms to consume them. Compare *decomposer*.

diffusion The movement of a substance from a region of higher concentration to a region of lower concentration.

dihybrid cross A controlled mating experiment involving organisms that are heterozygous for two traits.

diploid Possessing a double set of genetic information, represented by 2n. Somatic cells are diploid. Compare *haploid*.

directional selection The most common pattern of natural selection, in which individuals at one extreme of an inherited phenotypic trait have an advantage over other individuals in the population. Compare *disruptive selection* and *stabilizing selection*.

disruptive selection The least common pattern of natural selection, in which individuals with either extreme of an inherited trait have an advantage over individuals with an intermediate phenotype. Compare *directional selection* and *stabilizing selection*.

DNA Deoxyribonucleic acid, the genetic code of life, consisting of two parallel strands of nucleotides twisted into a double helix. DNA is the genetic material that transfers information from parents to offspring.

DNA polymerase The enzyme that builds new strands of DNA in DNA replication.

DNA replication The duplication of a DNA molecule.

DNA sequence similarity The degree to which the sequences of two different DNA molecules are the same—a measure of how closely related two DNA molecules are to each other.

domain The highest hierarchical level in the organization of life, describing the most basic and ancient divisions among living organisms. The three domains of life are Bacteria, Archaea, and Eukarya.

dominant allele An allele that prevents a second allele from affecting the phenotype when the two alleles are paired together. Compare *recessive allele*.

dominant genetic disorder A genetic disorder that is inherited as a dominant trait on an autosome. Compare *recessive genetic disorder*.

double helix The spiral formed by two complementary strands of nucleotides that is the backbone of DNA.

down-regulation The slowing down of gene expression. Compare *up-regulation*.

E

Earth equivalent The number of planet Earths needed to provide the resources we use and absorb the wastes we produce.

ecological community An association of species that live in the same area.

ecological footprint The area of biologically productive land and water that an individual or a population requires to produce the resources it consumes and to absorb the waste it produces.

ecological isolation The condition in which closely related species in the same territory are reproductively isolated by slight differences in habitat. Compare *geographic isolation* and *reproductive isolation*.

ecological niche The set of conditions and resources that a population needs in order to survive and reproduce in its habitat.

ecology The scientific study of interactions between organisms and their environment, where the environment of an organism includes both biotic factors (other living organisms) and abiotic (nonliving) factors.

ecosystem A particular physical environment and all the communities in it.

ecosystem process Any of four processes—nutrient cycling, energy flow, water cycling, and succession—that link the biotic and abiotic worlds in an ecosystem.

electron A negatively charged particle found outside the nucleus of an atom. Compare *neutron* and *proton*.

electron transport chain An elaborate chain of chemical events in which electrons and protons (H⁺) are handed over to other molecules that ultimately generates ATP and NADPH.

element A pure substance that has distinctive physical and chemical properties, and that cannot be broken down into other substances by ordinary chemical methods.

embryonic development The process by which an embryo develops. Common patterns of embryonic development across species provide evidence of evolution.

endocytosis The process by which materials are transported into a cell via vesicles. Compare *exocytosis*.

endoplasmic reticulum (ER) An extensive and interconnected network of sacs made of a single membrane that is continuous with the outer membrane of the nuclear envelope. See also **rough ER** and **smooth ER**.

energy The capacity of any object to do work, which is the capacity to bring about a change in a defined system.

energy capture The trapping and storing of solar energy by the producers at the base of an ecosystem's energy pyramid.

energy carrier A molecule that can store and deliver usable energy.

energy pyramid A pyramid-shaped representation of the amount of energy available to organisms in a food chain.

enzyme Any of a class of small molecules that speed up chemical reactions.

epistasis A form of inheritance in which the phenotypic effect of the alleles of one gene depends on the presence of alleles for another, independently inherited gene.

ER See **endoplasmic reticulum**.

estuary An aquatic biome characterized by shallow depth and high productivity. It is the tidal ecosystem where a river flows into the ocean and consequently has a constant ebb and flow of fresh and salt water. It is dominated by grasses and sedges.

Eukarya One of the three domains of life, including all the living organisms that do not fit into the domains *Archaea* or *Bacteria*, from amoebas to plants to fungi to animals.

eukaryote An organism that belongs to the Eukarya. Animals, plants, fungi, and protists are all eukaryotes. Compare *prokaryote*.

eutherians One of three main groups of mammals, whose members have a placenta and produce offspring that are born in a well-developed state. Compare *marsupials* and *monotremes*.

evaporation The transition from liquid to gas. Compare *condensation*.

evolution A change in the overall inherited characteristics of a group of organisms over multiple generations.

evolutionary tree A model of evolutionary relationships among groups of organisms that is based on similarities and differences in their DNA, physical features, biochemical characteristics, or some combination of these. It maps the relationships between ancestral groups and their descendants, and it clusters the most closely related groups on neighboring branches.

exocytosis The process by which materials are exported out of a cell via vesicles. Compare *endocytosis*.

exon A stretch of DNA that carries instructions for building a protein. Compare *intron*.

experiment A repeatable manipulation of one or more aspects of the natural world.

experimental group See **treatment group**.

exploitative competition Competition between species in which the two species indirectly compete for shared resources, such as food. Compare *interference competition*.

exponential growth A pattern of population growth in which the population increases by a constant proportion over a constant time interval, such as one year. Exponential growth occurs when there are no constraints on resources and is represented by a J-shaped curve. Compare *logistic growth*.

F

F₁ generation The first generation of offspring in a series of genetic crosses. Compare *F₂ generation* and *P generation*.

F₂ generation The second generation of offspring in a series of genetic crosses. Compare *F₁ generation* and *P generation*.

facilitated diffusion Diffusion that requires transport proteins. Compare *simple diffusion*.

fact A direct and repeatable observation of any aspect of the natural world. Compare *theory*.

falsifiable Able to be refuted.

family The unit of classification in the Linnaean hierarchy above genus and below order.

fermentation A metabolic pathway by which most anaerobic organisms extract energy from organic molecules. It begins with glycolysis and is followed by a special set of reactions whose only role is to help perpetuate glycolysis. Fermentation enables organisms to generate ATP anaerobically.

ferns A group of vascular, nonflowering plants that reproduce via spores.

fertilization The fusion of two gametes, resulting in a zygote.

flagellum (pl. flagella) A long, whiplike structure that assists bacteria in locomotion.

flower A structure in angiosperms that enhances sexual reproduction by bringing male gametes (sperm cells) to the female gametes (egg cells) in highly efficient ways, by attracting animal pollinators through scent, shape, and color.

food chain A single direct line of who eats whom among species in a community. Compare *food web*.

food web The way in which various and often overlapping *food chains* of a community are connected.

fossil The mineralized remains or impression of a formerly living organism.

founder effect A form of genetic drift that occurs when a small group of individuals establishes a new population isolated from its original, larger population.

freshwater biome An aquatic biome whose character is defined by the terrestrial biomes that it borders or through which its water flows. Lakes, rivers, and wetlands are all part of the freshwater biome.

fruiting body In fungi, the structure resulting from mating that releases offspring as sexual spores.

Fungi The fungi. One of the six kingdoms of life, in the domain Eukarya, distinguished by their modes of reproduction. Fungi are absorptive heterotrophs.

G

G₀ phase A nondividing state of the cell.

G₁ phase "Gap 1," the first phase in the life of a newborn cell.

G₂ phase "Gap 2," the phase of the cell cycle between the S phase and cell division.

gamete A sex cell. Male gametes are sperm; female gametes are eggs. Compare *somatic cell*.

gene The basic unit of information, consisting of a stretch of DNA, that codes for a distinct genetic characteristic.

gene expression The process by which genes are transcribed into RNA and then translated to make proteins.

gene flow The exchange of alleles between populations.

gene regulation The changing of the genes that are expressed in response to internal signals or external cues that allows organisms to adapt to their surroundings by producing different proteins as needed.

gene therapy A genetic engineering technique for correcting defective genes responsible for disease development.

genetic bottleneck A form of genetic drift that occurs when a drop in the size of a population causes a loss of genetic variation.

genetic carrier An individual who has only one copy of a recessive allele for a particular disease and therefore can pass on the disorder allele but does not have the disease.

genetic code The information specified by each of the 64 possible codons.

genetic cross A controlled mating experiment performed to examine how a particular trait may be inherited.

genetic disorder A disease caused by an inherited mutation, passed down from a parent to a child.

genetic divergence The presence of differences in the DNA sequences of genes.

genetic drift A change in allele frequencies produced by random differences in survival and reproduction among the individuals in a population.

genetic engineering The permanent introduction of one or more genes into a cell, tissue, or organism.

genetic modification (GM) Altering the genes of an organism for a specific purpose.

genetic recombination The exchange of DNA between homologous chromosomes brought about by *crossing-over*, contributing to variation in gametes.

genetic trait Any inherited characteristic of an organism that can be observed or detected in some manner.

genome The complete set of genes of an organism.

genotype The allelic makeup of a specific individual with respect to a specific genetic trait. Compare *phenotype*.

genus (pl. genera) The unit of classification in the Linnaean hierarchy above species and below family.

geographic isolation The condition in which populations are separated by physical barriers. Compare *ecological isolation* and *reproductive isolation*.

global warming A significant increase in the average surface temperature of Earth over decades or more. Compare *climate change*.

glycolysis The first of three stages of cellular respiration. During glycolysis, sugars (mainly glucose) are split to make the three-carbon compound pyruvate. For each glucose molecule that is split, two molecules of ATP and two molecules of NADH are released. Compare *Krebs cycle* and *oxidative phosphorylation*.

GM See **genetic modification**.

Golgi apparatus A collection of flattened membranes that packages and directs proteins and lipids produced by the ER to their final destinations either inside or outside the cell.

grassland A terrestrial biome characterized by low moisture levels (but not as low as in deserts), and dominated by grasses and herbaceous plants.

greenhouse effect The process by which greenhouse gases let in sunlight and trap heat.

greenhouse gas A gas in Earth's atmosphere that absorbs heat that radiates away from Earth's surface. Examples include carbon dioxide (CO_2), water vapor (H_2O), methane (CH_4), and nitrous oxide (N_2O).

guanine (G) One of the four nucleotides that make up DNA. The other three are adenine (A), thymine (T), and cytosine (C).

gymnosperms One of two main groups of seed-bearing plants, characterized by naked seeds. Compare *angiosperms*.

H

halophile A prokaryote, usually archaean, that can live in extremely salty environments.

haploid Possessing a single set of genetic information, represented by $2n$. Gametes are haploid. Compare *diploid*.

herbivore An animal that eats plants. Compare *carnivore* and *omnivore*.

herd immunity Protection against disease that is brought about by vaccination of a critical portion of a population.

heterotroph A metabolic consumer, an organism that obtains energy by taking it from other sources. *Chemoheterotrophs* consume organic molecules as a source of energy and carbon. *Photoheterotrophs* absorb the energy of sunlight but require an organic source of carbon. Compare *autotroph*.

heterozygous Carrying two different alleles for a given phenotype (*Bb*). Compare *homozygous*.

histone protein One of a class of specific proteins that, together with DNA, form nucleosomes.

homeostasis The process of maintaining constant internal conditions.

hominids The ape family, which includes humans and chimpanzees. All hominids are capable of tool use, symbolic language, and deliberate acts of deception. Compare *hominins*.

hominins The "human" branch of the *hominids*, including modern humans and extinct relatives such as Neanderthals.

homologous pair A pair of chromosomes consisting of one chromosome received from the father and one from the mother.

homologous trait A feature that is similar across species because of common descent. Homologous traits may begin to look different from one another over time. Compare *analogous trait*.

homozygous Carrying two copies of the same allele (such as *BB* or *bb*) for a particular gene. Compare *heterozygous*.

horizontal gene transfer The transfer of genes on plasmids from one bacterium to another.

host An organism in which a *parasite* lives.

human microbiome The complete collection of microbes that live in and on our cells and bodies.

hydrogen bond The weak electrical attraction between a hydrogen atom with a partial positive charge and a neighboring atom with a partial negative charge. Compare *covalent bond* and *ionic bond*.

hydrologic cycle The movement of water as it circulates from the land to the sky and back again.

hydrophilic Literally, "water-loving." Soluble in water. Compare *hydrophobic*.

hydrophobic Literally, "water-fearing." Excluded from water. Compare *hydrophilic*.

hypertonic Describing a fluid that has a solute concentration higher than that of the cell it surrounds. Compare *hypotonic* and *isotonic*.

hyphae (sing. hypha) The fine, branching threads of fungi that absorb nutrients from the environment.

hypothesis (pl. hypotheses) An informed, logical, and plausible explanation for observations of the natural world.

hypotonic Describing a fluid that has a solute concentration lower than that of the cell it surrounds. Compare *hypertonic* and *isotonic*.

I

in vitro fertilization (IVF) Fertilization of an egg by a sperm in a petri dish, followed by implantation of one or more embryos into a woman's uterus.

incomplete dominance An interaction between two alleles of a gene in which neither one can exert its full effect, causing a heterozygote to display an intermediate phenotype. Compare *codominance*.

independent assortment The random distribution of the homologous chromosomes into daughter cells during meiosis I.

independent variable The variable that is manipulated by the researcher in a scientific experiment. Compare *dependent variable*.

induced fit The way an enzyme changes shape when molecules bind to its active site.

insertion A mutation in which a base is inserted into the DNA sequence of a gene. Compare *deletion* and *substitution*.

interference competition Competition between species in which one organism directly excludes another from the use of a resource. Compare *exploitative competition*.

intermediate fossil A fossil that displays physical characteristics in between those of two known fossils in a family tree.

interphase The longest stage of the cell cycle. Most cells spend 90 percent or more of their life span in interphase.

intron A stretch of DNA that does not code for anything. Compare *exon*.

invariant trait A trait that is the same in all individuals of a species. Compare *variable trait*.

ion An atom that has lost or gained electrons and therefore is either negatively or positively charged.

ionic bond The chemical attraction between a negatively charged ion and a positively charged ion. Compare *covalent bond* and *hydrogen bond*.

irregular fluctuation An unpredictable pattern of change in population size. Compare *cyclical fluctuation*.

isotonic Describing a fluid that has a solute concentration equal to that of the cell it surrounds. Compare *hypertonic* and *hypotonic*.

isotopes Two or more forms of an element that have the same number of protons but different numbers of neutrons.

IVF See **in vitro fertilization**.

J

J-shaped growth curve The type of graphical curve that represents exponential growth. Compare *S-shaped growth curve*.

K

karyotype A depiction showing all the chromosomes of a particular individual or species arranged in homologous pairs.

keystone species A species that has a disproportionately large effect on a community, relative to the species' abundance.

kingdom The second-highest hierarchical level in the organization of life; the unit of classification in the Linnaean hierarchy above phylum and below domain. The six kingdoms of life are Bacteria, Archaea, Protista, Plantae, Fungi, and Animalia.

Krebs cycle Also called *citric acid cycle*. The second of three stages of cellular respiration. In this sequence of enzyme-driven reactions, the pyruvate made in glycolysis is broken down, releasing CO_2 and producing large amounts of energy carriers, including ATP, NADH, and $FADH_2$. Compare *glycolysis* and *oxidative phosphorylation*.

L

law of independent assortment The law, proposed by Gregor Mendel, stating (in modern terms) that when gametes form, the two alleles of any given gene segregate during meiosis independently of any two alleles of other genes. Compare *law of segregation*.

law of segregation The law, proposed by Gregor Mendel, stating (in modern terms) that the two alleles of a gene are separated during meiosis and end up in different gametes. Compare *law of independent assortment*.

lichen A mutualistic association between a photosynthetic microbe (usually a green alga or cyanobacterium) and a fungus.

light-independent reactions Also called the *Calvin cycle* or *carbon fixation*. The second of two principal stages of photosynthesis, in which a series of enzyme-catalyzed chemical reactions converts carbon dioxide (CO_2) into sugar, using energy delivered by ATP and electrons and hydrogen ions donated by NADPH. Compare *light reactions*.

light reactions The first of two principal stages of photosynthesis, in which chlorophyll molecules absorb energy from sunlight and use that energy for the splitting of water, which in turn produces oxygen gas (O_2) as a by-product that is released into the atmosphere. Compare *light-independent reactions*.

lignin A strengthening substance that links together *cellulose* fibers in plant cells to create a rigid network.

lineage A single line of descent.

Linnaean hierarchy A system of biological classification devised by the Swedish naturalist Carolus Linnaeus in the eighteenth century.

lipid Any of a major class of biomolecules built of fatty acids and insoluble in water.

liposome A sphere formed by a phospholipid bilayer.

locus (pl. loci) The physical location of a gene on a chromosome.

logistic growth A pattern of population growth in which the population grows nearly exponentially at first but then stabilizes at the maximum population size that can be supported indefinitely by the environment. Logistic growth is represented by an S-shaped curve. Compare *exponential growth*.

lysosome An organelle in animal cells that acts as a garbage or recycling center. Compare *vacuole*.

M

macromolecule See **biomolecule**.

malignant cell See **cancer cell**.

mammals A large class of animals that have body hair, sweat glands, and milk produced by mammary glands.

marine biome An aquatic biome characterized by salt water and encompassing both the coastal regions of all continents and the open ocean.

marsupials One of three main groups of mammals, whose members have a simple placenta and produce offspring that complete development in their mother's pouch. Compare *eutherians* and *monotremes*.

mass extinction A period of time during which a great number of Earth's species goes extinct. The fossil record shows that there have been five mass extinctions in the history of Earth.

matter Anything that has mass and occupies a volume of space.

meiosis A specialized type of cell division that kicks off sexual reproduction. It occurs in two stages: meiosis I and meiosis II, each involving one round of nuclear division followed by cytokinesis. Compare *mitosis*.

meiosis I The first stage of meiosis, in which the chromosome set is reduced by the separation of each homologous pair into two different daughter cells. Each homologous chromosome lines up with its partner and then separates to the two ends of the cells. Compare *meiosis II*.

meiosis II The second stage of meiosis, in which sister chromatids are separated into two new daughter cells. Compare *meiosis I*.

Mendelian trait A trait that is controlled by a single gene and unaffected by environmental conditions.

messenger RNA (mRNA) A type of RNA that is complementary to a DNA template strand. Compare *ribosomal RNA* and *transfer RNA*.

meta-analysis Work that combines results from different studies.

metabolic heat The heat released as a by-product of chemical reactions within a cell, typically during cellular respiration.

metabolic pathway Any of various chains of linked events that produce key biological molecules in a cell, including important chemical building blocks like amino acids and nucleotides.

metabolism All the chemical reactions that occur inside living cells, including those that release and store energy.

metastasis The spread of a disease from one organ to another.

methanogen An anaerobic archaean that feeds on hydrogen and produces methane gas as a by-product of its metabolism.

microbe A microscopic, single-celled organism.

mimicry Coloration of a nonpoisonous animal that resembles the coloration of a toxic species. Compare *camouflage* and *warning coloration*.

mitochondrial-DNA inheritance The passing down of DNA from the mitochondria in an egg cell to a new generation. Mitochondrial DNA passes virtually unchanged from mother to child, so it can be tracked from one generation, or one species, to another. Sequencing of mitochondrial DNA can determine how related an individual is to its female ancestors on its mother's side. Compare *nuclear-DNA inheritance*.

mitochondrion (pl. mitochondria) An organelle that is a tiny power plant fueling cellular activities. Mitochondria are the main source of energy in eukaryotic cells.

mitosis Division of the nucleus, the first step of mitotic division. Mitosis is divided into four main phases: prophase, metaphase, anaphase, and telophase. Compare *cytokinesis* and *meiosis*.

mitotic division A type of cell division that generates two genetically identical daughter cells from a single parent cell in eukaryotes. It consists of two steps: mitosis and cytokinesis.

molecule An association of atoms held together by chemical bonds.

monomer A small molecule that is the repeating unit of a *polymer*. For example, amino acids are the monomers that make up protein polymers.

monotremes One of three main groups of mammals, whose members lack a placenta and lay eggs. Compare *eutherians* and *marsupials*.

morphology An organism's physical characteristics.

most recent common ancestor The most immediate ancestor that two lineages share.

mRNA See **messenger RNA**.

mutation A random change to the sequence of bases in an organism's DNA.

mutualism A species interaction in which both species benefit. Compare *commensalism, competition,* and *predation.*

mycelium (pl. mycelia) The entire bundle of hyphae that composes the main body of a fungus.

mycorrhizal fungi Fungi that form mutualistic associations with the root systems of plants, that help the plants absorb more water and nutrients from the soil.

N

natural selection The process by which individuals with advantageous genetic characteristics for a particular environment survive and reproduce at a higher rate than do individuals with other, less useful characteristics. Compare *artificial selection.*

net primary productivity (NPP) The energy acquired through photosynthesis that is available for the growth and reproduction of producers within an ecosystem. It is the amount of energy captured by photosynthetic organisms, minus the amount they expend on cellular respiration and other maintenance processes.

neutron An electrically neutral particle found in the nucleus of an atom. Compare *electron* and *proton.*

nitrogen fixation The process, carried out by bacteria, of taking nitrogen gas from the air and converting it to ammonia, making it available for plants.

node The point on an evolutionary tree indicating the moment in time when an ancestral group split, or diverged, into two separate lineages. The node represents the most recent common ancestor of the two lineages in question.

noncoding DNA DNA that does not code for any kind of functional RNA.

notochord In chordates, a flexible yet rigid rod along the center of the body that is critical for development.

NPP See **net primary productivity**.

nuclear-DNA inheritance The passing down of DNA from the nucleus in an egg or sperm cell to a new generation. Sequencing of nuclear DNA can determine how related an individual is to all of its ancestors, both male and female. Compare *mitochondrial-DNA inheritance.*

nuclear envelope The boundary of a cell's nucleus, consisting of two concentric phospholipid bilayers.

nuclear pore Any of many small openings in the nuclear envelope that allow chemical messages to enter and exit the nucleus.

nucleic acid Any of a major class of biomolecules, including DNA and RNA, built of chains of nucleotides.

nucleosome DNA wrapped around histone proteins that, in multiples, form the beads-on-a-string complex in a chromatin fiber.

nucleotide The basic repeating subunit of DNA, composed of the sugar deoxyribose, a phosphate group, and one of four bases: adenine (A), cytosine (C), guanine (G), or thymine (T).

nucleotide pair See **base pair**.

nucleus (pl. nuclei) 1. The dense core of an atom, which contains protons and neutrons. 2. The control center of the eukaryotic cell, containing all of the cell's DNA and occupying up to 10 percent of the space inside the cell.

nutrient A chemical element that is required by a living organism.

nutrient cycling The process by which decomposers break down dead organisms or waste products, release the chemical elements locked in the biological material, and return them to the environment.

O

observation A description, measurement, or record of any object or phenomenon.

omnivore An animal that eats both animals and plants. Compare *carnivore* and *herbivore*.

opposable Able to be placed opposite other digits of the hand or foot. For example, opposable thumbs can be placed opposite each of the other four fingers.

order The unit of classification in the Linnaean hierarchy above family and below class.

organ A collection of different types of tissues that form a functional unit with a distinctive shape and location in the body.

organ system A network of organs that work in a closely coordinated manner to perform a distinct set of functions in the body.

organelle Any of the membrane-enclosed subcellular compartments found in eukaryotic cells.

organic molecule A molecule that includes at least one carbon-hydrogen bond.

organism An individual living thing composed of interdependent parts.

origin of replication A DNA sequence where DNA replication is initiated.

osmosis A form of simple diffusion in which water moves in and out of cells (and compartments inside cells).

ovule The egg-bearing structure in plants.

oxidative phosphorylation The third of three stages of cellular respiration. During this process, the chemical energy of NADH and $FADH_2$ is converted into the chemical energy of ATP, while electrons and hydrogen atoms removed from NADH and $FADH_2$ are handed over to molecular O_2, creating water (H_2O). A large amount of ATP is generated. Compare *glycolysis* and *Krebs cycle*.

P

P generation The first set of parents in a series of genetic crosses. Compare F_1 *generation* and F_2 *generation*.

parasite An organism that lives in or on another species and harms it by stealing resources. For example, some parasites suck blood or live off the food in our intestines. Compare *host*.

passive transport The movement of a substance without the addition of energy. Compare *active transport*.

PCR See **polymerase chain reaction**.

pedigree A chart similar to a family tree that shows genetic relationships among family members over two or more generations of a family's medical history.

peer-reviewed publication The publishing of original research only after it has passed the scrutiny of experts who have no direct involvement in the research under review, or a scientific journal that follows this standard.

PGD See **preimplantation genetic diagnosis**.

pH scale A logarithmic scale that indicates the concentration of hydrogen ions. The scale goes from 0 to 14, with 0 representing an extremely high concentration of free H^+ ions and 14 representing the lowest concentration.

phagocytosis Literally, "cellular eating." A large-scale version of endocytosis in which particles considerably larger than biomolecules are ingested. Compare *pinocytosis*.

phenotype The physical, biochemical, or behavioral expression of a particular version of a trait. Compare *genotype*.

phospholipid An organic molecule with a hydrophilic head and a hydrophobic tail.

phospholipid bilayer A double layer of phospholipids in which the heads face out and the tails face in. Plasma membranes are phospholipid bilayers.

photoautotroph See **autotroph**.

photoheterotroph See **heterotroph**.

photosynthesis The process by which organisms capture energy from the sun and use it to create sugars from carbon dioxide and water, thereby transforming light energy into chemical energy stored in the covalent bonds of sugar molecules. Compare *cellular respiration*.

phylum (pl. phyla) The unit of classification in the Linnaean hierarchy above class and below kingdom.

physical trait An anatomical or physiological characteristic of an individual, such as the shape of an animal's head. Compare *behavioral trait* and *biochemical trait*.

pili (sing. pilus) Short, hairlike projections that cover the surface of many bacteria.

pinocytosis Literally, "cellular drinking." A large-scale version of endocytosis in which fluids are ingested. Compare *phagocytosis*.

Plantae The plants. One of the six kingdoms of life, in the domain Eukarya, encompassing all plants, which are multicellular photosynthetic autotrophs.

plasma membrane A barrier consisting of a phospholipid bilayer that separates a cell from its external environment.

pleiotropy The pattern of inheritance in which a single gene influences multiple different traits. Compare *polygenic trait*.

point mutation A mutation in which only a single base is altered.

polar molecule A molecule whose electrical charge is shared unevenly, with some regions being electrically negative and others electrically positive.

pollen In plants, a microscopic structure containing sperm cells that can be lofted into the air in massive quantities.

polygenic trait A genetic trait that is governed by the action of more than one gene. Compare *pleiotropy*.

polymer A long strand of repeating units of small molecules called *monomers*. For example, proteins are polymers made up of amino acid monomers.

polymerase chain reaction (PCR) A technique for replicating DNA that can produce millions of copies of a DNA sequence in just a few hours from a small initial amount of DNA.

polyploidy The condition in which an individual's somatic cells have more than two sets of chromosomes.

population A group of individuals of the same species living and interacting in a shared environment.

population density The number of individuals per unit of area.

population doubling time The time it takes a population to double in size, as a measure of how fast a population is growing.

population ecology The study of the number of organisms in a particular place.

population size The total number of individuals in a population.

postzygotic barrier A barrier that prevents a zygote from developing into a healthy and fertile individual—that is, a reproductive barrier that acts after a zygote exists. Compare *prezygotic barrier*.

predation A species interaction in which one species benefits and the other species is harmed. Compare *commensalism, competition*, and *mutualism*.

predator A consumer that eats either plants or animals. Compare *prey*.

preimplantation genetic diagnosis (PGD) The removal of one or two cells from an embryo developing in a petri dish, usually 3 days after fertilization, followed by testing for genetic disorders. One or more embryos that are free of disorders are then implanted into a woman's uterus.

prey An animal that is eaten by a *predator*.

prezygotic barrier A barrier that prevents a male gamete and a female gamete from fusing to form a zygote—that is, a reproductive barrier that acts before a zygote exists. Compare *postzygotic barrier*.

primary consumer An organism that eats producers. Compare *secondary consumer, tertiary consumer*, and *quaternary consumer*.

primary literature Scientific literature in which research is first published. Compare *secondary literature*.

primary succession Succession that occurs in a newly created habitat, usually from bare rock or sand. Compare *secondary succession*.

primates The order of mammals to which humans belong. All primates have flexible shoulder and elbow joints, five functional fingers and toes, opposable thumbs, flat nails, and brains that are large in relation to the body.

primer A short stretch of RNA or DNA that is complementary-base-paired to a DNA template to provide a 3′-hydroxyl group for polymerase to initiate DNA replication or the polymerase chain reaction (PCR), respectively.

process of science See **scientific method**.

producer An organism at the bottom of a food chain that uses energy from the sun to produce its own food. Compare *consumer*.

product A substance that results from a chemical reaction. Compare *reactant*.

prokaryote An organism that belongs to either the Bacteria or the Archaea. Compare *eukaryote*.

promoter A segment of DNA near the beginning of a gene that RNA polymerase recognizes and binds to begin transcription. Compare *terminator*.

protein Any of a major class of biomolecules built of amino acids.

Protista The protists. One of the six kingdoms of life, in the domain Eukarya; a diverse group that is composed of mainly single-celled, microscopic organisms grouped together simply because they are not plants, animals, or fungi, and that includes amoebas and algae.

proton A positively charged particle found in the nucleus of an atom. Compare *electron* and *neutron*.

protozoans One of two main groups of protists, whose members are nonphotosynthetic and motile. Compare *algae*.

pseudoscience Scientific-sounding statements, beliefs, or practices that are not actually based on the scientific method.

Punnett square A grid-like diagram showing all possible ways that two alleles can be brought together through fertilization.

Q

quaternary consumer An organism that eats *tertiary consumers*. Compare *primary consumer* and *secondary consumer*.

quorum sensing A system of cell-to-cell communication used by prokaryotes that enables them to sense and respond to other bacteria in the area in accordance with the density of the population.

R

radial symmetry An animal body plan in which the body can be sliced symmetrically along any number of vertical planes that pass through the center of the animal. Compare *bilateral symmetry*.

reactant A substance that undergoes change in a chemical reaction. Compare *product*.

receptor-mediated endocytosis A form of specific endocytosis in which receptor proteins embedded in the membrane recognize specific surface characteristics of substances that will be incorporated into the cell.

receptor protein A site where a molecule from another cell can bind.

recessive allele An allele that has no effect on the phenotype when paired with a *dominant allele*.

recessive genetic disorder A genetic disorder that is inherited as a recessive trait on an autosome. Compare *dominant genetic disorder*.

relative species abundance How common one species is when compared to another.

reproduction The making of a new individual like oneself.

reproductive barrier A barrier that prevents two species from interbreeding, making them reproductively isolated.

reproductive isolation The condition in which barriers prevent populations from interbreeding. Compare *ecological isolation* and *geographic isolation*.

ribonucleic acid See **RNA**.

ribosomal RNA (rRNA) A type of RNA that is an important component of ribosomes. Compare *messenger RNA* and *transfer RNA*.

ribosome The site of protein synthesis (translation) in the cytoplasm. Ribosomes are embedded in the rough endoplasmic reticulum.

RNA Ribonucleic acid, a single-stranded nucleic acid transcribed from DNA and consisting of the ribonucleotides adenine, guanine, cytosine, and uracil.

RNA polymerase An enzyme that recognizes and binds a gene's promoter sequence and then separates the two strands of DNA during transcription.

RNA splicing Processing of mRNA in which the introns are snipped out of a pre-mRNA and the remaining pieces of mRNA—the exons—are joined to generate the mature mRNA.

rough ER A part of the endoplasmic reticulum, having a knobby appearance because of embedded ribosomes, where proteins are assembled. Compare *smooth ER*.

rRNA See **ribosomal RNA**.

rubisco The enzyme that catalyzes the first step in the light-independent reactions of photosynthesis, fixing a carbon molecule from CO_2.

S

S phase The "synthesis" phase of the cell cycle, in which preparations for cell division begin. A critical event during this phase is the replication of all the cell's DNA molecules.

S-shaped growth curve The type of graphical curve that represents logistic growth. Compare *J-shaped growth curve*.

scavenger An animal that eats dead or dying animals and plants. Scavengers are categorized as either *decomposers* or *detritivores*.

science A body of knowledge about the natural world, and an evidence-based process for acquiring that knowledge.

scientific claim A statement about how the world works that can be tested using the scientific method.

scientific literacy An understanding of the basics of science and the scientific process.

scientific method Also called *process of science*. The practices that produce scientific knowledge.

scientific name The unique two-word Latin name, consisting of the genus and species names, that is assigned to a species in the Linnaean hierarchy.

secondary consumer An organism that eats *primary consumers*. Compare *tertiary consumer* and *quaternary consumer*.

secondary literature Scientific literature that summarizes and synthesizes an area of research. Compare *primary literature*.

secondary succession Succession that occurs after a disturbance in a community. Compare *primary succession*.

seed In plants, the embryo and a supply of stored food, all encased in a protective covering.

selective breeding The process by which humans allow only individuals with certain inherited characteristics to mate.

selective permeability The quality of plasma membranes by which some substances are allowed to cross the membrane at all times, others are excluded at all times, and still others can pass through the membrane when they are aided by transport proteins.

semiconservative replication The mode of replication by which DNA is duplicated, where one "old" strand (the template strand) is retained (conserved) in each new double helix.

sense To perceive the world through a sensory system such as sight, touch, or smell.

serotype See **viral strain**.

sex chromosome One of the two chromosomes (X and Y) that determine gender. Compare *autosome*.

sex-linked Found solely on the X or Y chromosome. See also **X-linked** and **Y-linked**.

sexual dimorphism A distinct difference in appearance between the males and females of a species.

sexual reproduction The process by which genetic information from two individuals is combined to produce offspring. It has two steps: cell division through meiosis, followed by fertilization. Compare *asexual reproduction*.

sexual selection Natural selection in which a trait increases an individual's chance of mating even if it decreases the individual's chance of survival.

shared derived trait A unique feature common to all members of a group that originated in the group's most recent common ancestor and then were passed down in the group.

simple diffusion Diffusion in which substances such as the small, uncharged molecules of water, oxygen, or carbon dioxide, slip between the large molecules in the phospholipid bilayer without much hindrance. Compare *facilitated diffusion*.

sister chromatids The two identical DNA molecules produced by the replication of a chromosome.

smooth ER A part of the endoplasmic reticulum, having a smooth appearance, where lipids and hormones are manufactured. Compare *rough ER*.

soluble Able to mix completely with water.

solute A dissolved substance, such as sugar in water. Compare *solvent*.

solution Any combination of a solute and a solvent.

solvent The fluid, such as water, into which a substance has dissolved. Compare *solute*.

somatic cell A non–sex cell. Compare *gamete*.

speciation The process by which one species splits to form two species or more.

species 1. Members of a group that can mate with one another to produce fertile offspring.

2. The smallest unit of classification in the Linnaean hierarchy.

species interaction Any of four ecological interactions—mutualism, commensalism, predation, and competition—that occur between different species.

species richness The total number of different species that live in an ecological community.

spore In fungi, a reproductive structure that can survive for long periods of time in a dormant state and will sprout under favorable conditions to produce the body of the organism.

sporulation The formation of thick-walled dormant structures called spores.

stabilizing selection The pattern of natural selection in which individuals with intermediate values of an inherited phenotypic trait have an advantage over other individuals in the population. Compare *directional selection* and *disruptive selection*.

start codon The codon AUG; the point on an mRNA strand at which the ribosomes begin translation. Compare *stop codon*.

statistics A branch of mathematics that estimates the reliability of data.

stop codon The codon UAA, UAG, or UGA; the point on an mRNA strand at which the ribosomes end translation. Compare *start codon*.

substitution A mutation in which one base is substituted for another in the DNA sequence of a gene. Compare *deletion* and *insertion*.

substrate A molecule that will react to form a new product.

succession The process by which the species in a community change over time.

sustainable Able to be continued indefinitely without causing serious damage to the environment.

sympatric speciation The formation of new species in the absence of geographic isolation. Compare *allopatric speciation*.

T

temperate forest A terrestrial biome characterized by snowy winters and moist, warm summers and dominated by trees and shrubs adapted to relatively rich soil.

template strand The strand of DNA that is used as a template to make a new strand of DNA.

terminator A segment of DNA that, when reached by RNA polymerase, stops transcription. Compare *promoter*.

tertiary consumer An organism that eats *secondary consumers*. Compare *primary consumer* and *quaternary consumer*.

theory A hypothesis, or a group of related hypotheses, that has received substantial confirmation through diverse lines of investigation by independent researchers. Compare *fact*.

thermophile A prokaryote, usually archaean, that can live in extremely hot environments, such as geysers, hot springs, and hydrothermal vents.

thymine (T) One of the four nucleotides that make up DNA. The other three are adenine (A), guanine (G), and cytosine (C).

tissue A group of cells that function in an integrated manner to perform a unique set of tasks in the body.

transcription The synthesis of RNA based on a DNA template. Compare *translation*.

transfer RNA (tRNA) A type of RNA that facilitates translation by delivering specific amino acids to the ribosomes as codons are read off of an mRNA. Compare *messenger RNA* and *ribosomal RNA*.

translation The process by which ribosomes convert the information in mRNA into proteins. Compare *transcription*.

transpiration The process of plants absorbing water through their roots and releasing this water through their leaves into the atmosphere.

transport protein A protein that acts like a gate, channel, or pump that allows molecules to move into and out of a cell.

treatment group Also called *experimental group*. The group of subjects in an experiment that is maintained under the same standard set of conditions as the control group but is subjected to manipulation of the independent variable. Compare *control group*.

tRNA See **transfer RNA**.

trophic level Each level of the energy pyramid, corresponding to a step in a food chain.

tropical forest A terrestrial biome characterized by warm temperatures, about 12 hours of daylight, and seasonally heavy or year-round rains, and containing a rich diversity of organisms.

tumor A cell mass that results from runaway cell division.

tundra A terrestrial biome containing permafrost that is found at the poles and on mountaintops. Tundra is dominated by low-growing flowering plants and a lack of trees.

U

up-regulation The speeding up of gene expression. Compare *down-regulation*.

V

vacuole An organelle in plant cells that acts as a garbage or recycling center and that stores water. Compare *lysosome*.

variable A characteristic of any object or individual organism that can change.

variable trait A trait that is different in different individuals of a species. Compare *invariant trait*.

vascular system In plants, a network of tissues that is made up of tubelike structures specialized for transporting fluids.

vertebrae (sing. vertebra) The strong, hollow sections of the backbone, or vertebral column.

vertebrates Chordates that possess a backbone.

vesicle A sac, formed by the bulging inward or outward of a section of the plasma membrane, that moves molecules from place to place inside a cell but also may transport substances into and out of the cell.

vestigial trait A feature that is inherited from a common ancestor but no longer used. Vestigial traits may appear as reduced or degenerated parts whose function is hard to discern.

viral strain Also called *serotype*. Any of the variant forms of a particular type of virus.

virus A small, infectious agent that can replicate only inside a living cell.

W

warning coloration Bright coloring of an animal that alerts a potential predator to dangerous chemicals in the animal's tissues. Compare *camouflage* and *mimicry*.

weather Short-term atmospheric conditions, such as today's temperature, precipitation, wind, humidity, and cloud cover. Compare *climate*.

X

X-linked Found solely on the X chromosome. Compare *Y-linked*.

Y

Y-linked Found solely on the Y chromosome. Compare *X-linked*.

yeasts Single-celled fungi that belong in the group zygomycetes and are important in the rising of bread, the brewing of beer, and the fermenting of wine.

Z

zygomycetes One of three main groups of fungi, containing many species of molds. Compare *ascomycetes* and *basidiomycetes*.

zygote The single cell that results from fertilization.

Credits

Photos and Text

Front Matter

Page vii top: Jonathan Mays, Wildlife Biologist, Maine Department of Inland Fisheries and Wildlife/USFWS; bottom: Alexandra Grablewski/DigitalVision/Getty Images; **p. viii** top: Peter Zurek/Shutterstock; bottom: (puzzle) George Diebold/Getty Images; (cell) Mopic/Shutterstock; **p. ix** top: Michael Melford/National Geographic/Getty Images; bottom: Tsuneo Yamashita/Getty Images; **p. x** top: (background) Iaroslav Neliubov/Shutterstock; (dogs) Eric Isselee/Shutterstock; bottom: (background) bitterfly/Thinkstock/Getty Images Plus; (toy) Gaby Kooijman/Thinkstock/Getty Images Plus; **p. xi** top: Mike Kemp/Rubberball/Getty Images; center: (background) shironosov/iStock/Getty Images Plus; (tobacco) joannawnuk/Shutterstock; bottom: Andrey Nekrasov/Alamy; **p. xii** top: (background) muss/Shutterstock; (IV bag) Thinkstock/Getty Images; (VSRA) Science Source; center: (lizard) MattiaATH/Shutterstock; (diver) CyberEak/Shutterstock; bottom: Lou Linwei/Alamy; **p. xiii** top: Elke Van de Velde/Photodisc/Getty Images Plus; center: JimmyWrangles/iStock/Getty Images Plus; bottom: (karyotype) Leonard Lessin/Science Source/Getty Images; (background) thomaslenne/iStock/Getty Images Plus; (table) Piotr Adamowicz/iStock/Getty Images Plus (scientist) LattaPictures/iStock/Getty Images Plus; (left skull) Leemage/Universal Images Group/Getty Images; (right skull) Sabena Jane Blackbird/Alamy Stock Photo; **p. xiv** top: (background) Franz Lanting/Offset/Shutterstock; (smoke) Shutterstock; (charcoal edges) Anan Kaewkhammul/Shutterstock; (pre-burned edges) Sylwia Brataniec/Shutterstock; bottom: Paul Starosta/Corbis Documentary/Getty Images; **p. xv** top: Holly Kuchera/Shutterstock; bottom: SECCHI Disk; (water) TFoxFoto/Shutterstock; **p. xvii:** (Houtman) Will Prouty; (Scudellari): Sarah Der Photography; (Malone) courtesy Cindy Malone.

Chapter 1

Pages 2–3: Jonathan Mays, Wildlife Biologist, Maine Department of Inland Fisheries and Wildlife/USFWS; **p. 4:** Gerrit Vyn; **p. 6** (left to right): Judith Collins/Alamy, Scott Camazine/Science Source, AS Food studio/Shutterstock, Lev Kropotov/Shutterstock, Rosa Jay/Shutterstock; **p. 7** top: Photo by Dr. Kimberli Miller, courtesy USGS National Wildlife Health Center; bottom: Stephen Alvarez/National Geographic Creative;

p. 8 left: Photo courtesy Ryan von Linden. Photo used with permission from New York State Department of Environmental Conservation. All rights reserved; right top: Kevin Wenner/Pennsylvania Game Commission; right bottom: David S. Blehert, U.S. Geological Survey—National Wildlife Health Center; **p. 9** top: Image Courtesy of The Advertising Archives; bottom: photograph by Steve Grodsky; **p. 10** (all) Amy Smotherman Burgess/Knoxville News Sentinel/ZUMAPRESS.com; **p. 13** top left: Photo used with permission from New York State Department of Environmental Conservation; center left: Greg Turner/Pennsylvania Game Commission; bottom left: Photo courtesy Ryan von Linden. Photo used with permission from New York State Department of Environmental Conservation. All rights reserved; top right: Deborah Springer; center right: Design Pics Inc/Alamy; bottom right: Dr. Mary Hausbeck; **p. 15** top (left to right): Valeriy Vladimirovich Kirsanov/Shutterstock, Valeriy Vladimirovich Kirsanov/Shutterstock, Rosa Jay/Shutterstock, Vitalii Hulai/Shutterstock; bottom right: nico99/Shutterstock; bottom left: ASA-Carlos Asanuma/Getty Images; **p. 18:** Lindsey Heffernan, PA Game Commission/Patty Stevens, USGS.

Chapter 2

Pages 20–21: Alexandra Grablewski/Digital Vision/Getty Images; **p. 22:** Dr. Anna K. Eaton; **p. 23** top left: Gohil JR, Shah BA, Parekh AN. Diphtheria WebmedCentral INFECTIOUS DISEASES 2011;2(12): WMC002594 doi: 10.9754/journal.wmc.2011.002594. CC BY 3.0; center: Science Source/Colorization by Mary Madsen; top right: South West News Service; bottom left: AP Photo; bottom right: CDC; **p. 24:** Rachel Torres/Alamy Stock Photo; **p. 25** left: Dr. Anna K. Eaton; right: Ryan Collerd/The New York Times/Redux Pictures; **p. 28:** Graph (Fig. 2.6): "The Real Cause of Increasing Autism Prevalence?" by J. Emory Parker. Reprinted by permission of the author; **p. 31:** Republished with permission of Elsevier Science and Technology Journals, from "Ileal-Lymphoid-Nodular Hyperplasia, Non-specific Colitis, and Pervasive Developmental Disorder in Children." Wakefield, AJ, et al. *Lancet*, v351(9103) 1998; permission conveyed through Copyright Clearance Center, Inc. **p. 34:** Mara008/Shutterstock; **p. 35:** Dr. Anna K. Eaton.

Chapter 3

Pages 38–39: Peter Zurek/Shutterstock; **p. 40:** David McNew/Reuters/Newscom; **p. 41:** Time

& Life Pictures/Getty Images; **p. 43** top: Scripps Institution of Oceanography, UC San Diego; bottom: Courtesy of Henderson James Cleaves III; **p. 45:** Scripps Institution of Oceanography, UC San Diego; **p. 46:** Photolukacs/Shutterstock; **p. 47:** Kerstin Langenberger/imagebroker/Alamy; **p. 48** top: NASA, ESA, and D. Jewitt (UCLA); bottom: NASA Ames Research Center/SETI/Peter Jenniskens; **p. 49** top: Helen Sessions/Alamy; center: Shcherbinator/Shutterstock; bottom: Helen Sessions/Alamy; **p. 52** top: Africa Studio/Shutterstock; center: Now Foods; bottom: Kateryna Kon/Shutterstock; **p. 53** top left: Deep OV/Shutterstock; top center: designelements/Shutterstock; center right: Photo Melon/Shutterstock; center left: iofoto/Shutterstock; top right: Andrey Starostin/Shutterstock; center: Abramova Elena/Shutterstock; center bottom right: tryton2011/Shutterstock; bottom right: Jacqui Hurst/Corbis Documentary via Getty Images.

Chapter 4

Pages 58–59 (puzzle): George Diebold/Getty Images, (cell) Mopic/Shutterstock; **p. 60** top left: A. Barry Dowsett/Science Source; top right: Biophoto Associates/Science Source/Colorization by Mary Martin; bottom left: EMBL/Johanna Höög; bottom right: CNRI/Science Source; **p. 61:** Jessica Rinaldi/Reuters/Newscom; **p. 62:** Photos courtesy J. Craig Venter Institute. From Gibson, DG et al., "Creation of a Bacterial Cell Controlled by a Chemically Synthesized Genome." *Science*. 2010 Jul 2; 329(5987):52–6. doi: 10.1126/science.1190719; **p. 63:** Courtesy of Neal Devaraj; **p. 65:** Photo Researchers, Inc./Composition by: Eric Cohen/Science Source; **p. 66:** Roger Harris/Science Source; **p. 67** top left: Don W. Fawcett/Science Source; top right: SPL/Science Source; bottom left and right: Dennis Kunkel Microscopy/Science Photo Library/Science Source; **p. 69** top: Hybrid Medical Animation/Science Source; center and bottom: Biophoto Associates/Science Source; **p. 70** top: Dennis Kunkel Microscopy/Science Photo Library/Science Source; top left; center left and bottom: Russell Kightley; bottom left: Credit: Biophoto Associates/Science Source; Colorization by: Mary Martin; **p. 71** top: Biophoto Associates/Science Source; center top: Dr. Torsten Wittmann/Science Source; center bottom: Dr. George Chapman/Visuals Unlimited, Inc.; bottom: Dr. David Furness, Keele University/Science Source; **p. 74:** Photo courtesy J. Craig Venter Institute; **p. 76** top: Susumu

p. **258** top left: Roland Birke/Getty Images; top center: Rainer Fuhrmann/Shutterstock; top right: Dr. Peter Siver/Visuals Unlimited; center left: Robert and Jean Pollock/Science Source; center: Chris Mattison/FLPA/Science Source; bottom left: Dennis Flaherty/Science Source; bottom right: Schmitz Olaf/Getty Images; p. **259** top left: Andrew Syred/Science Source; top center: Fletcher & Baylis/Science Source; top right: Ingeborg Knol/imageBROKER/Corbis; center left: Borut Furlan/Getty Images; center: Stuart Wilson/Science Source; center right: Andrew J. Martinez/Science Source; bottom left: Andrew J. Martinez/Science Source; bottom center: Marius Dobilas/Shutterstock; bottom right: Martin Shields/Alamy; p. **260** 1st image in each row: Hans Kim/Shutterstock; row 1 (left to right): Science Source (Colorization by: Eric Cohen), BSIP/Science Source, Georges Antoni/Hemis/Alamy, Gary Meszaros/Science Source, Picture Partners/Science Source; row 2 (left to right): Gary Meszaros/Science Source, Irochka_T/iStock/Getty Images Plus, Anthony Mercieca/Science Source, James DeBoer/Shutterstock; row 3 (left to right): Valeriy Vladimirovich Kirsanov/Shutterstock, FLPA/Alamy, James Hager/Robert Harding/Alamy; row 4 (left to right): P. Wegner/Arco Images/Alamy, Bernd Rohrschneider/FLPA/Science Source; row 5 (left to right): Image Source/Alamy Stock Photo, Eric Isselee/Shutterstock; row 6 Sabena Jane Blackbird/Alamy Stock Photo; p. **261**: The Royal Belgian Institute of Natural Sciences, Brussels; p. **262**: upper left and right: The Royal Belgian Institute of Natural Sciences, Brussels; top: Emily Willoughby/Stocktrek Images/Alamy; center top: Masato Hattori; center: Natural History Museum, London; center bottom: Masato Hattori; bottom: Walt Anderson/Visuals Unlimited/GettyImages.

Chapter 15

Pages **268–69**: Elke Van de Velde/Photodisc/Getty Images Plus; p. **270**: image courtesy Allen Nutman, University of Wollongong. A.P. Nutman et al. "Rapid Emergence of Life Shown by Discovery of 3,700-Million-Year-Old Microbial Structures," *Nature*. Published online August 31, 2016. doi: 10.1038/nature19355; p. **271**: photo by Amanda Ward/Dunn Lab; p. **272** top left: Gillian Conquest, Extreme Citizen Science (ExCiteS) research group, University College London, top right: Photograph courtesy of Gardenista; bottom: Jim Davis/The Boston Globe via Getty Images; p. **273** all: Dunn Lab; p. **274** top left: Eye of Science/Science Source; center left: David McCarthy/Science Source; bottom left: Michael Abbey/Science Source; top right: Scimat/Science Source; bottom right: Eye of Science/Science Source; p. **275** top left: David Henderson/caia image/Alamy Stock Photo; top left inset: Eye of Science/Science Source; top right: Jose Arcos Aguilar/Shutterstock; top right inset: image courtesy Luis R. Comolli. Comolli, L. R., Baker, B., Downing, K.H., Siegerist, C. E., and Banfield, J. (2008). "Three-Dimensional Analysis of the Structure and Ecology of a Novel, Ultra-Small Archaeon." *ISME J*, 3(2), 159–167; bottom: IFE, URI-IAO, UW, Lost City Science Party; NOAA/OAR/OER; The Lost City 2005 Expedition; bottom inset: MichaelTaylor3d/Shutterstock; p. **277**: Photo courtesy of Lauren Nichols; p. **278** top left: Dr. Ron Dengler/

Visuals Unlimited/Getty Images; top right: Dennis Kunkel Microscopy/Science Source; bottom left: Eye of Science/Science Source; bottom right: CDC/James Archer, Illustrators: Alissa Eckert and Jennifer Oosthuizen; p. **279**: Universal History Archive/REX/Shutterstock; p. **280** background: Michael Heim/EyeEm/Getty Images; top right; top left and bottom: Dennis Kunkel Microscopy/Science Source; p. **281**: photo by Lauren Nichols, Your Wild Life/Dunn Lab.

Chapter 16

Pages **284–85**: JimmyWrangles/iStock/Getty Images Plus; p. **286** left: photo courtesy of Kumar Paudel; right: Jacob Phelps; p. **287** top: Wim van Egmond/Science Photo Library via Science Source; bottom: Plant-Success; p. **290**: Leon Neal/AFP/Getty Images; p. **291** left: divedog/Shutterstock; right: Casther/Shutterstock; p. **292**: Lance Cheung/USDA; p. **293**: Carroll & Carroll/AgStock/Design Pics Inc/Alamy Stock Photo, inset: Shawn Poynter/The New York Times/Redux; p. **295** top left: Stefano Rellandini/Reuters/Newscom; right: Laszlo Podor/Moment Open/Getty images; bottom left: Stefano Rellandini/Reuters/Newscom; p. **296** left: Le Do/Shutterstock; center: Suzifoo/iStock/Getty Images Plus; right: Egor Rodynchenko/Shutterstock; bottom (left to right): Amazing snapshot/Shutterstock, xpixel/Shutterstock, kzww/Shutterstock, tarapong srichaiyos/Shutterstock, Scott Sinklier/Alamy Stock Photo, Vitaly Korovin/Shutterstock; p. **297** top: Glass and Nature/Shutterstock; bottom: Frank Kohn/USFWS.

Chapter 17

Pages **300–301**: (background) thomaslenne/iStock/Getty Images Plus; (table) Piotr Adamowicz/iStock/Getty Images Plus; (karyotype) Leonard Lessin/Science Source/Getty Images; (scientist) LattaPictures/iStock/Getty Images Plus; (left skull) Leemage/Universal Images Group/Getty Images; (right skull) Sabena Jane Blackbird/Alamy Stock Photo; p. **302**: Dr. Mirjana Roksandic; p. **306**: Edward Westmacott/Shutterstock; p. **307** left: Steve Bloom Images/Alamy; center: John W Banagan/Getty Images; right: Dave Watts/Nature Picture Library/Alamy Stock Photo; p. **311**: John Reader/Science Source; p. **313**: Frank Vinken; p. **318** top: Sabena Jane Blackbird/Alamy Stock Photo; bottom: Silvana Condemi; p. **319**: Mark Thiessen/National Geographic Creative.

Chapter 18

Pages **322–23**: (background) Franz Lanting/Offset/Shutterstock; (smoke) Shutterstock; (charcoal edges) Anan Kaewkhammul/Shutterstock; (pre-burned edges) Sylwia Brataniec/Shutterstock; p. **324** top and bottom: Jennifer Balch; p. **325** left: NASA; right: Fotos593/Shutterstock; p. **326** top: Paul Lefebvre, Woods Hole Research Center; bottom: Dr. Michael T. Coe; p. **327** top left: David Greedy/Getty Images; top right: John SommersII/Reuters/Newscom; center: Greenpeace/Ges.oek.Forschung; bottom left: James Steinberg/Science Source; bottom right: AP Photo/Nick Ut; p. **330** top: Hervé Collart/Sygma via Getty Images; bottom: Ricky Rogers/Reuters/Newscom; p. **331**: (leaves) Valentyn

Volkov/Shutterstock, (tree section 1) My Life Graphic/Shutterstock, (tree section 2) andersphoto/Shutterstock; p. **337**: Andrew Orlemann/Shutterstock.

Chapter 19

Pages **340–41**: Paul Starosta/Corbis Documentary/Getty Images; p. **343** top: Estefan Radoviez/Agencia o Dia/AGENCIA ESTADO/Xinhua/Alamy Live News; bottom: photo courtesy of Dr. Matthew DeGennaro; p. **344**: CDC; p. **345**: CDC/Cynthia Goldsmith; p. **346**: 7th Son Studio/Shutterstock; p. **349** left: Dyakova Yulia/Shutterstock; right: Mark Newman/Tom Stack & Associates; p. **350**: Tom Brakefield/Getty Images; p. **351** top: Alan G. Nelson/Dembinsky Photo Associates; bottom: photo courtesy of Dr. Margareth Capurro; p. **352**: Angel Valentin/The New York Times/Redux; p. **353**: (mosquito) Antagain/E+/Getty Images Plus, (animal silhouettes) nikiteev_konstantin/Shutterstock; p. **355**: Graph: "Oxitec Approach Suppresses Mosquitoes in Piracicaba, Brazil." Appeared in "No GM Mosquitoes Didn't Start the Zika Outbreak," Discovermagazine.com, January 31, 2016. Reprinted by permission of Oxitec Ltd.

Chapter 20

Pages **356–57**: Holly Kuchera/Shutterstock; p. **358** top: photo by Dr. William Ripple; bottom: Pat & Chuck Blackley/Alamy; p. **359**: Dr. William Ripple; p. **361**: Don Johnston_IH/Alamy Stock Photo, inset: Kevin Ebi/Alamy Stock Photo; p. **362**: Ksanawo/Shutterstock; p. **363** left: Martin Harvey/Corbis via Getty Images; right: Luc Novovitch/Alamy Stock Photo; p. **364**: Daniel Stahler/NPS; p. **365** top: Mark Conlin/Oxford Scientific/Getty Images; bottom: David Fleetham/Visuals Unlimited/Getty Images; p. **366**: (wolves) rachisan alexandra/Shutterstock, (deer) KatarinaF/Shutterstock, (trees) Vertyr/Shutterstock; p. **367**: PhotoStock-Israel/Alamy; p. **368** top: Joe McDonald/Visuals Unlimited; center left: Nancy Nehring/Getty Images; center right: CHAINFOTO24/Shutterstock; bottom: Gerry Bishop/Visuals Unlimited; p. **369**: Norbert Rosing/National Geographic Creative; p. **371**: Anne-Marie Kalus; p. **372**: Adam Burton/Alamy; p. **373** left: Stan Osolinski/Dembinsky Photo Associates; center: Howard Garrett/Dembinsky Photo Associates; right: Walt Anderson/Visuals Unlimited.

Chapter 21

Pages **376–77**: (water) TFoxFoto/Shutterstock; SECCHI Disk; p. **378**: Gary Corbett/Alamy; p. **379** top: Jacques Descloitres, MODIS Rapid Response Team, NASA/GSFC, inset: D.P. Wilson/FLPA/Science Source; bottom left: Tania Wong; bottom right: Dr. Boris Worm; p. **382**: NASA image by Jesse Allen & Robert Simmon; p. **383**: Chris Linder/Visuals Unlimited; p. **384** top left: Bernd Zoller/imagebroker/Alamy; top right: John E Marriott/Alamy; center: Ken Lucas/Visuals Unlimited; bottom left: Pixtal Images/Media Bakery; bottom right: Design Pics Inc/Photolibrary/Getty Images; p. **385** top left: Michael P. Gadomski/Earth Sciences/Animals Animals; top right: Mark De Fraeye/SPL/Science Source; center: Willard Clay/Dembinsky Photo Associates; bottom left:

Mark Goodreau/Alamy; bottom right: Jim Zipp/ National Audubon Society Collection/Science Source; **p. 387:** Abigail McQuatters-Gollop.

Infographics

Chapter 1

Page 15, "Bug Zappers": Data from Justin G. Boyles, Paul M. Cryan, Gary F. McCracken, and Thomas H. Kunz, "Economic Importance of Bats in Agriculture," *Science*, April 1, 2001, www.sciencemag.org/content/332/6025/41.

Chapter 2

Page 34, "Safety in Numbers": Data from Caroline L. Trotter and Martin C. J. Maiden, "Meningococcal Vaccines and Herd Immunity: Lessons Learned from Serogroup C Conjugate Vaccination Programs," Expert Review of Vaccines, July 2009, www.ncbi.nlm.nih.gov /pubmed/19538112.

Chapter 3

Page 54, "What's It All Made Of?": Earth's Atmosphere data from Wikipedia, en.wikipedia .org/wiki/Abundance_of_the_chemical_ elements. The Universe data from WebElements, www.webelements.com/periodicity/abundance _universe/. The Human Body data from Wikipedia, en.wikipedia.org/wiki/Chemical _makeup_of_the_human_body. Earth's Crust data from Wikipedia, en.wikipedia.org/wiki /Abundance_of_elements_in_Earth%27s_crust.

Chapter 4

Page 73, "Sizing Up Life": Data from Genetic Science Learning Center at the University of Utah, "Cell Size and Scale," learn.genetics.utah .edu/content/cells/scale/.

Chapter 5

Page 92, "Making Way for Renewables": Data from REN21, "Renewables 2016 Global Status Report," www.ren21.net/status-of-renewables /global-status-report/.

Chapter 6

Page 112, "Cancer's Big 10": Data from Centers for Disease Control and Prevention, "United States Cancer Statistics: 2013 Top Ten Cancers," nccd.cdc.gov/uscs/toptencancers.aspx.

Chapter 7

Page 130, "Does Bigger Mean Better?": Data from National Center for Biotechnology Information, U.S. National Library of Medicine, "Genome Information by Organism," www.ncbi .nlm.nih.gov/genome/browse/.

Chapter 8

Page 143, "Genetic Diseases Affecting Americans": Data from Cold Spring Harbor Laboratory, DNA Learning Center, "Your Genes, Your Health," www.ygyh.org.

Chapter 9

Page 168, "The Meteoric Rise of CRISPR": PubMed search results data from National Center for Biotechnology Information, U.S. National Library of Medicine, www.ncbi.nlm .nih.gov/pubmed/?term=crispr. Timeline data from Broad Institute, "CRISPR Timeline," www.broadinstitute.org/what-broad/areas -focus/project-spotlight/crispr-timeline, and from CRISPR Update, "CRISPR Timeline," www.crisprupdate.com/crispr-timeline/.

Chapter 10

Page 185, "The Deadly Price of a Pandemic": Data from U.S. Department of Health and Human Services, "Pandemic Flu History," www.flu.gov/pandemic/history/index.html, and from World Health Organization, "Global Influenza Programme," www.who.int/influenza /en/.

Chapter 11

Page 206, "Watching Evolution Happen": Data from Oliver Tenaillon, Alejandra Rodríguez -Verdugo, Rebecca L. Gaut, Pamela McDonald, Albert F. Bennett, Anthony D. Long, and Brandon S. Gaut, "The Molecular Diversity of Adaptive Convergence," Science, January 27, 2012, www.ncbi.nlm.nih.gov /pubmed/22282810.

Chapter 12

Page 227, "Race against Resistance": Antibacterial drugs data from Infectious Diseases Society of America, "Bad Bugs, No Drugs: As Antibiotic Discovery Stagnates... A Public Health Crisis Brews," www.idsociety.org/uploadedfiles/idsa /policy_and_advocacy/current_topics_and _issues/antimicrobial_resistance/10x20/images /bad%20bugs%20no%20drugs.pdf, and from Dalia Deak, Kevin Outterson, John H. Powers, and Aaron S. Kesselheim, Annals of Internal Medicine, September 6, 2016, "Progress in the Fight Against Multidrug-Resistant Bacteria? A Review of U.S. Food and Drug Administration– Approved Antibiotics, 2010–2015," annals.org /aim/article/2526197/progress-fight-against -multidrug-resistant-bacteria-review-u-s-food/. Timeline data from Centers for Disease Control and Prevention, "Antibiotic Resistance Threats in the United States, 2013," www.cdc.gov /drugresistance/threat-report-2013/pdf/ar -threats-2013-508.pdf.

Chapter 13

Page 244, "On the Diversity of Species": Data from A.D. Chapman, "Numbers of Living Species in Australia and the World, 2nd Edition, A Report for the Australian Biological Resources Study, September 2009," www.environment.gov .au/node/13875.

Chapter 14

Page 264, "The Sixth Extinction": Data from International Union for Conservation of Nature and Natural Resources, The IUCN Red List of Threatened Species, Summary Statistics, www.iucnredlist.org/about/summary -statistics.

Chapter 15

Page 280, "The Bugs in Your Belly Button": Data from Robert R. Dunn, Julie Horvath-Roth, Sarah P. Council, Holly Menninger, and Matthew Fitzpatrick, Your Wild Life, November 25, 2013, "The Belly Buttons Will be Revealed, Slowly," yourwildlife.org/2013/11 /the-belly-buttons-will-be-revealed-slowly/.

Chapter 16

Page 296, "Food Banks": Data from Food and Agriculture Organization of the United Nations, "Biodiversity for Food Security and Nutrition: 30 Years of the Commission on Genetic Resources for Food and Agriculture," www.fao .org/resources/infographics/infographics -details/en/c/174199/, and "Commission on Genetic Resources for Food and Agriculture: Plant Genetic Resources," www.fao.org/nr/cgrfa /cthemes/plants/en/.

Chapter 17

Page 309, "Hereditary Heirlooms": Data from National Geographic, "Genes Are Us. And Them," ngm.nationalgeographic.com/2013/07 /125-explore/shared-genes, and from Stefan Lovgren, *National Geographic News*, August 31, 2005, "Chimps, Humans 96 Percent the Same, Gene Study Finds," news.nationalgeographic .com/news/2005/08/0831_050831_chimp _genes.html.

Chapter 18

Page 331, "Forest Devastation": Deforestation and forest degradation data from G. Kissinger, M. Herold, and V. De Sy, "Drivers of Deforestation and Forest Degradation: A Synthesis Report for REDD+ Policymakers," Lexeme Consulting, Vancouver, Canada, August 2012, www.gov.uk/government/uploads/system /uploads/attachment_data/file/65505/6316 -drivers-deforestation-report.pdf. Top 5 country data from Food and Agriculture Organization of the United Nations, "Change in Extent of Forest and Other Wooded Land 1990–2005," www.fao .org/forestry/32033/en/.

Chapter 19

Page 353, "World's Deadliest Animals": Data collected from multiple sources at Bill Gates, GatesNotes, April 25, 2014, "The Deadliest Animal in the World," www.gatesnotes.com /Health/Most-Lethal-Animal-Mosquito-Week/.

Chapter 20

Page 366, "Cause and Effect": Data from William J. Ripple and Robert L. Beschta, "Restoring Yellowstone's Aspen with Wolves," *Biological Conservation* 138 (2007), www.cof .orst.edu/leopold/papers/Restoring%20 Yellowstone%20aspen%20with%20wolves.pdf.

Chapter 21

Page 388, "Productive Plants": Data from Oak Ridge National Laboratory Distributed Active Archive Center, "NPP and the Global Carbon Cycle," daac.ornl.gov/NPP/html_docs /npp_ccycle.html.

Index

Page numbers set in *italic* type refer to materials in figures or tables.